Lecture Notes in Mathematics

Volume 2327

This series reports on new developments in all areas of mathematics and their applications - quickly, informally and at a high level. Mathematical texts analysing new developments in modelling and numerical simulation are welcome. The type of material considered for publication includes:

1. Research monographs
2. Lectures on a new field or presentations of a new angle in a classical field
3. Summer schools and intensive courses on topics of current research.

Texts which are out of print but still in demand may also be considered if they fall within these categories. The timeliness of a manuscript is sometimes more important than its form, which may be preliminary or tentative.

Titles from this series are indexed by Scopus, Web of Science, Mathematical Reviews, and zbMATH.

Ronen Eldan • Bo'az Klartag • Alexander Litvak •
Emanuel Milman

Editors

Geometric Aspects
of Functional Analysis

Israel Seminar (GAFA) 2020-2022

 Springer

Editors

Ronen Eldan
Department of Mathematics
Weizmann Institute of Science
Rehovot, Israel

Bo'az Klartag
Department of Mathematics
Weizmann Institute of Science
Rehovot, Israel

Alexander Litvak
Department of Mathematical
and Statistical Sciences
University of Alberta
Edmonton, AB, Canada

Emanuel Milman
Department of Mathematics
Technion - Israel Institute of Technology
Haifa, Israel

ISSN 0075-8434   ISSN 1617-9692   (electronic)
Lecture Notes in Mathematics
ISBN 978-3-031-26299-9   ISBN 978-3-031-26300-2   (eBook)
https://doi.org/10.1007/978-3-031-26300-2

Mathematics Subject Classification: 26D10, 47D07, 60G10, 60J60, 52A39, 52A40; Primary: 60E, 60F, 60E15; Seconday: 52A40

This Springer imprint is published by the registered company Springer Nature Switzerland AG
The registered company address is: Gewerbestrasse 11, 6330 Cham, Switzerland

Paper in this product is recyclable.

Preface

Since the mid-1980s, the following volumes containing collections of papers reflecting the activity of the Israel Seminar in Geometric Aspects of Functional Analysis have appeared:

1983–1984	Published privately by Tel Aviv University
1985–1986	Springer Lecture Notes in Mathematics, vol. 1267
1986–1987	Springer Lecture Notes in Mathematics, vol. 1317
1987–1988	Springer Lecture Notes in Mathematics, vol. 1376
1989–1990	Springer Lecture Notes in Mathematics, vol. 1469
1992–1994	Operator Theory: Advances and Applications, vol. 77, Birkhäuser
1994–1996	MSRI Publications, vol. 34, Cambridge University Press
1996–2000	Springer Lecture Notes in Mathematics, vol. 1745
2001–2002	Springer Lecture Notes in Mathematics, vol. 1807
2002–2003	Springer Lecture Notes in Mathematics, vol. 1850
2004–2005	Springer Lecture Notes in Mathematics, vol. 1910
2006–2010	Springer Lecture Notes in Mathematics, vol. 2050
2011–2013	Springer Lecture Notes in Mathematics, vol. 2116
2014–2016	Springer Lecture Notes in Mathematics, vol. 2169
2017–2019 (I)	Springer Lecture Notes in Mathematics, vol. 2256
2017–2019 (II)	Springer Lecture Notes in Mathematics, vol. 2266

The first six were edited by J. Lindenstrauss and V. Milman, the seventh by K. Ball and V. Milman, the subsequent four by V. Milman and G. Schechtman, the subsequent one by B. Klartag, S. Mendelson and V. Milman, and the last four by B. Klartag and E. Milman.

As in the previous Seminar Notes, the current volume reflects general trends in the study of Geometric Aspects of Functional Analysis, understood in a broad sense. A classical theme in the Local Theory of Banach Spaces is the Concentration of Measure Phenomenon, and the study of probability measures in high dimension. Several chapters study this phenomenon from different angles, through analysis on the Hamming cube, or via quantitative estimates in the Central Limit Theorem under thin-shell and related assumptions. Additionally, this volume includes contributions discussing the interactions of this circle of ideas with Linear Programming and Sampling Algorithms, e.g. solving a question in online learning algorithms by using a classical convexity construction from the nineteenth century. Classical convexity theory plays a central role in this volume, as well as the study of geometric inequalities. Such inequalities, which are somewhat in spirit of the Brunn-Minkowski inequality, shed light on convexity and on the geometry of Euclidean space. Probability measures with convexity or curvature properties, such as log-concave distributions, occupy a central role, as well as the Gaussian measures and non-trivial properties of the heat flow in Euclidean spaces. All contributions are original research papers and were subject to the usual refereeing standards.

The present volume also contains an extended introduction by Vitali Milman, outlining some main current directions and problems in the field of Asymptotic Geometric Analysis.

Rehovot, Israel   Ronen Eldan
Rehovot, Israel   Bo'az Klartag
Edmonton, AB, Canada   Alexander Litvak
Haifa, Israel   Emanuel Milman

Contents

Asymptotic Geometric Analysis: Achievements and Perspective

Vitali Milman

Abstract The reader will have noticed the non-standard appearance of this piece. Indeed, we are used to reading papers which are either survey papers or research ones (or a mixture of both). However, it seems to be beneficial for a given field to sometimes take a pause and to observe the picture of the field's development in a broad sense, from "above", and to examine the directions in which things are progressing, in various directions, at the same time. This collection of short essays on some particular subdirections of the theory is an attempt to present such an overview. In recent years Asymptotic Geometric Analysis has grown enormously in its areas of interest, directions and results. Trying to understand and digest the picture of this development, I asked several experts who are close to me and represent the centers of various directions, to write for me a very concise and short description of their present central interests. More precisely, that part of their interests that relates to Asymptotic Geometric Analysis. Many agreed, and I am posting below the short texts I received. After each of them, I will place my comments, as well as some problems that arise when reading these texts. Of course, I know that a few promising and interesting directions are missing. Some articles I expected, I did not receive, and some directions are not active at present around me. It is my hope that such a presentation will add curiosity to some questions for experts, but more important, it will have, hopefully, a positive influence on the young generation joining this field.

1 A Few Words About Asymptotic Geometric Analysis

Asymptotic Geometric Analysis (AGA) studies properties of geometric objects, such as normed spaces, convex bodies, or convex functions on finite dimensional domains, when the dimensions of these objects increase to infinity.

V. Milman (✉)
Department of Mathematics, Tel Aviv University, Tel Aviv, Israel
e-mail: milman@tauex.tau.ac.il

The asymptotic approach reveals many very novel phenomena which influence other fields in mathematics, especially where a large data set is of main concern, or the number of parameters becomes uncontrollably large. One of the important features of this relatively new theory is in developing tools which allow studying high-parametric families. Among the tools developed in this theory are measure concentration, thin-shell estimates, stochastic localization, the geometry of Gaussian measures, volume inequalities for convex bodies, symmetrizations, and functional versions of geometric notions and inequalities (see [24] and [25]).

This field started on the border between geometry and functional analysis in the 1980s and 1990s. In this field, isometric problems that are typical for geometry in low dimensions are substituted by an "isomorphic" point of view, and an asymptotic approach (as dimension tends to infinity) is introduced. Geometry and analysis meet here in a non-trivial way. One central theme of this subject is the interaction of randomness and pattern. At first glance, life in a high dimension seems to mean the existence of multiple "possibilities", so one may expect an increase in the diversity and complexity as dimension increases. However, the concentration of measure and effects caused by convexity show that this diversity is compensated, and order and patterns are created for arbitrary convex bodies in the mixture caused by high dimensionality.

As mentioned in the abstract, in recent years Asymptotic Geometric Analysis has grown enormously in its areas of interest, directions and results. The following pieces, together, will hopefully help to understand and digest the picture of most of these development. After each short note, I am placing my comments, as well as some problems that arise when reading these texts.

Throughout my comments I will use the following two references [24] and [25].

2   Semyon Alesker, Valuations

2.1   Valuations on Convex Sets

1 For a finite dimensional real vector space V let us denote by $\mathcal{K}(V)$ the family of all convex compact subsets of V. A valuation is a functional $\phi: \mathcal{K}(V) \to \mathbb{C}$ satisfying the following additivity property

$$\phi(K \cup L) = \phi(K) + \phi(L) - \phi(K \cap L),$$

whenever $K, L, K \cup L \in \mathcal{K}(V)$. Valuation is a very classical object which goes back at least to M. Dehn's solution of the 3rd Hilbert problem in 1900.

2 The class of all valuations is too large to control. It was realized by Hadwiger that there is a fruitful restriction of analytic nature: continuity of valuations in the Hausdorff metric on $\mathcal{K}(V)$. It is roughly analogous to the condition of countable additivity of finitely additive measures. However the set of examples and most

of intuition in the valuations theory and in the classical measure theory are quite different.

There are many geometrically interesting examples of continuous valuations coming from convexity (e.g. mixed volumes), integral geometry, geometric measure theory (integration with respect to the normal cycle), and Monge-Ampère operators.

3 Fundamental results in the theory were obtained by Hadwiger in 1940s and 1950s and P. McMullen in 1970s. In 1995 there was a breakthrough in the theory: Klain [88] and Schneider [135] have obtained an explicit classification of translation invariant continuous so called simple (i.e. vanishing on convex sets with empty interior) valuations. Thus Klain showed that any such valuation which is also even is proportional to a Lebesgue measure; the odd case is due to Schneider and is more technical to state. The Klain-Schneider theorem turned out to be a basis for most of the subsequent developments. In particular it was used in the proof of Alesker's irreducibility theorem [7] which says that the natural representation of the group $GL(V)$ in the space of translation invariant continuous valuations of given degree of homogeneity and parity is topologically irreducible, i.e. has no closed proper $GL(V)$-invariant subspaces. The irreducibility theorem also played a role in many of the subsequent developments.

In the last two decades several new structures with non-trivial properties on valuations have been discovered. Some of them are discussed below.

It was also realized that the notion of valuation can be extended beyond convexity, and one can imitate the space of valuations on an arbitrary smooth manifold [9]. This space retains some of the structures on valuations on convex sets. It has applications to integral geometry [38].

The main applications of valuations belong to integral geometry so far, see [89] for classical applications and [36, 38] for more recent ones. Most recently there were obtained applications of valuations to pseudo-Riemannian geometry [39], and geometric inequalities (discussed below). In the more recent applications the use of new structures plays an important role.

4 In this note we discuss the recently discovered connection between valuations and geometric inequalities. Very recently Kotrbatý [99] has discovered a new fundamental property of valuations—the Hodge-Riemann (HR) type relations. He proved them for even valuations, and most recently Kotrbatý and Wannerer [100] proved HR for odd valuations. These results, in combination with previously developed theory, opened a way for applications of valuations to geometric inequalities.

Furthermore Kotrbatý has formulated a conjectural more general (mixed) version of HR relations (MHR). We are going to formulate this conjecture as well as two new geometric inequalities which follow from Kotrbatý's theorem.

5 Let us introduce some background. Let $Val^\infty(V)$ denote the space of smooth translation invariant valuations on an n-dimensional vector space V equipped with the Garding topology. We omit here the formal definition of a smooth valuation as well as of the Garding topology: they are special cases of the standard notions of smooth vectors and the Garding topology on smooth vectors in the

representation theory (see e.g. [156], p. 33). Let us only mention that $Val^\infty(V)$ is dense in the space of continuous translation invariant valuations. Valuations of the form $\phi(K) = vol(K + A)$ are smooth provided A has a smooth boundary with positive Gauss curvature. Denote

$$Val_j^\infty(V) := \{\phi \in Val^\infty(V) | \phi(\lambda K) = \lambda^j \phi(K) \,\forall \lambda > 0, \forall K \in \mathcal{K}(V)\}.$$

We have McMullen's decomposition with respect to degrees of homogeneity

$$Val^\infty(V) = \oplus_{i=0}^n Val_i^\infty(V).$$

Set $Val^{i,\infty}(V) := Val_{n-i}^\infty(V)$.

6 The following result summarizes the first properties of the Bernig-Fu convolution.

Theorem 2.1 (Bernig-Fu [36]) *Fix a Lebesgue measure vol on V.*

(1) There exists a unique continuous (in the Garding topology) map called convolution

$$*: Val^\infty \times Val^\infty \to Val^\infty$$

such that if $\phi(\bullet) = vol(\bullet + A)$, $\psi(\bullet) = vol(\bullet + B)$ then

$$(\psi * \psi)(\bullet) = vol(\bullet + A + B).$$

(2) $(Val^\infty, *)$ is a commutative associative algebra with a unit $(= vol)$.
(3) $Val^{i,\infty} * Val^{j,\infty} \subset Val^{i+j,\infty}$.

7 Moreover $(Val^\infty, *)$ satisfies the Poincaré duality:

$$Val^{i,\infty} \times Val^{n-i,\infty} \xrightarrow{*} Val^{n,\infty} = \mathbb{C} \cdot \chi$$

is a perfect paring, i.e. for any non-zero valuation $\phi \in Val^{i,\infty}$ there exists $\psi \in Val^{n-i,\infty}$ such that $\phi * \psi \neq 0$.

8 In following conjecture all bodies A_i are assumed to have smooth positively curved boundary. Denote by V_A the mixed volume $V_A(\bullet) := V(\bullet[n-1], A)$.

Conjecture 2.2 (Kotrbatý [99])

(1) (MHL) Let $i < n/2$ then the map $Val^{i,\infty} \to Val^{n-i,\infty}$ given by

$$\phi \mapsto \phi * V_{A_1} * \cdots * V_{A_{n-2i}}$$

is an isomorphism.
(2) (MHR) Let $i \le n/2$. Define primitive subspace

$$P^i = \{\phi \in Val^{i,\infty} | \phi * V_{A_1} * \cdots * V_{A_{n-2i}} * V_{A_{n-2i+1}} = 0\}.$$

Then on the subspace P_i the following Hermitian form is positive definite:

$$\phi \mapsto (-1)^i \phi * \bar{\phi} * V_{A_1} * \cdots * V_{A_{n-2i}} \geq 0$$

with equality iff $\phi = 0$.

Remark 2.3

(1) In the special case when all $A_i = B$ are the Euclidean balls part (1) was proved by Alesker [8] (even valuations) and Bernig-Bröcker [35] (general case).

(2) In the special case when all $A_i = B$ part (2) was proved by Kotrbatý [99] for even valuations, and for odd ones by Kotrbatý-Wannerer [101].

(3) The case $i = 1$ of part (2) implies easily the Alexandrov-Fenchel inequalities. This case was proved by Kotrbatý-Wannerer [100] using the method of the proof of the Alexandrov-Fenchel inequality.

9 As we have said, in the case when all $A_i = B$ are Euclidean balls the Conjecture 2.2 is proven. In this case it has an equivalent version on the language of the product on valuations which is obtained by applying the Fourier type transform (this also was observed by Kotrbatý [99]). From the latter version Alesker obtained two new inequalities for mixed volumes as follows. We denote by $\Delta, \iota_1, \iota_2 \colon \mathbb{R}^n \to \mathbb{R}^n \times \mathbb{R}^n = \mathbb{R}^{2n}$ the imbeddings $\Delta(x) = (x, x), \iota_1(x) = (x, 0), \iota_2(x) = (0, x)$.

Theorem 2.4 (Alesker [12]) *Let $n \geq 2$. Let $A_1, \ldots, A_{n-1} \subset \mathbb{R}^n$ be convex compact sets. Then we have two inequalities for mixed volumes (below V_n denotes the mixed volume in $\mathbb{R}^n$ and V_{2n} denotes the mixed volume in $\mathbb{R}^{2n}$)*

$$(a) \quad V_{2n}(\iota_1(A_1), \ldots, \iota_1(A_{n-1}); \iota_2(A_1), \ldots, \iota_2(A_{n-1}); \Delta(B)[2]) \geq$$
$$V_{2n}(\iota_1(A_1), \ldots, \iota_1(A_{n-1}); -\iota_2(A_1), \ldots, -\iota_2(A_{n-1}); \Delta(B)[2]),$$

$$(b) \quad V_{2n}(\iota_1(A_1), \ldots, \iota_1(A_{n-1}); \iota_2(A_1), \ldots, \iota_2(A_{n-1}); \Delta(B)[2]) +$$
$$V_{2n}(\iota_1(A_1), \ldots, \iota_1(A_{n-1}); -\iota_2(A_1), \ldots, -\iota_2(A_{n-1}); \Delta(B)[2]) \leq$$
$$\gamma_n V_n(A_1, \ldots A_{n-1}, B)^2,$$

where γ_n is such a constant that the equality is achieved for $A_1 = \cdots = A_{n-1} = B$.

2.2 Valuations on Functions

1 While theory of valuations on convex sets is a classical part of convexity, the notion of valuation on a class of functions is recent. It is being mostly developed by M. Ludwig with her collaborators and students. This text is not supposed to be a survey of these developments but rather a glimpse into the subject. We focus on valuations on convex functions: examples and two recent results due to Colesanti, Ludwig, and Mussnig which are analogues of the classical McMullen's decomposition [115] and Hadwiger's characterization [78] on valuations on convex bodies. The choice of material reflects author's taste.

2 Let V denote a finite dimensional real vector space. Let $\mathcal{F}$ be a set of real valued functions on V ($\mathcal{F}$ does not have to be a linear space), e.g. the set of convex function, a Sobolev space etc.

Definition 2.5 A function $Z \colon \mathcal{F} \to \mathbb{R}$ is called a *valuation* if

$$Z(\max\{f, g\}) + Z(\min\{f, g\}) = Z(f) + Z(g) \tag{1}$$

for any $f, g \in \mathcal{F}$ such that $\max\{f, g\}, \min\{f, g\} \in \mathcal{F}$.

3 Let us give two simplest examples of valuations on functions.

Example 2.6

(1) $Z(f) = 1$ is a valuation on any class $\mathcal{F}$.
(2) Let Z be a linear functional on $\mathcal{F}$ (note that $\mathcal{F}$ does not have to be a vector space; the linearity of Z is understood whenever it makes sense). Then Z is a valuation on $\mathcal{F}$. Indeed this follows from the identity

$$\max\{f, g\} + \min\{f, g\} = f + g.$$

More examples will be discussed below.

4 Let us compare explicitly the notions of valuations on functions vs on convex bodies. This comparison can be considered as a motivation of Definition 2.5, but I am not sure that it was the historical one. Recall that a valuation ϕ on convex bodies is a functional on the class $\mathcal{K}(V)$ of all convex compact sets $\phi \colon \mathcal{K}(V) \to \mathbb{R}$ such that

$$\phi(K \cup L) + \phi(K \cap L) = \phi(K) + \phi(L) \tag{2}$$

for any $K, L \in \mathcal{K}(V)$ such that $K \cup L \in \mathcal{K}(V)$.

To see the relation between (1) and (2), recall that any convex compact set K is characterized uniquely by its supporting functional $h_K \colon V^* \to \mathbb{R}$ defined by

$$h_K(\xi) = \sup_{x \in K} \xi(x).$$

The following properties are well known:

(a) the map $K \mapsto h_K$ induces a bijection between $\mathcal{K}(V)$ and the class of convex 1-homogeneous functions on the dual space V^*;

(b) for $K, L \in \mathcal{K}(V)$ one has $K \cup L \in \mathcal{K}(V)$ if and only if $\min\{h_K, h_L\}$ is convex;

(c) if $K \cup L$ is convex then $h_{K \cup L} = \max\{h_K, h_L\}$ and $h_{K \cap L} = \min\{h_K, h_L\}$;

(d) a sequence $\{K_i\} \subset \mathcal{K}(V)$ converges to K in the Hausdorff metric if and only if $h_{K_i} \to h_K$ in $C^0(V^*)$, i.e. uniformly on compact subsets of V^*.

Thus we see that valuations on $\mathcal{K}(V)$ in the sense of (2) are identified with valuations on the class

$$\mathcal{F} = \{\text{convex 1-homogeneous funcions on } V^*\}.$$

Moreover valuations on $\mathcal{K}(V)$ continuous in the Hausdorff metric correspond to valuations on $\mathcal{F}$ continuous with respect to $C^0(V^*)$-convergence.

Translation invariant valuations on $\mathcal{K}(V)$ correspond to valuations on this $\mathcal{F}$ satisfying

$$Z(f + l) = Z(f) \quad \forall f \in \mathcal{F} \text{ and } \forall \text{ linear functional } l \text{ on } V^*, \qquad (3)$$

i.e. $l \in V$.

5 We have seen that valuations on $\mathcal{K}(V)$ are the same as valuations on the class of convex 1-homogeneous functions on V^*. Hence Definition 2.5 is more general than the definition of valuations on convex bodies since more general classes of functions $\mathcal{F}$ can be considered. In this note we restrict the discussion to the space of convex (not necessarily 1-homogeneous) functions. For valuations on Sobolev spaces see [108, 109, 113]. For valuation on L^p-spaces see [146].

6 The classical Legendre transform provides a general method to obtain new examples of valuations out of the known ones. To state it precisely, let us denote

$$Conv(V, \mathbb{R}) := \{f \colon V \to \mathbb{R} |\ f \text{ is convex}\},$$

$$Conv_{sc}(V) := \{f \colon V \to \mathbb{R} \cup \{+\infty\} |\ f \text{ is convex, l.s.c., and}$$

$$\lim_{|x| \to \infty} \frac{f(x)}{|x|} = \infty, \ f \not\equiv +\infty\}.$$

In the latter space the subscript sc stays for super-coersive which is the terminology used in the field. The l.s.c. means lower semi-continuous, i.e. for any x_0 one has $\lim_{x \to x_0} f(x) \leq f(x_0)$.

Recall that the Legendre transform is defined by

$$f^*(\xi) := \sup_{x \in V}(\xi(x) - f(x)).$$

It establishes a bijection between $Conv(V, \mathbb{R})$ and $Conv_{sc}(V^*)$, and $f^{**} = f$. Moreover Z is a valuation on $Conv(V, \mathbb{R})$ if and only if $[f \mapsto Z(f^*)]$ is a valuation on $Conv_{sc}(V^*)$ (see [57], Section 3.2, and references therein for these facts). $C^0(V)$-continuity corresponds to another kind of convergence which can be explicitly described; we omit the description.

7 Let us remind the linear algebraic notion of mixed determinant. Let $\mathcal{H}_n(\mathbb{R})$ denote the space of real symmetric $n \times n$ matrices. The determinant $\det\colon \mathcal{H}_n(\mathbb{R}) \to \mathbb{R}$ is a homogenous polynomial of degree n. Then there exists a unique n-linear symmetric map (also denoted by det)

$$\det\colon (\mathcal{H}_n(\mathbb{R}))^n \to \mathbb{R}$$

such that $\det(A, \ldots, A) = \det A$ for any $A \in \mathcal{H}_n(\mathbb{R})$. This map is called mixed determinant.

Example 2.7 Let $0 \le k \le n$. Let I_n denote the identity matrix of size n. For any $A \in \mathcal{H}_n(\mathbb{R})$ the mixed determinant

$$\det(\underbrace{A \ldots, A}_{k \text{ times}}, \underbrace{I_n \ldots, I_n}_{n-k \text{ times}}) = \binom{n}{k}^{-1} [A]_k,$$

where $[A]_k$ denotes the kth elementary symmetric polynomial in the eigenvalues of A.

8 The following result [11] provides some examples of $C^0(\mathbb{R}^n)$-continuous valuations on $Conv(\mathbb{R}^n, \mathbb{R})$.

Theorem 2.8 *Fix $0 \le k \le n$. Fix a sequence $A_1, \ldots, A_{n-k}\colon \mathbb{R}^n \to \mathcal{H}_n(\mathbb{R})$ of continuous functions, and a continuous function $b\colon \mathbb{R}^n \to \mathbb{R}$. Assume that at least one of the functions $b, A_1, \ldots, A_{n-k}$ has a compact support.*

(a) *Then there is a unique $C^0(\mathbb{R}^n)$-continuous valuation Z on $Conv(\mathbb{R}^n, \mathbb{R})$ such that its value on any C^2-smooth convex function f is equal to*

$$Z(f) := \int_{\mathbb{R}^n} b(x) \det(\underbrace{Hess_{\mathbb{R}} f \ldots, Hess_{\mathbb{R}} f}_{k \text{ times}}, A_1, \ldots, A_{n-k}) dx.$$

(b) *Furthermore*

$$Z(f+p) = Z(f) \; \forall \; polynomial\; p\; of\; degree\; 0\; or\; 1\; and\; \forall f \in Conv(\mathbb{R}^n, \mathbb{R}).$$
$$(4)$$

Remark 2.9

(1) This theorem is a relatively straightforward consequence of two general facts on expressions of the Monge-Ampère type. The continuity of Z follows from the Alexandrov's theorem [6]. The valuation property of Z follows from Blocki's theorem [42].

(2) The class of valuations in Theorem 2.8 is not sufficient to get the Hadwiger type classification of valuations on convex functions obtained by Colesanti, Ludwig, and Mussnig [56]. We will see that the function b there might have a singularity at 0.

9 Colesanti, Ludwig, and Mussnig [55] obtained an analogue of P. McMullen's decomposition [115] of a valuation into homogeneous components. To state their result let us introduce a definition.

Definition 2.10 A valuation $Z \colon Conv(\mathbb{R}^n, \mathbb{R}) \to \mathbb{R}$ is called α-homogeneous if

$$Z(\lambda f) = \lambda^\alpha Z(f)$$

for any $f \in Conv(\mathbb{R}^n, \mathbb{R})$ and any $\lambda > 0$.

Theorem 2.11 ([55]) *Let* $Z \colon Conv(\mathbb{R}^n, \mathbb{R}) \to \mathbb{R}$ *be a* $C^0(\mathbb{R}^n)$*-continuous valuation satisfying condition (4).*

(a) Then there exist unique $C^0(\mathbb{R}^n)$*-continuous valuations satisfying (4)*

$$Z_0, Z_1, \ldots, Z_n \colon Conv(\mathbb{R}^n, \mathbb{R}) \to \mathbb{R}$$

such that each Z_i *is i-homogeneous and* $Z = Z_0 + Z_1 + \cdots + Z_n$.
(b) Z is 0-homogeneous if and only if it is constant, i.e. $Z(f)$ *is independent of* f.
(c) Z is n-homogeneous if and only if there is a compactly supported continuous function $b \colon \mathbb{R}^n \to \mathbb{R}$ *such that*

$$Z(f) = \int_{\mathbb{R}^n} b(x) \det Hess_{\mathbb{R}}(f) dx$$

for any convex C^2*-smooth function* f.

10 Following [56] let us introduce analogues of intrinsic volumes. Let us denote $D_n^n := C_c([0, \infty))$. Let us denote by D_0^n the space of continuous functions $\zeta \colon (0, \infty) \to \mathbb{R}$ with bounded support such that there exists $\lim_{s \to +0} \int_s^\infty t^{n-1} \zeta(t) dt < \infty$. Finally for $1 \le j \le n - 1$ denote

$$D_j^n :=$$

$$\{\zeta \in C((0, \infty)) | \, supp(\zeta) \text{ is bounded and}$$

$$\lim_{s \to +0} s^{n-j} \zeta(s) = 0, \exists \lim_{s \to +0} \int_s^\infty t^{n-j-1} \zeta(t) dt < \infty\}.$$

Theorem 2.12 ([56]) *Let* $0 \le j \le n$. *Let* $\zeta \in D_j^n$. *There exists unique* $C^0(\mathbb{R}^n)$-*continuous valuation* $V_{j,\zeta}^* : Conv(\mathbb{R}^n, \mathbb{R}) \to \mathbb{R}$ *such that its value on any* C^2-*smooth convex function* f *is equal to*

$$V_{j,\zeta}^*(f) = \int_{\mathbb{R}^n} \zeta(|x|)[Hess_{\mathbb{R}}(f)]_j dx.$$

Clearly $V_{j,\zeta}^*$ is j-homogeneous, satisfies (4), and is rotation invariant, i.e. $V_{j,\zeta}^*(f \circ \mathcal{O}) = V_{j,\zeta}^*(f)$ for any convex function f and any transformation $\mathcal{O} \in SO(n)$.

11 Now let us state the Hadwiger type theorem due to Colesanti, Ludwig, and Mussnig [56].

Theorem 2.13 ([56]) *A functional* $Z : Conv(\mathbb{R}^n, \mathbb{R}) \to \mathbb{R}$ *is a* $C^0(\mathbb{R}^n)$-*continuous rotation invariant valuation satisfying condition (4) if and only if*

$$Z = \sum_{j=0}^{n} c_j V_{j,\zeta}^*$$

for some constants $c_j \in \mathbb{R}$ *and functions* $\zeta_j \in D_j^n$.

2.3   Comments by V.M.

Let us start with some comments to 2.1. I would like to explain to people of Functional Analysis and AGA how interesting and important the Theory of Valuation is. For this goal I will very compactly return to our discussion in [24] (see subsection "Valuation" in section B6). The definition of a function we call "valuation" is introduced in the first point 2.1.1. of the first Alesker's article. Such a function may always be extended to finite unions of convex sets as a finitely additive function. Of course, after such an extension a valuation is a finitely additive functions on (special) subsets of a started set, and in modern terminology, finitely additive measure. The importance to study such measures became evident in the end of the nineteenth century, and Lebesgue and his student that time Emile Borel intensively studied this object. However, the class of such measures is too large to control. In order to have a more restricted class of measures which still covers interesting examples from geometry and analysis, one could try to look for an extra condition of analytic nature which would determine how the measures of sets behave with respect to limits. In the classical measure theory of Lebesgue this condition is countable additivity. To introduce such novel condition instead of more natural that time some kind of continuity condition was a huge step in Mathematics. Any kind of continuity was too restrictive for problems which occupied both, Lebesgue and Borel. As is well known, this condition, sigma additive measures, turned out to be extremely useful and is one of the pillars which created Functional Analysis. However, in classical convexity theory in $\mathbb{R}^n$ a different condition was necessary. One of the most interesting examples like mixed volumes were NOT, in general, sigma additive. And then the mathematical mind had a step back, and a "new-old" very useful condition of continuity of the valuation V on the class of convex compact sets with respect to the Hausdorff metric was introduced. To the best of our knowledge, this condition of continuity was first introduced and systematically studied by Hadwiger in 1957 (see [24] for references). To-day a continuity condition is a part of the definition of a valuation.

Of course, "enough smooth" countably additive measures are the simplest examples of valuations. But usually valuations cannot be defined on too broad a class of sets, say on Borel sets. So, very different functionals, which are not measures are appearing when the closure of finite additive measures is taken in Hausdorff sense and not in the sense of sigma additivity. Let me provide only one simple example of valuation which is obviously extremely important. It is the surface area of a convex body as a functional on the class of all convex compact bodies in $\mathbb{R}^n$. Obviously, there is no sigma additive measure on $\mathbb{R}^n$ which applying to a given convex body will provide its surface area. However, it is a valuation ON THIS CLASS. One short remark on quermassintegrals. As we noted they are not defined by measures on $\mathbb{R}^n$ (beside the trivial one, namely volume). However, one may change the space on which the measure "lives", and they will be defined by such new spaces with measures on them. For example, consider the space of all affine one dimensional lines with the uniform measure. Then for every convex body its measure in this space is, after proper normalization, the surface area of this body (this is simply "Crofton's formula", see for example [24, page 245]). Similarly, the mean-width can be defined by another "affine Grassmannian": the uniform measure on all affine $(n-1)$-dimensional subspaces of $\mathbb{R}^n$. There is an even more general fact proved by Alesker, see e.g. [24], Theorem B.6.4.

And now some remarks on the part 2.2. The introduction and study of valuations in the functional spaces is a very significant change of setting, especially because of its infinite dimensional background. Exactly this fact attracts me very much. I recall what is written above: the approach of valuation is a complimentary to another approach in the study of finitely additive functions/measures, which led Lebesgue and Borel to the notion of measure. However, the notion of a measure on the infinite dimensional spaces are not satisfactory. The experts who study and use Wiener measures, or, alternatively, Wiener process (also called Brownian motion) will violently disagree with me. However, I cannot feel comfortably with a measure which has zero value on any bounded subset of a Hilbert space. In the same time, we see through the work of Ludwig and then Colesanti , Ludwig and others (see Alesker's article) that valuations may be defined in the infinite dimensional setting on important subclasses and behave very nicely. It provides some light, although weak, at the "end of the tunnel". Of course, they are functionals, but not necessary linear functionals, and à-la-measure functionals. Of course, we may represent some subsets by their characteristic functions, and discuss valuation type property as a property of these functions. In point 2.2.4 of his article Alesker suggested, for the family of convex sets, another representation of these sets by functions, taking supporting functions as the representative family. Note, that a measure is also a (linear) functional on some classes of functions. But functions which are valuations highly increase the possibilities to play with them.

Let me note that the use of measures in all developments in AGA was very crucial, but the main thing we were always interested is to know that some objects of interest have positive measure, because the positivity of the measure plays the role of existing results. We want to know that some objects exist, but we are unable to visualized them, and we prove their existence by proving that they exist with a positive measure (and usually we are able to prove that this measure is very close to one; one should correct and tell here that we are unable to prove a positivity of measure without proving that it is almost one). Very long ago, still before the valuation in the functional setting appeared, I dreamed to prove existence of some objects we are searching for in AGA, by proving that some (non-trivial) valuation is positive. For this goal I even "pushed" Alesker to study integration along valuations (not just volumes). And he created such very successful theory, but an application for AGA is still missing. Now, with the notion of valuation was extended to infinite dimensional setting, some very new applications became more visible.

3   Shiri Artstein-Avidan, A Playground of Dualities

In the progress of functionalization of duality [118], the importance of the Legendre transform [18], as the functional analogue of polarity for sets, became apparent. This brought about many functional theorems, as well as the framing of known functional inequalities as pure functional analogues of known geometric inequalities see [25, Chapter 9]. By this is meant more than just "a functional inequality from which the geometric one may be recovered" of which many were known, for example Bobkov's functional isoperimetric inequality, but some more "geometric" way of considering all of these structures (something quite hard to define).

In this process, a new duality transform (which we called the $\mathcal{A}$ transform) emerged [17, 19], which was, up to linear terms, the only other order reversing involution on the class of geometric convex functions (non-negative lower semi continuous convex functions which vanish at the origin) except for the Legendre transform. The $\mathcal{A}$ transform had its own "differential analysis" [20], where the gradient (and gradient map) which is closely associated with Legendre transform, is replaced by the "polar gradient" which points in the same direction but whose monotonicity is opposite in a sense (for example, in one dimension the gradient of a convex function is increasing but its polar gradient, when defined, is decreasing).

The gradient map of a convex function and its properties are a key element in many proofs in convex geometry (for example Brenier map can be used to prove Brascamp-Lieb inequalities and their reverse). Understanding whether these properties are linked with the fact that it is the optimal transport with respect to the quadratic cost, in such a way that the inverse map is associated with the Legendre transform is an important goal. To this end, we rediscovered, and further developed [26] and [27], some of the optimal transport techniques for other costs. It turns out that the $\mathcal{A}$ transform is linked with a different cost (in the same way that Legendre is linked with the quadratic cost), this cost being $-\log(\langle x, y \rangle - 1)_+$. Further, it turns out that to every choice of cost, there is associated a class of sets (or functions, under some restrictions on the cost) on which the cost-transform serves as an order reversing involution [28]. In this sense, the specialty of $\mathcal{L}$ and $\mathcal{A}$ is that the associated class is the class of convex functions (or geometric convex functions) which was of interest to begin with. However, other costs can bring about other "convexity-type" classes, some of which were known to be of interest for other reasons (such as the class of so-called "flowers", and the class of "reciprocal convex bodies") and some of which are at the moment a curiosity, but might eventually lead us to new structures of interest expanding well about currently studied areas.

In the same vein, any such cost and duality bring about their own "gradient map" and differential analysis, as well as other interesting directions of study, such as invariant sets ($\mathcal{L}$ and polarity have only one, $\mathcal{A}$ has many, other costs have none), corresponding Santaló-type inequalities, associated measure-concentration estimates, and many more. At the moment this serves as a big "playground" of possibilities, and the main challenge is to focus on the most useful members of this playground, which may serve to help solve and advance other problems and directions.

3.1 Comments by V.M.

It is a remarkable development to discover that the polarity-type transforms are a natural byproduct of the optimal transport way of thinking. Actually, it is shown in a number of publications by Shiri Artstein-Avidan and her collaborators (see, e.g. [28], Chapter 1, [25]) that any cost transform is also a polarity transform.

In this connection I recall some question which I tried to promote around 10 years ago, but did not succeed, and which, may be, just inside the above development:

Can the standard polarity transform K to K° (for convex compact K with 0 in their interior) be realized as the solution of some extremal problem on the set of all n-dimensional (dim $K = n$) convex compact bodies with 0 in their interior?

Perhaps, it is just an interpretation of the above results. But it would be good to formulate it in such a way.

4   Ronen Eldan, Dimension-Free Concentration in High-Dimensional Distributions

A functional inequality is said to be *dimension-free* if it has no explicit dependence on the dimension. This usually comes together with the fact that the extremizers of this functional inequality are sets defined by a constant number of directions in space. Perhaps the most canonical example of a distribution which exhibits dimension-free functional inequalities is the Gaussian space, which is just the space $\mathbb{R}^n$ equipped with the standard Gaussian measure, whose density is

$$\frac{d\gamma}{dx} := (2\pi)^{-n/2} \exp(-|x|^2/2).$$

While the extremizers in the isoperimetric inequality in Euclidean space equipped with the Lebesgue measures are metric balls, in Gaussian space, the extremizers turn out to be halfspaces.

Theorem 4.1 (Borell, Sudakov-Tsirelson [43, 140]) *If $A \subset \mathbb{R}^n$ is a measurable set and $H \subset \mathbb{R}^n$ is a set of the form $\{x;\ x \cdot \theta > t\}$ with $\gamma(A) = \gamma(H)$ then for all $\epsilon > 0$ we have*

$$\gamma(A_\epsilon) \geq \gamma(H_\epsilon)$$

where $A_\epsilon := \{x;\ \exists y \in A,\ |y - x| < \epsilon\}$.

Since the Gaussian measure is a product measure, the function $\epsilon \to \gamma(H_\epsilon)$ is independent of the dimension. A weaker fact is that the Gaussian measure satisfies a dimension-free Poincaré inequality.

Corollary 4.2 *For any differentiable $f : \mathbb{R}^n \to \mathbb{R}$,*

$$\mathrm{Var}_{X \sim \gamma}[f(X)] \leq \mathbb{E}_{X \sim \gamma}|\nabla f(X)|^2,$$

and equality holds if and only if f is a linear function.

In words, a function which does not typically vary locally in space, cannot vary in a global sense.

Both linear functions and half-spaces essentially depend on one direction in space, in this sense the two inequalities are dictated "one-dimensional" objects.

It is natural to look for other settings which exhibit the same behavior. Below we consider two such settings.

4.1 Concentration Under a Convexity Condition

A natural family of distributions, which contains the Gaussian distribution as a special case, are *logarithmically-concave* distributions. These are measures whose density is of the form $\frac{d\mu(x)}{dx} = \exp(-V(x))$ where $V : \mathbb{R}^n \to \mathbb{R}$ is convex. An important subclass of these distributions are uniform measures over compact, convex sets with nonempty interior. Define the covariance matrix of a measure μ on $\mathbb{R}^n$ by

$$\mathrm{Cov}(\mu)_{i,j} := \mathbb{E}[X_i X_j] - \mathbb{E}[X_i]\mathbb{E}[X_j],$$

where $X \sim \mu$. The following conjecture, by Kannan, Lovász and Simonovits ([84]) asserts that logarithmically concave measures exhibit dimension-free concentration.

Conjecture 4.3 (KLS Conjecture [84]) There exists a universal constant $C > 0$ such that

$$\mathrm{Var}_\mu[f] \leq C \mathbb{E}_\mu \left[|\mathrm{Cov}(\mu)^{-1/2} \nabla f|^2 \right].$$

The conjecture is known to imply the slicing problem by Bourgain [95].

A series of works, starting from [84], followed by works of Klartag, Guédon-Milman and the author have been able to obtain increasigly better polynomial dependence on the dimension. A recent breakthrough by Chen [52] shows that the conjecture is true if C is replaced by $n^{o(1)}$, and very recently, Klartag and Lehec [93] have been able to replace C by a polylogarithmic factor. See [93] for history and references.

4.2 The Discrete Hypercube

Consider the discrete hypercube $\mathcal{C}_n := \{-1, 1\}^n$ equipped with a probability measure ν. A natural counterpart for the Dirichlet form $\mathbb{E}|\nabla f|^2$ is

$$\mathcal{E}_\nu(f, f) := \sum_{x \sim y} \frac{\nu(\{x\})}{\nu(\{x\}) + \nu(\{y\})} (f(x) - f(y))^2,$$

where the summation is over all neighboring pairs of vertices on the discrete hypercube. A well-known fact is that if ν is taken to be the uniform measure, then one has the Poincaré-type inequality

$$\mathrm{Var}_\nu[f] \leq C \mathcal{E}_\nu(f, f), \tag{5}$$

where $C = 1/4$. This is one of many senses in which the uniform measure on the discrete hypercube can be thought of as a discrete analogue of the Gaussian measure. An inequality of the above form has direct implications on the existence of sampling algorithms from the measure v, which is one motivation for studying the following question.

Question 4.4 Find sufficient conditions on v under which (5) holds with a constant C independent of the dimension n.

A recent line of works, starting with the work of Anari, Liu and Oveis Gharan [16] gives an interesting sufficient condition for a polynomial dependence on the dimension. For $w \in \{-1, \star, 1\}^n$ define $\mathcal{R}_w v = v|_{S(u)}$ where $S(u) := \{x; \ u_i \in \{x_i, \star\}, \ \forall i \in [n]\}$, which we can understand as a "pinning" of some of the coordinates to given values. Moreover, define

$$\Psi(v)_{i,j} := \mathbb{E}_{X \sim v}[X_i | X_j = 1] - \mathbb{E}_{X \sim v}[X_i | X_j = -1],$$

sometimes called the "influence matrix" of the measure. Finally, set $\eta_i(v) = \max_{\|w\|=i} \|\Psi(\mathcal{R}_w v)\|_{\mathrm{OP}}$ where $\|w\| = \#\{j, w_j = \pm 1\}$. We have the following sufficient condition for an inequality of type (5).

Theorem 4.5 ([16]) *One has for all v and every test function f,*

$$\mathrm{Var}_v[f] \leq \mathcal{E}_v(f, f) \prod_{i=0}^{n-1} \left(1 - \frac{\eta_i(v)}{n-i}\right)^{-1}.$$

This theorem ensures that a Poincaré inequality holds when the covariance structure of restrictions of the measures are well-behaved.

A related condition for concentration was given in [62] in terms of the logarithmic Laplace transform of the measure v. Set

$$\mathcal{L}[v](\theta) := \log \int \exp(\langle x, \theta \rangle) dv(x).$$

Theorem 4.6 ([62]) *Suppose that v satisfies $\|\nabla^2 \mathcal{L}[v](\theta)\|_{\mathrm{OP}} \leq \beta$ for all $\theta \in \mathbb{R}^n$. Then for any function f satisfying $|f(x) - f(y)| \leq \|x - y\|_1$, one has*

$$\mathrm{Var}_v[f] \leq n^{2-c/(1+\beta)}$$

The above theorem is just a small improvement over the trivial bound $\mathrm{Var}_v[f] \leq n^2$ which follows from the diameter of $\mathcal{C}_n$. We conjecture that a much stronger concentration, in the form of a dimension-free spectral gap should follow from the same assumption.

Thus, the two above theorems are examples for (arguably) natural conditions which imply concentration inequalities, however at this point it is not quite well-understood what conditions should imply *dimension-free* concentration inequalities (which can easily be shown to be satisfied by the uniform measure).

4.3   Comments by V.M.

The goal of Eldan's note, as I understand it, is how to increase efficiency of the Concentration Phenomenon approach to some problems, especially discrete problems. To achieve this goal, Eldan discusses different possibilities to improve concentration estimates.

In my comments here I will suggest a different approach. The idea, as I will explain below, is taken from an old paper by M. Karpovsky and V. Milman ([87]). Our setting to demonstrate the idea will be an n-dimensional space E_q^n over a final field of q elements. Let $M \subset E_q^n \backslash 0$, $|M| = t$ and $|E_q^n \backslash 0| = q^n - 1$. Then (uniform) measure M in the set $E(n, q) = E_q^n \backslash 0$ is $\lambda(M, E_q^n) := \lambda = |M|/(q^n - 1) = t/(q^n - 1)$. Denote $G_q(n, k)$ the discrete Grassmanian of all k-dimensional subspaces of the E_q^n. The question we ask is the following: how to find the number $T = T(n, k_0, q)$ such that any set $M \subset E_q^n \backslash 0$ of at least T elements will contain some k_0-dimensional subspace $E \backslash 0 \in G_q(n, k_0)$. The exact answer may be found in that paper, but it is not the problem which interests us here, but the method to find it. So, we want to find E such that $\lambda(M \cap E, E) = 1$, but the original measure of M inside $E_q^n \backslash 0$ is $\lambda = t/(q^n - 1)$.

Note that if we average $\lambda(M \cap E, E)$, E is k-dimensional, over $G_q(n, k)$ we will receive the same number λ. However, because $\lambda(M \cap E, E) \cdot (q^k - 1)$ must be an integer, we may find a subspace E such that $\lambda(M \cap E, E) \geq]\lambda(q^k - 1)[$ (]a[is the smallest integer not less than a), and the measure of our set inside E is increased! One may think that this increase is insignificant and the exact answer cannot be achieved using such increases. However, what is shown in that paper that iterating this argument many times, reducing the dimension on every step only by 1 (i.e. for the starting dimension $\dim n$ to go down to dimension $n - 1$, then $n - 2, \ldots$, till the final dimension of k_0) we arrive at $\lambda(M \cap E, E) = 1$ exactly for the optimal dimension k. So, there is nothing lost at all if we "are not in a hurry" and repeat the trivial integration procedure maximal number of times.

I would think that a similar idea should be tried, and may be useful, in the continuous case.

The model problem I have in mind is the estimate in the Quotient of Subspace Theorem (QS-theorem):

Theorem (Milman) *Let X be an n-dimensional normed space. For every $1 \leq k < n$ there exists a subspace $Y \in QS(X)$ with $\dim(Y) = n - k$ and*

$$d(Y, \ell_2^{n-k}) \leq C \frac{n}{k} \log\left(\frac{C_n}{k}\right),$$

where $C > 0$ is an absolute constant (see [24], th. 7.9.1, page 241).

We would like to consider small codim. and write $k = \alpha n$. Then the estimate on the distance is

$$C \frac{1}{\alpha} \log \frac{C}{\alpha}.$$

Note that for $k = 1$ (i.e. $\alpha = 1/n$) the distance which the estimate provides is of the order n (I ignore here the logarithmic factor). However, we know that the worst possible distance of m-dim. space to the euclidean is of the order $\sqrt{m}$. So, it is suspected that in the above estimate the correct answer should be $\sim C(1/\sqrt{\alpha}) \log C/\alpha$. So, where could we lose this $\sqrt{}$?

Recall that the original proof of the theorem used iteration of the M^*-estimate of subspaces and quotient spaces (see Chapter 7 in [24]). However, we jumped in one step for a significant loss of dimension. But we can actually lose one dimension (or a few more) in one step (i.e. consider 1-codim. subspaces). This was impossible thinking when the original proof was discovered, but Gordon's estimate on M^* which appeared later (see formula (7.3.9), p. 241, in the same book) provides such a possibility.

5   Dmitry Faifman, the Weyl Principle in Valuation Theory, and Projective Geometries

5.1   Intrinsic Volumes and the Weyl Principle

The intrinsic volumes μ_j in Euclidean space, also known as quermassintegrals, go back to Steiner's formula. They are given by

$$\text{vol}(K + \epsilon B^n) = \sum_{j=0}^{n} \mu_j(K) \omega_{n-j} \epsilon^{n-j},$$

where K is any convex body, B^k the k-dimensional Euclidean ball, and ω_k its volume. The intrinsic volumes have many remarkable properties. Notably, Hadwiger has shown in 1957 [78] that they span the space of all continuous, rigid-motion invariant valuations on convex bodies.

The following remarkable property first observed by Weyl in 1939 [157] extended the intrinsic volumes from convexity to Riemannian geometry. Here we state it in a form taking advantage of the Nash embedding theorem.

Theorem 5.1 (Weyl) *Let M be a Riemannian manifold, and $A^k \subset M$ a submanifold. Fix an isometric embedding $M \hookrightarrow \mathbb{R}^n$, and denote $A_\epsilon = \{x \in \mathbb{R}^n : \text{dist}(x, A) \leq \epsilon\}$. Then for $\epsilon \ll 1$,*

$$\text{vol}(A_\epsilon) = \sum_{j=0} \mu_j(A)\omega_{n-j}\epsilon^{n-j},$$

with coefficients $\mu_j(A)$ that are independent of the isometric embedding.

The statement remains true for rather general subsets A, such as compact differentiable polyhedra. When $A = M$, the values $\mu_j(M)$ are known as the intrinsic volumes of M. They are given by integrals of the various Lipschitz-Killing curvatures; the most well-known is $\mu_0(M)$, which is just the Euler characteristic of M. Its integral representation $\mu_0(M) = \int_M LK_0$ is the contents of the Chern-Gauss-Bonnet theorem.

Weyl's theorem took a more refined form with the introduction of the theory of valuations on smooth manifolds by Alesker in the early 2000s [13]. Namely, the k-th intrinsic volume assigns to a Riemannian manifold (M, g) a valuation $\mu_k^M \in \mathcal{V}^\infty(M)$ which is k-homogeneous in $\sqrt{g}$, and commutes with isometric embeddings. Furthermore, the μ_k generate a canonical finite dimensional subalgebra in $\mathcal{V}^\infty(M)$ with respect to the Alesker product, which in many cases can be related to the integral geometry of M.

The proof of Weyl's theorem is a rather technical but mostly straightforward computation; a substantial part of the computation can be sidestepped by exploiting the invariant theory of the orthogonal group, which was also developed by Weyl. However it is fair to say that a conceptual proof of Weyl's theorem remains to be found.

It is therefore natural to look for generalizations of Weyl's theorem to other geometric settings, both for their own sake and in search of a unifying principle. Such extensions came to be known as the Weyl principle.

In the Finsler setting, a linear Weyl principle holds.

Theorem 5.2 (Alvarez Paiva-Fernandes [14], Bernig [34]) *In a normed space $(\mathbb{R}^n, F)$ there exist intrinsic volumes $\mu_j^F \in \text{Val}_j^+(\mathbb{R}^n)$ which restrict to the Holmes-Thompson k-volume on any given k-dimensional subspace. Moreover, their span is closed under the Alesker product.*

In fact, a multiplicative family of Holmes-Thompson intrinsic volumes is constructed for all reversible projective Finsler metrics.

Contrasting the Euclidean model, it was shown recently that a full-fledged Finsleriean extension of the Weyl principle is not possible.

Theorem 5.3 (Faifman-Wannerer [67]) *There exists a 3-dimensional submanifold $M \subset \mathbb{R}^6$ and two different norms F_1, F_2 on $\mathbb{R}^6$ that induce the same Finsler structure on M, but the restrictions of the corresponding first intrinsic volumes $\mu_1^{F_1}|_M, \mu_1^{F_2}|_M$ are distinct valuations.*

In light of those results, it is natural to ask how far the flat case can be extended.

Question 5.4 For which families of Finsler metrics can Holmes-Thompson intrinsic volumes be defined?

A particularly interesting case to understand would be that of the non-symmetric Funk metric, where invariance under projective polarity could be expected. For more on volume in Funk geometry, see the next section.

In pseudo-Riemannian geometry, Weyl's principle holds. The invariant valuations in general have singularities.

Theorem 5.5 (Bernig-Faifman-Solanes[40]) *There is for each $k \geq 0$ a natural functor μ_k, which assigning to a pseudo-Riemannian manifold a complex-valued generalized valuation $\mu_k \in \mathcal{W}_k^{-\infty}(M)$. The μ_k extend the corresponding Riemannian intrinsic volumes, and are compatible with isometric embeddings.*

A curious setting outside of metric geometry where the Weyl principle applies is that of contact manifolds. A contact structure is a maximally non-integrable hyperplane distribution on a manifold.

Theorem 5.6 (Faifman[64]) *A contact manifold M^{2n+1} admits a canonical family of generalized valuations $\mu_{2k} \in \mathcal{W}_{2k}^{-\infty}(M)$, $0 \leq k \leq n$, which are compatible with contact embeddings.*

Finally, in a yet unpublished work, Bernig-Fu-Solanes-Wannerer construct a family of valuations on general Kähler manifolds that generalize the hermitian intrinsic volumes [37] in the same way Weyl's theorem extends Euclidean intrinsic volumes. This last construction deviates from the Euclidean model, as the general Kähler manifold cannot be embedded into hermitian flat space even locally. Nevertheless, embedding is possible in a weaker sense, which proves sufficient.

It is worthwhile to notice that the holonomy group of a Kähler manifold is restricted to the unitary group. This led to the following question of Alesker.

Question 5.7 Which valuations, or valuation subalgebras, can naturally be defined on a Riemannian manifold with a specified holonomy group?

Let us finally mention an intriguing conjecture of Alesker [10], that asserts the extendibility of intrinsic volumes of Riemannian manifolds to certain Alexandrov spaces with curvature bounded from below.

5.2 Volume in Funk and Hilbert Geometries

The Funk and Hilbert metrics are natural examples of projective metrics, namely metrics that have straight segments as geodesics. Given a convex body $K \subset \mathbb{R}^n$, the *Funk metric* d^F arises from the Finsler structure that has K as the unit tangent ball at each $x \in \text{int}(K)$, with origin at x. It amounts to the non-symmetric distance function

$$d_K^F(x, y) = \log \frac{|xz|}{|yz|}, \quad z = (x + \mathbb{R}_+(y - x)) \cap \partial K.$$

The *Hilbert metic* is its symmetrization: $d_K^H(x, y) = \frac{1}{2}(d_K^F(x, y) + d_K^F(y, x))$. The Funk geometry is naturally affine-invariant, while Hilbert geometry is projectively invariant. Hilbert geometry has been studied quite extensively. One reason is that it generalizes hyperbolic geometry, and so contains hints to negative curvature phenomena in more general metric spaces. Another is the study of convex projective structures on manifolds, generalizing hyperbolic manifolds, where the Hilbert metric naturally replaces hyperbolic metric. Funk geometry received less attention.

One class of problems in those geometries concerns the volume of metric balls. Let $B_K^{F/H}(x, R)$ denote the ball of radius R around x in $K \subset \mathbb{R}^n$ in the Funk/Hilbert geometry. A well-known question in Hilbert geometry, attributed to Colbois-Verovic, asks to maximize $Ent_H(K) = \limsup_{R \to \infty} \frac{\log \text{vol}(B(x,R))}{R}$, the *volume growth entropy*.

Theorem 5.8 (Tholozan [144], Vernicos-Walsh [151]) *For any K, $Ent_H(K) \leq n - 1$.*

The maximum is achieved by ellipsoids (corresponding to the hyperbolic metric), and in fact by all C^2 bodies K with positive gaussian curvature. The conjecture has been resolved recently, by two very different approaches: Tholozan used a difficult result from PDE of Cheng-Yau [54], effectively replacing the Hilbert metric with the Riemannian Blaschke metric; Vernicos-Walsh used subdivisions into flag simplices and approximated general convex sets by polyhedra.

The question admits several refinements, which are in turn connected to some open questions in convex geometry.

Question 5.9 For a fixed radius $0 < R < \infty$, find the extremal values of $M_R(K) := \inf_{x \in \text{int}(K)} \text{vol}(B_K(x, R))$ and describe all equality cases.

Both Funk and Hilbert metrics can be addressed, and the definition of volume has to be specified. The connection to convex geometry is most pronounced for the Funk metric, with the Holmes-Thompson volume definition, which is the setting we will consider.

Conjecturally, $M_R(K)$ is maximized by ellipsoids for any R. This would strengthen and extend the Blaschke-Santaló inequality, which is recovered in the limit $R \to 0$. For $R \to \infty$, the centro-affine isoperimetric inequality is obtained. Presently ellipsoids are only known to maximize $M_R(K)$ among unconditional convex bodies [65].

Minimizing $M_R(K)$ for fixed R similarly extends the Mahler conjectured inequality $|K||K^o| \geq |B^n|^2$, which is recovered in the regime $R \to 0$, with the same conjectured minimizers, namely the Hanner polytopes. Of particular interest is the regime $R \to \infty$. The general case is well understood, as follows.

Theorem 5.10 (Vernicos [150], Vernicos-Walsh [151]) $c(K) := \limsup_{R\to\infty}$ $\frac{M_R(K)}{R^n}$ *is only finite for polyhedra, and is minimized by simplices.*

In fact, $c(K)$ is proportional to a combinatorial invariant known as the full flag number—it is the number of chains vertex$\subset \cdots \subset k$-dimensional face$\subset \cdots \subset$ K. Under the additional assumption of central symmetry, it is an old and open conjecture of Kalai [83] that the full flag number of a polytope is minimized by Hanner polytopes. A related conjecture, also of Kalai, is the minimality of the total face number of all dimensions among all centrally symmetric polytopes, again conjectured to be achieved by Hanner polytopes.

It's also worthwhile to note that the full flag number of a polyhedron recently appeared as a limiting value of another interesting affine construction in convex geometry, that of floating bodies and affine surface area [41]. More precisely, letting K_δ denote the δ-floating body of K, we have

Theorem 5.11 (Besau-Schuett-Werner) *For a polyhedron* $P \subset \mathbb{R}^n$,

$$\lim_{\delta\to 0^+} \frac{|P| - |P_\delta|}{\delta(\log\frac{1}{\delta})^{n-1}} = \frac{\text{flag}(P)}{n!n^{n-1}}.$$

Let us finally mention a surprising property of the Funk metric. While the Funk metric in $\text{int}(K)$ is only an affine invariant, many of the invariants associated to it are in fact projective invariants, most notably the Holmes-Thompson volume of a domain $\Omega \subset K$, and the billiard dynamics it induces on Ω. Moreover when Ω is itself convex, both are invariant under the projective duality $(K, \Omega) \leftrightarrow (\Omega^\vee, K^\vee)$. This further suggests the intriguing possibility of relating Funk geometry to the Mahler and Viterbo conjectures

5.3   Comments by V.M.

I don't feel Funk and Hilbert geometries in Convex bodies, Faifman's article is about volume in these geometries. But I would like to understand the geometries themselves. So, the first, introductory question is: Does, say, Hilbert geometry uniquely define euclidean balls?

I presume the answer is yes, and it may even be known.

But, also, how stable is it? How does the distance of the (Funk) Hilbert metric of K from this distance of the euclidean ball (of some radius) influence the distance of K from the euclidean ball?

What kind of Dvorestsky-type theorem is possible in this metric, or QS-theorem?

6 Bo'az Klartag, Recent Developments Towards Understanding the Distribution of Volume in High Dimensions

Upon first encountering geometry in high dimensions, one might think that the diversity and the rapid increase of the number of configurations would make it impossible to formulate general, interesting theorems that apply to large classes of high-dimensional geometric objects. However, it turns out that the contrary is often true. One example is the classical Central Limit Theorem, stating that product measures in high dimensions are rather regular and well-behaved, in the sense that they have approximately Gaussian marginals. Another example is Dvoretzky's theorem which demonstrates that any high-dimensional convex body has nearly-Euclidean sections of a high dimension.

There is a strong motif in high-dimensional geometry which compensates for the vast amount of different possibilities. This is the concentration of measure phenomenon, put forth by Milman starting with his proof of Dvoretzky's theorem. Quite unexpectedly, a scalar Lipschitz function on a high-dimensional space behaves in many cases as if it were a constant function. For example, if we sample 10 random points from the n-dimensional unit sphere, for large n, and substitute them into a 1-Lipschitz function, then we will almost certainly obtain 10 numbers that are very close to one another. This phenomenon is reminiscent of the well-known geometric property that in the high-dimensional Euclidean sphere, "most of the mass is close to the equator, for any equator". This geometric property, which may be strengthened using the spherical isoperimetric inequality, is unthinkable in, say, three dimensions.

Convexity assumptions fit naturally with high dimensionality, and enable us to harness this motif in order to formulate non-trivial theorems. For example, using concentration of measure ideas, we established the Central Limit Theorem for Convex Bodies, stating that if X is a random vector that is distributed uniformly in some convex body in $\mathbb{R}^n$, then there exists $0 \neq \theta \in \mathbb{R}^n$ such that

$$\langle X, \theta \rangle$$

is approximately a Gaussian random variable of mean zero and variance one; the total variation distance between $\langle X, \theta \rangle$ and the Gaussian is at most C/n^α where $C, \alpha > 0$ are universal constants. The general understanding that has emerged is that geometric conditions such as convexity, or positive Ricci curvature, may replace strong regularity assumptions such as symmetry or independence of the random variables.

Still, one of the central questions in high-dimensional convex geometry remains unsolved. This is the "slicing problem" originating from Bourgain's work in the 1980s. In its simplest formulation, the question is whether for any convex body $K \subseteq \mathbb{R}^n$ of volume one, there exists a hyperplane $H \subseteq \mathbb{R}^n$ such that

$$\text{Vol}_{n-1}(K \cap H) > c,$$

where $c > 0$ is a universal constant, and Vol_{n-1} stands for $(n-1)$-dimensional volume. Surprisingly, the answer to this simple-looking question is still unknown. It has several implications for convexity and quite a few equivalent formulations. Let us mention just one of these equivalent formulations: Suppose that $K \subseteq \mathbb{R}^n$ is a convex body of volume one. Does there exist an ellipsoid $\mathcal{E} \subseteq \mathbb{R}^n$ of volume one, such that the volume of $K \cap C\mathcal{E}$ is at least $1/2$, where $C > 0$ is a universal constant?

For many years, the best estimate for Bourgain's slicing problem has been $n^{1/4}$. That is, we knew that a convex body of volume one in $\mathbb{R}^n$ has a hyperplane section whose $(n-1)$-dimensional volume is at least $c \cdot n^{-1/4}$. A trivial bound is $c/\sqrt{n}$, which follows from Fubini theorem (since a convex body of volume one has a direction in which its width is at most $C\sqrt{n}$; recall that a Euclidean ball of volume one has a radius of the order of magnitude $\sqrt{n}$). There were reasons to believe that $n^{1/4}$ is the right answer, and in fact, there were three completely different proofs that led to this $n^{1/4}$:

1. The original proof by Bourgain, from the 1980s, relying on comparison with Gaussian processes and Pisier's deep theorem on the K-convexity constant. This proof yields an unnecessary logarithmic factor [47].
2. The author's argument from 2005 going through random "tilts" of the original convex body, through the log-Laplace transform, which utilizes Paouris' large deviations estimate.
3. The Lee-Vempala approach from 2016, that used the heat flow and Eldan's stochastic localization, in order to bound the thin-shell constant by $Cn^{1/4}$, which implies the same bound for the slicing problem.

At the time, the impression was that one perhaps needs to construct a counter-example. An analogous problem for projections of convex bodies was settled in the negative by Ball in the 1980s. Experts tried to construct a counter example by using random matrices: after all, if we work in dimension $n \times n = n^2$, then we are aiming at estimates involving $(n^2)^{1/4} = \sqrt{n}$, which is a prevalent scale in random matrix theory. Surprisingly, towards the end of 2020, Yuansi Chen improved the $n^{1/4}$ bound to $C_\epsilon n^\epsilon$, for all $\epsilon > 0$. His proof relied on familiar techniques, such as stochastic localization, with an ingenius regularity lemma that was proved by clever manipulations of 3-tensors. It feels as if we had a psychological block, a fixation on $n^{1/4}$, and a young newcomer showed us the light.

We now switch gears and move to a more technical description. For $n \geq 1$ define

$$\frac{1}{L_n} := \inf_{K \subseteq \mathbb{R}^n} \sup_{H \subseteq \mathbb{R}^n} \mathrm{Vol}_{n-1}(K \cap H),$$

where the infimum runs over all convex bodies $K \subseteq \mathbb{R}^n$ of volume one, and the supremum runs over all hyperplanes $H \subseteq \mathbb{R}^n$. Thus for years we knew that $L_n \leq Cn^{1/4}$, and then Chen proved that for any $\epsilon > 0$,

$$L_n \leq C_1 \exp\left(C_2\sqrt{\log n} \cdot \sqrt{\log\log(3n)}\right) \leq C_\epsilon n^\epsilon. \tag{6}$$

Chen arrives at (6) by exploiting the relation between the slicing problem and the thin-shell problem, due to Eldan and Klartag, who proved that

$$L_n \leq C\sigma_n, \tag{7}$$

where σ_n is the thin-shell constant which we will describe shortly. A probability density ρ in $\mathbb{R}^n$ is *log-concave* if the set $\{\rho > 0\} = \{x \in \mathbb{R}^n ; \rho(x) > 0\}$ is convex, and $\log \rho$ is concave in $\{\rho > 0\}$. A probability measure in $\mathbb{R}^n$ (or a random vector in $\mathbb{R}^n$) is log-concave if it is supported in an affine subspace of $\mathbb{R}^n$ and it has a log-concave density in this subspace. For instance, the uniform probability measure on any compact, convex set is log-concave, as well as all Gaussian measures. We say that a log-concave probability measure μ on $\mathbb{R}^n$ is *isotropic* if

$$\int_{\mathbb{R}^n} x_i d\mu(x) = 0 \quad \text{and} \quad \int_{\mathbb{R}^n} x_i x_j d\mu(x) = \delta_{ij} \qquad (i, j = 1, \dots, n),$$
$$\tag{8}$$

where δ_{ij} is Kronecker's delta. Thus, a log-concave probability measure is isotropic when it has mean zero and identity covariance. A log-concave probability measure has moments of all orders and the convolution of two log-concave probability measures is again log-concave. The relevance of the class of log-concave distributions to the slicing problem was realized by Ball in the 1980s. The thin-shell constant σ_μ of an isotropic, log-concave probability measure μ in $\mathbb{R}^n$ is defined via

$$n\sigma_\mu^2 = Var_\mu(|x|^2), \tag{9}$$

where $Var_\mu(f) = \int f^2 d\mu - \left(\int f d\mu\right)^2$. It may be shown that most of the mass of the measure μ is located in a thin spherical shell whose width is at most $C\sigma_\mu$, and this estimate for the width is always tight, hence the name *thin-shell constant*. The thin-shell constant is crucial for establishing the Central Limit Theorem for convex sets. The parameter σ_n mentioned above is defined as

$$\sigma_n = \sup_{\mu \text{ in } \mathbb{R}^n} \sigma_\mu$$

where the supremum runs over all isotropic, log-concave probability measures μ in $\mathbb{R}^n$. Earlier bounds for σ_n utilized the Concentration of Measure Phenomenon, but more recent advances, due to Eldan, Lee and Vempala and to Chen, deal with the Poincaré constant. The Poincaré constant $C_P(\mu)$ of a Borel probability measure μ in $\mathbb{R}^n$ is defined as the smallest constant $C \geq 0$ such that for any locally-Lipschitz function $f \in L^2(\mu)$,

$$Var_\mu(f) \leq C \cdot \int_{\mathbb{R}^n} |\nabla f|^2 d\mu. \tag{10}$$

The fact that $\sigma_\mu \leq 2C_P(\mu)$ for an isotropic, log-concave probability measure μ is easily proven:

$$n\sigma_\mu^2 = Var_\mu(|x|^2) \leq 4C_P(\mu) \int_{\mathbb{R}^n} |x|^2 d\mu(x) = 4nC_P(\mu). \tag{11}$$

The Poincaré constant is closely related to the *isoperimetric constant* or the *Cheeger constant* of μ. Given a probability measure μ in $\mathbb{R}^n$ with log-concave density ρ, its isoperimetric constant is

$$\frac{1}{\psi_\mu} = \inf_{A \subseteq \mathbb{R}^n} \frac{\int_{\partial A} \rho}{\min\{\mu(A), 1 - \mu(A)\}}$$

where the infimum runs over all open sets $A \subseteq \mathbb{R}^n$ with smooth boundary. By the Cheeger inequality and the Buser-Ledoux inequality, for any absolutely-continuous, log-concave probability measure μ in $\mathbb{R}^n$,

$$\frac{1}{4} \leq \frac{C_P(\mu)}{\psi_\mu^2} \leq \pi, \tag{12}$$

where the inequality on the left—Cheeger's inequality—is rather general and does not require log-concavity. Define

$$\psi_n := \sup_{\mu \text{ in } \mathbb{R}^n} \psi_\mu \tag{13}$$

where the supremum runs over all isotropic, log-concave probability measures μ in $\mathbb{R}^n$. The Kannan-Lovász-Simonovits (KLS) conjecture from the 1990s suggests that ψ_n is bounded by a universal constant. Thanks to (7), (11) and (12) we have the chain of inequalities

$$L_n \leq C\sigma_n \leq \tilde{C}\psi_n, \tag{14}$$

where $C, \tilde{C} > 0$ are universal constants. The right-hand side inequality in (14) may be reversed, up to a logarithmic factor. A deep theorem by Eldan from 2012 (Stochastic Localization was invented for its proof) states that

$$\psi_n^2 \leq C \log n \cdot \sum_{k=1}^{n} \frac{\sigma_k^2}{k} \leq \tilde{C} \log^2 n \cdot \sigma_n^2, \tag{15}$$

where the second inequality follows from the fact that $\sigma_{n+1} \geq \sigma_n$, which follows from the fact that $Var(|X|^2 + Z^2) \geq Var(|X|^2)$ when X is a log-concave random vector in $\mathbb{R}^n$ and Z is any real-valued, log-concave random variable that is independent of X. Chen uses Eldan's stochastic localization and the analysis of Lee and Vempala in order to show that

$$\psi_n \leq C_1 \exp\left(C_2\sqrt{\log n} \cdot \sqrt{\log\log(3n)}\right), \tag{16}$$

where $C_1, C_2 > 0$ are universal constants. This bound implies (6), in view of (14). The bound in (16) grows slower than any power law, and it is natural to expect that this bound for ψ_n may be improved to a polylogarithmic one. This is indeed true, as we recently showed with Lehec: for any $n \geq 2$,

$$\psi_n \leq C(\log n)^{\alpha} \tag{17}$$

for some universal constants $C, \alpha > 0$. Our proof yields $\alpha \leq 5$, though this exponent is probably non-optimal. For slicing, we obtain $L_n \leq (\log n)^4$. As of March 2022, it is the logarithmic factor that is yet to be understood in the slicing problem and the KLS isoperimetric conjecture.

For more thorough information about the interaction between high-dimensionality and convexity, we may refer the reader for instance to the recent book "Asymptotic Geometric Analysis I+II" by Artstein-Avidan, Giannopoulos and Milman, and to references therein.

6.1 Comments by V.M.

1. For the integers $1 \leq k < n$, define the number $C(k, n)$ as the smallest number C s.t. $\forall K$ and T in $\mathcal{K}_s^n$ (i.e. centrally symmetric convex compact bodies) the inequalities

$$\mathrm{Vol}_k(K \cap E) \leq \mathrm{Vol}_k(T \cap E),$$

for every subspace $E \in \mathcal{G}_{k,n}$, imply

$$\mathrm{Vol}\, K \leq C\, \mathrm{Vol}\, T.$$

The result of Klartag–Lehec discussed in Klartag's article shows that $C(n-1, n) \leq C(\log n)^4$ for some universal $C > 0$. This is a great result which follows 40 years of non-trivial development. Obviously, $C(1, n) = 1$ (as in this case $K \subseteq T$). Just recently Klartag improved this to $C(n-1, 1) \leq C\sqrt{\log n}$ where $C > 0$ is a universal constant, see [92].

Question Does $C(k, n)$ grow monotonically for a fixed n and $1 \leq k \leq n/2$?

It is an open question if $C(2, n) = 1$ and $C(3, n) = 1$. However, it is known after Bourgain and G. Zhang [49] that $C(k, n) > 1$ for any $k > 3$. (See Koldobsky [96].)

But the behavior of $C(k, n)$ for a fixed $k > 3$ and $n \to \infty$ is not known. Assuming that $C(n - 1, n)$ is not bounded and, as I expect, has logarithmic growth, I would expect that $C(k, n)$ has polylogarithmic growth by k (and $n \gg k$).

2. Now, when we have an upper estimate for the slicing problem of the order of a multi-logarithmic of the dimension, the search for counter-examples becomes different. We may expect only logarithmic low bound. I will describe below some series of convex bodies which I suspect may have isotropic constants which increase with dimension. I thought about these kind of examples already 20 years ago, but the second proof by Klartag provided power-type upper bound again as in Bourgain's results $n^{1/4}$, and this created the feeling that this may be the actual answer. This stopped considering examples which may provide only logarithmic low bounds.

We describe first the special embeddings of the sphere S^{k-1} to higher-dimensional spheres S^{N-1}. These embeddings $\varphi : S^{k-1} \to S^{N-1}$ will be actually a factor of isometry:

$$\text{denote} \qquad \varphi(S^{k-1}) = M \subset S^{N-1} \qquad \text{and}$$

$d_M(u, w) :=$ the length of the shortest curve inside M joining $u \in M$ with $w \in M$. Then

$$d_M\big(\varphi(x), \varphi(y)\big) = C \cdot \rho(x, y).$$

where ρ is the geodesic distance on S^{k-1}.

Of course, such embeddings exist only for very special N and explicitly constructed using irreducible representations of $SO(k)$ of so called Class 1. I suggest the excellent book by Vilenkin [154] Chapter 9, to consult on this subject. Such explicit embeddings use Gegenbauer polynomials.

The dimensions when such embeddings are possible we describe by the following formula: For any integer $d = 1, 2, \ldots$, consider

$$N := N(k, d) = \binom{k + d - 2}{d} \frac{2d + k - 2}{d + k - 2}.$$

For any such N the maps as we described above exist; there is $\varphi_d : S^{k-1} \to S^{N-1}$. And the constants $C := C(d)$ have the order around d. So, the manifold $M(k, d) = \varphi_d(S^{k-1})$ is, in some sense, highly "oscillated" on $S^{N(k,d)-1}$. Now, take

$$K_{k,d} = \text{Conv } M(k, d) \subset \mathbb{R}^N.$$

These are "my suspected" bodies for high slicing constants for any fixed k and increasing d to infinity. I suggest even to consider the case $k = 3$.

Again, very long ago Klartag pointed out to me that the case $K_{k,2}$ creates the 1-Shatten class. But what happens for d large is absolutely not clear. Note that all bodies $K_{k,d}$ are already in isotropic positions. The group theory creates them, which leads to a lot of symmetries and nice properties.

PS I would like to thank Boaz Klartag for reminding me about all this information from our past discussions.

7 Alexander E. Litvak, Random Matrices and Related Topics

7.1 Random Matrices in Asymptotic Geometric Analysis

Many phenomena in Asymptotic Geometric Analysis (**AGA**) are closely related to the behavior of singular values of random matrices, i.e., matrices whose entries are random variables from a certain class. Questions on distributions of singular values of such matrices are of major importance due to applications in pure and applied mathematics, statistics, computer sciences, electrical engineering, among others. Classical Random Matrix Theory has extensively studied corresponding *limiting* distributions for a long time. In sharp contrast, **AGA** interest concentrates on a *non-limiting* regime, although some results can be used to obtain limiting distributions. We consider a high dimensional random matrix and seek asymptotically sharp bounds for several parameters such as the largest and the smallest singular values which hold with an overwhelming probability, i.e., probability tending to 1 as dimensions grow to infinity (usually we ask even more—that probability tends to one exponentially fast in dimensions). While the Gaussian case was well understood for a long time (see survey [59] and references therein), till recently almost nothing was known for other classes of matrices (except bounds on norms). In [106] first results were obtained in this direction. From that time significant progress has been made in understanding the (non-limiting) behavior of singular values of matrices with independent entries [68, 130, 132, 133, 142, 143, 152]. Recall that the largest singular number, also called the *spectral norm*, is the operator norm of the matrix considered as a linear operator between corresponding Euclidean spaces, while the smallest singular number is the reciprocal of the spectral norm of the inverse (from the image) operator.

7.1.1. Approximation of the Covariance Matrix
In the works mentioned above the authors used assumptions of independence on entries. However in many problems coming from **AGA** a random matrix appears as a collection of independent vectors uniformly distributed on a given convex body. Among such questions are various questions related to random polytopes [72, 75, 76, 106] and questions on approximation of the covariance matrix. In the case of such vectors coordinates of the vector are not independent anymore. The problem of approximation of

covariance matrix (CM) by the empirical covariance matrix (ECM) is one of fundamental problems in Statistics. CM of a random n-dimensional vector X is

$$S = \mathbb{E}XX^t,$$

where X^t denotes the transpose. The ECM is obtained by random sampling:

$$S_N = (1/N) \sum_{i=1}^{N} X_i X_i^t,$$

where X_i's are independent copies of X. By the law of large numbers S_N converges to S. The quantitative question related to this convergence is:

Question What is $N = N(n)$ that guarantees a good approximation with high probability?

In the context of log-concave distributions this question was raised by three renown combinatorists, Kannan, Lovasz, and Simonovits, who studied random walks and needed to construct a volume computing algorithm with good bounds on the complexity [85]. They roughly estimated N as n^2. In 1996 Bourgain [48] obtained almost linear dependence $n \ln^3 n$. Then several strong mathematicians worked on this problem, improving the power of logarithm. In 2010 in [3] it was proved that proportionally many vectors are enough, i.e. N of the order n works. Moreover, the approximation works with high probability. After this the main interest of researchers shifted to the question: *under what minimal assumptions on a random vector X can one obtain approximation of CM?* One of natural assumptions is the boundedness of q-sth moments of one dimensional marginals of X together with a bound on maximum of norms of X_i. More precisely, we can formulate the questions as follows.

Question (Covariance Matrix) Let X be an n-dimensional random vector in the isotropic position, that is $\mathbb{E}XX^t = I$, the identity matrix. Let X_i, $i \leq N$ be independent copies of X and $S_N = \frac{1}{N} \sum_{i=1}^{N} X_i X_i^t$ denote the sample covariance matrix. Assume that there exist $q \geq 4, C_1, C_2 > 1$ such that

$$\max_{i \leq N} \|X_i\|^2 \leq C_1 n \qquad \text{with high probability} \tag{18}$$

and

$$\forall u \in S^{n-1} \quad \mathbb{E}|\langle X_i, u \rangle|^q \leq C_2^q. \tag{19}$$

Is it true that $\|S_N - I\| \leq C(n/N)^\alpha$, for an $\alpha = \alpha(q)$ and $C = C(q, C_1, C_2)$?

The setting in this question corresponds to well-known Bay–Yin [29] limiting result for matrices with independent identically distributed (*i.i.d.*) random entries with bounded 4-th moment (with $\alpha = 1/2$, see [124] for the non-limiting version). Several attempts were done to solve this question [98, 128, 139, 148, 153, 158], however either much stronger additional conditions were imposed or additional logarithmic factor appeared. Then in [125] this question was solved for $q > 8$ with $\alpha = 1/2$. In [74] it was proved with $q > 4$ and some $\alpha(q)$. Finally Tikhomirov [145] showed it with $\alpha = 1/2$ and, moreover, for $q = 4$ with $\sqrt{n/N} \ln^4(N/n)$ in place of $(n/N)^\alpha$ (he also proved for $q > 2$, however it is known that in the absence of 4-th moment the condition (1) does not hold even for matrices with i.i.d. entries and $N = n$ [105, 137]).

Question 7.1 (Optimality) Solve the above question with $q = 4$ and $\alpha = 1/2$ without logarithmic factor (or prove that this logarithmic factor is necessary).

Question 7.2 (Optimality) Let $q \geq 4$, $\beta \in (0, 1)$, and $n = \lfloor \beta N \rfloor$. Does (19) imply (18) with C_1 depending only on β? If not, what are (minimal) assumptions on the distribution required in order to get (18)?

7.1.2. Nonhomogeneous Gaussian Matrix and Sparse Matrices
A remarkable result of Seginer [136] states that for a square random matrix with i.i.d. symmetric random variables the expectation of its norm is of the order of the largest Euclidean norm of its rows and columns. The lower bound is obvious, while the upper bound is counter-intuitive. Surprisingly, a similar question about a random matrix with independent but not identically distributed entries was open till very recently even in the Gaussian case. After several steps toward the solution [31, 102, 131, 149] it was recently solved by Latała, van Handel, and Youssef [103]—they confirmed that Seginer's result holds in the case of independent centered Gaussian variables (note that in the case of Gaussian variables the expectation of the largest Euclidean norm of rows and columns can be computed in terms of variations of the variables). However the corresponding question about the smallest singular value is still open and seems much harder. To describe formally, let $B = \{b_{ij}\}_{i,j \leq n}$ be a fixed matrix with non-negative entries and let Γ be a random $n \times n$ matrix with $\mathcal{N}(0, b_{ij})$ independent entries. Let $s_n(\Gamma)$ denote the smallest singular value of Γ. Note, that for the standard Gaussian matrix, that is, when all $b_{ij} = 1$, one has $s_n(\Gamma) \approx 1/\sqrt{n}$. The following question is a long-term program in this direction.

Question 7.3 (Smallest Singular Value) Describe the behaviour of $s_n(\Gamma)$ in terms of B.

A weaker interesting question is

Question 7.4 Under what minimal assumptions on B, $s_n(\Gamma)$ is of the order $1/\sqrt{n}$?

Some initial steps in this direction were done in [104, 107, 134].

7.2 Approximation of Convex Bodies by Polytopes with a Small Number of Vertices or Faces

Approximation of convex bodies by polytopes is one of central subjects in Convex Geometry. In our setting it can be formulated as follows. Let $n, d > 1$. What is the minimal $N = N(d, n)$ such that every convex body can be approximated up to an error d (in the sense of the Banach–Mazur distance) by a convex polytope P_N with N vertices? More precisely,

Question 7.5 (Approximation of Convex Bodies by Polytopes) Given $n, d > 1$ what is the minimal N such that for every convex body $K \subset \mathbb{R}^n$ there exist $x_1, ..., x_N \in K$ satisfying $P_N \subset K \subset dP_N$, where P_N is the convex hull of these x_i's?

Equivalently, one can investigate the inverse function, that is, given $1 \leq n \leq N$, one can ask for the minimal $d = d(N, n)$ with the above property. The case when the distance d is close to 1, say $d = 1 + \epsilon, \epsilon \in (0, 1)$, is known. After several works [33, 50, 51, 60] an almost sharp bound

$$N(1 + \epsilon, n) \leq C^n \epsilon^{-(n-1)/2}$$

was obtained in [127]. The case of large d is less studied. In the symmetric case, that is, when $K = -K$, it follows from the John decomposition that Cn points already give the distance of the order of $\sqrt{n}$, namely, $N(3\sqrt{n}, n) \leq 8n$ [33, 73]. For $2 \leq d \leq \sqrt{n}$ and $K = -K$, Barvinok [33] obtained the bound

$$N(d, n) \leq \exp(C(n/d^2) \ln d).$$

In the general case of not necessarily symmetric bodies, considering the simplex of the maximal volume in K, one has $N(n + 2, n) \leq n + 1$. Brazitikos, Chasapis, and Hioni [50], Naszodi [126] and Szarek [141] independently proved that for $2 \leq d \leq n$

$$N(d, n) \leq n \exp(Cn/d).$$

The latter bound does not look to be sharp. In particular, it is not clear when (for what values d) $N(d, n)$ changes the behaviour from polynomial (in n) to exponential (in n). The approaches in these three papers are completely different—while Szarek used a greedy algorithm to construct "good" points, the authors of two other papers used a random choice of points and either certain tools from combinatorics [126] or properties of L_p-centroid bodies [50].

7.3 Comments by V.M.

I would like just explicitly state an idea which is actually implicitly presented in the Litvak's text.

One of unexpected consequences of the introduction and study of the isotropicity in AGA was the following observation. Recall first that convex compact K is in the isotropic position if a random vector in K with respect to volume measure is isotropic (as it is defined in Litvak's article). It was studied in AGA in [121] and the authors were inspired by the earlier work by F. John [81] (see [121] and references there). Fix the standard Euclidean scalar product $\langle ., . \rangle$ in $\mathbb{R}^n$ and let $K \subset \mathbb{R}^n$ be in the isotropic position in this Euclidean norm. Let X be randomly selected in K (with respect to the volume, and let the volume of K be equal 1). Then the coordinates $\{x_j \in \mathbb{R}\}$ of $X = (x_1, ..., x_n)$ (in the standard Euclidean decomposition) behave similarly to independent random variables. And this similarity became more precise when n tends to infinity.

Still back in 1995 I suggested to Semyon Alesker as one of a possible interesting direction to study works of M. Marcus and G. Pisier on uniform convergence of random trigonometric series (see the book [114]). They considered trigonometric series and put random signs for i-th coefficient. And very interesting results were following for i.i.d. variables $\{\epsilon_i = \pm 1\}$. So, the suggestion was to consider a family of isotropic convex bodies $\{K_j\}$, dim K_j tends to infinity, and use instead of $\{\epsilon_i\}$ the coordinates of random vector $X = (x_1, ..., x_{n_j})$ uniformly distributed in K_j, $n_j = \dim K_j$. However, a problem was necessary to be still formulated, as the results of Marcus-Pisier were for infinite series, but dimensions of K_j were finite.

Unfortunately, the problem was not moved as Alesker became involved in the valuation theory and was very successful there.

The ideology was used later, say in O. Friedland [70] and A. Pajor and L. Pastur [129], but not written very explicitly, although the authors definitely understood this point.

It would be nice to develop it in a more explicit form.

I would also like also to mention that once in Paris, in the late 1990s or the first decade of our millennium, I attended the talk by K. Ball where he presented the explicit dictionary of correspondence between isotropic convex bodies in high dimension and the language of independent random variables. However, I don't know his results on this subject to mention.

8 Emanuel Milman, Isomorphic Version of the Log-Brunn–Minkowski Inequality

One important question in contemporary Brunn–Minkowski theory is that of existence and uniqueness in the L^p-Minkowski problem for $p \in (-\infty, 1)$: given a finite non-negative Borel measure μ on the Euclidean unit-sphere $\mathbb{S} = S^{n-1}$, determine conditions on μ which ensure the existence and/or uniqueness of a convex body K in $\mathbb{R}^n$ so that:

$$S_p K := h_K^{1-p} S_K = \mu. \tag{20}$$

Here h_K and S_K denote the support function and surface-area measure of K, respectively. When $h_K \in C^2(\mathbb{S})$,

$$S_K = \det(D^2 h_K)\mathfrak{m},$$

where $\mathfrak{m}$ is the induced Lebesgue measure on $\mathbb{S}$, $D^2 h_K = \nabla_{\mathbb{S}}^2 h_K + h_K \delta_{\mathbb{S}}$ and $\nabla_{\mathbb{S}}$ is the Levi-Civita connection on $\mathbb{S}$ with its standard Riemannian metric $\delta_{\mathbb{S}}$. Consequently, (20) is a Monge–Ampère-type equation.

The case $p = 1$ above corresponds to the classical Minkowski problem of finding a convex body with prescribed surface-area measure; when μ is not concentrated on any hemisphere and its barycenter is at the origin, existence and uniqueness (up to translation) of K were established by Minkowski, Alexandrov and Fenchel–Jessen. The extension to general p was put forth and publicized by E. Lutwak [110] as an L^p-analog of the Minkowski problem for the L^p surface-area measure $S_p K = h_K^{1-p} S_K$ which he introduced. Existence and uniqueness in the class of origin-symmetric convex bodies, when the measure μ is even and not concentrated in a hemisphere, was established for $n \neq p > 1$ by Lutwak [110] and for $p = n$ by Lutwak–Yang–Zhang [112]. A key tool in the range $p \geq 1$ is the prolific L^p-Brunn–Minkowski theory, initiated by Lutwak [110, 111] following Firey [69], and developed by Lutwak–Yang–Zhang and others, which extends the classical $p = 1$ case. Recall that the L^p-Minkowski sum $a \cdot K_0 +_p b \cdot K_1$ of $K_0, K_1 \in \mathcal{K}$ ($a, b \geq 0$) was defined by Firey for $p \geq 1$ [69], and extended by Böröczky–Lutwak–Yang–Zhang [44, 45] to all $p \in \mathbb{R}$, as the largest convex body (with respect to inclusion) L so that:

$$h_L \leq \left(a h_{K_0}^p + b h_{K_1}^p \right)^{1/p}$$

(with the case $p = 0$ interpreted as $h_{K_0}^a h_{K_1}^b$ when $a+b = 1$). Note that for $p \geq 1$ one has equality above, that the case $p = 1$ coincides with the usual Minkowski sum, and that for $p < 1$ the resulting convex body $a \cdot K_0 +_p b \cdot K_1$ is the Alexandrov body associated to the continuous function on the right-hand-side.

The case $p < 1$ turns out to be more challenging because of the lack of an appropriate L^p-Brunn–Minkowski theory. Existence, (non-)uniqueness and regularity under various conditions on μ were studied by numerous authors when $p < 1$ (from either side of the critical exponent $p = -n$). The case $p = 0$ is of particular importance as it corresponds to the *log-Minkowski problem* for the cone-volume measure

$$V_K := \frac{1}{n} h_K S_K = \frac{1}{n} S_0 K,$$

obtained as the push-forward of the cone-measure on ∂K onto $\mathbb{S}$ via the Gauss map, and having total mass $V(K)$, the volume of K. Being a self-similar solution to the isotropic Gauss curvature flow, the case $p = 0$ and $\mu = \mathfrak{m}$ of (20) describes the ultimate fate of a worn stone in a model proposed by Firey.

Let $\mathcal{K}$ denote the collection of convex bodies in $\mathbb{R}^n$ containing the origin in their interior, and let $\mathcal{K}_e$ denote the subset of origin-symmetric elements. In [45], Böröczky–Lutwak–Yang–Zhang showed that an *even* measure μ is the cone-volume measure V_K of an *origin-symmetric* convex body $K \in \mathcal{K}_e$ if and only if it satisfies a certain subspace concentration condition, thereby completely resolving the existence part of the *even* log-Minkowski problem. As put forth by Böröczky–Lutwak–Yang–Zhang in their influential work [44, 45] and further developed in [97], the uniqueness question is intimately related to the validity of a conjectured L^0- (or log-)Brunn–Minkowski inequality for origin-symmetric convex bodies $K, L \in \mathcal{K}_e$, which would constitute a remarkable strengthening of the classical $p = 1$ case.

Specifically, the following equivalence may be shown by following the arguments of [44, 45]. We denote by $\mathcal{K}_{+,e}^{2,\alpha}$ the subset of $\mathcal{K}_e$ having $C^{2,\alpha}$-smooth boundary and strictly positive curvature.

Theorem 8.1 (After Böröczky–Lutwak–Yang–Zhang) *The following statements are equivalent for any fixed $p \in (-n, 1)$:*

1. *For any $q \in (p, 1)$, uniqueness holds in the even L^q-Minkowski problem for any $K \in \mathcal{K}_{+,e}^{2,\alpha}$:*

$$\forall L \in \mathcal{K}_e, \quad S_q L = S_q K \implies L = K. \tag{21}$$

2. *The even L^p-Brunn–Minkowski inequality holds:*

$$\forall \lambda \in [0, 1] \quad \forall K, L \in \mathcal{K}_e \quad V((1 - \lambda) \cdot K +_p \lambda \cdot L)$$
$$\geq \left((1 - \lambda) V(K)^{\frac{p}{n}} + \lambda V(L)^{\frac{p}{n}} \right)^{\frac{n}{p}}. \tag{22}$$

The case $p = 0$, called the even log-Brunn–Minkowski inequality, is interpreted in the limiting sense as:

$$V((1 - \lambda) \cdot K +_0 \lambda \cdot L) \geq V(K)^{1-\lambda} V(L)^{\lambda}. \tag{23}$$

3. *The even L^p-Minkowski inequality holds:*

$$\forall K, L \in \mathcal{K}_e \quad \frac{1}{p} \int_{\mathbb{S}} h_L^p dS_p K \geq \frac{n}{p} V(K)^{1-\frac{p}{n}} V(L)^{\frac{p}{n}}. \tag{24}$$

The case $p = 0$, called the even log-Minkowski inequality, is interpreted in the limiting sense as:

$$\frac{1}{V(K)} \int_{\mathbb{S}} \log \frac{h_L}{h_K} dV_K \geq \frac{1}{n} \log \frac{V(L)}{V(K)}.$$

Using Jensen's inequality in formulation (24) (or (22)), it is immediate to check that the above (equivalent) statements become stronger as p decreases. The restriction to origin-symmetric bodies is natural, and necessitated by the fact that no L^p-Brunn–Minkowski inequality nor uniqueness in the L^p-Minkowski problem can hold for general convex bodies when $p < 1$. Even when restricting to origin-symmetric bodies, it is easy to show that (22) or (24) are false for any $p < 0$, and that uniqueness in (21) does not hold for general $K, L \in \mathcal{K}_e$ and $q = 0$, as may be verified by testing two different centered parallelepipeds with appropriately chosen parallel facets.

Conjecture 8.2 (Böröczky–Lutwak–Yang–Zhang, "Even log-Brunn–Minkowski Conjecture") Any (and hence all) of the above statements hold for origin-symmetric convex bodies in the "logarithmic case" $p = 0$ (and hence for all $p \in [0, 1)$ as well).

A confirmation of this conjecture would constitute a dramatic improvement over the classical Brunn–Minkowski theory for the subfamily of origin-symmetric convex bodies, which had gone unnoticed for over a century. The conjecture is known to hold in the plane [44], but remains open in general for $n \geq 3$.

Various partial results are known regarding the BLYZ conjecture (see e.g. [97, 120]). The main result in [97] confirmed the *local* uniqueness in the even L^p-Minkowski problem (21) for all $K \in \mathcal{K}_{+,e}^{2,\alpha}$ and $p \in (p_0, 1)$ for $p_0 := 1 - \frac{c}{n^{3/2}}$. In [53], Chen–Huang–Li–Liu established a local-to-global principle for the uniqueness question, and deduced (21) and (24) for all $K \in \mathcal{K}_{+,e}^{2,\alpha}$ and $p \in (p_0, 1)$. In fact, thanks to recent progress on the KLS conjecture due to Y. Chen [52], the estimate from [97] immediately improves to $p_0 = 1 - \frac{c_\epsilon}{n^{1+\epsilon}}$ for any $\epsilon > 0$.

In [120], the following isomorphic version of the conjecture regarding uniqueness in the even log-Minkowski problem was recently resolved. We denote by $d_G(K, L)$ the geometric distance between two origin-symmetric bodies K, L, namely $d_G(K, L) := \inf\{ab > 0 \; ; \; \frac{1}{b}K \subset L \subset aK\}$.

Theorem 8.3 (Isomorphic Log-Minkowski) *For any $\bar{K} \in \mathcal{K}_e$, there exists $\tilde{K} \in \mathcal{K}_{+,e}^{\infty}$ with:*

$$d_G(\bar{K}, \tilde{K}) \leq 8,$$

so that for any $T \in GL_n$, the even log-Minkowski problem for $K = T(\tilde{K})$ has a unique solution:

$$\forall L \in \mathcal{K}_e \, , \quad V_L = V_K \Rightarrow L = K,$$

and the even log-Minkowski inequality holds for K:

$$\forall L \in \mathcal{K}_e \quad \frac{1}{V(K)} \int_{\mathbb{S}} \log \frac{h_L}{h_K} dV_K \geq \frac{1}{n} \log \frac{V(L)}{V(K)},$$

with equality if and only if $L = cK$ for some $c > 0$.

The constant 8 obtained in the isomorphic version above is the worst case behavior for a general $\bar{K} \in \mathcal{K}_e$, when $D = d_{BM}(\bar{K}, B_2^n)$ may be as large as John's upper bound $\sqrt{n}$. However, whenever $D \ll \sqrt{n}$, a slightly finer analysis yields an *isometric* version of the above results, where one only perturbs $\bar{K}$ by at most $\gamma = 1 + \epsilon$, with $\epsilon = C \frac{\sqrt{D}}{\sqrt[4]{n}}$.

Theorem 8.3 is a result about existence of an isomorphic position in a localized version of the log-Brunn-Minkowski inequality problem by BLYZ (Conjecture 8.2 above). At the same time we do not know if the following problem has a positive solution:

Problem 8.4 *There is a universal constant $C > 0$ such that for every origin-symmetric convex body K there is an isomorphic version K' such that $d_G(K, K') < C$, and such that for any two bodies K and L, their isomorphic versions K' and L' satisfy the log-Brunn–Minkowski inequality (23).*

8.1 Comments by V.M.

8.1.1 Isomorphic Position of Convex Body

I will consider only centrally symmetric convex bodies in the n-dimensional real space, i.e. such convex compact K that $K = -K$ and with non-empty interior. Of course, we may think of such K as the unit ball of some normed space X and, to emphasize this, we will write $K := K(X)$. For any non-degenerated linear map u in $\mathbb{R}^n$, of course, uK is the unit ball of isometrically the same normed space. However, geometrically it is a different body in $\mathbb{R}^n$. We call such a body a position of K. With every K there is associated with it a family of very interesting ellipsoids, which reflect, actually, different hidden symmetries in K. In the asymptotic study of the normed spaces and convex bodies, by increasing dimension to infinity the role of selected positions is crucial. Different remarkable properties of convex bodies (one may call them hidden symmetries) are recovered by considering them in different positions specially selected for different goals. We understand this part of the theory very well now.

However, many very central problems of the Asymptotic Geometry of high-dimensional convex bodies are still open and I would suggest here an additional "step of freedom" in attacking them. These reflections are inspired by two results. One of them has been known for a relatively long time. This is the result of Klartag from 2006. Another result is by Emanuel Milman presented in the contribution above.

To solve some specific problem (let us call it Problem X) of the Asymptotic Theory, we will ask if there is a universal constant C such that for every dimension n and every convex body K (from our family) in $\mathbb{R}^n$ one may find another body T (from the family, i.e. centrally symmetric convex body) and the Banach–Mazur distance at most C from K and such that the Problem X would have a solution for T. Such a T we will call now isomorphic position of K.

And now we will recall the remarkable result of Klartag from 2006 called the isomorphic version of the Bourgain's slicing problem. We suggest the books [24, 25] to see the details.

Problem 8.5 *Let* $K, T \subseteq \mathbb{R}^n$ *be centrally-symmetric convex bodies such that* $\mathrm{Vol}_{n-1}(K \cap \theta^\perp) \leq \mathrm{Vol}_{n-1}(T \cap \theta^\perp)$ *for all* $\theta \in S^{n-1}$. *Does it follow that* $\mathrm{Vol}_n K \leq C \cdot \mathrm{Vol}_n T$ *for some universal constant* C?

This is known as Bourgain's slicing problem (from 1985, see [46]). A positive answer would have important consequences in convex geometry. In some sense the slicing problem, also called the hyperplane conjecture, is the "opening gate" to a better understanding of uniform measures in high dimensions. The problem is still open. See Klartag's contribution in this collection.

The problem may be reduced to estimating the isotropic constants L_K of the convex bodies K (see [24] for the definition of L_K; and for the present state of knowledge see Klartag's note in this collection).

However, Klartag found another approach to the problem slightly modifying the question (see [90]).

Theorem 8.6 (B. Klartag (2006)) *Let* $K \subset \mathbb{R}^n$ *be a convex body and* $0 < \epsilon < 1$. *Then there exists a convex body* $T \subset \mathbb{R}^n$ *such that*

(i) $(1 - \epsilon)T \subseteq K \subseteq (1 + \epsilon)T$.
(ii) $L_T < C/\sqrt{\epsilon}$, *where* $C > 0$ *is a universal constant.*

Later, in 2018, Klartag additionally proved that the body T from the theorem can be assumed to be a projective image of K [91].

The problem has a positive solution but in an isomorphic sense: there is an isomorphic position for which the Problem 8.5 is solved.

So we now have two problems solved in the isomorphic form: one of Klartag in 2006 and another in this article by E. Milman, which is the isomorphic version of the Log-Brunn–Minkowski inequality.

More isomorphic versions of well-known problems of AGA.
In the article by B. Klartag and V. Milman [95] we listed a number of problems which are connected with the slicing problem of Bourgain and either follow from it,

in the case of the positive solution, or implied it if they would be positively solved. In every of these problems one may ask if their isomorphic versions would be correct. Let me list some of them:

(i) the thin shell conjecture would imply the hyperplane conjecture (see [61])
(ii) Kannan, Lovasz, Simonovich (KLS) isoperimetric conjecture [84]
(iii) Mahler conjecture on the low bound for the product of the volumes of the convex body and its polar (see [91] for connections with slicing problem)
(iv) problems on "quick Steiner symmetrizations" (see [94]).

I am not introducing these problems here. I refer the reader to [95] for their exact formulations and to the books [24] and [25] for detail discussions of these major problems of the theory. As an example, let me just formulate some of these problems in the isomorphic form.

(ii) Isomorphic KLS Problem: Do universal constants C and C' exist such that for every centrally-symmetric convex body K there exist another centrally-symmetric body K' such that Banach-Mazur distance $d(K, K') < C$ and KLS conjecture is correct for K' with a constant C' ?

(iii) Isomorphic Mahler problem: Does a universal constant C exist such that for any centrally-symmetric convex body K there is a body K' such that $d(K, K') < C$ and the Mahler volume of K' is more (or equal) to the Mahler volume of the cube (of the same dimension).

Many other problems of AGA may be reformulated the same way, and all of them, beside what is written above, are open.

9 Yaron Ostrover, Convex Bodies in the Classical Phase Space

Consider a convex body K in the phase space of classical mechanics $\mathbb{R}^{2n}$, where the latter is equipped with the standard symplectic form $\omega = \sum dx_i \wedge dy_i$. There is a natural dynamical system (known as the "Reeb Dynamics") associated with the body K. Indeed, the restriction of ω to the boundary ∂K canonically defines a one-dimensional sub-bundle, $\ker(\omega|_{\partial K})$, whose integral curves comprise the characteristic foliation of ∂K. In local coordinates, this dynamics, which is the classical Hamiltonian dynamical system on the energy surface ∂K, is given by $\dot{z} = J \nabla g_K(z)$, where g_K is the gauge function of K, and $J^2 = -\mathrm{Id}$ is the standard complex structure on $\mathbb{R}^{2n} \simeq \mathbb{C}^n$.

A classical result proven independently by Rabinowitz, Weinstein, and Clarke asserts that the Reeb dynamics associated with a convex body always posses a periodic orbit. One can naturally assign a number ("action") to a periodic orbit γ via $A(\gamma) = \int_\gamma \lambda$, where λ is a primitive of ω, i.e., $\omega = d\lambda$. The quantity $A(\gamma)$ is the symplectic area of a disc spanned by the loop γ. Let $c(K)$ be the minimal action among all closed characteristics on ∂K.

Fact It is known that $c(K)$ is a symplectic invariant, monotone with respect to inclusion, 2-homogenuous, and moreover satisfies a Brunn-Minkowski type inequality [21] i.e.,

$$c^{1/2}(K + T) \geq c^{1/2}(K) + c^{1/2}(T).$$

A weak version of a famous conjecture by C. Viterbo [155] states that

Conjecture (Viterbo) For a convex body $K \subset \mathbb{R}^{2n}$ one has

$$c^n(K) \leq n!\text{Vol}(K).$$

In [22] Viterbo's conjecture was proved up to a universal constant, i.e., there is $A \in \mathbb{R}$ so that $c^n(K) \leq A^n n!\text{Vol}(K)$. By considering convex domains of the form $T \times T^\circ \subset \mathbb{R}^{2n}$, where $T \subset \mathbb{R}^n$ is a centrally symmetric convex body, and interpreting the Reeb dynamics associated with $T \times T^\circ$ in terms of Minkowski billiard dynamics, it was proved in [23] that $c(T \times T^\circ) = 4$ for every centrally symmetric convex $T \subset \mathbb{R}^n$. This immediately implies

Theorem (Artstein-Avidan, Karasev, Ostrover) *Viterbo's volume-capacity conjecture implies the symmetric Mahler Conjecture.*

Remark Using a "tensor power trick" (based on 2-products), one can show [80] that it is enough to prove Viterbo's conjecture in the asymptotic regime, i.e., when $n \to \infty$.

Remark It was observed in [4] that one can not use the approach of [23] to relate Viterbo's conjecture with the general Mahler conjecture. Still, it might be interesting to find a "symplectic approach" to the Mahler conjecture also for non-symmetric convex bodies.

Remark Viterbo's conjecture is still open in general. Recent progress includes a proof of a "local version" in a C^3-small neighborhood of the ball (see [2], and Corollary 2 in [1]), and a proof of the conjecture in some special cases (see Theorem 2.1 in [86], and Theorem 1.11 in [77]).

In [79], P. Haim-Kislev studied Reeb dynamics on convex polytopes, and proved the following beautiful "combinatorial formula":

Theorem (Haim-Kislev) *Let $P \subset \mathbb{R}^{2n}$ be a convex polytope. Then,*

$$c(P) = \frac{1}{2}\left[\max_{\sigma \in S_{P_F}, (\beta_i) \in M(P)} \sum_{1 \leq j \leq i \leq P_F} \beta_{\sigma(i)} \beta_{\sigma(j)} \omega(n_{\sigma(i)}, n_{\sigma(j)})\right]^{-1},$$

where P_F is the number of facets of P, S_m is the symmetric group of m letters, and

$$M(P) = \left\{(\beta_i)_{i=1}^{P_F} \mid \beta_i \geq 0, \sum_{i=1}^{P_F} \beta_i h_i = 1, \sum_{i=1}^{P_F} \beta_i n_i = 0\right\},$$

where n_i is the unit outer normal to the facet F_i, and $h_i = h_P(n_i)$ is the "oriented height".

This leads naturally to the following

Question 9.1 Can one use Haim-Kislev's theorem to prove Viterbo's Conjecture?

There are other natural questions regarding the quantity $c(K)$. For example,

Question 9.2 For a convex body K in $\mathbb{R}^{2n}$, study the position $A \in SL(2n)$ which maximizes the capacity $c(AK)$.

In [5], motivated by Bang's problem, the authors conjectured that the symplectic capacity of a convex body $c(K)$ is subadditive. More precisely,

Conjecture (Akopyan, Karasev, Petrov) If a convex body $K \subset \mathbb{R}^{2n}$ if covered by a finite set of convex bodies $\{K_i\}_{i=1}^M$, then one has

$$c(K) \le \sum_{i=1}^{M} c(K_i).$$

A partial answer for hyperplane cuts is given in [79]. Another research direction concerns a "symplectic characterisation" of the Euclidean ball among convex domains in $\mathbb{R}^{2n}$. More precisely, we say that a smooth convex body K is "Zoll" if for every point $x \in \partial K$ one has a closed characteristic (of the Reeb dynamics) passing through x, and moreover all the closed characteristics have the same action. It was proved in [2] that every smooth Zoll body in $\mathbb{R}^4$ is a symplectic ball. It is an open question whether this is true in higher dimensions. For a convex polytope P in $\mathbb{R}^{2n}$, it would be interesting to study combinatoricals/dynamical characterizations that imply that P is symplectomorphic to a Euclidean ball up to an ε-neighborhood. Moreover, one can ask the following very general question:

Question 9.3 How much information on a convex body K is encoded in the Reeb dynamics on its boundary?

More precisely, there are several infinite families of numerical symplectic invariants based on the Reeb dynamics (such as the Ekeland-Hofer capacities or the ECH capacities). It was proved in [58] that one can relate the asymptotics of the ECH capacities of a convex $K \subseteq \mathbb{R}^4$ to the volume of K. It is very interesting to generalize this result, and understand exactly how much information about a convex body $K \subset \mathbb{R}^{2n}$ (e.g., volume, surface area, inradius, etc.) can be recovered via these invariants?

Symplectic Capacities and Billiards There is a natural connection between the classical billiard dynamics in a convex body K in the configuration space $\mathbb{R}_q^n$, and the Reeb dynamics of $K \times B$ in the phase space $\mathbb{R}_q^n \times \mathbb{R}_p^n$, where B is the Euclidean ball (see e.g., Section 2 in [23]). Roughly speaking, this is a consequence of the fact that $K \times B$ can be considered as the cotangent bundle of K. More precisely, the projection of any closed characteristic γ on $\partial(K \times B)$ is a billiard trajectory inside K, and vice versa, any billiard trajectory of K can be lifted to a closed characteristic of $K \times B$. In particular, the symplectic capacity $c(K \times B)$ is the length of the shortest periodic billiard trajectory in K.

Remark This connection implies in particular that the length of the shortest periodic billiard trajectory in a convex body $K \subset \mathbb{R}^n$ is monotone with respect to inclusion, and that it satisfies a Brunn-Minkowski type inequality [21]. Moreover, if one replaces the Euclidean ball B above with a convex body T, then the same holds for the so-called "Minkowski billiards" (i.e., Billiard dynamic in normed spaces, which model the propagation of waves in a homogeneous, anisotropic medium that contains perfectly reflecting mirrors).

Remark The symplectic point of view on billiard dynamics immediately implies the following "duality result": the billiard dynamics in a (centrally symmetric) convex body K with respect to the norm associated with a (centrally symmetric) convex body T, is equivalent to the billiard dynamics inside T govern by the norm associated with the body K. We remark that the centrally symmetry assumption here is not really necessary.

There are many open questions regarding the shortest periodic billiard trajectory in a convex body. For example, already in $\mathbb{R}^2$, the following is unknown:

Question 9.4 Which $K \subset \mathbb{R}^2$ with $Area(K) = 1$ maximizes $c(K \times B)$? In other words, for which convex body with a given area the shortest periodic billiard orbit is maximal?

Very recently, D. Tsodikovich proved a Blaschke–Santaló type inequality for the shortest periodic billiard trajectory in a convex body $K \subset \mathbb{R}^n$ (see [147]). It would be interesting to further explore the properties of the length of the shortest periodic billiard trajectory in a convex body $K \subset \mathbb{R}^{2n}$, and its interrelations with other classical quantities associated with the convex body K. I end this note with the following question about billiard dynamics:

Question 9.5 What is the length of the shortest periodic billiard trajectory in a "random" convex body?

9.1   *Comments by V.M.*

There are many unexpected connections between Convex Geometry, and particularly AGA, and Symplectic Geometry and its methods. The first instance which I noted long ago was a remarkable proof by Alvarez–Paiva [15] of Schaffer's conjecture. This conjecture is a purely Convex Geometry statement but its proof uses very essentially symplectic methods. It was later extended by D. Faifman in [63].

Also in the opposite direction some methods and point of view of AGA influenced some proofs and questions which arose in Symplectic Geometry.

I would like to discuss the following fact from one of the Remarks in the Note by Y. Ostrover: "the length of the shortest periodic billiard trajectory in a convex body $K \subset \mathbb{R}^n$ is monotone with respect to inclusion, and that it satisfies a Brunn-Minkowski type inequality [21]." This was observed by Artstein-Avidan and Ostrover and considered to be "an easy consequence" of some other symplectic results. Although, please, stop for a second and think about this statement. It is absolutely not intuitive. Even knowing it, I find difficult it to believe. There are more such parameters associated to convex bodies which are monotone with respect to embedding for some reasons which are not obvious. There are very classical examples, e.g. monotonicity of mixed volumes observed by Minkowski. I find it very interesting to collect more examples. They tell us something in Convexity which exists behind the scenes, invisible at our first glance.

Below is a different kind of example by Vladimir Kadets [82]:

Theorem (V. Kadets) *Let H be a Hilbert space. For a closed convex body A denote by $r_i(A)$, the supremum of the radiuses of balls is contained in A. Then $\sum_{n=1}^{\infty} r(A_n) \geq r(A)$ for every covering of a convex closed body $A \subset H$ by a sequence of convex closed bodies A_n, $n \in N$.*

This fact is actually similar to Bang's theorem which solved the so-called "plank problem" [32]. Indeed, let sets A_i be planks, i.e.

$$A_i = \left\{ x \in H : |(x - x_0), e)| \leq r_i \right\}.$$

For some e and $x_0 \in H$, $r_i > 0$. This is a plank of the width $2r_i$. And we see that Kadets' theorem is parallel in this case to Bang's plank theorem which states that

$$2 \sum r_i \geq \text{width}(A).$$

This is a very interesting turn of the problem which changes the spirit and meaning of the original question. Let us note that there is another generalization of the plank problem by K. Ball [30] in which the width is computed with respect to an arbitrary norm, not necessary the Euclidean norm. Such a generalization is still unknown for Kadets' result. Although some progress was recently recorded (see [71])

To be precise, in his paper "Coverings by convex bodies and inscribed balls", Kadets makes the following conjecture.

Conjecture (Kadets) If $K, K_1, \ldots, K_N$ are convex bodies in $\mathbb{R}^n$ such that $K \subset \cup_{i=1}^{N} K_i$, and $B \subset \mathbb{R}^n$ defines a (non-symmetric) norm $\| \bullet \|_B$, then the inradius function r_B is subadditive:

$$r_B(K) \leq \sum r_B(K_i).$$

He proves it when B is an ellipsoid.

This conjecture, if proved, will already imply K. Ball's result (and, this means the result of Bang as well). Actually, D. Faifman [66] noted that the Kadets conjecture has the following equivalent form:

Conjecture (Faifman [66]) Let f be a monotone under inclusion, translation-invariant, 1-homogeneous, non-negative functional on convex bodies. Then it is subadditive: whenever $K \subset \cup K_i$ are all convex, one has

$$f(K) \leq \sum f(K_i).$$

In particular, the Euclidean case that Kadets proves implies that Conjecture of Faifman holds for any f and K_i whenever K is an ellipsoid.

10 Liran Rotem, Flowers and Convex Bodies

We describe a class of subsets of $\mathbb{R}^n$ which we call flowers. Flowers are not necessarily convex, but they are intimately related to convex bodies and can help shed a light on their properties. The most direct way to define this class is as follows:

Definition 10.1 A flower $F \subseteq \mathbb{R}^n$ is a compact set which is an arbitrary union of closed Euclidean balls, $F = \bigcup_{i \in I} B_i$, such that each ball B_i contains the origin.

(one may also study non-bounded flowers, but let us restrict ourselves to the compact case)

Flowers correspond to convex bodies in at least two fundamental ways. For the first, let us write B_x for the ball with center $\frac{x}{2}$ and radius $\frac{|x|}{2}$, where $|\cdot|$ denotes the Euclidean norm. In other words B_x is the unique ball with the interval $[0, x]$ as a diameter. Every ball B containing 0 can be written as the union of balls of the form B_x, and therefore the same is true for all flowers. If we consider

$$K = \left\{ x \in \mathbb{R}^n : B_x \subseteq \mathbb{R}^n \right\}$$

then K is a compact convex set with containing the origin. Conversely, every such convex body is obtained from a unique flower in such a way, and we write $F = K^{\clubsuit}$.

Alternatively, one may consider the spherical inversion map $\phi : \mathbb{R}^n \setminus \{0\} \to \mathbb{R}^n \setminus \{0\}$ defined by $\phi(x) = \frac{x}{|x|^2}$. Given a flower F, there exists a unique convex body T such that $\partial F = \phi(\partial T)$ and vice versa. Note that $\phi(T)$ is not F but the complement of F (up to the boundary), so we summarize this relationship by writing $F = (\mathrm{co}\phi)(T)$.

The two convex bodies which are associated to a flower are of course intimately related. If $F = K^{\clubsuit} = (\mathrm{co}\phi)(T)$ then K and T are *polar* to each other, which means that

$$T = K^\circ = \left\{ y \in \mathbb{R}^n : \langle x, y \rangle \leq 1 \text{ for all } x \in K \right\}.$$

Since the polarity map is an involution, i.e. $K^{\circ\circ} = K$, we also have $K = T^{\circ}$.

This is a good point to stop for a brief remark about nomenclature: The flower $K^{\clubsuit}$ of a convex body K was studied in stochastic geometry for its applications to Voronoi tessellations, so it was named the Voronoi flower of K (see e.g. [138]). Since we are interested in flowers for more geometric and less probabilistic reasons, we omit the name of Voronoi from the definition. The content of this note is based on the papers [123] and [122].

It is noteworthy that the class of flowers is closed under many operations. Most of the following Proposition is immediate from the definition, but this does not reduce its usefulness:

Proposition 10.2

1. Let F be a flower. Then its convex hull conv F is also a flower.
2. Let F be a flower and $E \subseteq \mathbb{R}^n$ a linear subspace. Then the intersection $F \cap E$ and the orthogonal projection $P_E F$ are also flowers.
3. Let F_1, F_2 be flowers. Then their Minkowski sum

$$F_1 + F_2 = \{x + y : \ x \in F_1 \ and \ y \in F_2\}$$

and their radial sum

$$F_1 \widetilde{+} F_2 = \{x + y : \ x \in F_1, \ y \in F_2 \ and \ x = \lambda y \ for \ \lambda \geq 0\}$$

are also flowers.

As flowers correspond to convex bodies, every operation on flowers induces an operation on convex bodies. In some cases this resulting operation is well-known. For example, one may check that $(K_1 + K_2)^{\clubsuit} = K_1^{\clubsuit} \widetilde{+} K_2^{\clubsuit}$ and that $(P_E K)^{\clubsuit} = K^{\clubsuit} \cap E$, so in these cases we do not obtain anything new. However, we can also define a new addition $\oplus$ on convex bodies by

$$(K_1 \oplus K_2)^{\clubsuit} = K_1^{\clubsuit} + K_2^{\clubsuit},$$

and a new type of "projection" $Q_E K$ by $(Q_E K)^{\clubsuit} = P_E \left(K^{\clubsuit}\right)$. These operations are still somewhat mysterious, and it is not clear how to define them directly without the use of flowers. As one concrete open problem, one may ask for lower bounds on the volume of $K_1 \oplus K_2$ which are better than the trivial bound that one obtains from the inclusion $K_1 \oplus K_2 \supseteq K_1 + K_2$ and the Brunn–Minkowski inequality.

As for the convex hull, we do know to describe (somewhat) explicitly the convex body T which satisfies $T^{\clubsuit} = \text{conv}\left(K^{\clubsuit}\right)$. Recall that the support function h_K of a convex body K is a function on the unit sphere, $h_K : \mathbb{S}^{n-1} \to \mathbb{R}$, defined by $h_K(\theta) = \max_{x \in K} \langle x, \theta \rangle$. We may then define the reciprocal body K' as the largest convex body such that $h_{K'} \leq \frac{1}{h_K}$. We then have:

Theorem 10.3 *For every convex body K we have* $\left(K^{\nu}\right)^{\clubsuit} = \text{conv}\left(K^{\clubsuit}\right)$. *In particular $K = K''$ if and only if $K^{\clubsuit}$ is convex.*

Flowers are also very interesting from the point of view of high dimensional phenomena. Given two star bodies $A, B \subseteq \mathbb{R}^n$ we define their distance as

$$d(A, B) = \inf \left\{ a \cdot b : \frac{1}{a} A \subseteq B \subseteq b \cdot A \right\}.$$

We recall the celebrated Dvoretzky's theorem (See [116], or e.g. [24]): Let $K \subseteq \mathbb{R}^n$ be a convex body such that $K = -K$. Then for every $\epsilon > 0$ there exists a subspace $E \subseteq \mathbb{R}^n$ of dimension $\dim E \geq c(\epsilon) \cdot \log n$ such that $d\left(P_E K, B_2^E\right) \leq 1 + \epsilon$. Here B_2^E denotes the unit ball in the subspace E.

The dependence of $\dim E$ on the dimension n in Dvoretzky's theorem is sharp. If for example

$$K = B_1^n = \left\{ x = (x_1, x_2, \ldots, x_n) \in \mathbb{R}^n : \sum_{i=1}^{n} |x_i| \leq 1 \right\},$$

then $d\left(P_E K, B_2^E\right) \leq 1 + \epsilon$ is only possible when $\dim E \geq \tilde{c}(\epsilon) \cdot \log n$. Surprisingly however, it turns out that Dvoretzky's theorem also holds for flowers with a much better dependency on n:

Theorem 10.4 *Let $F \subseteq \mathbb{R}^n$ be a flower with $F = -F$. Then for every $\epsilon > 0$ there exists a subspace E of dimension $\dim E = c(\epsilon) \cdot n$ such that*

$$d\left(P_E F, B_2^E\right) \leq 1 + \epsilon.$$

However, if one replaces the projection $P_E F$ with the section $F \cap E$, no better dimension bound can be achieved then the standard bound $\dim E \geq c(\epsilon) \cdot \log n$.

Other than this result not a lot is known about the asymptotic theory of flowers, and there are undoubtedly many results waiting to be discovered.

10.1   Comments by V.M.

As Liran Rotem wrote in his short description of Flowers, the family $\mathcal{F}$ of flowers is in one-to-one correspondence with the family $\mathcal{K}_0$ of convex bodies containing 0. Actually, there are two such natural correspondences: $F \in \mathcal{F}$ corresponds to $K \in \mathcal{K}_0$, $K^{\clubsuit} = F$ and $K^{\circ} = \varphi(F)$, where φ is the spherical inversion. So, F positions between K and K°. This by itself should lead to many interesting consequences.

(i) But I would mention the connection between some deep results in AGA with results in $\mathcal{F}$.

One result (from our joint paper with Liran Rotem [122]).

Theorem *Let $r_F(\theta)$ be the radial function of a flower $F \in \mathcal{F}$ and fix $\epsilon > 0$.*
Then there exists $N \leq c \cdot \frac{n}{\epsilon^2}$ rotations $\{u_i\}_{i=1}^N$ such that for some $r > 0$, and all $\theta \in S^{n-1}$,

$$(1-\epsilon)r \leq \frac{1}{N} \sum_{i=1}^N r_{u_i F}(\theta) \leq (1+\epsilon)r,$$

for some universal consant $c > 0$.

Think also about a partial case where F is a single petal. This is a very nontrivial statement.

Both cases are highly non-trivial but follow from some AGA results just as an interpretation (see [122]).

The above result leads to a similar question about Lipschitz functions on the sphere which are mostly not flowers. And Faifman and Klartag proved an interesting statement in this direction.

Proposition (Faifman–Klartag) *Let $f : S^{n-1} \rightarrow \mathbb{R}$ be a 1-Lipschitz function of mean zero and $\epsilon > 0$. Assume that $k \geq C\epsilon^{-2}|\log \epsilon|$. Let $U_1, \ldots, U_k \in O(n)$ be random, uniformly-distributed, i.i.d matrices. Then with probability of at least $1 - C \exp(-ckn\epsilon^2) \geq 1 - (C\epsilon)^n$ of choosing the matrices,*

$$\left| \frac{1}{k} \sum_{j=1}^k f(U_j(x)) \right| \leq \epsilon \qquad\qquad \text{for all } x \in S^{n-1}.$$

Here, $c, C > 0$ are universal constants.

Of course, to regularize a given function as we see in the Theorem and the Proposition above, is a major goal. However, it is interesting to observe that different functions from the same class have "the same problems" (I mean the same type of "bad regions" which should be regularized), and they may regularize each other. Let me put it precisely in the case of Faifman–Klartag.

Proposition *Let $f_j : S^{n-1} \rightarrow \mathbb{R}$ be 1-Lipschitz functions of mean zero and $\epsilon > 0$. Assume that $k \geq C\epsilon^{-2}|\log \epsilon|$. Let $U_1, \ldots, U_k \in O(n)$ be random, uniformly-distributed, i.i.d matrices. Then with probability of at least $1 - C \exp(-ckn\epsilon^2) \geq 1 - (C\epsilon)^n$ of choosing the matrices,*

$$\left| \frac{1}{k} \sum_{j=1}^k f_j(U_j(x)) \right| \leq \epsilon \qquad\qquad \text{for all } x \in S^{n-1}.$$

Here, $c, C > 0$ are universal constants.

(The proof of the statement is exactly the same as for a single function f).

I did not check this type of generalization for the (Milman–Rotem) Theorem above, but I am sure it is also correct.

This last point of "the sameness" of differences between different bodies (or functions, as it was above) may look more surprising on the results for convex bodies.

Let $\mathcal{D}$ be the standard euclidean ball, and $|K|$ means the volume of K.

Theorem *For any four centrally-symmetric convex bodies $K_i \subset \mathbb{R}^n$, $i = 1, 2, 3, 4$, $|K_i| = |\mathcal{D}|$, there are $\{u_i\}_1^4 \subset SL_n$ such that if $P_1 = u_1 K_1 \cap u_2 K_2$, $P_2 = u_3 K_3 \cap u_4 K_4$ and $Q = \mathrm{Conv}\, P_1 \cup P_2$, then $\frac{1}{C}\mathcal{D} \subset Q \subset C\mathcal{D}$ for some universal constant C independent of the dimension n and $\{K_i\}_1^4$.*

Moreover, if all K_i, $i = 1, 2, 3, 4$, are in M-position (see [24], Chapter 8), then operators $\{u_i\}_{i=1}^4$ are orthogonal rotations and what we call in AGA "random rotations", i.e. there is exponentially (by dimension n) close to 1 probability to select $\{u_i\}$. This fact for the case of all K_i is the same body K is well-known to experts (see [117]).

(ii) Let us return to the theory of Flowers and bring another reformulation of a result from AGA to the language of Flowers.

For the flower $F \subset \mathbb{R}^n$ define the number

$$r_\ell(F) = \max\left\{ r \,\middle|\, r\mathcal{D} \subset \frac{1}{\ell}\left(\tilde{+} \sum_{i=1}^{\ell} u_i F \right), \, u_i \in SO(n) \right\}.$$

Here $\tilde{+}$ means the radial sum of star-sets, as in Rotem's article.

Then Theorem 2 of [117] may be reformulated in the language of flowers as

Theorem *There is a number $c > 0$ such that for any integer n, any centrally symmetric compact flower $F \subset \mathbb{R}^n$ with $0 \in \overset{\circ}{F}$ we have*

$$c \leq r_2(F) \cdot r_3(F^*),$$

where F^* is the dual flower, i.e. if F is the flower of the convex body K, then F^* is the flower of the polar K°.

Moreover one may take $c = \frac{1}{21}\sqrt{\frac{2}{3}}$.

(iii) My next remark is about a flower of a quasi-concave function $f : \mathbb{R}^n \to [0, \infty)$. The next definition was suggested by S. Bobkov

For $x \in \mathbb{R}^n$ let $H_x = \{y \in \mathbb{R}^n : (y, x) \geq |x|^2\}$.

Then a definition of a flower for such a (quasi-concave) function is

$$f^\clubsuit(x) = \sup\{ f(y) \mid y \in H_x \}.$$

Fact $\forall t > 0$ we have $\{f^{\clubsuit} \geq t\} = \{f \geq t\}^{\clubsuit}$.

(iv) Shiri Artstein-Avidan explained to me that flowers also have a cost-duality description. For example cost-duality leads to reciprocity.

Recall $K \in \mathcal{K}_{\circ} : K' = (K^{\clubsuit})^{\circ} = A\left[\frac{1}{h_K}\right]$.

Let cost-function $c(x, y) : \mathbb{R}^n \times \mathbb{R}^n \rightarrow (-\infty, \infty]$. For $K \subset \mathbb{R}^n$ define (c)-duality

$$K^c = \bigcap_{x \in K} \{y \in \mathbb{R}^n : c(x, y) \geq 0\} = \{y \in \mathbb{R}^n : \inf c(x, y) \geq 0\}.$$

Examples

1. $c_2(x, y) := 1 - \langle x, y \rangle$.
   Then $K^{c_2} = K^{\circ}$ (the standard duality is created by this cost-function through the cost-transform).
2. $c(x, y) := 1 - \sup_{\theta \in S^{n-1}} (\langle x, \theta \rangle \cdot \langle y, \theta \rangle)$.
   Then computation shows that

$$K^c = A\left[\frac{1}{h_K(\theta)}\right] = K',$$

i.e. the reciprocal is also created by some cost-transform.

Acknowledgments I am very thankful to every author who contributed his/her notes to this collection. In addition, I would like to thank Liran Rotem for his useful advice and help in creating the right format for this article. I would like also to note that the article could not have appeared without invaluable help in typesetting and editing I received from Miriam Hercberg. I am very grateful to her.

The author is partially supported by ISF grant 519/17 and BSF grant 2016050.

References

1. A. Abbondandolo, G. Benedetti, On the local systolic optimality of Zoll contact forms. Geom. Funct. Anal. **33**(2), 299–363 (2023)
2. A., Abbondandolo, B. Bramham, U.L. Hryniewicz, P. Salomao, Sharp systolic inequalities for Reeb flows on the three-sphere. Invent. Math. **211**, 687–778 (2018)
3. R. Adamczak, A.E. Litvak, A. Pajor, N. Tomczak-Jaegermann, Quantitative estimates of the convergence of the empirical covariance matrix in Log-concave Ensembles. J. Am. Math. Soc. **234**, 535–561 (2010)
4. A. Akopyan, A. Balitskiy, R. Karasev, A. Sharipova, Elementary approach to closed billiard trajectories in asymmetric normed spaces. Proc. Am. Math. Soc. **144**, 4501–4513 (2016)
5. A. Akopyan, R. Karasev, F. Petrov, Bang's problem and symplectic invariants. J. Symplectic Geom. **17**(6), 1579–1611 (2019)

6. A.D. Aleksandrov, Dirichlet's problem for the equation $\det \|z_{ij}\|$ = $\phi(z_1, \ldots, z_n, z, x_1, \ldots, x_n)$. I. (Russian) Vestnik Leningrad. Univ. Ser. Mat. Meh. Astr. **13**(1), 5–24 (1958). See also in: A.D. Alexandrov, Selected Works, Part I (edited by Yu.G. Reshetnyak and S.S. Kutateladze, translated from the Russian by P.S.V. Naidu), Gordon and Breach Publishers, 1996, pp. 251–288

7. S. Alesker, Description of translation invariant valuations on convex sets with solution of P. McMullen's conjecture. Geom. Funct. Anal. **11**(2), 244–272 (2001)

8. S. Alesker, Hard Lefschetz theorem for valuations, complex integral geometry, and unitarily invariant valuations. J. Differ. Geom. **63**(1), 63–95 (2003)

9. S. Alesker, Theory of valuations on manifolds: a survey. Geom. Funct. Anal. **17**(4), 1321–1341 (2007)

10. S. Alesker, Some conjectures on intrinsic volumes of Riemannian manifolds and Alexandrov spaces. Arnold Math. J. **4**(1), 1–17 (2018)

11. S. Alesker, Valuations on convex functions and convex sets and Monge-Ampère operators. Adv. Geom. **19**(3), 313–322 (2019)

12. S. Alesker, Kotrbatý's theorem on valuations and geometric inequalities for convex bodies. Israel J. Math. **247**(1), 361–378 (2022)

13. S. Alesker, J.H.G. Fu, *Integral Geometry and Valuations*. Advanced Courses in Mathematics. CRM Barcelona (Birkhäuser, Basel, 2014). Lectures from the Advanced Course on Integral Geometry and Valuation Theory held at the Centre de Recerca Matemàtica (CRM), Barcelona, September 6–10, 2010, Edited by Eduardo Gallego and Gil Solanes

14. J.C. Álvarez Paiva, E. Fernandes. Crofton formulas in projective Finsler spaces. Electron. Res. Announc. Am. Math. Soc. **4**, 91–100 (1998)

15. J.C. Álvarez Paiva, Dual spheres have the same girth. Am. J. Math. **128**(2), 361–371 (2006)

16. N. Anari, K. Liu, S. Oveis Gharan, Spectral independence in high-dimensional expanders and applications to the hardcore model, in *2020 IEEE 61st Annual Symposium on Foundations of Computer Science* (IEEE Computer Society, Los Alamitos, 2020), pp. 1319–1330

17. S. Artstein-Avidan, V. Milman, A new duality transform. C. R. Math. Acad. Sci. Paris **346**(21-22), 1143–1148 (2008)

18. S. Artstein-Avidan, V. Milman, The concept of duality in convex analysis, and the characterization of the Legendre transform. Ann. Math. (2) **169**(2), 661–674 (2009)

19. S. Artstein-Avidan, V. Milman, Hidden structures in the class of convex functions and a new duality transform. J. Eur. Math. Soc. (JEMS) **13**(4), 975–1004 (2011)

20. S. Artstein-Avidan, Y.A. Rubinstein, Differential analysis of polarity: polar Hamilton-Jacobi, conservation laws, and Monge Ampére equations. J. Anal. Math. **132**, 133–156 (2017)

21. S. Artstein-Avidan, Y. Ostrover, Brunn-Minkowski inequality for symplectic capacities of convex domains. Int. Math. Res. Not. **2008**, rnn044 (2008)

22. S. Artstein-Avidan, V. Milman, Y. Ostrover, The M-ellipsoid, symplectic capacities and volume. Commentarii Mathematici Helvetici **83**(2), 359–369 (2008)

23. Artstein-Avidan, S., Karasev, R., Ostrover, Y. From symplectic measurements to the mahler conjecture. Duke Math. J. **163**, 2003–2022 (2014)

24. S. Artstein-Avidan, A. Giannopoulos, V. Milman, *Asymptotic Geometric Analysis, Part I*. Mathematical Surveys and Monographs, vol. 202 (American Mathematical Society, Providence, 2015)

25. S. Artstein-Avidan, A. Giannopoulos, V. Milman, *Asymptotic Geometric Analysis. Part II*. Mathematical Surveys and Monographs, vol. 261 (American Mathematical Society, Providence, RI, 2021), xx+645pp.

26. S. Artstein-Avidan, S. Sadovsky, K. Wyczesany, A Rockafellar-type theorem for non-traditional costs. Adv. Math. **395**, 108157 (2022)

27. S. Artstein-Avidan, S. Sadovsky, K. Wyczesany, Optimal measure transportation with respect to non-traditional costs. Calc. Var. Partial Differential Equations **62**(1), 35 (2023)

28. S. Artstein-Avidan, S. Sadovsky, K. Wyczesany, A zoo of dualities. The Journal of Geometric Analysis **33**, 238 (2023)

29. Z.D. Bai, Y.Q. Yin, Limit of the smallest eigenvalue of a large dimensional sample covariance matrix. Ann. Probab. **21**, 1275–1294 (1993)
30. K. Ball, The plank problem for symmetric bodies. Invent. Math. **104**(3), 535–543 (1991)
31. A.S. Bandeira, R. van Handel, Sharp nonasymptotic bounds on the norm of random matrices with independent entries. Ann. Prob. **44**, 2479–2506 (2016)
32. T. Bang, A solution of the "Plank problem". Proc. Am. Math. Soc. **2**, 900–993 (1951)
33. A. Barvinok, Thrifty approximations of convex bodies by polytopes. Int. Math. Res. Not. **2014/16**, 4341–4356 (2014)
34. A. Bernig, Valuations with Crofton formula and Finsler geometry. Adv. Math. **210**(2), 733–753 (2007)
35. A. Bernig, L. Bröcker, Valuations on manifolds and Rumin cohomology. J. Differ. Geom. **75**(3), 433–457 (2007)
36. A. Bernig, J.H.G. Fu, Convolution of convex valuations. Geom. Ded. **123**, 153–169 (2006)
37. A. Bernig, J.H.G. Fu, G.Solanes, T.Wannerer, The Weyl tube theorem for Kähler manifolds. arXiv:2209.05806 [math.DG], preprint
38. A. Bernig, J.H.G. Fu, G. Solanes, Integral geometry of complex space forms. Geom. Funct. Anal. **24**(2), 403–492 (2014)
39. A. Bernig, D. Faifman, G. Solanes, Uniqueness of curvature measures in pseudo-Riemannian geometry. J. Geom. Anal. **31**(12), 11819–11848 (2021)
40. A. Bernig, D. Faifman, G. Solanes, Curvature measures of pseudo-Riemannian manifolds. J. Reine Angew. Math. **788**, 77–127 (2022)
41. F. Besau, C. Schütt, E.M. Werner, Flag numbers and floating bodies. Adv. Math. **338**, 912–952 (2018)
42. Z. Blocki, Equilibrium measure of a product subset of $\mathbb{C}^n$. Proc. Am. Math. Soc. **128**(12), 3595–3599 (2000)
43. C. Borell, The Brunn-Minkowski inequality in Gauss space. Invent. Math. **30**(2), 207–216 (1975)
44. K.J. Böröczky, E. Lutwak, D. Yang, G. Zhang, The log-Brunn-Minkowski inequality. Adv. Math. **231**(3–4), 1974–1997 (2012)
45. K.J. Böröczky, E. Lutwak, D. Yang, G. Zhang, The logarithmic Minkowski problem. J. Am. Math. Soc. **26**(3), 831–852 (2013)
46. J. Bourgain, On high-dimensional maximal functions associated to convex bodies. Am. J. Math. **108**(6), 1467–1476 (1986)
47. J. Bourgain, On the distributions of polynomials on high-dimensional functions çonvex sets, in *Geometric Aspects of Functional Analysis, Israel Seminar (1988–90)*. Springer Lecture Notes in Mathematics, vol. 1469 (Springer, Berlin, 1991), pp. 127–137
48. J. Bourgain, Random points in isotropic convex sets, in *Convex Geometric Analysis, Berkeley, CA, 1996*. Mathematical Sciences Research Institute Publications, vol. 34 (Cambridge University Press, Cambridge, 1999), pp. 53–58
49. J. Bourgain, G. Zhang, *On a Generalization of the Busemann-Petty Problem, Convex Geometric Analysis* (Berkeley, CA, 1996). Mathematical Sciences Research Institute Publications, vol. 34 (Cambridge University Press, Cambridge, 1999). (with a little correction in B. Rubin and Gaoyong Zhang, Generalizations of the Busemann-Petty problem for sections of convex bodies, J. Funct. Anal. 213 (2004), 473–501)
50. S. Brazitikos, G. Chasapis, L. Hioni, Random approximation and the vertex index of convex bodies. Arch. Math. **108**, 209–221 (2017)
51. E.M. Bronshtein, L.D. Ivanov, The approximation of convex sets by polyhedra (in Russian). Sibirskii Mat. Zhurnal **16/5**, 1110–1112 (1975). Translation in Siberian Math. J. **16** (1976), 852–853
52. Y. Chen, An almost constant lower bound of the isoperimetric coefficient in the KLS conjecture. Geom. Funct. Anal. **31**(1), 34–61 (2021)
53. S. Chen, Y. Huang, Q.-R. Li. J. Liu, The L_p-Brunn–Minkowski inequality for $p < 1$. Adv. Math. **368**, 107–166 (2020)

54. S.Y. Cheng, S.T. Yau, On the regularity of the Monge-Ampère equation $\det(\partial^2 u/\partial x_i \partial s x_j) = F(x, u)$. Comm. Pure Appl. Math. **30**(1), 41–68 (1977)
55. A. Colesanti, M. Ludwig, F. Mussnig, A homogeneous decomposition theorem for valuations on convex functions. J. Funct. Anal. **279**(5), 108573, 25 pp. (2020)
56. A. Colesanti, M. Ludwig, F. Mussnig, The Hadwiger theorem on convex functions. I arXiv:2009.03702
57. A. Colesanti, M. Ludwig, The Hadwiger theorem on convex functions, III: Steiner formulas and mixed Monge-Ampère measures. Calc. Var. Partial Differential Equations 61(5), Paper No. 181, 37 pp. (2022)
58. D. Cristofaro-Gardiner, M. Hutchings, V.G.B. Ramos, The asymptotics of ECH capacities. Invent. Math. **199**(1), 187–214 (2015)
59. K.R. Davidson, S.J. Szarek, Local operator theory, random matrices and Banach spaces, in *Handbook on the Geometry of Banach Spaces*, vol. 1 (Elsevier, Amsterdam, 2001), pp. 317–366
60. R.M. Dudley, Metric entropy of some classes of sets with differentiable boundaries. J. Approx. Theory **10/3**, 227–236 (1974)
61. R. Eldan, B. Klartag, Approximately Gaussian marginals and the hyperplane conjecture, in *Proceedings of a Workshop on "Concentration, Functional Inequalities and Isoperimetry"*. Contemporary Mathematics, vol. 545 (American Mathematical Society, Providence, 2011), pp. 55–68
62. R. Eldan, O. Shamir, Log concavity and concentration of Lipschitz functions on the Boolean hypercube. J. Funct. Anal. **282**(8), (2022)
63. D. Faifman, An extension of Schäffer's dual girth conjecture to Grassmannians. J. Differ. Geom. **92**(2), 201–22 (2012)
64. D. Faifman, Contact integral geometry and the Heisenberg algebra. Geom. Topol. **23**(6), 3041–3110 (2019)
65. D. Faifman, A funk perspective on billiards, projective geometry and mahler volume. J. Differential Geom., (2022) (to appear.)
66. D. Faifman, Private Communication (2021)
67. D. Faifman, T. Wannerer, The Weyl principle on the Finsler frontier. Sel. Math. **27**(2), 27, 30 (2021)
68. O. Feldheim, S. Sodin, A universality result for the smallest eigenvalues of certain sample covariance matrices. Geom. Funct. Anal. **20**, 88–123 (2010)
69. W.J. Firey, *p*-means of convex bodies. Math. Scand. **10**, 17–24 (1962)
70. O. Friedland, Kahane-Khinchin type averages. Proc. Am. Math. Soc. **136**(10), 3639–3645 (2008)
71. A. Gergely, A generalization of Bang's Lemma. https://arxiv.org/pdf/2201.08823.pdf
72. A. Giannopoulos, M. Hartzoulaki, Random spaces generated by vertices of the cube. Discrete Comp. Geom. **28**, 255–273 (2002)
73. E.D. Gluskin, A.E. Litvak, *A Remark on Vertex Index of the Convex Bodies*. GAFA, Lecture Notes in Mathematics, vol. 2050 (Springer, Berlin, 2012), pp. 255–265
74. O. Guedon, A.E. Litvak, A. Pajor, N. Tomczak-Jaegermann, On the interval of fluctuation of the singular values of random matrices. J. Eur. Math. Soc. **19**, 1469–1505 (2017)
75. O. Guedon, A.E. Litvak, K. Tatarko, Random polytopes obtained by matrices with heavy tailed entries. Commun. Contemp. Math. **22**, 1950027 (2020)
76. O. Guedon, F. Krahmer, C. Kummerle, S. Mendelson, H. Rauhut, On the geometry of polytopes generated by heavy-tailed random vectors. Commun. Contemp. Math. **24**, 2150056 (2022)
77. J. Gutt, M. Hutchings, V.G.B. Ramos, Examples around the strong Viterbo conjecture. J. Fixed Point Theory Appl. **24**(2), 22 (2022)
78. H. Hadwiger, *Vorlesungen über Inhalt, Oberfläche und Isoperimetrie* (in German) (Springer, Berlin, 1957)
79. P. Haim-Kislev, On the symplectic size of convex polytopes. Geom. Funct. Anal. **29**, 440–463 (2019)

80. P. Haim-Kislev, Y. Ostrover, Remarks on symplectic capacities of p-products. Internat. J. Math. **34**(4), 21 (2023)
81. F. John, Polar correspondence with respect to a convex region. Duke Math. J. **3**(2), 355–369 (1937)
82. V. Kadets, Coverings by convex bodies and inscribed balls. Proc. Am. Math. Soc. **133**(5), 1491–1495 (2005)
83. G. Kalai, The number of faces of centrally-symmetric polytopes. Graphs Combin. **5**(1), 389–391 (1989)
84. R. Kannan, L. Lovász, M. Simonovits, Isoperimetric problems for convex bodies and a localization lemma. Discrete Comput. Geom. **13**, 541–559 (1995)
85. R. Kannan, L. Lovász, M. Simonovits, Random walks and $O^*(n^5)$ volume algorithm for convex bodies. Random Struct. Algorithms **2**(1), 1–50 (1997)
86. R. Karasev, A. Sharipova, Viterbo's conjecture for certain hamiltonians in classical mechanics. Arnold Math. J. **5**(10), 1–18 (2019)
87. M. Karpovsky, V. Milman, On subspaces contained in subsets of finite homogeneous spaces. Discrete Math. **22**, 273–280 (1978)
88. D.A. Klain, A short proof of Hadwiger's characterization theorem. Mathematika **42**(2), 329–339 (1995)
89. D.A. Klain, G.-C. Rota, *Introduction to Geometric Probability*. Lezioni Lincee. [Lincei Lectures] (Cambridge University Press, Cambridge, 1997)
90. B. Klartag, On convex perturbations with a bounded isotropic constant. Geom. Func. Anal. **16**(6), 1274–1290 (2006)
91. B. Klartag, Isotropic constants and the Mahler volumes. Adv. Math. **330**, 74–180 (2018)
92. B. Klartag, Logarithmic bounds for isoperimetry and slices of convex sets. To appear in Ars Inveniendi Analytica, (2023)
93. B. Klartag, J. Lehec, Bourgain's slicing problem and KLS isoperimetry up to polylog (preprint)
94. B. Klartag, V. Milman, Rapid Steiner symmetrization of most of the convex and the slicing problem. Combin. Probab. Comput. **14**(5–6), 829–843 (2005)
95. B. Klartag, V. Milman, The slicing problem by Bourgain, in *Ananysis at Large*, ed. by A. Avila, M. Rassias, Y. Sinai (Springer, Berlin, 2022). A collection of articles in memory of Jean Bourgain
96. A. Koldobsky, *Fourier Analysis in Convex Geometry* (American Mathematical Society, Providence, 2005)
97. A.V. Kolesnikov, E. Milman, Local L^p-Brunn–Minkowski inequalities for $p < 1$. Mem. Amer. Math. Soc. **277**, 1360 v+78 (2022)
98. V. Koltchinskii, S. Mendelson, Bounding the smallest singular value of a random matrix without concentration. Int. Math. Res. Not. **23**, 12991–13008 (2015)
99. J. Kotrbatý, On Hodge-Riemann relations for translation-invariant valuations. Adv. Math. **390**, 107914, 28pp. (2021)
100. J. Kotrbatý, T. Wannerer, On mixed Hodge-Riemann relations for translation-invariant valuations and Aleksandrov-Fenchel inequalities. Commun. Contemp. Math. 24(7), Paper No. 2150049, 24 pp. (2022)
101. J. Kotrbatý, T. Wannerer, From harmonic analysis of translation-invariant valuations to geometric inequalities for convex bodies. Geom. Funct. Anal. **33**(2), 541–592 (2023)
102. R. Latała, Some estimates of norms of random matrices. Proc. Am. Math. Soc. **133**, 1273–1282 (2005)
103. R. Latała, R. van Handel, P. Youssef, The dimension-free structure of nonhomogeneous random matrices. Inv. Math. **214**, 1031–1080 (2018)
104. A.E. Litvak, O. Rivasplata, Smallest singular value of sparse random matrices. Stud. Math. **212**, 195–218 (2012)
105. A.E. Litvak, S. Spektor, Quantitative version of a Silverstein's result, in *Geometric Aspects of Functional Analysis*. Lecture Notes in Mathematics, vol. 2116, 335–340 (2014)

106. A.E. Litvak, A. Pajor, M. Rudelson, N. Tomczak-Jaegermann, Smallest singular value of random matrices and geometry of random polytopes. Adv. Math. **195**, 491–523 (2005)

107. G.V. Livshyts, K. Tikhomirov, R. Vershynin, The smallest singular value of inhomogeneous square random matrices. Ann. Prob. **49**, 1286–1309 (2021)

108. M. Ludwig, Fisher information and matrix-valued valuations. Adv. Math. **226**(3), 2700–2711 (2011)

109. M. Ludwig, Valuations on Sobolev spaces. Am. J. Math. **134**(3), 827–842 (2012)

110. E. Lutwak, The Brunn-Minkowski-Firey theory. I. Mixed volumes and the Minkowski problem. J. Differ. Geom. **38**(1), 131–150 (1993)

111. E. Lutwak, The Brunn-Minkowski-Firey theory. II. Affine and geominimal surface areas. Adv. Math. **118**(2), 244–294 (1996)

112. E. Lutwak, D. Yang, G. Zhang, On the L_p-Minkowski problem. Trans. Am. Math. Soc. **356**(11), 4359–4370 (2004)

113. D. Ma, Real-valued valuations on Sobolev spaces. Sci. China Math. **59**(5), 921–934 (2016)

114. M.B. Marcus, G. Pisier, *Random Fourier Series with Applications to Harmonic Analysis*. Annals of Mathematics Studies, vol. 101 (Princeton University Press, Princeton/University of Tokyo Press, Tokyo, 1981)

115. P. McMullen, Valuations and Euler-type relations on certain classes of convex polytopes. Proc. Lond. Math. Soc. **35**(1), 113–135 (1977)

116. V. Milman, New proof of the theorem of A. Dvoretzky on intersections of convex bodies. Funktsional'nyi Analiz i Ego Prilozheniya **5**(4), 28–37 (1971)

117. V. Milman, Some applications of duality relations, in *Geometric Aspects of Functional analysis (1989–90)*. Lecture Notes in Mathematics, vol. 1469 (Springer, Berlin, 1991). pp. 13–40

118. V. Milman, Geometrization of probability, in *Geometry and Dynamics of Groups and Spaces*, Progr. Math., vol. 265 (Birkhäuser, Basel, 2008), pp. 647–667

119. E. Milman, On the role of convexity in isoperimetry, spectral-gap and concentration. Invent. Math. **177** (1), 1–43 (2009)

120. E. Milman, Centro-affine differential geometry and the log-Minkowski problem. J. Eur. Math. Soc. (2021, to appear)

121. V.D. Milman, A. Pajor, Isotropic position and inertia ellipsoids and zonoids of the unit ball of a normed n-dimensional space, in *Geometric Aspects of Functional Analysis (1987–88)*. Lecture Notes in Mathematics, vol. 1376 (Springer, Berlin, 1989), pp. 64–104

122. V. Milman, L. Rotem, Novel View on Classical Convexity Theory. J. Math. Phys. Anal. Geom. **16**(3), 291–311 (2020)

123. E. Milman, V. Milman, L. Rotem, Reciprocals and flowers in convexity, in *Geometric Aspects of Functional Analysis, Israel Seminar 2017–2019 Volume II*, ed. by B. Klartag, E. Milman. Lecture Notes in Mathematics (Springer, Cham, 2020), pp. 199–227

124. S. Mendelson, G. Paouris, On generic chaining and the smallest singular value of random matrices with heavy tails. J. Funct. Anal. **262**, 3775–3811 (2012)

125. S. Mendelson, G. Paouris, On the singular values of random matrices. J. Eur. Math. Soc. **16**, 823–834 (2014)

126. M. Naszódi, Approximating a convex body by a polytope using the epsilon-net theorem. Disc. Comput. Geom. **61/3**, 686–693 (2019)

127. M. Naszódi, F. Nazarov, D. Ryabogin, Fine approximation of convex bodies by polytopes. Am. J. Math. **142/3**, 809–820 (2020)

128. R.I. Oliveira, The lower tail of random quadratic forms with applications to ordinary least squares and restricted eigenvalue properties. Probab. Theory Relat. Fields **166**, 1175–1194 (2016)

129. A. Pajor, L. Pastur, On the limiting empirical measure of eigenvalues of the sum of rank one matrices with log-concave distribution. Stud. Math. **195**(1), 11–29 (2009)

130. E. Rebrova, K. Tikhomirov, Coverings of random ellipsoids and invertibility of matrices with i.i.d. heavy-tailed entries. Isr. J. Math. **227**, 507–544 (2018)

131. S. Riemer, C. Schuett, On the expectation of the norm of random matrices with non-identically distributed entries. Electr. J. Prob. **18**, 1–13 (2013)

132. M. Rudelson, R. Vershynin, The Littlewood-Offord Problem and invertibility of random matrices. Adv. Math. **218**, 600–633 (2008)
133. M. Rudelson, R. Vershynin, Smallest singular value of a random rectangular matrix. Commun. Pure Appl. Math. **62**, 1707–1739 (2009)
134. M. Rudelson, O. Zeitouni, Singular values of Gaussian matrices and permanent estimators. Rand. Struct. Algorithm **48**, 183–212 (2016)
135. R. Schneider, Simple valuations on convex bodies. Mathematika **43**(1), 32–39 (1996)
136. Y. Seginer, The expected norm of random matrices. Combin. Probab. Comput. **9**, 149–166 (2000)
137. J. Silverstein, On the weak limit of the largest eigenvalue of a large dimensional sample covariance matrix. J. Mult. Anal. **30**, 307–311 (1989)
138. E. Spodarev (Ed.), *Stochastic Geometry, Spatial Statistics and Random Fields*. Lecture Notes in Mathematics, vol. 2068 (Springer, Berlin, 2013)
139. N. Srivastava, R. Vershynin, Covariance estimation for distribution with $2 + \epsilon$ moments. Ann. Prob. **41**, 3081–3111 (2013)
140. V.N. Sudakov, B.S. Cirel'son, Extremal properties of half-spaces for spherically invariant measures. J. Math. Sci. **9**(1), 9–18 (1978)
141. S.J. Szarek, Approximation by polytopes (2014)
142. T. Tao, V. Vu, Random matrices: the circular law. Commun. Contemp. Math. **10**, 261–307 (2008)
143. T. Tao, V. Vu, Inverse Littlewood-Offord theorems and the condition number of random discrete matrices. Ann. Math. **169**, 595–632 (2009)
144. N. Tholozan, Volume entropy of Hilbert metrics and length spectrum of Hitchin representations into PSL(3, $\mathbb{R}$). Duke Math. J. **166**(7), 1377–1403 (2017)
145. K. Tikhomirov, Sample covariance matrices of heavy-tailed distributions. Int. Math. Res. Not. **20**, 6254–6289 (2018)
146. A. Tsang, Valuations on L^p-spaces. Int. Math. Res. Not. IMRN **2010**(20), 3993–4023 (2010)
147. D. Tsodikovich, An analogue of the Blaschke-Santaló inequality for billiard dynamics (preprint). arXiv:2204.06209
148. S. van de Geer, A. Muro, On higher order isotropy conditions and lower bounds for sparse quadratic forms. Elect. J. Stat. **8**, 3031–3061 (2014)
149. R. van Handel, On the spectral norm of Gaussian random matrices. Trans. Am. Math. Soc. **369**, 8161–8178 (2017)
150. C. Vernicos, Asymptotic volume in Hilbert geometries. Ind. Univ. Math. J. **62**(5), 1431–1441 (2013)
151. C. Vernicos, C. Walsh, Flag-approximability of convex bodies and volume growth of hilbert geometry. Ann. Sci. Éc. Norm. Supér. **54**(5), 1297–1314 (2021)
152. R. Vershynin, Spectral norm of products of random and deterministic matrices. Prob. Theory Rel. Fields **150**, 471–509 (2011)
153. R. Vershynin, How close is the sample covariance matrix to the actual covariance matrix? J. Theor. Prob. **25**, 655–686 (2012)
154. N.J. Vilenkin, *Special Functions and the Theory of Group Representations*. Translated from the Russian by V. N. Singh Translations of Mathematical Monographs, vol. 22 (American Mathematical Society, Providence, 1968), x+613pp.
155. C. Viterbo, Metric and isoperimetric problems in symplectic geometry. J. Am. Math. Soc. **13**(2), 411–431 (2000)
156. N.R. Wallach, *Real Reductive Groups. I*. Pure and Applied Mathematics, vol. 132 (Academic Press, Boston, 1988)
157. H. Weyl, On the volume of tubes. Am. J. Math. **61**(2), 461–472 (1939)
158. P. Yaskov, Lower bounds on the smallest eigenvalue of a sample covariance matrix. Electron. Commun. Prob. **19**, 1–10 (2014)

On the Gaussian Surface Area
of Spectrahedra

Srinivasan Arunachalam, Oded Regev, and Penghui Yao

Abstract We show that for sufficiently large $n \geq 1$ and $d = Cn^{3/4}$ for some universal constant $C > 0$, a random spectrahedron with matrices drawn from Gaussian orthogonal ensemble has Gaussian surface area $\Theta(n^{1/8})$ with high probability.

1 Introduction

A *spectrahedron* $S \subseteq \mathbb{R}^n$ is a set of the form

$$S = \left\{ x \in \mathbb{R}^n : \sum_i x_i A^{(i)} \preceq B \right\},$$

for some $d \times d$ symmetric matrices $A^{(1)}, \ldots, A^{(n)}, B \in \mathrm{Sym}_d$. Here we will be concerned with the *Gaussian surface area* of S, defined as

$$\mathrm{GSA}(S) = \liminf_{\delta \to 0} \frac{\mathcal{G}^n(S_\delta^{\mathrm{out}})}{\delta}, \tag{1}$$

where $S_\delta^{\mathrm{out}} = \{ x \notin S : \mathrm{dist}(x, S) \leq \delta \}$ denotes the outer δ-neighborhood of S under Euclidean distance and $\mathcal{G}^n(\cdot)$ denotes the standard Gaussian measure on $\mathbb{R}^n$ whose

S. Arunachalam
IBM T.J. Watson Research Center, New York, NY, USA
e-mail: Srinivasan.Arunachalam@ibm.com

O. Regev (✉)
Courant Institute of Mathematical Sciences, New York University, New York, NY, USA
e-mail: regev@cims.nyu.edu

P. Yao
State Key Laboratory for Novel Software Technology, Nanjing University, Nanjing, Jiangsu, China
e-mail: pyao@nju.edu.cn

© The Author(s), under exclusive license to Springer Nature Switzerland AG 2023
R. Eldan et al. (eds.), *Geometric Aspects of Functional Analysis*, Lecture Notes in Mathematics 2327, https://doi.org/10.1007/978-3-031-26300-2_2

density is $(2\pi)^{-n/2} \exp(-\|x\|^2/2)$. Ball showed that the GSA of any convex body in $\mathbb{R}^n$ is $O(n^{1/4})$ [3], which was later shown to be tight by Nazarov [11]. Moreover, Nazarov [8] showed that the GSA of a d-facet polytope[1] in $\mathbb{R}^n$ is $O(\sqrt{\log d})$ and this fact has found application in learning theory and constructing pseudorandom generators for polytopes [4, 5, 8, 13]. We refer the interested reader to [5, 8] for more details. Motivated by recent work [2], this raises the question of whether the GSA of spectrahedra is also small. In this note we answer this question in the negative. Recall that a matrix A drawn from the Gaussian orthogonal ensemble is a symmetric matrix whose entries $\{A_{i,j}\}_{i \le j}$ are all independent normal random variables of mean 0 having variance 1 if $i < j$ and variance 2 if $i = j$.

Theorem 1 *For a universal constant $C > 0$ and any integers $n, d \ge 1$ satisfying $d \le n/C$ the following hold. If $A^{(1)}, \ldots, A^{(n)}$ are i.i.d. drawn from the $d \times d$ Gaussian orthogonal ensemble, then the spectrahedron*

$$\mathcal{T} = \left\{ x \in \mathbb{R}^n : \sum_i x_i A^{(i)} \preceq 2\sqrt{nd} \cdot \mathbb{I} \right\} \tag{2}$$

satisfies $\mathsf{GSA}(\mathcal{T}) \ge c \cdot \sqrt{n/d}$ *for some absolute constant $c > 0$ with probability at least $1 - C \exp(-dn^{-3/4}/C)$. Moreover, for any integer d satisfying $d \le n/C$, $\mathsf{GSA}(\mathcal{T}) \le 2\sqrt{n}/(\sqrt{\pi d})$ holds with probability at least $1 - \exp(-n/50)$.*

The theorem shows the existence of spectrahedra with GSA of $\Omega(n^{1/8})$. (In fact, a random spectrahedron as above satisfies this with constant probability.) This lower bound can be contrasted with the GSA upper bound of Ball [3] of $O(n^{1/4})$ for *arbitrary* convex bodies. Moreover, the lower bound shows that in contrast to the case of polytopes, the GSA of spectrahedra can depend polynomially on d. A natural open question is how large the GSA of arbitrary spectrahedra can be; can spectrahedra with small d (say, polynomial in n) achieve a GSA of $\Theta(n^{1/4})$?

2 Preliminaries

For a matrix A, $\lambda_{\max}(A)$ is the maximum eigenvalue of A. We use g, x, A to denote random variables. We let $\mathcal{G}(0, \sigma^2)$ be the normal distribution with mean 0 and variance σ^2. We denote by $\mathcal{H}_d$ the $d \times d$ Gaussian orthogonal ensemble (GOE). Namely, $A \sim \mathcal{H}_d$ if it is a symmetric matrix with entries $\{A_{i,j}\}_{i \le j}$ independently distributed satisfying $A_{i,j} \sim \mathcal{G}(0, 1)$ for $i < j$ and $A_{i,i} \sim \mathcal{G}(0, 2)$. To keep notations short, for $b \ge 0$ we use $[a \pm b]$ to represent the interval $[a - b, a + b]$. For every $c \ge 0$, we use $c \cdot [a \pm b]$ to represent the interval $[ac \pm bc]$. We denote the set of n-dimensional unit vectors by S^{n-1}. Finally, we let χ_n be the χ distribution

---

[1] A d-facet polytope is the special case of a spectrahedron when the matrices, $A^{(1)}, \ldots, A^{(n)}, B$ are *diagonal*.

with n degrees of freedom, which is the square root of the sum of the squares of n independent standard normal variables. The following are some simple facts about the χ distribution.

Fact 2 *Let $n \in \mathbb{Z}_{>0}$ and $h(\cdot)$ be the pdf of χ_n. Then the following hold.*

1. $h(x) \geq c$ for $x \in [\sqrt{n} \pm c]$, where $c > 0$ is an absolute constant.
2. $h(x) \leq \sqrt{n}/(\sqrt{\pi} \cdot |x|)$ for $x \in \mathbb{R}$.

Proof Recall that by definition

$$h(x) = \frac{1}{2^{\frac{n}{2}-1}\Gamma(\frac{n}{2})} x^{n-1} e^{-x^2/2}$$

for $x \geq 0$, where $\Gamma(\cdot)$ denotes the gamma function, and $h(x) = 0$ otherwise. By elementary calculus, $x^{n-1}e^{-x^2/2}$ monotonically increases for $0 \leq x < \sqrt{n-1}$ and monotonically decreases for $x > \sqrt{n-1}$. We therefore have

$$x^{n-1}e^{-x^2/2} \geq \min\left\{(\sqrt{n}+c)^{n-1}e^{-(\sqrt{n}+c)^2/2}, (\sqrt{n}-c)^{n-1}e^{-(\sqrt{n}-c)^2/2}\right\} \tag{3}$$

for $0 < c \leq 1$ and $x \in [\sqrt{n} \pm c]$. Item 1 now follows from Eq. (3) and the fact that $\Gamma(z) \leq \sqrt{2\pi} z^{z-1/2} e^{-z+1/(12z)}$ for all $z > 0$ [1, 7].
   Item 2 is trivial for $x \leq 0$. For $x > 0$, it follows from the inequalities $\Gamma(z) \geq \sqrt{2\pi} z^{z-1/2} e^{-z}$ for all $z > 0$ [1, 7] and $x^n e^{-x^2/2} \leq n^{n/2} e^{-n/2}$, which follows from the same argument as above. $\qquad\square$

Lemma 3 ([9, Comment Below Lemma 1]) *For $n \geq 1$, let r be a random variable distributed according to χ_n. Then for every $x > 0$, we have*

$$\Pr\left[n - 2\sqrt{nx} \leq r^2 \leq n + 2\sqrt{nx} + 2x\right] \geq 1 - 2e^{-x}.$$

For our purposes, it will be convenient to use an alternative definition of Gaussian surface area in terms of the *inner* surface area. Namely, for $S_\delta^{\text{in}} = \{x \in S : \text{dist}(x, S^c) \leq \delta\}$ where S^c is the complement of the body S, we define,

$$\text{GSA}(S) = \lim_{\delta \to 0} \frac{\mathcal{G}^n(S_\delta^{\text{in}})}{\delta}. \tag{4}$$

It follows from Huang et al. [6, Theorem 3.3] that this definition is equivalent to the one in Eq. (1) when S is a convex body that contains the origin, which is sufficient for our purposes.
   To prove our main theorem, we use the following facts, starting with a well known bound on the size of an ε-net of the n-dimensional sphere.

Fact 4 ([14, Lemma 2.3.4]) *For every $d \geq 1$ and any $0 < \varepsilon < 1/2$ there exists an ε-net of the sphere S^{d-1} of cardinality at most $(3/\varepsilon)^d$.*

The following claim gives a formula for the pdf of the product of two real-valued random variables.

Claim 5 ([12, p. 134, Theorem 3]) *Let x, y be two real-valued random variables and f be the pdf of (x, y). Then the pdf of $z = x \cdot y$ is given by*

$$g(z) = \int_{-\infty}^{\infty} f\left(x, \frac{z}{x}\right) \cdot \frac{1}{|x|} dx.$$

Theorem 6 ([10, Theorem 1]) *Let $A \sim \mathcal{H}_d$. For every $0 < \eta < 1$, it holds that*

$$\Pr\left[\lambda_{\max}(A) \in 2\sqrt{d}\left[1 \pm \eta\right]\right] \geq 1 - C \cdot e^{-d\eta^{3/2}/C},$$

for some absolute constant $C > 0$.

3   Proof of Main Theorem

The core of the argument is in the following lemma, bounding $q(2\sqrt{nd})$ where q is the pdf of the largest eigenvalue of the matrix showing up in Eq. (2). We will later show that this value is essentially the same as $\mathsf{GSA}(\mathcal{T})$, where $\mathcal{T}$ is the spectrahedron in the statement of the theorem.

Lemma 7 *For $n, d \geq 1$ and $A^{(1)}, \ldots, A^{(n)} \in \mathrm{Sym}_d$, let $q(\cdot)$ be the probability density function of*

$$\lambda_{\max}\left(\sum_i x_i A^{(i)}\right),$$

where $x = (x_1, \ldots, x_n)$ is a random vector and each entry is i.i.d. drawn from $\mathcal{G}(0, 1)$. If $A^{(1)}, \ldots, A^{(n)}$ are i.i.d. drawn from the $d \times d$ Gaussian orthogonal ensemble, then $q(2\sqrt{nd}) \geq c \cdot \sqrt{1/d}$ with probability at least $1 - C \exp(-dn^{-3/4}/C)$ (over the choice of $A^{(1)}, \ldots, A^{(n)}$) where $c, C > 0$ are universal constants. Moreover, for any integer d and any $d \times d$ matrices $A^{(1)}, \ldots, A^{(n)}$, $q(2\sqrt{nd}) \leq 1/(2\sqrt{\pi d})$.

Proof Let $y \sim S^{n-1}$ be chosen uniformly from the unit sphere and for matrices $A^{(1)}, \ldots, A^{(n)}$, denote by p the pdf of $\lambda_{\max}(\sum_i y_i A^{(i)})$. Let $r \sim \chi_n$ and notice that ry is distributed like x (since both are spherically symmetric and by definition, have equally distributed norms). Denote by h the pdf of r. By Claim 5, we have

$$q\left(2\sqrt{nd}\right) = \int_{-\infty}^{\infty} h\left(2\sqrt{nd}/z\right) p(z) \frac{1}{|z|} dz. \tag{5}$$

Using Item 2 of Fact 2, $h(2\sqrt{nd}/z)/|z| \leq 1/(2\sqrt{\pi d})$ for all z. Hence Eq. (5) can be bounded as $(1/(2\sqrt{\pi d})) \cdot \int_{-\infty}^{\infty} p(z)dz = 1/(2\sqrt{\pi d})$, establishing the claimed upper bound on q.

To prove the lower bound on q, let $A^{(1)}, \ldots, A^{(n)} \sim \mathcal{H}_d$ be n matrices chosen i.i.d. from the Gaussian orthogonal ensemble. Observe that by Theorem 6, we have

$$\Pr\left[\lambda_{\max}\left(\sum_{i=1}^{n} y_i A^{(i)}\right) \in I\right] \geq 1 - C\exp(-dn^{-3/4}/C)\,, \tag{6}$$

where

$$I = 2\sqrt{d} \cdot [1 \pm c/\sqrt{n}]\,,$$

for some universal constants $C, c > 0$. Define the set of matrices

$$G = \left\{\left(A^{(1)}, \ldots, A^{(n)}\right) : \Pr\left[\lambda_{\max}\left(\sum_{i=1}^{n} y_i A^{(i)}\right) \in I\right] \geq \frac{1}{2}\right\}.$$

Then, using the definition of G and Eq. (6), we have

$$\Pr\left[\left(A^{(1)}, \ldots, A^{(n)}\right) \in G\right] \geq 1 - 2C\exp(-dn^{-3/4}/C)\,.$$

Now fix any $(A^{(1)}, \ldots, A^{(n)}) \in G$. By definition of G, $\int_I p(z)dz \geq 1/2$, and therefore the right-hand side of Eq. (5) is at least

$$\int_I h\left(2\sqrt{nd}/z\right)p(z)\frac{1}{z}dz \geq c \cdot \int_I p(z)\frac{1}{z}dz$$

$$\geq \frac{c}{2\sqrt{d}(1 + c/\sqrt{n})} \cdot \int_I p(z)dz \geq \frac{c}{5\sqrt{d}}\,, \tag{7}$$

for some absolute constant $c > 0$, where we used Item 1 of Fact 2 to conclude that $h(2\sqrt{nd}/z) \geq c$ for all $z \in I$. $\square$

We next relate $q(2\sqrt{nd})$ to $\mathsf{GSA}(\mathcal{T})$. For a vector $v \in S^{d-1}$, and $d \times d$ symmetric matrices $A^{(1)}, \ldots, A^{(n)}$, define the vector

$$W_v = \left(v^T A^{(1)} v, v^T A^{(2)} v, \ldots, v^T A^{(n)} v\right) \in \mathbb{R}^n\,. \tag{8}$$

Notice that $\mathcal{T}$ can be written as

$$\mathcal{T} = \left\{ x \in \mathbb{R}^n : \sum_i x_i A^{(i)} \preceq 2\sqrt{nd} \cdot \mathbb{I} \right\}$$

$$= \left\{ x \in \mathbb{R}^n : \forall v \in S^{d-1}, \ \langle x, W_v \rangle \leq 2\sqrt{nd} \right\}.$$

We say that $A^{(1)}, \ldots, A^{(n)}$ are *good* if

$$\forall v \in S^{d-1}, \ \frac{1}{2}\sqrt{n} \leq \|W_v\| \leq 2\sqrt{n} \ .$$

Lemma 8 *There exists a constant $C \geq 1$ such that for all integers n and $d \leq n/C$, random matrices $A^{(1)}, \ldots, A^{(n)}$ drawn i.i.d. from $\mathcal{H}_d$ are good with probability at least $1 - \exp(-n/50)$.*

Proof For a fixed $v \in S^{d-1}$, we claim that

$$\Pr[n \leq \|W_v\|^2 \leq 3n] \geq 1 - 2\exp(-n/40). \tag{9}$$

To see this, observe that by definition of the Gaussian orthogonal ensemble, for $A \sim \mathcal{H}_d$ and unit vector $v \in \mathbb{R}^d$, $v^T A v = \sum_{i,j} v_i v_j A_{i,j}$ is distributed according to

$$\left(4 \sum_{i<j} v_i^2 v_j^2 + 2 \sum_i v_i^4 \right)^{1/2} \cdot \mathcal{G}(0,1) = \sqrt{2} \cdot \mathcal{G}(0,1).$$

Therefore, each entry in W_v is distributed according to $\mathcal{G}(0,2)$, and Lemma 3 implies Eq. (9). We next prove that with high probability (over the $A^{(i)}$s), for *every* unit vector z, $\|W_z\|$ is large. First, by Fact 4, there exists a set $\mathcal{V} = \{v_1, \ldots, v_{10^{5d}}\} \subseteq \mathbb{R}^d$ of unit vectors that form a 10^{-4}-net of the unit Euclidean sphere. Applying a union bound on $\mathcal{V}$, we have

$$\Pr[\forall v \in \mathcal{V} : n \leq \|W_v\|^2 \leq 3n] \geq 1 - 2\exp(-n/40) \cdot 10^{5d} \geq 1 - \exp(-n/50) , \tag{10}$$

here we used that $d \leq n/C$ for a sufficiently large C.

To conclude the proof, it suffices to show that if $A^{(1)}, \ldots, A^{(n)}$ are such that

$$\forall v \in \mathcal{V}, \ n \leq \|W_v\|^2 \leq 3n \ ,$$

then also

$$\forall z \in S^{d-1}, \ \|W_z\| \geq \frac{1}{2}\sqrt{n} \ .$$

Let $b_{max} = \max_{z \in S^{d-1}} \|W_z\|$ and $b_{min} = \min_{z \in S^{d-1}} \|W_z\|$. Let z_{max} and z_{min} be the vectors achieving the maximum and the minimum respectively. Let v_{max} and v_{min} be the vectors in $\mathcal{V}$ that are closest to z_{max} and z_{min}, respectively. For any vectors $z, v \in S^{d-1}$ with $\|z - v\| \leq 10^{-4}$, applying the spectral decomposition of $zz^T - vv^T$, there exist unit vectors u_1, u_2 and $0 \leq \lambda \leq \frac{1}{100}$ such that

$$zz^T - vv^T = \lambda \cdot \left(u_1 u_1^T - u_2 u_2^T \right). \tag{11}$$

Hence

$$\|W_z - W_v\|^2 = \sum_{i=1}^n \left(z^T A^{(i)} z - v^T A^{(i)} v \right)^2$$

$$= \sum_{i=1}^n \left(\mathrm{Tr}\left(A^{(i)} \left(zz^T - vv^T \right) \right) \right)^2$$

$$\leq \frac{1}{10^4} \sum_{i=1}^n \left(u_1^T A^{(i)} u_1 - u_2^T A^{(i)} u_2 \right)^2$$

$$\leq \frac{1}{5000} \sum_{i=1}^n \left(\left(u_1^T A^{(i)} u_1 \right)^2 + \left(u_2^T A^{(i)} u_2 \right)^2 \right) \ .$$

$$\leq \frac{b_{max}^2}{2500} \ .$$

Choosing $z = z_{max}$ and $v = v_{max}$, we have

$$\|W_{z_{max}}\| \leq \|W_{v_{max}}\| + \frac{b_{max}}{50} \ .$$

Now, since $\|W_{z_{max}}\| = b_{max}$, we have

$$b_{max} \leq \frac{50}{49} \|W_{v_{max}}\| \leq \frac{50}{49} \sqrt{3n} \leq 2\sqrt{n} \ .$$

Similarly, we set $z = z_{min}$ and $v = v_{min}$ and obtain

$$b_{min} \geq \|W_{v_{min}}\| - \frac{b_{max}}{50} \geq \sqrt{n} - \frac{1}{25}\sqrt{n} > \frac{1}{2}\sqrt{n} \ .$$

This concludes the result. $\qquad\square$

For the following claim, we define the inner and outer shells of $\mathcal{T}$ as

$$\mathcal{D}_\delta^{\mathrm{in}} = \left\{ x : \lambda_{\max}\left(\sum_i x_i A^{(i)}\right) \in \sqrt{n} \cdot \left[2\sqrt{d} - \delta, 2\sqrt{d}\right]\right\},$$

$$\mathcal{D}_\delta^{\mathrm{out}} = \left\{ x : \lambda_{\max}\left(\sum_i x_i A^{(i)}\right) \in \sqrt{n} \cdot \left[2\sqrt{d}, 2\sqrt{d} + \delta\right]\right\}.$$

Also recall the inner and outer neighborhoods of $\mathcal{T}$, defined as

$$\mathcal{T}_\delta^{\mathrm{in}} = \{x \in \mathcal{T} : \exists y \notin \mathcal{T} : \|x - y\| \le \delta\},$$

$$\mathcal{T}_\delta^{\mathrm{out}} = \{x \notin \mathcal{T} : \exists y \in \mathcal{T} : \|x - y\| \le \delta\}.$$

Claim 9 *For sufficiently small $\delta > 0$ and any good $A^{(1)}, \ldots, A^{(n)}$, we have $\mathcal{D}_\delta^{\mathrm{in}} \subseteq \mathcal{T}_{4\delta}^{\mathrm{in}}$ and $\mathcal{T}_\delta^{\mathrm{out}} \subseteq \mathcal{D}_{2\delta}^{\mathrm{out}}$.*

Proof For every $x \in \mathcal{D}_\delta^{\mathrm{in}}$, let v be a unit eigenvector of $\sum_i x_i A^{(i)}$ with the eigenvalue $\lambda_{\max}(\sum_i x_i A^{(i)})$. Therefore,

$$\langle x, W_v \rangle = v^T \left(\sum_i x_i A^{(i)}\right) v \ge (2\sqrt{d} - \delta)\sqrt{n}.$$

Setting $y = 2\delta\sqrt{n} W_v / \|W_v\|^2$, we have

$$\langle x + y, W_v \rangle = \langle x, W_v \rangle + 2\delta\sqrt{n} \ge \left(2\sqrt{d} - \delta\right)\sqrt{n} + 2\delta\sqrt{n} = \left(2\sqrt{d} + \delta\right)\sqrt{n},$$

and so $x + y \notin \mathcal{T}$. Moreover, since $A^{(1)}, \ldots, A^{(n)}$ are good, $\|y\| = 2\delta\sqrt{n}/\|W_v\| \le 4\delta$ and therefore $x \in \mathcal{T}_{4\delta}^{\mathrm{in}}$, as desired. For the other containment, let $x \in \mathcal{T}_\delta^{\mathrm{out}}$. Then for any unit vector v, by Cauchy-Schwarz and using $\|W_v\| \le 2\sqrt{n}$,

$$\langle x, W_v \rangle \le 2\sqrt{nd} + 2\delta\sqrt{n},$$

implying that $x \in \mathcal{D}_{2\delta}^{\mathrm{out}}$, as desired. $\square$

We now prove our main theorem.

Proof of Theorem 1 By Lemmas 7 and 8, if $A^{(1)}, \ldots, A^{(n)}$ are i.i.d. drawn from the $d \times d$ Gaussian orthogonal ensemble, then with probability at least $1 - C\exp(-dn^{-3/4}/C)$, we have that $q(2\sqrt{nd}) \ge c \cdot \sqrt{1/d}$ (where $q(\cdot)$ is as defined in Lemma 7) and that $A^{(1)}, \ldots, A^{(n)}$ are good, where $c, C > 0$ are some constants. Since $q(\cdot)$ is continuous, the former implies that $\mathcal{G}^n(\mathcal{D}_\delta^{\mathrm{in}}) \ge c\delta\sqrt{n/(2d)}$ for sufficiently small $\delta > 0$. Thus, $\mathcal{G}^n(\mathcal{T}_{4\delta}^{\mathrm{in}}) \ge c\delta\sqrt{n/(2d)}$ by Claim 9. By definition of $\mathsf{GSA}(S) = \lim_{\delta \to 0} \mathcal{G}^n(S_\delta^{\mathrm{in}})/\delta$, we obtain the desired lower bound on $\mathsf{GSA}(\mathcal{T})$. Similarly, by Lemmas 7 and 8, if $A^{(1)}, \ldots, A^{(n)}$ are i.i.d. drawn from the $d \times d$ Gaussian orthogonal ensemble, then with probability at least

$1 - \exp(-n/50)$, $\mathcal{G}^n(\mathcal{D}_\delta^{\text{out}}) \leq \delta\sqrt{n}/(\sqrt{\pi d})$ for sufficiently small $\delta > 0$. Thus, $\mathcal{G}^n(\mathcal{T}_{\delta/2}^{\text{out}}) \leq \delta\sqrt{n}/(\sqrt{\pi d})$ by Claim 9. We complete the proof using $\mathsf{GSA}(S) = \lim_{\delta \to 0} \mathcal{G}^n(S_\delta^{\text{out}})/\delta$. $\qquad\square$

Acknowledgments We thank Daniel Kane, Assaf Naor, Fedor Nazarov, and Yiming Zhao for useful correspondence. O.R. is supported by the Simons Collaboration on Algorithms and Geometry, a Simons Investigator Award, and by the National Science Foundation (NSF) under Grant No. CCF-1814524. P.Y. is supported by the National Key R&D Program of China 2018YFB1003202, National Natural Science Foundation of China (Grant No. 61972191), the Program for Innovative Talents and Entrepreneur in Jiangsu and Anhui Initiative in Quantum Information Technologies Grant No. AHY150100.

References

1. G.E. Andrews, R. Askey, R. Roy. *The Gamma and Beta Functions*. Encyclopedia of Mathematics and Its Applications (Cambridge University Press, Cambridge, 1999), pp. 1–60
2. S. Arunachalam, P. Yao, Positive spectrahedra: invariance principles and pseudorandom generators. arXiv:2101.08141 (2021)
3. K. Ball, The reverse isoperimetric problem for Gaussian measure. Discrete Comput. Geom. **10**(4), 411–420 (1993)
4. E. Chattopadhyay, A. De, R.A. Servedio, Simple and efficient pseudorandom generators from Gaussian processes, in *34th Computational Complexity Conference (CCC 2019)*. Schloss Dagstuhl-Leibniz-Zentrum fuer Informatik (2019)
5. P. Harsha, A. Klivans, R. Meka, An invariance principle for polytopes. J. ACM **59**(6), 1–25 (2013)
6. Y. Huang, D. Xi, Y. Zhao, The Minkowski problem in Gaussian probability space. Adv. Math. **385**, 107769 (2021)
7. G.J.O. Jameson, A simple proof of Stirling's formula for the gamma function. Math. Gaz. **99**(544), 68–74 (2015)
8. A.R. Klivans, R. O'Donnell, R.A. Servedio, Learning geometric concepts via Gaussian surface area, in *49th Annual IEEE Symposium on Foundations of Computer Science* (IEEE, 2008), pp. 541–550
9. B. Laurent, P. Massart, Adaptive estimation of a quadratic functional by model selection. Ann. Stat. **28**(5), 1302–1338 (2000)
10. M. Ledoux, B. Rider, Small deviations for beta ensembles. Electron. J. Probab. **15**, 1319–1343 (2010)
11. F. Nazarov, On the maximal perimeter of a convex set in $\mathbb{R}^n$ with respect to a Gaussian measure, in *Geometric Aspects of Functional Analysis*, vol. 1807. Lecture Notes in Mathematics (Springer, Berlin, 2003), pp. 169–187
12. V.K. Rohatgi, E. Saleh, *An Introduction to Probability and Statistics* (Wiley, New York, 2015)
13. R.A. Servedio, L.-Y. Tan, Fooling intersections of low-weight halfspaces, in *2017 IEEE 58th Annual Symposium on Foundations of Computer Science (FOCS)* (IEEE, 2017), pp. 824–835
14. T. Tao, *Topics in Random Matrix Theory*, vol. 132. Graduate Studies in Mathematics (American Mathematical Society, Providence, 2012)

Asymptotic Expansions and Two-Sided Bounds in Randomized Central Limit Theorems

Sergey G. Bobkov, Gennadiy P. Chistyakov, and Friedrich Götze

Abstract Lower and upper bounds are explored for the uniform (Kolmogorov) and L^2-distances between the distributions of weighted sums of dependent summands and the normal law. The results are illustrated for several classes of random variables whose joint distributions are supported on Euclidean spheres. We also survey several results on improved rates of normal approximation in randomized central limit theorems.

Keywords Typical distributions · Normal approximation · Central limit theorem

1 Introduction

A random vector $X = (X_1, \ldots, X_n)$ in $\mathbb{R}^n$ ($n \geq 2$) defined on the probability space $(\Omega, \mathfrak{F}, \mathbb{P})$ is called isotropic, if

$$\mathbb{E} X_i X_j = \delta_{ij} \quad \text{for all } i, j \leq n,$$

where δ_{ij} is the Kronecker symbol. Equivalently, all weighted sums

$$S_\theta = \theta_1 X_1 + \cdots + \theta_n X_n, \qquad \theta = (\theta_1, \ldots, \theta_n), \quad \theta_1^2 + \cdots + \theta_n^2 = 1,$$

with coefficients from the unit sphere $\mathbb{S}^{n-1}$ in $\mathbb{R}^n$ have a second moment $\mathbb{E} S_\theta^2 = 1$. In this case, provided that the Euclidean norm $|X|$ is almost constant, and if n is large, a theorem due to Sudakov [27] asserts that the distribution functions

$$F_\theta(x) = \mathbb{P}\{S_\theta \leq x\}, \quad x \in \mathbb{R},$$

S. G. Bobkov (✉)
School of Mathematics, University of Minnesota, Minneapolis, MN, USA
e-mail: bobko001@umn.edu

G. P. Chistyakov · F. Götze
Faculty of Mathematics, University of Bielefeld, Bielefeld, Germany

© The Author(s), under exclusive license to Springer Nature Switzerland AG 2023
R. Eldan et al. (eds.), *Geometric Aspects of Functional Analysis*, Lecture Notes in Mathematics 2327, https://doi.org/10.1007/978-3-031-26300-2_3

are well approximated for most of $\theta \in \mathbb{S}^{n-1}$ by the standard normal distribution function

$$\Phi(x) = \frac{1}{\sqrt{2\pi}} \int_{-\infty}^{x} e^{-y^2/2} \, dy.$$

Here, "most" should refer to the normalized Lebesgue measure $\mathfrak{s}_{n-1}$ on the sphere. This property may be quantified, for example, in terms of the Kolmogorov distance

$$\rho(F_\theta, \Phi) = \sup_x |F_\theta(x) - \Phi(x)|.$$

Being rather universal (since no independence of the components X_k is required), randomized central limit theorems of such type have received considerable interest in recent years. For the history, bibliography, and interesting connections with other concentration problems we refer an interested reader to [8, 9, 12]. Let us mention one general upper bound

$$\mathbb{E}_\theta \, \rho(F_\theta, \Phi) \le c \, (1 + \sigma_4) \frac{\log n}{\sqrt{n}}, \tag{1.1}$$

which holds true with an absolute constant $c > 0$ for any isotropic random vector X (cf. Theorem 1.2 in [8]). Here and elsewhere, $\mathbb{E}_\theta$ denotes an integral over $\mathbb{S}^{n-1}$ with respect to the measure $\mathfrak{s}_{n-1}$, and the bound involves the variance-type functional

$$\sigma_4^2 = \sigma_4^2(X) = \frac{1}{n} \text{Var}(|X|^2) \quad (\sigma_4 \ge 0).$$

Modulo a logarithmic factor, the bound (1.1) exhibits a standard rate of normal approximation for F_θ, in analogy with the classical case of independent identically distributed (iid) summands with equal coefficients. It turns out, however, that in the model with arbitrary $\theta \in \mathbb{S}^{n-1}$ and independent components X_k, the standard rate for $\rho(F_\theta, \Phi)$ is dramatically improved to the order $1/n$ on average and actually for most of θ. Motivated by the seminal paper of Klartag and Sodin [20], this interesting phenomenon was recently studied in [9, 10] for dependent data under certain correlation-type conditions. The last chapters of this paper provide a short account of these improved rates of normal approximation.

One of the main aims of this work is to develop lower bounds with a similar standard rate as in (1.1) (modulo logarithmic factors) and to illustrate them with a number of examples of random variables X_k often appearing in Functional Analysis. These results rely on a careful examination of the closely related L^2-distance

$$\omega(F_\theta, \Phi) = \left(\int_{-\infty}^{\infty} (F_\theta(x) - \Phi(x))^2 \, dx \right)^{1/2}.$$

Similarly to (1.1), it can be shown that for the class of isotropic random vectors the inequality

$$\mathbb{E}_\theta \, \omega^2(F_\theta, \Phi) \leq c \, (1 + \sigma_4^2) \, \frac{1}{n} \tag{1.2}$$

holds without an unnecessary logarithmic term. However, in order to explore the real behavior of the average L^2-distance, some other characteristics of the distribution of X are required. For example, assuming that the distribution is supported on the sphere $\sqrt{n} \, \mathbb{S}^{n-1}$, the L^2-distance admits an asymptotic expansion in terms of the moment functionals (normalized L^p-norms)

$$m_p = m_p(X) = \frac{1}{\sqrt{n}} \left(\mathbb{E} \, \langle X, Y \rangle^p \right)^{1/p} = \frac{1}{\sqrt{n}} \left(\sum (\mathbb{E} X_{i_1} \dots X_{i_p})^2 \right)^{1/p}.$$

Here, Y is an independent copy of X, and the summation is performed over all indices $1 \leq i_1, \dots, i_p \leq n$. The second representation shows that these functionals are non-negative for any integer $p \geq 1$. Note that $m_1 = 0$ if X has mean zero, $m_2 = 1$ if X is isotropic, and $m_p = 0$ with odd p when the distribution of X is symmetric about the origin. The following expansion involves the moments m_p up to order 4.

Theorem 1.1 *Let X be an isotropic random vector in $\mathbb{R}^n$ with mean zero and such that $|X|^2 = n$ a.s. We have*

$$\mathbb{E}_\theta \, \omega^2(F_\theta, \Phi) = \frac{c}{n^{3/2}} \, m_3^3 + O\left(\frac{1}{n^2} \, m_4^4 \right) \tag{1.3}$$

with $c = \frac{1}{16\sqrt{\pi}}$. Similarly, with some absolute constants $c_1, c_2 > 0$,

$$\mathbb{E}_\theta \, \rho^2(F_\theta, \Phi) \leq \frac{c_1 \log n}{n^{3/2}} \, m_3^3 + \frac{c_2 (\log n)^2}{n^2} \, m_4^4. \tag{1.4}$$

As we will see, in the general isotropic case without the support assumption, but with bounded σ_4, the average L^2-distance is described by a more complicated formula

$$\mathbb{E}_\theta \, \omega^2(F_\theta, \Phi) = \frac{1}{\sqrt{2\pi n}} \left(1 + \frac{1}{8n} \right) \mathbb{E}\sqrt{|X|^2 + |Y|^2}$$
$$- \frac{1}{\sqrt{2\pi n}} \left(1 + \frac{1}{4n} \right) \mathbb{E} \, |X - Y| + O\left(\frac{1 + \sigma_4^2}{n^2} \right), \tag{1.5}$$

which holds whenever $\mathbb{E} \, |X|^2 = n$.

In the setting of Theorem 1.1, using the pointwise bound $|\langle X, Y \rangle| \leq n$ together with the isotropy assumption, we have $\mathbb{E} \langle X, Y \rangle^3 \leq n^2$ and $\mathbb{E} \langle X, Y \rangle^4 \leq n^3$.

Therefore, the inequalities (1.3) and (1.4) yield with some absolute constant $c > 0$

$$\mathbb{E}_\theta \, \omega^2(F_\theta, \Phi) \leq \frac{c}{n}, \quad \mathbb{E}_\theta \, \rho^2(F_\theta, \Phi) \leq \frac{c \, (\log n)^2}{n}, \tag{1.6}$$

thus recovering the upper bounds (1.1) and (1.2) for this particular case (since $\sigma_4 = 0$). On the other hand, for a large variety of examples, such bounds turn out to be optimal and may be reversed modulo a logarithmic factor (for large n). To see this, one may use the following lower bound which will be derived from a slightly modified variant of (1.5).

Theorem 1.2 *Let X be a random vector in $\mathbb{R}^n$ satisfying $\mathbb{E} \, |X|^2 = n$, and let Y be its independent copy. For some absolute constants $c_1, c_2 > 0$, we have*

$$\mathbb{E}_\theta \, \omega^2(F_\theta, \Phi) \geq c_1 \, \mathbb{P}\left\{ |X - Y| \leq \frac{1}{2}\sqrt{n} \right\} - c_2 \frac{1 + \sigma_4^4}{n^2}. \tag{1.7}$$

Thus, if the probability in (1.7) is of order at least $1/n$, and σ_4 is bounded, the right-hand side of this bound will be of the same order. If, for example, $|X| = \sqrt{n}$ a.s., we then obtain that $\mathbb{E}_\theta \, \omega^2(F_\theta, \Phi) \sim 1/n$. In order to derive a similar conclusion for the Kolmogorov distance, one may refer to the next statement.

Theorem 1.3 *Let X be an isotropic random vector in $\mathbb{R}^n$ such that $|X| \leq b\sqrt{n}$ a.s. Suppose that we have a lower bound at the standard rate*

$$\mathbb{E}_\theta \, \omega^2(F_\theta, \Phi) \geq \frac{D}{n}$$

with some $D > 0$. Then with some absolute constants $c_0, c_1 > 0$

$$\mathbb{E}_\theta \, \rho(F_\theta, F) \geq \frac{c_0}{(1 + \sigma_4)^3 \, b^2} \frac{D^2}{(\log n)^4 \, \sqrt{n}} - \frac{c_1 \, (1 + \sigma_4^2)}{n}.$$

These estimates may be employed to arrive at the two-sided bounds of the form

$$\frac{c_0}{n} \leq \mathbb{E}_\theta \, \omega^2(F_\theta, \Phi) \leq \frac{c_1}{n}, \quad \frac{c_0}{(\log n)^4 \, \sqrt{n}} \leq \mathbb{E}_\theta \, \rho(F_\theta, \Phi) \leq \frac{c_1 \log n}{\sqrt{n}} \tag{1.8}$$

with some absolute constants $c_0 > 0$ and $c_1 > 0$. Examples where both inequalities in (1.8) are fulfilled include the following uniformly bounded orthonormal systems in $L^2(\Omega, \mathfrak{F}, \mathbb{P})$:

(i) The trigonometric system $X = (X_1, \ldots, X_n)$ with components

$$X_{2k-1}(t) = \sqrt{2} \, \cos(kt),$$
$$X_{2k}(t) = \sqrt{2} \, \sin(kt) \quad (-\pi < t < \pi, \, k = 1, \ldots, n/2, \, n \text{ even})$$

on the interval $\Omega = (-\pi, \pi)$ equipped with the normalized Lebesgue measure $\mathbb{P}$.

(ii) The cosine trigonometric system $X = (X_1, \ldots, X_n)$ with

$$X_k(t) = \sqrt{2}\,\cos(kt)$$

on the interval $\Omega = (0, \pi)$ equipped with the normalized Lebesgue measure $\mathbb{P}$.

(iii) The normalized Chebyshev polynomials $X_1, \ldots, X_n$ defined by

$$X_k(t) = \sqrt{2}\cos(k \arccos t)$$

$$= \sqrt{2}\left[t^n - \binom{n}{2} t^{n-2}(1 - t^2) + \binom{n}{4} t^{n-4}(1 - t^2)^2 - \cdots \right]$$

on $\Omega = (-1, 1)$ equipped with the probability measure $d\mathbb{P}(t) = \dfrac{1}{\pi\sqrt{1-t^2}}\,dt$, $|t| < 1$.

(iv) The systems of functions of the form

$$X_k(t, s) = \Psi(kt + s), \quad k = 1, \ldots, n \ \ (0 < t, s < 1)$$

on the square $\Omega = (0, 1) \times (0, 1)$ equipped with the Lebesgue measure $\mathbb{P}$. In this case, (1.8) holds true for any 1-periodic Lipschitz function Ψ on the real line such that $\int_0^1 \Psi(x)\,dx = 0$ and $\int_0^1 \Psi(x)^2\,dx = 1$ with constants c_0 and c_1 depending on Ψ only.

(v) The Walsh system

$$X = \{X_\tau\}_{\tau \neq \emptyset}, \quad \tau \subset \{1, \ldots, d\},$$

of dimension $n = 2^d - 1$ on the discrete cube $\Omega = \{-1, 1\}^d$ (the ordering of the components does not play any role). Here, $\mathbb{P}$ denotes the normalized counting measure, and

$$X_\tau(t) = \prod_{k \in \tau} t_k \quad \text{for } t = (t_1, \ldots, t_d) \in \Omega.$$

(vi) Random vectors X with associated empirical distribution functions F_θ based on the "observations" $X_k = \sqrt{n}\,\theta_k$ $(1 \leq k \leq n)$.

The paper is organized as follows. We start in Sect. 2 with a review of several results on the so-called typical distributions F which serve as main approximations for F_θ (in general, they do not need to be normal, or even nearly normal). Sections 3–7 deal with the L^2-distances $\omega(F_\theta, F)$ only, while Sects. 8–12 are mostly focused on the Kolmogorov distances $\rho(F_\theta, F)$. In Sect. 13, the examples described in items (i)–(vi) illustrate the applicability of Theorems 1.1–1.3, thus with a standard rate of normal approximation. In Sect. 14 we consider lacunary trigonometric systems and show that the typical rate is improved to the order $1/n$. Similar improved rates are also reviewed in the last section in presence of certain correlation-type conditions. Thus an outline of all sections reads as:

1. Introduction
2. Typical distributions
3. Upper bound for the L^2-distance at standard rate
4. General approximations for the L^2-distance with error of order at most $1/n$
5. Proof of Theorem 1.1 for the L^2-distance
6. General lower bounds for the L^2-distance. Proof of Theorem 1.2
7. Lipschitz systems
8. Berry-Esseen-type bounds
9. Quantitative forms of Sudakov's theorem for the Kolmogorov distance
10. Proof of Theorem 1.1 for the Kolmogorov Distance
11. Relations between L^1, L^2 and Kolmogorov distances
12. Lower bounds. Proof of Theorem 1.3
13. Functional examples
14. The Walsh system; Empirical measures
15. Improved rates for lacunary systems
16. Improved rates for independent and log-concave summands
17. Improved rates under correlation-type conditions

As usual, the Euclidean space $\mathbb{R}^n$ is endowed with the canonical norm $|\cdot|$ and the inner product $\langle\cdot,\cdot\rangle$. In the sequel, we denote by $\mathbb{E}_\theta$ an integral over $\mathbb{S}^{n-1}$ with respect to the measure $\mathfrak{s}_{n-1}$. By $c, c_1, c_2, \ldots$, we denote positive absolute constants which may vary from place to place (if not stated explicitly that c depends on some parameter). Similarly C will denote a quantity bounded by an absolute constant. Throughout, we assume that X is a given random vector in $\mathbb{R}^n$ ($n \geq 2$) and Y is its independent copy.

2 Typical Distributions

In the sequel, we denote by

$$F(x) = \mathbb{E}_\theta F_\theta(x) = \mathbb{E}_\theta \mathbb{P}\{S_\theta \leq x\}, \quad x \in \mathbb{R},$$

the mean distribution function of the weighted sums $S_\theta = \langle X, \theta\rangle$ with respect to the uniform measure $\mathfrak{s}_{n-1}$. It is also called a typical distribution function using the terminology of [27]. Indeed, according to Sudakov's theorem, if X is isotropic, then most of F_θ are concentrated about F in a weak sense (cf. [1, 2, 8] for quantitative statements).

However, whether or not F itself is close to the normal distribution function Φ is determined by the concentration properties of the distribution of $|X|$. Note that, due to the rotational invariance of $\mathfrak{s}_{n-1}$, the typical distribution can be described as the distribution of the product $\theta_1 |X|$, assuming that $\theta = (\theta_1, \dots, \theta_n)$ is a random vector which is independent of X and has distribution $\mathfrak{s}_{n-1}$. In this product, $\theta_1 \sqrt{n}$ is almost standard normal, so that F is almost standard normal, if and only if $\frac{1}{\sqrt{n}} |X|$ is almost 1 (like in the weak law of large numbers). This assertion can be quantified in terms of the weighted total variation distance by virtue of the following upper bound derived in [7].

Proposition 2.1 *If* $\mathbb{E} |X|^2 = n$ *(in particular, when X is isotropic), then*

$$\int_{-\infty}^{\infty} (1 + x^2) |F(dx) - \Phi(dx)| \leq \frac{c}{n} (1 + \mathrm{Var}(|X|)).$$

In particular, this gives a non-uniform bound for the normal approximation, namely

$$|F(x) - \Phi(x)| \leq \frac{c}{n(1 + x^2)} (1 + \mathrm{Var}(|X|)), \quad x \in \mathbb{R}. \tag{2.1}$$

In these bounds we shall rely on the following monotone functionals (of p)

$$\sigma_{2p} = \sqrt{n} \left(\mathbb{E} \left| \frac{|X|^2}{n} - 1 \right|^p \right)^{1/p}, \quad p \geq 1, \tag{2.2}$$

where the particular cases $p = 1$ and $p = 2$ will be most important. If $\mathbb{E} |X|^2 = n$, we thus deal with a more tractable quantity

$$\sigma_4^2 = \frac{1}{n} \mathrm{Var}(|X|^2).$$

Using an elementary inequality $\mathrm{Var}(\xi) \, \mathbb{E}\xi^2 \leq \mathrm{Var}(\xi^2)$ (which is true for any random random variable $\xi \geq 0$), we have $\mathrm{Var}(|X|) \leq \sigma_4^2$. Another similar relation

$$\frac{1}{4} \sigma_2^2 \leq \mathrm{Var}(|X|) \leq \sqrt{n} \, \sigma_2$$

can be found in [8]. From (2.1), we therefore obtain the following bounds for the normal approximation in all L^p-norms

$$\|F - \Phi\|_p = \left(\int_{-\infty}^{\infty} |F(x) - \Phi(x)|^p \, dx \right)^{1/p},$$

including the limit case

$$\|F - \Phi\|_\infty = \rho(F, \Phi) = \sup_x |F(x) - \Phi(x)|.$$

Corollary 2.2 *If* $\mathbb{E}|X|^2 = n$, *then, for all* $p \geq 1$,

$$\|F - \Phi\|_p \leq c\,\frac{1 + \sigma_2}{\sqrt{n}}, \qquad \|F - \Phi\|_p \leq c\,\frac{1 + \sigma_4^2}{n}. \tag{2.3}$$

Note that the characteristic function associated to F is given by

$$f(t) = \mathbb{E}_\theta\,\mathbb{E}\,e^{it\langle X, \theta\rangle} = \mathbb{E}_\theta\,\mathbb{E}\,e^{it|X|\,\theta_1} = \mathbb{E}\,J_n(t|X|), \qquad t \in \mathbb{R}, \tag{2.4}$$

where J_n denotes the characteristic function of the first coordinate θ_1 of θ under $\mathfrak{s}_{n-1}$. Hence, by the Plancherel theorem,

$$\omega^2(F, \Phi) = \frac{1}{2\pi}\int_{-\infty}^{\infty}\left(\mathbb{E}\,J_n(t|X|) - e^{-t^2/2}\right)^2\frac{dt}{t^2}. \tag{2.5}$$

For $p = 2$, the relations in (2.3) can also be derived by means of (2.5) and by virtue of the following Edgeworth-type approximations derived in [8] and [10].

Lemma 2.3 *For all* $t \in \mathbb{R}$,

$$\left|J_n\left(t\sqrt{n}\right) - e^{-t^2/2}\right| \leq \frac{c}{n}\,\min\{1, t^2\}. \tag{2.6}$$

Moreover,

$$\left|J_n\left(t\sqrt{n}\right) - \left(1 - \frac{t^4}{4n}\right)e^{-t^2/2}\right| \leq \frac{c}{n^2}\,\min\{1, t^4\}. \tag{2.7}$$

The functions J_n have a subgaussian (although oscillatory) decay on a long interval of the real line. In particular, as was shown in [8],

$$\left|J_n\left(t\sqrt{n}\right)\right| \leq 5\,e^{-t^2/2} + 4\,e^{-n/12}, \qquad t \in \mathbb{R}. \tag{2.8}$$

This bound can be used for the estimation of the characteristic function of the typical distribution, by involving the variance-type functionals σ_{2p}.

Lemma 2.4 *The characteristic function of the typical distribution satisfies, for all* $t \in \mathbb{R}$,

$$c_p\,|f(t)| \leq e^{-t^2/4} + \frac{1 + \sigma_{2p}^p}{n^{p/2}}$$

with constants $c_p > 0$ *depending on* $p \geq 1$ *only. Consequently, for all* $T > 0$,

$$\frac{c_p}{T}\int_0^T|f(t)|\,dt \leq \frac{1}{T} + \frac{1 + \sigma_{2p}^p}{n^{p/2}}.$$

Proof One may split the expectation in (2.4) to the event $A = \{|X|^2 \le \lambda n\}$ and its complement $B = \{|X|^2 > \lambda n\}$, $0 < \lambda < 1$. By (2.8),

$$\mathbb{E}\,|J_n(t|X|)|\,1_B \le \mathbb{E}\left(5\,e^{-t^2|X|^2/2n} + 4\,e^{-n/12}\right)1_B$$

$$\le 5\,e^{-\lambda t^2/2} + 4\,e^{-n/12}.$$

On the other hand, recalling the definition (2.2), we have

$$\mathbb{P}(A) = \mathbb{P}\left\{n - |X|^2 \ge (1-\lambda)n\right\}$$

$$\le \frac{1}{((1-\lambda)n)^p}\,\mathbb{E}\,|n - |X|^2|^p \;=\; \frac{\sigma_{2p}^p}{(1-\lambda)^p\,n^{p/2}}. \tag{2.9}$$

Choosing $\lambda = \frac{1}{2}$, and since $|J_n(s)| \le 1$ for all $s \in \mathbb{R}$, we get

$$\mathbb{E}\,|J_n(t|X|)|\,1_A \;\le\; (2\sigma_{2p})^p\,n^{-p/2},$$

thus implying that

$$|f(t)| \;\le\; 5\,e^{-t^2/4} + 4\,e^{-n/12} + (2\sigma_{2p})^p\,n^{-p/2}.$$

This readily yields the desired pointwise and integral bounds of the lemma.   □

If $|X| = \sqrt{n}$ a.s., the typical distribution F is just the distribution of $\sqrt{n}\,\theta_1$, the normalized first coordinate of a point on the unit sphere under $\mathfrak{s}_{n-1}$, whose characteristic function is $J_n(t\sqrt{n})$. In this case, the subgaussian character of F manifests itself in corresponding deviation and moment inequalities such as the following.

Lemma 2.5 *For all $p > 0$,*

$$\mathbb{E}_\theta\,|\theta_1|^p \;\le\; 2\left(\frac{p}{n}\right)^{p/2}. \tag{2.10}$$

This inequality can be derived from the well-known bound on the Laplace transform

$$\mathbb{E}_\theta\,e^{t\theta_1} \;\le\; \exp\left\{\frac{t^2}{2(n-1)}\right\}, \quad t \in \mathbb{R},$$

which follows from the fact that the logarithmic Sobolev constant for the unit sphere is equal to $n - 1$ (cf. [21]). Using $x^p \leq (\frac{p}{e})^p e^x$, $x \geq 0$, we have $|x|^p \leq 2 (\frac{p}{e})^p \cosh(x)$, $x \in \mathbb{R}$, and the above bound implies

$$t^p \, \mathbb{E}_\theta \, |\theta_1|^p \leq 2 \left(\frac{p}{e}\right)^p e^{\frac{t^2}{2(n-1)}} \quad \text{for all } t \geq 0.$$

The latter can be optimized over t, which leads to (2.10), even in a sharper form.

In this connection, let us emphasize that rates for the normal approximation for F that are better than $1/n$ cannot be obtained under the support assumption as above.

Proposition 2.6 *For any random vector X in $\mathbb{R}^n$ such that $|X|^2 = n$ a.s., we have*

$$\mathbb{E}_\theta \, \rho(F, \Phi) \geq \frac{c}{n}.$$

Proof One may apply the following lower bound

$$\rho(F, \Phi) \geq \frac{1}{3T} \left| \int_0^T (f(t) - e^{-t^2/2}) \left(1 - \frac{t}{T}\right) dt \right|, \tag{2.11}$$

which holds for any $T > 0$ (cf. [3]). Since $|X|^2 = n$ a.s., we have $f(t) = J_n(t\sqrt{n})$. Choosing $T = 1$ and applying (2.7), it follows from (2.11) that $\rho(F, \Phi) \geq \frac{c}{n}$ for all $n \geq n_0$ where n_0 is determined by c only. But, a similar bound also holds for $n < n_0$ since F is supported on the interval $[-\sqrt{n}, \sqrt{n}]$. □

3 Upper Bound for the L^2-Distance at Standard Rate

Like in the problem of normal approximation for the typical distribution function $F = \mathbb{E}_\theta F_\theta$, the closeness of distribution functions F_θ of the weighted sums $S_\theta = \langle X, \theta \rangle$ $(\theta \in \mathbb{S}^{n-1})$ to F in the metric ω can also be explored in terms of the associated characteristic functions (the Fourier-Stieltjes transforms)

$$f_\theta(t) = \mathbb{E} \, e^{it\langle X, \theta \rangle} = \int_{-\infty}^\infty e^{it\langle x, \theta \rangle} \, dF_\theta(x), \quad t \in \mathbb{R}. \tag{3.1}$$

Again, let us start with the identity

$$\omega^2(F_\theta, F) = \frac{1}{2\pi} \int_{-\infty}^\infty \frac{|f_\theta(t) - f(t)|^2}{t^2} \, dt. \tag{3.2}$$

Here, the mean value of the numerator represents the variance $\mathbb{E}_\theta \, |f_\theta(t)|^2 - |f(t)|^2$ with respect to $\mathfrak{s}_{n-1}$. Moreover, using an independent copy Y of X, we have

$$\mathbb{E}_\theta \, |f_\theta(t)|^2 = \mathbb{E}_\theta \, \mathbb{E} \, e^{it\langle X-Y, \theta \rangle} = \mathbb{E} J_n(t|X - Y|). \tag{3.3}$$

Hence, the Plancherel formula (3.2) together with (2.4) yields

$$\mathbb{E}_\theta\, \omega^2(F_\theta, F) = \frac{1}{2\pi} \int_{-\infty}^{\infty} \left(\mathbb{E} J_n(t|X-Y|) - \left(\mathbb{E} J_n(t|X|) \right)^2 \right) \frac{dt}{t^2}. \tag{3.4}$$

In this section our aim is to show that the above expression is of order at most $O(1/n)$ provided that the mean $a = \mathbb{E}X$, $m_2 = m_2(X)$ and $\sigma_4^2 = \sigma_4^2(X)$ are of order 1. The next statement contains the upper bound (1.2) as a partial case.

Proposition 3.1 *Given a random vector X in $\mathbb{R}^n$ with $\mathbb{E}X = a$ and $\mathbb{E}\,|X|^2 = n$, we have*

$$\mathbb{E}_\theta\, \omega^2(F_\theta, F) \le \frac{cA}{n} \tag{3.5}$$

with $A = 1 + |a|^2 + m_2^2 + \sigma_4^2$. A similar inequality continues to hold with the normal distribution function Φ in place of F.

If X is isotropic, then $m_2 = 1$, while $|a| \le 1$ (by Bessel's inequality). Hence, both characteristics m_2 and a may be removed from the parameter A in this case. However, in the general case, it may happen that m_2 and σ_4 are bounded, while $|a|$ is large. The example in Remark 3.2 shows that this parameter can not be removed.

Proof Note that, for any $\eta > 0$,

$$\int_{-\infty}^{\infty} \frac{\min\{1, t^2\eta^2\}}{t^2}\, dt = 4\eta, \tag{3.6}$$

Hence, in the formula (3.4), the expectation $\mathbb{E} J_n(t|X-Y|)$ can be replaced using the normal approximation (2.6) at the expense of an error not exceeding

$$\frac{c}{n}\, \mathbb{E} \int_{-\infty}^{\infty} \min\left\{1, \frac{t^2|X-Y|^2}{n}\right\} \frac{dt}{t^2} = \frac{4c}{n}\, \mathbb{E}\, \frac{|X-Y|}{\sqrt{n}} \le \frac{8c}{n},$$

where we used that $\mathbb{E}\,|X| \le \sqrt{n}$. Similarly, by (2.6) and (3.6),

$$\int_{-\infty}^{\infty} \left| \left(\mathbb{E} J_n(t|X|) \right)^2 - \left(\mathbb{E}\, e^{-t^2|X|^2/2n} \right)^2 \right| \frac{dt}{t^2} \le 2\mathbb{E} \int_{-\infty}^{\infty} \left| J_n(t|X|) - e^{-t^2|X|^2/2n} \right| \frac{dt}{t^2}$$

$$\le \frac{2c}{n}\, \mathbb{E} \int_{-\infty}^{\infty} \min\left\{1, \frac{t^2|X|^2}{n}\right\} \frac{dt}{t^2}$$

$$= \frac{8c}{n}\, \mathbb{E}\, \frac{|X|}{\sqrt{n}} \le \frac{8c}{n}.$$

Hence, using these bounds in (3.4), we arrive at the general approximation

$$\mathbb{E}_\theta \, \omega^2(F_\theta, F) = \frac{1}{2\pi} \int_{-\infty}^{\infty} \left(\mathbb{E} \, e^{-t^2|X-Y|^2/2n} - \left(\mathbb{E} \, e^{-t^2|X|^2/2n} \right)^2 \right) \frac{dt}{t^2} + \frac{C}{n}, \quad (3.7)$$

where we recall that C denotes a quantity bounded by an absolute constant.

Introduce the random variable

$$\rho^2 = \frac{|X-Y|^2}{2n} \quad (\rho \geq 0).$$

By Jensen's inequality, $\mathbb{E} \, e^{-t^2|X|^2/2n} \geq e^{-t^2/2}$, so that, by (3.7),

$$\mathbb{E}_\theta \, \omega^2(F_\theta, F) \leq \frac{1}{2\pi} \mathbb{E} \int_{-\infty}^{\infty} \frac{e^{-\rho^2 t^2} - e^{-t^2}}{t^2} \, dt + \frac{c}{n}.$$

The above integral is easily evaluated (by differentiating with respect to the variable "ρ^2"), and we arrive at the bound

$$\mathbb{E}_\theta \, \omega^2(F_\theta, F) \leq \frac{1}{\sqrt{\pi}} (1 - \mathbb{E}\rho) + \frac{c}{n}. \quad (3.8)$$

To further simplify, one may apply an elementary inequality $1 - x \leq \frac{1}{2}(1 - x^2) + (1 - x^2)^2$ ($x \geq 0$), which gives

$$\mathbb{E}_\theta \, \omega^2(F_\theta, F) \leq \frac{1}{2\sqrt{\pi}} \mathbb{E} (1 - \rho^2) + \frac{1}{\sqrt{\pi}} \mathbb{E} (1 - \rho^2)^2 + \frac{c}{n}.$$

Since

$$1 - \rho^2 = \frac{n - |X|^2}{2n} + \frac{n - |Y|^2}{2n} + \frac{\langle X, Y \rangle}{n},$$

we have

$$1 - \mathbb{E}\rho^2 = \frac{1}{n} \mathbb{E} \langle X, Y \rangle = \frac{1}{n} |\mathbb{E}X|^2 = \frac{1}{n} |a|^2.$$

In addition,

$$(1 - \rho^2)^2 \leq 2 \left(\frac{n - |X|^2}{2n} + \frac{n - |Y|^2}{2n} \right)^2 + 2 \frac{\langle X, Y \rangle^2}{n^2},$$

which implies

$$\mathbb{E}\,(1 - \rho^2)^2 \;\leq\; \frac{\mathrm{Var}(|X|^2)}{n^2} + 2\,\frac{\mathbb{E}\,\langle X, Y\rangle^2}{n^2} \;=\; \frac{\sigma_4^2 + 2m_2^2}{n}.$$

Using this estimate in (3.8), the inequality (3.5) follows immediately.
    For the second assertion, it remains to apply Corollary 2.2.   □

Remark 3.2 Let us illustrate the inequality (3.5) in the example where the random
vector X has a normal distribution with a large mean value. Given a standard normal
random vector $Z = (Z_1, \ldots, Z_{n-1})$ in $\mathbb{R}^{n-1}$ (which we identify with the space of
all points in $\mathbb{R}^n$ with zero last coordinate), define

$$X = \alpha Z + \lambda e_n \quad \text{with } 1 \leq \lambda \leq n^{1/4}, \ \alpha^2(n-1) + \lambda^2 = n,$$

where $e_n = (0, \ldots, 0, 1)$ is the last unit vector in the canonical basis of $\mathbb{R}^n$. Since
Z is orthogonal to e_n, so that $|X|^2 = \alpha^2 |Z|^2 + \lambda^2$, we have $\mathbb{E}\,|X|^2 = n$, and

$$\sigma_4^2 = \frac{\alpha^4}{n}\,\mathrm{Var}(|Z|^2) = \frac{2\alpha^4\,(n-1)}{n} = 2\,\frac{(n-\lambda^2)^2}{n(n-1)} < 2.$$

Let Z' be an independent copy of Z. Then $Y = \alpha Z' + \lambda e_n$ is an independent
copy of X, so that

$$m_2^2 = \frac{1}{n}\,\mathbb{E}\,\langle X, Y\rangle^2 = \frac{1}{n}\,(\alpha^4\,(n-1) + \lambda^4) < 2.$$

Thus, both m_2 and σ_4 are bounded, while the mean $a = \mathbb{E}X = \lambda e_n$ has the
Euclidean length $|a| = \lambda \geq 1$. Hence, the inequality (3.5) being stated for the
normal distribution function in place of F simplifies to

$$\mathbb{E}_\theta\, \omega^2(F_\theta, \Phi) \leq \frac{c\lambda^2}{n}.$$

Let us show that this bound may be reversed up to an absolute factor (which
would imply that $|a|^2$ may not be removed from A). For any unit vector $\theta = (\theta_1, \ldots, \theta_n)$, the linear form

$$S_\theta = \langle X, \theta\rangle = \alpha\theta_1 Z_1 + \cdots + \alpha\theta_{n-1} Z_{n-1} + \lambda\theta_n$$

has a normal distribution on the line with mean $\mathbb{E}S_\theta = \lambda\theta_n$ and variance $\mathrm{Var}(S_\theta) = \alpha^2(1 - \theta_n^2)$. Consider the normal distribution function $\Phi_{\mu,\sigma^2}(x) = \Phi(\frac{x-\mu}{\sigma})$ with
parameters $0 \leq \mu \leq 1$ and $\frac{1}{2} \leq \sigma^2 \leq 1$ ($\sigma > 0$). If $x \leq \frac{\mu}{1+\sigma}$, then $\frac{x-\mu}{\sigma} \leq x$, and on

the interval with these endpoints the standard normal density $\varphi(y)$ attains minimum at the left endpoint. Hence

$$|\Phi_{\mu,\sigma^2}(x) - \Phi(x)| = \int_{\frac{x-\mu}{\sigma}}^{x} \varphi(y)\,dy \geq \left(x - \frac{x-\mu}{\sigma}\right)\varphi\left(\frac{x-\mu}{\sigma}\right),$$

so that

$$\omega^2(\Phi_{\mu,\sigma^2}, \Phi) \geq \int_{-\infty}^{\frac{\mu}{1+\sigma}} \left(x - \frac{x-\mu}{\sigma}\right)^2 \varphi\left(\frac{x-\mu}{\sigma}\right)^2 dx$$

$$= \frac{\sigma}{2\pi} \int_{-\infty}^{-\frac{\mu}{1+\sigma}} (\mu - (1-\sigma)y)^2\, e^{-y^2/2}\,dy$$

$$\geq \frac{\sigma\mu^2}{2\pi} \int_{-\infty}^{-\frac{\mu}{1+\sigma}} e^{-y^2/2}\,dy \geq c\mu^2.$$

In our case, since $\lambda \leq n^{1/4}$ and

$$\alpha^2 = \frac{n - \lambda^2}{n - 1} \geq \frac{n - \sqrt{n}}{n - 1} \geq 1 - \frac{1}{\sqrt{n}},$$

we have $|\mathbb{E}S_\theta| \leq 1$ and $\mathrm{Var}(S_\theta) \geq \frac{1}{2}$ on the set $\Omega_n = \{\theta \in \mathbb{S}^{n-1} : |\theta_n| < \frac{\log n}{\sqrt{n}}\}$ with n large enough. It follows that

$$\mathbb{E}_\theta\, \omega^2(F_\theta, \Phi) \geq c\lambda^2\, \mathbb{E}\,\theta_n^2\, 1_{\{\theta \in \Omega_n\}} \geq \frac{c'\lambda^2}{n}.$$

4   General Approximations for the L^2-Distance with Error of Order at Most $1/n$

We now turn to general representations for the average L^2-distance between F_θ and the typical distribution function F with error of order at most $1/n$.

Proposition 4.1 *Suppose that* $\mathbb{E}\,|X| \leq b\sqrt{n}$ *for some* $b \geq 0$. *Then*

$$\mathbb{E}_\theta\, \omega^2(F_\theta, F) = \frac{1}{\sqrt{2\pi}}\, \mathbb{E}R + \frac{Cb}{n^2}, \tag{4.1}$$

where

$$R = \frac{(|X|^2 + |Y|^2)^{1/2}}{\sqrt{n}}\left(1 + \frac{1}{4n}\frac{|X|^4 + |Y|^4}{(|X|^2 + |Y|^2)^2}\right) - \frac{|X - Y|}{\sqrt{n}}\left(1 + \frac{1}{4n}\right). \tag{4.2}$$

We use the convention that $R = 0$ if $X = Y = 0$. Note that $|R| \leq 3 \frac{|X|+|Y|}{\sqrt{n}}$, so $\mathbb{E}R \leq 3b$.

Let us give a simpler expression by involving the functional $\sigma_4^2 = \frac{1}{n} \mathrm{Var}(|X|^2)$ and assuming that $\mathbb{E}|X|^2 = n$. Since

$$\frac{|X|^4 + |Y|^4}{(|X|^2 + |Y|^2)^2} - \frac{1}{2} = \frac{(|X|^2 - |Y|^2)^2}{2(|X|^2 + |Y|^2)^2},$$

we may write

$$R = \frac{1}{8n^{3/2}} \frac{(|X|^2 - |Y|^2)^2}{(|X|^2 + |Y|^2)^{3/2}} + \frac{(|X|^2 + |Y|^2)^{1/2}}{\sqrt{n}} \left(1 + \frac{1}{8n}\right) - \frac{|X - Y|}{\sqrt{n}} \left(1 + \frac{1}{4n}\right). \quad (4.3)$$

As we will see, the first term here is actually of order at most σ_4^2 / n^2. As a result, we arrive at the relation (1.5).

Proposition 4.2 *If $\mathbb{E}|X|^2 = n$, then*

$$\mathbb{E}_\theta \, \omega^2(F_\theta, F) = \frac{1}{\sqrt{2\pi}} \mathbb{E}R + C \frac{1 + \sigma_4^2}{n^2}, \quad (4.4)$$

where

$$R = \frac{(|X|^2 + |Y|^2)^{1/2}}{\sqrt{n}} \left(1 + \frac{1}{8n}\right) - \frac{|X - Y|}{\sqrt{n}} \left(1 + \frac{1}{4n}\right). \quad (4.5)$$

Proof of Proposition 4.1 Let us return to the Plancherel formula (3.4). To simplify the integrand therein, we apply the inequality (2.7) in Lemma 2.3, by replacing t^4 with t^2 in the remainder term. Using the equality (3.6), the expectation $\mathbb{E}J_n(t|X - Y|)$ in the formula (3.4) can be therefore replaced according to (2.7) at the expense of an error not exceeding

$$\frac{c}{n^2} \mathbb{E} \int_{-\infty}^{\infty} \min\left\{1, \frac{t^2|X - Y|^2}{n}\right\} \frac{dt}{t^2} = \frac{4c}{n^2} \mathbb{E} \frac{|X - Y|}{\sqrt{n}} \leq \frac{8cb}{n^2}.$$

As for the main term $(1 - \frac{t^4}{4n}) e^{-t^2/2}$ in (2.7), it is bounded by an absolute constant, which implies that

$$J_n(t\sqrt{n}) J_n(s\sqrt{n}) = \left(1 - \frac{t^4}{4n}\right)\left(1 - \frac{s^4}{4n}\right) e^{-(t^2+s^2)/2} + O\left(n^{-2} \min\{1, t^2 + s^2\}\right)$$

$$= \left(1 - \frac{t^4 + s^4}{4n}\right) e^{-(t^2+s^2)/2} + O\left(n^{-2} \min\{1, t^2 + s^2\}\right).$$

Hence

$$|\mathbb{E} J_n(t|X|)|^2 = \mathbb{E} J_n(t|X|) J_n(t|Y|) = \mathbb{E} \left(1 - \frac{t^4 (|X|^4 + |Y|^4)}{4n^3} \right) e^{-\frac{t^2 (|X|^2 + |Y|^2)}{2n}}$$

$$+ O\left(n^{-2} \min \left\{ 1, \frac{t^2 (|X|^2 + |Y|^2)}{n} \right\} \right).$$

As before, after integration in (3.4) the latter remainder term will produce a quantity not exceeding a multiple of b/n^2. As a preliminary step, we therefore obtain the representation

$$\mathbb{E}_\theta \, \omega^2(F_\theta, F) = \frac{1}{2\pi} I + \frac{Cb}{n^2} \qquad (4.6)$$

with

$$I = \mathbb{E} \int_{-\infty}^{\infty} \left[\left(1 - \frac{t^4 |X - Y|^4}{4n^3} \right) e^{-\frac{t^2 |X-Y|^2}{2n}} - \left(1 - \frac{t^4 (|X|^4 + |Y|^4)}{4n^3} \right) e^{-\frac{t^2 (|X|^2 + |Y|^2)}{2n}} \right] \frac{dt}{t^2}.$$

To evaluate the integrals of this type, consider the functions

$$\psi_r(\alpha) = \frac{1}{\sqrt{2\pi}} \int_{-\infty}^{\infty} \left((1 - rt^4) e^{-\alpha t^2/2} - e^{-t^2/2} \right) \frac{dt}{t^2} \qquad (\alpha > 0, \, r \in \mathbb{R}).$$

Clearly,

$$\psi_r(1) = -\frac{1}{\sqrt{2\pi}} \int_{-\infty}^{\infty} rt^2 e^{-t^2/2} \, dt = -r$$

and

$$\psi_r'(\alpha) = -\frac{1}{2\sqrt{2\pi}} \int_{-\infty}^{\infty} (1 - rt^4) e^{-\alpha t^2/2} \, dt$$

$$= -\frac{1}{2\sqrt{\alpha}} \frac{1}{\sqrt{2\pi}} \int_{-\infty}^{\infty} \left(1 - \frac{r}{\alpha^2} s^4 \right) e^{-s^2/2} \, ds = -\frac{1}{2\sqrt{\alpha}} \left(1 - \frac{3r}{\alpha^2} \right).$$

Hence

$$\psi_r(\alpha) - \psi_r(1) = \int_{1}^{\alpha} \left(-\frac{1}{2} z^{-1/2} + \frac{3r}{2} z^{-5/2} \right) dz$$

$$= (1 + r) - (\alpha^{1/2} + r\alpha^{-3/2}),$$

and we get

$$\psi_r(\alpha) = 1 - (\alpha^{1/2} + r\alpha^{-3/2}). \qquad (4.7)$$

Here, when α and r both approach zero subject to the relation $r = O(\alpha^2)$, we get in the limit $\psi_0(0) = 1$. From this,

$$\frac{1}{\sqrt{2\pi}} I = \mathbb{E}\left(\psi_{r_1}(\alpha_1) - \psi_{r_2}(\alpha_2)\right)$$

$$= \mathbb{E}\left(\alpha_2^{1/2} + r_2\alpha_2^{-3/2}\right) - \mathbb{E}\left(\alpha_1^{1/2} + r_1\alpha_1^{-3/2}\right),$$

which we need with

$$\alpha_1 = \frac{|X - Y|^2}{n}, \quad r_1 = \frac{|X - Y|^4}{4n^3},$$

$$\alpha_2 = \frac{|X|^2 + |Y|^2}{n}, \quad r_2 = \frac{|X|^4 + |Y|^4}{4n^3}.$$

It follows that

$$\alpha_2^{1/2} + r_2\alpha_2^{-3/2} = \left(\frac{|X|^2 + |Y|^2}{n}\right)^{1/2}\left(1 + \frac{1}{4n}\frac{|X|^4 + |Y|^4}{(|X|^2 + |Y|^2)^2}\right),$$

$$\alpha_1^{1/2} + r_1\alpha_1^{-3/2} = \left(\frac{|X - Y|^2}{n}\right)^{1/2}\left(1 + \frac{1}{4n}\right)$$

with the assumption that both expressions are equal to zero in the case $X = Y = 0$. As a result, (4.6) yields the desired representation (4.1) with quantity R described in (4.2). □

In order to modify (4.1) and (4.2) to the form (4.4) and (4.5), first let us verify the following general relation.

Lemma 4.3 *Let ξ be a non-negative random variable with finite second moment (not identically zero), and let η be its independent copy. Then*

$$\mathbb{E}\frac{(\xi - \eta)^2}{(\xi + \eta)^{3/2}} 1_{\{\xi+\eta>0\}} \leq 12\frac{\mathrm{Var}(\xi)}{(\mathbb{E}\xi)^{3/2}}.$$

Applying the lemma with $\xi = |X|^2$, $\eta = |Y|^2$ and assuming that $\mathbb{E}|X|^2 = n$, we get that

$$\mathbb{E}\frac{(|X|^2 - |Y|^2)^2}{(|X|^2 + |Y|^2)^{3/2}} \leq 12\frac{\mathrm{Var}(|X|^2)}{(\mathbb{E}|X|^2)^{3/2}} = 12\frac{\mathrm{Var}(|X|^2)}{n^{3/2}} = 12\frac{\sigma_4^2}{n^{1/2}}.$$

In view of (4.3), this proves Proposition 4.2.

Proof of Lemma 4.3 By homogeneity, we may assume that $\mathbb{E}\xi = 1$. In particular, $\mathbb{E}\,|\xi - \eta| \leq 2$. We have

$$\mathbb{E}\,\frac{(\xi - \eta)^2}{(\xi + \eta)^{3/2}}\,1_{\{\xi+\eta>1/2\}} \leq 2^{3/2}\,\mathbb{E}\,(\xi - \eta)^2\,1_{\{\xi+\eta>1/2\}}$$

$$\leq 2^{3/2}\,\mathbb{E}\,(\xi - \eta)^2 = 4\sqrt{2}\,\mathrm{Var}(\xi).$$

Also note that, by Chebyshev's inequality,

$$\mathbb{P}\{\xi \leq 1/2\} = \mathbb{P}\{1 - \xi \geq 1/2\} \leq 4\,\mathrm{Var}(\xi)^2,$$

so

$$\mathbb{P}\{\xi + \eta \leq 1/2\} \leq \mathbb{P}\{\xi \leq 1/2\}\,\mathbb{P}\{\eta \leq 1/2\} \leq 16\,\mathrm{Var}(\xi)^2.$$

Hence, since $\frac{|\xi-\eta|}{\xi+\eta} \leq 1$ for $\xi + \eta > 0$, we have, by Cauchy's inequality,

$$\mathbb{E}\,\frac{(\xi - \eta)^2}{(\xi + \eta)^{3/2}}\,1_{\{0<\xi+\eta\leq1/2\}} \leq \mathbb{E}\,\sqrt{|\xi - \eta|}\,1_{\{0<\xi+\eta\leq1/2\}}$$

$$\leq \sqrt{\mathbb{E}\,|\xi - \eta|}\,\sqrt{\mathbb{P}\{\xi + \eta \leq 1/2\}} \leq 4\sqrt{2}\,\mathrm{Var}(\xi).$$

It remains to combine both inequalities, which yield

$$\mathbb{E}\,\frac{(\xi - \eta)^2}{(\xi + \eta)^{3/2}}\,1_{\{\xi+\eta>0\}} \leq 8\sqrt{2}\,\mathrm{Var}(\xi) \leq 12\,\mathrm{Var}(\xi).$$

$\square$

5 Proof of Theorem 1.1 for the L^2-Distance

The expression (4.5) may be further simplified in the particular case where the distribution of X is supported on the sphere $\sqrt{n}\,\mathbb{S}^{n-1}$. Introduce the random variable

$$\xi = \frac{\langle X, Y \rangle}{n},$$

where Y is an independent copy of X. Since $|X - Y|^2 = 2n\,(1 - \xi)$, Proposition 4.2 yields:

Corollary 5.1 *If* $|X|^2 = n$ *a.s., then*

$$\sqrt{\pi}\,\mathbb{E}_\theta\,\omega^2(F_\theta, F) = \left(1 + \frac{1}{4n}\right)\mathbb{E}\left(1 - (1 - \xi)^{1/2}\right) - \frac{1}{8n} + O\left(\frac{1}{n^2}\right). \quad (5.1)$$

Note that $|\xi| \leq 1$. Therefore, the relation (5.1) suggests to develop an expansion in powers of ε for the function $w(\varepsilon) = 1 - \sqrt{1 - \varepsilon}$ near zero, which will be needed up to the term ε^4.

Lemma 5.2 *For all* $|\varepsilon| \leq 1$,

$$1 - \sqrt{1 - \varepsilon} \leq \frac{1}{2}\varepsilon + \frac{1}{8}\varepsilon^2 + \frac{1}{16}\varepsilon^3 + 3\varepsilon^4.$$

In addition,

$$1 - \sqrt{1 - \varepsilon} \geq \frac{1}{2}\varepsilon + \frac{1}{8}\varepsilon^2 + \frac{1}{16}\varepsilon^3 + 0.01\,\varepsilon^4.$$

Proof By Taylor's formula for the function $w(\varepsilon)$ around zero on the half-axis $\varepsilon < 1$,

$$1 - \sqrt{1 - \varepsilon} = \frac{1}{2}\varepsilon + \frac{1}{8}\varepsilon^2 + \frac{1}{16}\varepsilon^3 + \frac{5}{128}\varepsilon^4 + \frac{w^{(5)}(\varepsilon_1)}{120}\varepsilon^5$$

for some ε_1 between zero and ε. Since $w^{(5)}(\varepsilon) = \frac{105}{32}(1 - \varepsilon)^{-9/2} \geq 0$, we have an upper bound

$$1 - \sqrt{1 - \varepsilon} \leq \frac{1}{2}\varepsilon + \frac{1}{8}\varepsilon^2 + \frac{1}{16}\varepsilon^3 + \frac{5}{128}\varepsilon^4, \qquad \varepsilon \leq 0.$$

Also, $w^{(5)}(\varepsilon) \leq \frac{105}{32} 3^{9/2} < 461$ for $0 \leq \varepsilon \leq \frac{2}{3}$, so, in this interval

$$\frac{5}{128}\varepsilon^4 + \frac{w^{(5)}(\varepsilon_1)}{120}\varepsilon^5 \leq 3\varepsilon^4.$$

Thus, in both cases,

$$1 - \sqrt{1 - \varepsilon} \leq \frac{1}{2}\varepsilon + \frac{1}{8}\varepsilon^2 + \frac{1}{16}\varepsilon^3 + 3\varepsilon^4, \qquad \varepsilon \leq \frac{2}{3}.$$

To treat the remaining values $\frac{2}{3} \leq \varepsilon \leq 1$, it is sufficient to select a positive constant b such that the polynomial

$$Q(\varepsilon) = \frac{1}{2}\varepsilon + \frac{1}{8}\varepsilon^2 + \frac{1}{16}\varepsilon^3 + b\varepsilon^4$$

is greater than or equal to 1 for $\varepsilon \geq \frac{2}{3}$. On this half-axis, $Q(\varepsilon) \geq \frac{11}{27} + b\frac{16}{81} \geq 1$ for $b \geq 3$. Thus, the upper bound of the lemma is proved.

Now, from Taylor's formula we also get that

$$1 - \sqrt{1 - \varepsilon} \geq \frac{1}{2}\varepsilon + \frac{1}{8}\varepsilon^2 + \frac{1}{16}\varepsilon^3 + \frac{5}{128}\varepsilon^4, \qquad \varepsilon \geq 0.$$

In addition, if $-1 \le \varepsilon \le 0$, then $w^{(5)}(\varepsilon) \le \frac{105}{32}$, so

$$
\begin{aligned}
1 - \sqrt{1 - \varepsilon} &= \frac{1}{2}\varepsilon + \frac{1}{8}\varepsilon^2 + \frac{1}{16}\varepsilon^3 + \frac{5}{128}\varepsilon^4 \left(1 + \frac{w^{(5)}(\varepsilon_1)}{120}\varepsilon\right) \\
&\ge \frac{1}{2}\varepsilon + \frac{1}{8}\varepsilon^2 + \frac{1}{16}\varepsilon^3 + \varepsilon^4 \left(\frac{5}{128} - \frac{\frac{105}{32}}{120}\right) \\
&\ge \frac{1}{2}\varepsilon + \frac{1}{8}\varepsilon^2 + \frac{1}{16}\varepsilon^3 + 0.01\,\varepsilon^4.
\end{aligned}
$$

□

Proof of Theorem 1.1 *(First Part)* Using Lemma 5.2 with $\varepsilon = \xi$ and applying Corollary 5.1, we get an asymptotic representation

$$
\sqrt{\pi}\,\mathbb{E}_\theta\,\omega^2(F_\theta, F) = \left(1 + \frac{1}{4n}\right)\left(\frac{1}{8}\mathbb{E}\xi^2 + \frac{1}{16}\mathbb{E}\xi^3 + c\,\mathbb{E}\xi^4\right) - \frac{1}{8n} + O\left(\frac{1}{n^2}\right)
$$

for some quantity c such that $0.01 \le c \le 3$. If additionally X is isotropic, then $\mathbb{E}\,\langle X, Y\rangle^2 = n$, i.e. $\mathbb{E}\xi^2 = \frac{1}{n}$, and the representation is simplified to

$$
\sqrt{\pi}\,\mathbb{E}_\theta\,\omega^2(F_\theta, F) = \left(1 + \frac{1}{4n}\right)\left(\frac{1}{16}\mathbb{E}\xi^3 + c\,\mathbb{E}\xi^4\right) + O\left(\frac{1}{n^2}\right),
$$

thus removing the term of order $1/n$. Moreover, since $\mathbb{E}\xi^4 \le \mathbb{E}\,|\xi|^3 \le \mathbb{E}\xi^2 = \frac{1}{n}$, the fraction $\frac{1}{4n}$ may be removed from the brackets at the expense of the remainder term. Thus

$$
\sqrt{\pi}\,\mathbb{E}_\theta\,\omega^2(F_\theta, F) = \frac{1}{16}\mathbb{E}\xi^3 + c\,\mathbb{E}\xi^4 + O\left(\frac{1}{n^2}\right),
$$

which is exactly the expansion (1.3).

□

Remark 5.3 In the isotropic case with $|X|^2 = n$ a.s., but without the mean zero assumption, the above expansion takes the form

$$
\sqrt{\pi}\,\mathbb{E}_\theta\,\omega^2(F_\theta, F) = \frac{1}{2}\mathbb{E}\xi + \frac{1}{16}\mathbb{E}\xi^3 + c\,\mathbb{E}\xi^4 + O\left(\frac{1}{n^2}\right). \tag{5.2}
$$

Since the last two expectations are non-negative, this implies in particular that

$$
\mathbb{E}_\theta\,\omega^2(F_\theta, F) \ge \frac{1}{2\sqrt{\pi}}\mathbb{E}\xi + O\left(\frac{1}{n^2}\right). \tag{5.3}
$$

6 General Lower Bounds for the L^2-Distance: Proof of Theorem 1.2

Proposition 4.1 may be used to establish the following general lower bound which will be the first step in the proof of Theorem 1.2. Recall that Y denotes an independent copy of a random vector X in $\mathbb{R}^n$.

Proposition 6.1 *If* $\mathbb{E}\,|X| \le b\sqrt{n}$, *then*

$$\mathbb{E}_\theta\, \omega^2(F_\theta, F) \ge c_1\, \mathbb{E}\,\rho\,\xi^4 - c_2\frac{b}{n^2}, \tag{6.1}$$

where

$$\rho = \left(\frac{|X|^2 + |Y|^2}{2n}\right)^{1/2}, \qquad \xi = \frac{2\,\langle X, Y\rangle}{|X|^2 + |Y|^2}.$$

The argument employs two elementary lemmas.

Lemma 6.2 *If* $\mathbb{E}\,|X|^2$ *is finite, then*

$$\mathbb{E}\,\langle X, Y\rangle^2 \ge \frac{1}{n}\left(\mathbb{E}\,|X|^2\right)^2. \tag{6.2}$$

By the invariance of (6.2) under linear orthogonal transformations, we may assume that $\mathbb{E}X_iX_j = \lambda_i\delta_{ij}$ where λ_i's appear as eigenvalues of the covariance operator of X. Since

$$\mathbb{E}\,|X|^2 = \sum_{i=1}^n \lambda_i, \qquad \mathbb{E}\,\langle X, Y\rangle^2 = \sum_{i=1}^n \lambda_i^2,$$

the inequality (6.2) follows by applying Cauchy's inequality.

Lemma 6.3 *If* $\mathbb{E}\,|X|^p$ *is finite for an integer* $p \ge 1$, *then, for any real number* $0 \le \alpha \le p$,

$$\mathbb{E}\,\frac{\langle X, Y\rangle^p}{(|X|^2 + |Y|^2)^\alpha} \ge 0,$$

where the ratio is defined to be zero in case $X = Y = 0$. *In addition, for* $\alpha \in [0, 2]$,

$$\mathbb{E}\,\frac{\langle X, Y\rangle^2}{(|X|^2 + |Y|^2)^\alpha} \ge \frac{1}{n}\,\mathbb{E}\,\frac{|X|^2\,|Y|^2}{(|X|^2 + |Y|^2)^\alpha}.$$

Proof First, let us note that

$$\mathbb{E}\,\frac{|\langle X, Y\rangle|^p}{(|X|^2 + |Y|^2)^\alpha} \le \mathbb{E}\,\frac{(|X|\,|Y|)^p}{(|X|\,|Y|)^\alpha} = (\mathbb{E}\,|X|^{p-\alpha})^2,$$

so, the expectation on the left is finite. Without loss of generality, we may assume that $0 < \alpha \le p$ and $r = |X|^2 + |Y|^2 > 0$ with probability 1. We use the identity

$$\int_0^\infty e^{-rt^{1/\alpha}}\,dt = c_\alpha\,r^{-\alpha} \quad \text{where } c_\alpha = \int_0^\infty e^{-s^{1/\alpha}}\,ds,$$

which gives

$$c_\alpha\,\mathbb{E}\,\langle X, Y\rangle^p\,r^{-\alpha} = \int_0^\infty \mathbb{E}\,\langle X, Y\rangle^p\,e^{-rt^{1/\alpha}}\,dt.$$

Writing $X = (X_1, \ldots, X_n)$ and $Y = (Y_1, \ldots, Y_n)$, we have

$$\mathbb{E}\,\langle X, Y\rangle^p\,e^{-rt^{1/\alpha}} = \mathbb{E}\,\langle X, Y\rangle^p\,e^{-t^{1/\alpha}(|X|^2 + |Y|^2)}$$

$$= \sum_{i_1,\ldots,i_p=1}^{n} \left(\mathbb{E}\,X_{i_1}\ldots X_{i_p}\,e^{-t^{1/\alpha}|X|^2}\right)^2,$$

which shows that the left expectation is always non-negative. Integrating over $t > 0$, this proves the first assertion.

For the second assertion, write

$$c_\alpha\,\mathbb{E}\,\langle X, Y\rangle^2\,r^{-\alpha} = \int_0^\infty \mathbb{E}\,\langle X, Y\rangle^2\,e^{-t^{1/\alpha}(|X|^2 + |Y|^2)}\,dt = \int_0^\infty \mathbb{E}\,\langle X_t, Y_t\rangle^2\,dt,$$

where

$$X_t = e^{-t^{1/\alpha}|X|^2/2}\,X, \quad Y_t = e^{-t^{1/\alpha}|Y|^2/2}\,Y.$$

Since Y_t represents an independent copy of X_t, one may apply Lemma 6.2 which gives

$$\mathbb{E}\,\langle X_t, Y_t\rangle^2 \ge \frac{1}{n}\,\mathbb{E}\,|X_t|^2\,|Y_t|^2.$$

Hence,

$$\int_0^\infty \mathbb{E}\,\langle X_t, Y_t\rangle^2\,dt \ge \frac{1}{n}\int_0^\infty \mathbb{E}|X_t|^2\,|Y_t|^2\,dt$$

$$= \frac{1}{n}\int_0^\infty \mathbb{E}|X|^2\,|Y|^2\,e^{-t^{1/\alpha}(|X|^2 + |Y|^2)}\,dt = \frac{c_\alpha}{n}\,\mathbb{E}\,|X|^2\,|Y|^2\,r^{-\alpha}.$$

$\square$

Proof of Proposition 6.1 Let us return to the representation (4.3) in Proposition 4.1 and write

$$\mathbb{E}_\theta\, \omega^2(F_\theta, F) = \frac{1}{\sqrt{2\pi}}\, \mathbb{E}\,(R_0 + R_1) + \frac{Cb}{n^2},$$

where

$$R_0 = \frac{1}{8n^{3/2}} \frac{(|X|^2 - |Y|^2)^2}{(|X|^2 + |Y|^2)^{3/2}}$$

and

$$R_1 = \frac{(|X|^2 + |Y|^2)^{1/2}}{\sqrt{n}}\left(1 + \frac{1}{8n}\right) - \frac{|X - Y|}{\sqrt{n}}\left(1 + \frac{1}{4n}\right)$$

$$= \frac{(|X|^2 + |Y|^2)^{1/2}}{\sqrt{n}}\left[\left(1 + \frac{1}{4n}\right)\left(1 - \sqrt{1 - \xi}\,\right) - \frac{1}{8n}\right]$$

with the assumption that $R_0 = 0$ when $X = Y = 0$. Since $|\xi| \leq 1$, one may apply Lemma 5.2 which gives

$$R_1 \geq \frac{(|X|^2 + |Y|^2)^{1/2}}{\sqrt{n}}\left[\left(1 + \frac{1}{4n}\right)\left(\frac{1}{2}\xi + \frac{1}{8}\xi^2 + \frac{1}{16}\xi^3 + 0.01\,\xi^4\right) - \frac{1}{8n}\right].$$

The expectation of the terms on the right-hand side containing ξ and ξ^3 is non-negative according to Lemma 6.3 with $\alpha = \frac{1}{2}$, $p = 1$, and with $\alpha = \frac{5}{2}$, $p = 3$, respectively. Hence, removing the unnecessary factor $1 + \frac{1}{4n}$, we get

$$\mathbb{E}_\theta\, \omega^2(F_\theta, F) \geq \frac{1}{\sqrt{2\pi}}\, \mathbb{E}R_0 + \frac{1}{\sqrt{2\pi}}\, \mathbb{E}\, \frac{(|X|^2 + |Y|^2)^{1/2}}{8\sqrt{n}}\left(\xi^2 - \frac{1}{n}\right)$$

$$+ c_1\, \mathbb{E}\, \frac{(|X|^2 + |Y|^2)^{1/2}}{\sqrt{n}}\,\xi^4 - c_2 \frac{b}{n^2}. \tag{6.3}$$

Now, by the second inequality of Lemma 6.3 applied with $\alpha = 3/2$, $p = 2$, we have

$$\mathbb{E}\,(|X|^2 + |Y|^2)^{1/2}\,\xi^2 = 4\,\mathbb{E}\, \frac{\langle X, Y\rangle^2}{(|X|^2 + |Y|^2)^{3/2}}$$

$$\geq \frac{4}{n}\, \mathbb{E}\, \frac{|X|^2\,|Y|^2}{(|X|^2 + |Y|^2)^{3/2}}.$$

This gives

$$\mathbb{E}\,\frac{(|X|^2+|Y|^2)^{1/2}}{8\sqrt{n}}\left(\xi^2-\frac{1}{n}\right) \geq \frac{1}{8n^{3/2}}\,\mathbb{E}\left[\frac{4\,|X|^2\,|Y|^2}{(|X|^2+|Y|^2)^{3/2}}-(|X|^2+|Y|^2)^{1/2}\right]$$

$$=-\frac{1}{8n^{3/2}}\,\mathbb{E}\,\frac{(|X|^2-|Y|^2)^2}{(|X|^2+|Y|^2)^{3/2}}=-\mathbb{E}R_0.$$

Thus, the summand $\mathbb{E}R_0$ in (6.3) neutralizes the second expectation, and we are left with the term containing ξ^4. □

Proof of Theorem 1.2 We apply Proposition 6.1. By the assumption, $\mathbb{E}\rho^2=1$ and $\mathrm{Var}(\rho^2)=\frac{1}{2n}\sigma_4^2$, where $\sigma_4^2=\frac{1}{n}\,\mathrm{Var}(|X|^2)$. Using

$$2\,\langle X,Y\rangle=|X|^2+|Y|^2-|X-Y|^2,\qquad \xi=1-\frac{|X-Y|^2}{|X|^2+|Y|^2},$$

we have

$$\xi^4\geq(1-\alpha)^4\,1_{\{|X-Y|^2\leq\alpha\,(|X|^2+|Y|^2)\}}$$

$$\geq(1-\alpha)^4\,1_{\{|X-Y|^2\leq\alpha\lambda n,\,|X|^2+|Y|^2\geq\lambda n\}},\quad 0<\alpha,\lambda<1.$$

On the set $|X|^2+|Y|^2\geq\lambda n$, we necessarily have $\rho^2\geq\frac{\lambda}{2}$, so

$$\mathbb{E}\rho\,\xi^4\geq\frac{(1-\alpha)^4}{\sqrt{2}}\,\sqrt{\lambda}\,\mathbb{P}\left\{|X-Y|^2\leq\alpha\lambda n,\,|X|^2+|Y|^2\geq\lambda n\right\}$$

$$\geq\frac{(1-\alpha)^4}{\sqrt{2}}\,\sqrt{\lambda}\,\left(\mathbb{P}\{|X-Y|^2\leq\alpha\lambda n\}-\mathbb{P}\{|X|^2+|Y|^2\leq\lambda n\}\right).$$

But, by Chebyshev's inequality

$$\mathbb{P}\{|X|^2\leq\lambda n\}=\mathbb{P}\{n-|X|^2\geq(1-\lambda)\,n\}\leq\frac{\mathrm{Var}(|X|^2)}{(1-\lambda)^2\,n^2}=\frac{\sigma_4^2}{(1-\lambda)^2\,n},$$

implying

$$\mathbb{P}\{|X|^2+|Y|^2\leq\lambda n\}\leq\left(\mathbb{P}\{|X|^2\leq\lambda n\}\right)^2\leq\frac{1}{(1-\lambda)^4}\,\frac{\sigma_4^4}{n^2}.$$

Hence

$$\mathbb{E}\rho\,\xi^4\geq\frac{(1-\alpha)^4}{\sqrt{2}}\,\sqrt{\lambda}\,\left(\mathbb{P}\{|X-Y|^2\leq\alpha\lambda n\}-\frac{1}{(1-\lambda)^4}\,\frac{\sigma_4^4}{n^2}\right).$$

Choosing, for example, $\alpha = \lambda = \frac{1}{2}$, we get

$$\mathbb{E}\,\rho\,\xi^4 \geq \frac{1}{32}\,\mathbb{P}\left\{|X - Y|^2 \leq \frac{1}{4}\,n\right\} - \frac{\sigma_4^4}{2n^2}.$$

It remains to apply (6.1) with $b = 1$ and replace F with Φ on the basis of (2.3).   □

7 Lipschitz Systems

While upper bounds of order $n^{-1/2}$ for the L^2-distance $\omega(F_\theta, F)$ on average are provided in (1.2) and in the more general inequality (3.5) of Proposition 3.1, in this section we focus on the conditions that provide similar lower bounds, as a consequence of Theorem 1.2.

Let L be a fixed measurable function on the underlying probability space $(\Omega, \mathfrak{F}, \mathbb{P})$. We will say that the system $X_1, \ldots, X_n$ of random variables on $(\Omega, \mathfrak{F}, \mathbb{P})$, or the random vector $X = (X_1, \ldots, X_n)$ in $\mathbb{R}^n$ satisfies a Lipschitz condition with a parameter function L, if

$$\max_{1 \leq k \leq n} |X_k(t) - X_k(s)| \leq n\,|L(t) - L(s)|, \qquad t, s \in \Omega. \tag{7.1}$$

When Ω is an interval of the real line (finite or not), and $L(t) = Lt$, $L > 0$, this condition means that every function X_k in the system has a Lipschitz semi-norm at most Ln.

As before, we use the variance functional $\sigma_4^2 = \frac{1}{n}\,\mathrm{Var}(|X|^2)$.

Proposition 7.1 *Suppose that* $\mathbb{E}\,|X|^2 = n$. *If the random vector X satisfies the Lipschitz condition with a parameter function L, then*

$$\mathbb{E}_\theta\,\omega^2(F_\theta, F) \geq \frac{c_L}{n} - \frac{c_0\,(1 + \sigma_4^4)}{n^2} \tag{7.2}$$

with some absolute constant $c_0 > 0$ and with a constant c_L depending on the distribution of L only. Moreover, if L has finite second moment, then with some absolute constant $c_1 > 0$

$$\mathbb{E}_\theta\,\omega^2(F_\theta, F) \geq \frac{c_1}{n\,\sqrt{\mathrm{Var}(L)}} - \frac{c_0\,(1 + \sigma_4^4)}{n^2}. \tag{7.3}$$

Note that, if $X_1, \ldots, X_n$ form an orthonormal system in $L^2(\Omega, \mathfrak{F}, \mathbb{P})$, i.e., the random vector X is isotropic, and if L has finite second moment $\|L\|_2^2 = \mathbb{E}L^2$, then this moment has to be bounded from below by a multiple of $1/n^2$. Indeed, the projection of the function $\eta(t) = 1$ in $L^2(\Omega, \mathfrak{F}, \mathbb{P})$ to the linear hull H of

$X_1, \ldots, X_n$ has the form $\mathrm{Proj}_H(\eta) = \sum_{k=1}^n \langle \eta, X_k \rangle X_k$, and we have Bessel's inequality

$$1 = \|\eta\|_2^2 \geq \|\mathrm{Proj}_H(\eta)\|_2^2 = \sum_{k=1}^n \langle \eta, X_k \rangle^2 = \sum_{k=1}^n (\mathbb{E}X_k)^2$$

(where we used the canonical innde product $\langle \cdot, \cdot \rangle$ in $L^2(\Omega, \mathfrak{F}, \mathbb{P})$). By the Lipschitz assumption, $|X_k(t) - X_k(s)|^2 \leq n^2 |L(t) - L(s)|^2$. Integrating this inequality over the product measure $\mathbb{P}(dt) \otimes \mathbb{P}(ds)$, we obtain a lower bound

$$n^2 \mathrm{Var}(L) \geq \mathrm{Var}(X_k) = 1 - (\mathbb{E}X_k)^2.$$

One may now perform summation over $k = 1, \ldots, n$, which together with Bessel's inequality leads to

$$\mathrm{Var}(L) \geq \frac{n-1}{n^3} \geq \frac{1}{2n^2} \quad (n \geq 2).$$

The Lipschitz condition (7.1) guarantees the validity of the following property, which can be combined with Theorem 1.2 to obtain (7.2) and (7.3).

Lemma 7.2 *Suppose that the random vector* $X = (X_1, \ldots, X_n)$ *satisfies the Lipschitz condition with the parameter function L. If Y is an independent copy of X, then*

$$\mathbb{P}\{|X - Y|^2 \leq \lambda n\} \geq \frac{c\sqrt{\lambda}}{n}, \quad 0 \leq \lambda \leq 1,$$

where the constant $c > 0$ depends on the distribution of L only. Moreover, if L has finite second moment, then

$$\mathbb{P}\{|X - Y|^2 \leq \lambda n\} \geq \frac{\sqrt{\lambda}}{6n\sqrt{\mathrm{Var}(L)}}, \quad 0 \leq \lambda \leq n^2 \mathrm{Var}(L).$$

In turn, this lemma is based on the following general observation.

Lemma 7.3 *If η is an independent copy of a random variable ξ, then for any $\varepsilon_0 > 0$,*

$$\mathbb{P}\{|\xi - \eta| \leq \varepsilon\} \geq c\varepsilon, \quad 0 \leq \varepsilon \leq \varepsilon_0,$$

with some constant $c > 0$ independent of ε. Moreover, if the standard deviation $\sigma = \sqrt{\mathrm{Var}(\xi)}$ is finite, then

$$\mathbb{P}\{|\xi - \eta| \leq \varepsilon\} \geq \frac{1}{6\sigma}\varepsilon, \quad 0 \leq \varepsilon \leq \sigma.$$

Proof The difference $\xi - \eta$ has a non-negative characteristic function $h(t) = |\psi(t)|^2$, where ψ is the characteristic function of ξ. Denoting by H the distribution function of $\xi - \eta$, we start with a general identity

$$\int_{-\infty}^{\infty} \hat{p}(x)\, dH(x) = \int_{-\infty}^{\infty} p(t) h(t)\, dt, \tag{7.4}$$

which is valid for any integrable function $p(t)$ on the real line with Fourier transform $\hat{p}(x) = \int_{-\infty}^{\infty} e^{itx} p(t)\, dt$, $x \in \mathbb{R}$. Given $\varepsilon > 0$, here we take a standard pair

$$p(t) = \frac{1}{2\pi} \left(\frac{\sin \frac{\varepsilon t}{2}}{\frac{\varepsilon t}{2}} \right)^2, \qquad \hat{p}(x) = \frac{1}{\varepsilon} \left(1 - \frac{|x|}{\varepsilon} \right)^+,$$

where we use the notation $a^+ = \max\{a, 0\}$. In this case,

$$\int_{-\infty}^{\infty} \hat{p}(x)\, dH(x) \leq \frac{1}{\varepsilon} \int_{[-\varepsilon, \varepsilon]} dH(x) = \frac{1}{\varepsilon} \mathbb{P}\{|\xi - \eta| \leq \varepsilon\}.$$

On the other hand, since the function $\frac{\sin u}{u}$ is decreasing in $0 < u < \frac{\pi}{2}$, we have

$$\int_{-\infty}^{\infty} p(t) h(t)\, dt \geq \frac{1}{2\pi} \left(2\sin(1/2) \right)^2 \int_{-1/\varepsilon}^{1/\varepsilon} h(t)\, dt \geq \frac{1}{7} \int_{-1/\varepsilon}^{1/\varepsilon} h(t)\, dt.$$

Hence, whenever $0 < \varepsilon \leq \varepsilon_0$, by (7.4),

$$\mathbb{P}\{|\xi - \eta| \leq \varepsilon\} \geq \frac{\varepsilon}{7} \int_{-1/\varepsilon}^{1/\varepsilon} h(t)\, dt \geq \frac{\varepsilon}{7} \int_{-1/\varepsilon_0}^{1/\varepsilon_0} h(t)\, dt.$$

Since $h(t)$ is bounded away from zero near the origin, the first assertion follows.

One may quantify this statement in terms of the variance $\sigma^2 = \mathrm{Var}(\xi)$ by using Taylor's expansion for $h(t)$ about zero. Indeed, it gives $1 - h(t) \leq \sigma^2 t^2$, and thus for $\varepsilon \leq \varepsilon_0 = \sigma$,

$$\int_{-1/\varepsilon}^{1/\varepsilon} h(t)\, dt \geq \int_{-1/\sigma}^{1/\sigma} (1 - \sigma^2 t^2)\, dt = \frac{4}{3\sigma}.$$

Since $\frac{\varepsilon}{7} \cdot \frac{4}{3\sigma} \geq \frac{1}{6\sigma} \varepsilon$, the lemma is proved. $\square$

Proof of Lemma 7.2 Let us equip the product space $\Omega^2 = \Omega \times \Omega$ with the product measure $\mathbb{P}^2 = \mathbb{P} \otimes \mathbb{P}$ and redefine X on this new probability space as $X(t, s) =$

$X(t)$, $(t, s) \in \Omega^2$. Then one can introduce an independent copy of X in the form $Y(t, s) = X(s)$. By the Lipschitz condition,

$$|X(t, s) - Y(t, s)|^2 = \sum_{k=1}^{n} |X_k(t) - X_k(s)|^2 \leq n^3 |L(t) - L(s)|^2.$$

Hence, if η is an independent copy of the random variable $\xi = L$, then

$$\mathbb{P}\{|X - Y|^2 \leq \lambda n\} \geq \mathbb{P}\{n^3 |\xi - \eta|^2 \leq \lambda n\} = \mathbb{P}\left\{|\xi - \eta| \leq \frac{\sqrt{\lambda}}{n}\right\}.$$

But, by Lemma 7.3 with $\varepsilon_0 = 1$, the latter probability is at least $c \frac{\sqrt{\lambda}}{n}$, where the constant c depends on L only (via its distribution). An application of the second inequality of Lemma 7.3 yields the second assertion. $\square$

To include more examples, let us now give a bit more general form of Lemma 7.2, assuming that $(\Omega, \mathbb{P}) = (\Omega_1 \times \Omega_2, \mathbb{P}_1 \otimes \mathbb{P}_2)$ is a product probability space.

Lemma 7.4 *Let* $X = (X_1, \ldots, X_n) : \Omega \to \mathbb{R}^n$ *be a random vector such that, for some measurable functions* L_1 *and* L_2 *defined on* Ω_1 *and* Ω_2 *respectively,*

$$\max_{1 \leq k \leq n} |X_k(t_1, t_2) - X_k(s_1, s_2)| \leq n |L_1(t_1) - L_1(s_1)| + |L_2(t_2) - L_2(s_2)| \quad (7.5)$$

for all (t_1, t_2), $(s_1, s_2) \in \Omega$. *If* Y *is an independent copy of* X, *then*

$$\mathbb{P}\{|X - Y|^2 \leq \lambda n\} \geq \frac{c\lambda}{n}, \qquad 0 \leq \lambda \leq 1, \quad (7.6)$$

where the constant $c > 0$ *depends on the distributions of* L_1 *and* L_2 *only.*

Proof Again, let us equip the product space $\Omega^2 = \Omega \times \Omega$ with the product measure $\mathbb{P}^2 = \mathbb{P} \otimes \mathbb{P}$ and put $X(t, s) = X(t)$, $Y(t, s) = X(s)$ for $t = (t_1, t_2) \in \Omega$ and $s = (s_1, s_2) \in \Omega$, so that Y is an independent copy of X. By the Lipschitz condition (7.5), for any $k \leq n$,

$$|X_k(t) - X_k(s)|^2 \leq 2n^2 |L_1(t_1) - L_1(s_1)| + 2 |L_2(t_2) - L_2(s_2)|^2,$$

so

$$|X(t) - Y(s)|^2 = \sum_{k=1}^{n} |X_k(t) - X_k(s)|^2$$

$$\leq 2n^3 |L_1(t_1) - L_1(s_1)|^2 + 2n |L_2(t_2) - L_2(s_2)|^2.$$

Putting $L_1(t_1, t_2) = L_1(t_1)$ and $L_2(t_1, t_2) = L_2(t_2)$, one may treat L_1 and L_2 as independent random variables. If L_1' is an independent copy of L_1 and L_2' is an independent copy of L_2, we obtain that

$$\mathbb{P}\{|X - Y|^2 \leq \lambda n\} \geq \mathbb{P}\left\{n^2 |L_1 - L_1'|^2 + |L_2 - L_2'|^2 \leq \frac{\lambda}{2}\right\}$$

$$\geq \mathbb{P}\left\{n^2 |L_1 - L_1'|^2 \leq \frac{\lambda}{4}\right\} \mathbb{P}\left\{|L_2 - L_2'|^2 \leq \frac{\lambda}{4}\right\}$$

$$= \mathbb{P}\left\{|L_1 - L_1'| \leq \frac{1}{2n}\sqrt{\lambda}\right\} \mathbb{P}\left\{|L_2 - L_2'| \leq \frac{1}{2}\sqrt{\lambda}\right\}.$$

It remains to apply Lemma 7.3.   □

Let us now combine the inequality (1.8) of Theorem 1.2 with the inequality (7.6) applied with $\lambda = \frac{1}{4}$. Then we obtain the following generalization of Proposition 7.1.

Proposition 7.5 *Under the Lipschitz condition* (7.5), *we have*

$$\mathbb{E}_\theta \, \omega^2(F_\theta, F) \geq \frac{c}{n} - \frac{c_0 \, (1 + \sigma_4^4)}{n^2},$$

where $c_0 > 0$ is an absolute constant, while $c > 0$ depends on the distributions of L_1 and L_2. A similar estimate also holds when F is replaced with the normal distribution function Φ.

The last assertion follows from the inequality (2.3), cf. Corollary 2.2.

8   Berry-Esseen-Type Bounds

We now turn to the study of the Kolmogorov distance

$$\rho(F_\theta, F) = \sup_x |F_\theta(x) - F(x)|, \quad \theta \in \mathbb{S}^{n-1},$$

between the distribution functions F_θ of the weighted sums $S_\theta = \langle X, \theta \rangle$ and the typical distribution function $F = \mathbb{E}_\theta F_\theta$. We are mostly interested in bounding the second moment $\mathbb{E}_\theta \, \rho^2(F_\theta, F)$. As in the case of the L^2-distance, our basic tool will be a Fourier analytic approach relying upon a general Berry-Esseen-type bound

$$c \, \rho(U, V) \leq \int_0^T \frac{|\hat{U}(t) - \hat{V}(t)|}{t} \, dt + \frac{1}{T} \int_0^T |\hat{V}(t)| \, dt, \qquad T > 0, \qquad (8.1)$$

where U and V may be arbitrary distribution functions on the line with characteristic functions $\hat{U}$ and $\hat{V}$ respectively (cf. e.g. [3, 23, 24]).

As before, we denote by f_θ and f the characteristic functions associated to F_θ and F. Recall that σ_{2p}-functionals were defined in (2.2).

Lemma 8.1 *If $T \geq T_0 \geq 1$, then for all $p \geq 1$,*

$$c_p \, \mathbb{E}_\theta \, \rho^2(F_\theta, F) \leq \int_0^1 \frac{\mathbb{E}_\theta \, |f_\theta(t) - f(t)|^2}{t^2} \, dt + \log T \int_0^{T_0} \frac{\mathbb{E}_\theta \, |f_\theta(t) - f(t)|^2}{t} \, dt$$

$$+ \log T \int_{T_0}^T \frac{\mathbb{E}_\theta \, |f_\theta(t)|^2}{t} \, dt + \frac{1}{T^2} + \frac{1 + \sigma_{2p}^{2p}}{n^p}, \qquad (8.2)$$

where the constants $c_p > 0$ depend on p only.

Proof By (8.1), for any $\theta \in \mathbb{S}^{n-1}$,

$$c \, \rho(F_\theta, F) \leq \int_0^T \frac{|f_\theta(t) - f(t)|}{t} \, dt + \frac{1}{T} \int_0^T |f(t)| \, dt,$$

and squaring it, we get

$$c \, \rho^2(F_\theta, F) \leq \left(\int_0^T \frac{|f_\theta(t) - f(t)|}{t} \, dt \right)^2 + \frac{1}{T^2} \left(\int_0^T |f(t)| \, dt \right)^2.$$

Let us split integration in the first integral into the intervals $[0, 1]$ and $[1, T]$. By Cauchy's inequality,

$$\left(\int_0^1 \frac{|f_\theta(t) - f(t)|}{t} \, dt \right)^2 \leq \int_0^1 \frac{|f_\theta(t) - f(t)|^2}{t^2} \, dt,$$

while

$$\left(\int_1^T \frac{|f_\theta(t) - f(t)|}{t} \, dt \right)^2 \leq \log T \int_1^T \frac{|f_\theta(t) - f(t)|^2}{t} \, dt.$$

Hence

$$c \, \rho^2(F_\theta, F) \leq \int_0^1 \frac{|f_\theta(t) - f(t)|^2}{t^2} \, dt$$

$$+ \log T \int_1^T \frac{|f_\theta(t) - f(t)|^2}{t} \, dt + \frac{1}{T^2} \left(\int_0^T |f(t)| \, dt \right)^2.$$

Without an essential loss one may extend integration in the second integral to the larger interval $[0, T]$. Moreover, taking the expectation over θ, we then get

$$c\, \mathbb{E}_\theta\, \rho^2(F_\theta, F) \leq \int_0^1 \frac{\mathbb{E}_\theta\, |f_\theta(t) - f(t)|^2}{t^2}\, dt$$

$$+ \log T \int_0^T \frac{\mathbb{E}_\theta\, |f_\theta(t) - f(t)|^2}{t}\, dt + \frac{1}{T^2} \Big(\int_0^T |f(t)|\, dt \Big)^2.$$

Again, one may split integration in the second last integral to the two intervals $[0, T_0]$ and $[T_0, T]$, so that to consider separately sufficiently large values of t for which $|f_\theta(t)|$ is small enough (with high probability). More precisely, since $f(t) = \mathbb{E}_\theta\, f_\theta(t)$ and

$$|f_\theta(t) - f(t)|^2 \leq 2\, |f_\theta(t)|^2 + 2\, |f(t)|^2,$$

we have $|f(t)|^2 \leq \mathbb{E}_\theta\, |f_\theta(t)|^2$ and therefore

$$\mathbb{E}_\theta\, |f_\theta(t) - f(t)|^2 \leq 4\, \mathbb{E}_\theta\, |f_\theta(t)|^2.$$

It remains to apply Lemma 2.4.   □

In order to control the last integral in (8.2), one may apply the upper bound (2.8) on J_n in the representation (3.3) to get that, for all $t \in \mathbb{R}$,

$$\mathbb{E}_\theta\, |f_\theta(t)|^2 \leq 5\, \mathbb{E}\, e^{-t^2 |X-Y|^2/2n} + 4\, e^{-n/12},$$

where Y is an independent copy of the random vector X. Splitting the last expectation to the event $A = \{|X - Y|^2 \leq \frac{1}{4} n\}$ and its complement leads to

$$\mathbb{E}_\theta\, |f_\theta(t)|^2 \leq 5\, e^{-t^2/8} + 4\, e^{-n/12} + 5\, \mathbb{P}(A). \tag{8.3}$$

The latter probability may further be estimated by using the moment functionals such as m_p.

To recall the argument (cf. also [8], Proposition 2.5), first note that, by (2.9) with $\lambda = \frac{3}{4}$,

$$\mathbb{P}\Big\{ |X|^2 + |Y|^2 \leq \frac{3}{4} n \Big\} \leq \mathbb{P}\Big\{ |X|^2 \leq \frac{3}{4} n \Big\} \mathbb{P}\Big\{ |Y|^2 \leq \frac{3}{4} n \Big\} \leq \frac{(4\sigma_{2p})^{2p}}{n^p}.$$

On the other hand, by Markov's inequality, assuming that $p \geq 1$ is integer, we have

$$\mathbb{P}\Big\{ |\langle X, Y \rangle| \geq \frac{1}{4} n \Big\} \leq \frac{4^{2p}\, \mathbb{E}\, \langle X, Y \rangle^{2p}}{n^{2p}} = \frac{4^{2p}\, m_{2p}^{2p}}{n^p}.$$

Since $|X - Y|^2 = |X|^2 + |Y|^2 - 2\langle X, Y\rangle$, we have

$$\left\{ |X - Y|^2 \le \frac{1}{4} \right\} \subset \left\{ |X| + |Y|^2 \le \frac{1}{4} n \right\} \cup \left\{ \langle X, Y\rangle > \frac{1}{4} n \right\},$$

and it follows that

$$\mathbb{P}(A) \le \mathbb{P}\left\{ |X|^2 + |Y|^2 \le \frac{3}{4} n \right\} + \mathbb{P}\left\{ \langle X, Y\rangle > \frac{1}{4} n \right\} \le \frac{4^{2p}}{n^p} (m_{2p}^{2p} + \sigma_{2p}^{2p}).$$

Returning to (8.3) and noting that necessarily $m_{2p} \ge m_2 \ge 1$ under the assumption that $\mathbb{E}\,|X|^2 = n$, we thus obtain that

$$c_p\,\mathbb{E}_\theta\,|f_\theta(t)|^2 \le \frac{m_{2p}^{2p} + \sigma_{2p}^{2p}}{n^p} + e^{-t^2/8}.$$

Using this bound, the inequality (8.2) is simplified:

Lemma 8.2 *If the random vector X in $\mathbb{R}^n$ satisfies $\mathbb{E}\,|X|^2 = n$, then for all $T \ge T_0 \ge 1$ and any integer $p \ge 1$,*

$$c_p\,\mathbb{E}_\theta\,\rho^2(F_\theta, F) \le \int_0^1 \frac{\mathbb{E}_\theta\,|f_\theta(t) - f(t)|^2}{t^2}\,dt + \log T \int_0^{T_0} \frac{\mathbb{E}_\theta\,|f_\theta(t) - f(t)|^2}{t}\,dt$$

$$+ \frac{m_{2p}^{2p} + \sigma_{2p}^{2p}}{n^p}(1 + \log T)^2 + \frac{1}{T^2} + e^{-T_0^2/8}\log T \qquad (8.4)$$

with constants c_p depending on p only.

9 Quantitative Forms of Sudakov's Theorem for the Kolmogorov Distance

Let us specialize Lemma 8.2 to the value $p = 1$, assuming that the random vector X is isotropic in $\mathbb{R}^n$ (so that $m_2 = 1$). If σ_2 is bounded, then choosing

$$T = 4n, \quad T_0 = 4\sqrt{\log n},$$

the last three terms in (8.4) produce a quantity of order at most $(\log n)^2/n$. In order to bound the integrals in (8.4), one may apply the classical Poincaré inequality on the unit sphere $\mathbb{S}^{n-1}$

$$\mathbb{E}_\theta\,|u(\theta)|^2 \le \frac{1}{n-1}\,\mathbb{E}_\theta\,|\nabla u(\theta)|^2 \qquad (9.1)$$

to the mean zero functions $u_t(\theta) = f_\theta(t) - f(t)$. They are well defined and smooth on $\mathbb{R}^n$ for any fixed value $t \in \mathbb{R}$ and have gradients (by differentiating in (3.1)) given by

$$\langle \nabla u_t(\theta), w \rangle = it \, \mathbb{E} \, \langle X, w \rangle \, e^{it\langle X, \theta \rangle}, \quad w \in \mathbb{C}^n,$$

where we use the canonical inner product in the product complex space. By the isotropy assumption,

$$|\langle \nabla u_t(\theta), w \rangle| \le |t| \, \mathbb{E} \, |\langle X, w \rangle| \le |t| \, |w|$$

for all w. Hence $|\nabla u_t(\theta)|^2 \le t^2$ for any $\theta \in \mathbb{R}^n$, so that by (9.1),

$$\mathbb{E}_\theta \, |f_\theta(t) - f(t)|^2 \le \frac{t^2}{n-1}. \tag{9.2}$$

Applying this inequality in (8.4) together with the first bound in (2.3) in order to replace F with Φ, we obtain:

Proposition 9.1 *Given an isotropic random vector X in $\mathbb{R}^n$,*

$$\mathbb{E}_\theta \, \rho^2(F_\theta, \Phi) \le c \, (1 + \sigma_2^2) \, \frac{(\log n)^2}{n}.$$

Since $\sigma_2 \le \sigma_4$, we thus have

$$\left(\mathbb{E}_\theta \, \rho^2(F_\theta, \Phi) \right)^{1/2} \le c \, (1 + \sigma_4) \, \frac{\log n}{\sqrt{n}} \tag{9.3}$$

which sharpens (1.1). The latter bound will be an essential step in the proof of Theorem 1.3, while (1.1) is not strong enough.

Let us now consider another scenario in Lemma 8.2, where the distribution of X is supported on the sphere $\sqrt{n} \, \mathbb{S}^{n-1}$. In this case,

$$\mathbb{E}_\theta \, |f_\theta(t) - f(t)|^2 = \mathbb{E}_\theta \, |f_\theta(t)|^2 - |f(t)|^2$$
$$= \mathbb{E} J_n(t|X - Y|) - J_n(t\sqrt{n})^2$$

according to (3.3), while $\sigma_4 = 0$. Hence, in (8.4) with $p = 2$ we arrive at the following preliminary bound which is needed for the proof of Theorem 1.1 in its second part. Here we use again that $m_4 \ge m_2 \ge 1$.

Corollary 9.2 *Suppose that $|X| = \sqrt{n}$ a.s., and Y is an independent copy of X. Then*

$$c \, \mathbb{E}_\theta \, \rho^2(F_\theta, F) \le \int_0^1 \frac{\Delta_n(t)}{t^2} \, dt + \log n \int_0^{4\sqrt{\log n}} \frac{\Delta_n(t)}{t} \, dt + \frac{(\log n)^2}{n^2} \, m_4^4, \tag{9.4}$$

where

$$\Delta_n(t) = \mathbb{E}J_n(t|X - Y|) - J_n(t\sqrt{n})^2. \tag{9.5}$$

10 Proof of Theorem 1.1 for the Kolmogorov Distance

To study the integrals in (9.4), assume additionally that the random vector X in $\mathbb{R}^n$ is isotropic with mean zero and put

$$\xi = \frac{\langle X, Y \rangle}{n},$$

where Y is an independent copy of X. Note that $\frac{1}{n^2} m_4^4 = \mathbb{E}\xi^4$ which is present in the last term on the right-hand side of (9.4).

Focusing on the first integral, we need to develop an asymptotic bound on $\Delta_n(t)$ for $t \in [0, 1]$. Since $|X - Y|^2 = 2n(1 - \xi)$, (9.5) becomes

$$\Delta_n(t) = \mathbb{E}J_n\big(t\sqrt{2n(1 - \xi)}\big) - \big(J_n(t\sqrt{n})\big)^2.$$

We use the asymptotic formula (2.7),

$$J_n\big(t\sqrt{n}\big) = \Big(1 - \frac{t^4}{4n}\Big)e^{-t^2/2} + \varepsilon_n(t), \quad t \in \mathbb{R}, \tag{10.1}$$

where $\varepsilon_n(t)$ denotes a quantity of the form $O\big(n^{-2}\min(1, t^4)\big)$ with a universal constant in O. It implies a similar representation

$$\big(J_n(t\sqrt{n})\big)^2 = \Big(1 - \frac{t^4}{2n}\Big)e^{-t^2} + \varepsilon_n(t). \tag{10.2}$$

Since $|\xi| \le 1$ a.s., we also have

$$J_n\big(t\sqrt{2n(1 - \xi)}\big) = \Big(1 - \frac{t^4}{n}(1 - \xi)^2\Big)e^{-t^2(1-\xi)} + \varepsilon_n(t).$$

Hence, subtracting from $e^{t^2\xi}$ the linear term $1 + t^2\xi$ and adding, one may write

$$\Delta_n(t) = e^{-t^2}\,\mathbb{E}\Big(\Big(1 - \frac{t^4}{n}(1 - \xi)^2\Big)e^{t^2\xi} - \Big(1 - \frac{t^4}{2n}\Big)\Big) + \varepsilon_n(t)$$

$$= e^{-t^2}\,\mathbb{E}\,(U + V) + \varepsilon_n(t)$$

with

$$U = \frac{t^4}{n}\left(\frac{1}{2} - (1-\xi)^2\right) + \left(1 - \frac{t^4}{n}(1-\xi)^2\right) \cdot t^2\xi,$$

$$V = \left(1 - \frac{t^4}{n}(1-\xi)^2\right)(e^{t^2\xi} - 1 - t^2\xi).$$

Using $\mathbb{E}\xi = 0$, $\mathbb{E}\xi^2 = \frac{1}{n}$ and hence $\mathbb{E}\,|\xi|^3 \le \mathbb{E}\xi^2 \le \frac{1}{n}$, we find that in the interval $0 \le t \le 1$,

$$\mathbb{E}\,U = -\frac{t^4}{2n} - \frac{t^4}{n^2} + \frac{2t^6}{n^2} - \frac{t^6}{n}\,\mathbb{E}\xi^3 = -\frac{t^4}{2n} + \varepsilon_n(t).$$

Next write

$$V = W - \frac{t^4}{n}(1-\xi)^2\,W, \qquad W = e^{t^2\xi} - 1 - t^2\xi.$$

Using $|e^x - 1 - x| \le 2x^2$ for $|x| \le 1$, we have $|W| \le 2t^4\xi^2$. Hence, the expected value of the second term in the representation for V does not exceed $8t^8/n^2$. Moreover, by Taylor's expansion,

$$W = \frac{1}{2}t^4\xi^2 + \frac{1}{6}t^6\xi^3 + Rt^8\xi^4, \qquad R = \sum_{k=4}^{\infty} \frac{t^{2k-8}}{k!}\,\xi^{k-4},$$

implying that

$$\mathbb{E}\,W = \frac{t^4}{2n} + \frac{t^6}{6}\,\mathbb{E}\xi^3 + Ct^8\,\mathbb{E}\xi^4,$$

where C is bounded by an absolute constant. Summing the two expansions, we arrive at

$$\mathbb{E}\,(U + V) = \frac{t^6}{6}\,\mathbb{E}\xi^3 + Ct^8\,\mathbb{E}\xi^4 + \varepsilon_n(t)$$

and therefore

$$\int_0^1 \frac{\Delta_n(t)}{t^2}\,dt \le \mathbb{E}\xi^3 + c\,\mathbb{E}\xi^4 + O(n^{-2}).$$

Here $\mathbb{E}\xi^4 \ge (\mathbb{E}\xi^2)^2 = n^{-2}$, so the term $O(n^{-2})$ may be absorbed by the 4-th moment of ξ. Since $\mathbb{E}\xi^3 \ge 0$, the bound (9.4) may be simplified to

$$c\,\mathbb{E}_\theta\,\rho^2(F_\theta, F) \le \mathbb{E}\xi^3 + \mathbb{E}\xi^4 + \log n \int_0^{4\sqrt{\log n}} \frac{\Delta_n(t)}{t}\,dt + \frac{(\log n)^2}{n^2}\,m_4^4,$$

that is,

$$c \, \mathbb{E}_\theta \, \rho^2(F_\theta, F) \le \log n \int_0^{4\sqrt{\log n}} \frac{\Delta_n(t)}{t} \, dt + \mathbb{E}\xi^3 + (\log n)^2 \, \mathbb{E}\xi^4. \qquad (10.3)$$

Turning to the remaining integral (which is most important), let us express it in terms of the functions $g_n(t) = J_n(t\sqrt{2n})$ and

$$\psi(\alpha) = \int_0^T \frac{g_n(\alpha t) - g_n(t)}{t} \, dt, \qquad 0 \le \alpha \le \sqrt{2}, \ T > 1,$$

which will be needed with $T = 4\sqrt{\log n}$ and $\alpha = \sqrt{1 - \xi}$. Namely, we have

$$\int_0^T \frac{\Delta_n(t)}{t} \, dt = \mathbb{E} \, \psi(\sqrt{1 - \xi}) + \int_0^T \frac{J_n(t\sqrt{2n}) - (J_n(t\sqrt{n}))^2}{t} \, dt. \qquad (10.4)$$

To proceed, we need to develop a Taylor expansion for $\xi \to \psi(\sqrt{1 - \xi})$ around zero in powers of ξ. Recall that $g_n(t)$ represents the characteristic function of the random variable $\sqrt{2n} \, \theta_1$ on the probability space $(\mathbb{S}^{n-1}, \mathfrak{s}_{n-1})$. This already ensures that $|g_n(t)| \le 1$ and

$$|g_n'(t)| \le \sqrt{2n} \, \mathbb{E} \, |\theta_1| \le \sqrt{2n} \, (\mathbb{E} \, \theta_1^2)^{1/2} = \sqrt{2}$$

for all $t \in \mathbb{R}$. Hence

$$|g_n(\alpha t) - g_n(t)| \le \sqrt{2} \, |\alpha - 1| \, |t| \le 2 \, |t|,$$

so that

$$|\psi(\alpha)| \le \int_0^1 \frac{|g_n(\alpha t) - g_n(t)|}{t} \, dt + \int_1^T \frac{|g_n(\alpha t) - g_n(t)|}{t} \, dt$$
$$\le 2 + 2 \log T \ < \ 4 \log T \qquad (10.5)$$

(since $T > e$). In addition, $\psi(1) = 0$ and

$$\psi'(\alpha) = \int_0^T g_n'(\alpha t) \, dt = \frac{1}{\alpha} \, (g_n(\alpha T) - 1).$$

Therefore, we arrive at another expression

$$\psi(\alpha) = \int_1^\alpha \frac{g_n(Tx) - 1}{x} \, dx = \int_1^\alpha \frac{g_n(Tx)}{x} \, dx - \log \alpha.$$

For $|\varepsilon| \leq 1$, let

$$v(\varepsilon) = \int_1^{(1-\varepsilon)^{1/2}} \frac{g_n(Tx)}{x} \, dx,$$

$$u(\varepsilon) = \psi\left((1-\varepsilon)^{1/2}\right) = v(\varepsilon) - \frac{1}{2}\log(1-\varepsilon),$$

so that $\mathbb{E}\,\psi\left(\sqrt{1-\xi}\right) = \mathbb{E}\,u(\xi)$. Applying the non-uniform bound $|g_n(t)| \leq 5\left(e^{-t^2} + e^{-n/12}\right)$, cf. (2.8), we have that, for $-1 \leq \varepsilon \leq \frac{1}{2}$,

$$|v(\varepsilon)| \leq \sup_{\frac{1}{\sqrt{2}} \leq x \leq \sqrt{2}} |g_n(Tx)| \int_{\frac{1}{\sqrt{2}}}^{\sqrt{2}} \frac{1}{x} \, dx$$

$$\leq \sup_{z \geq T/\sqrt{2}} |g_n(z)| \log 2 \leq 5\log 2 \left(e^{-T^2/2} + e^{-n/12}\right) \leq \frac{c}{n^8},$$

where the last inequality is specialized to the choice $T = 4\sqrt{\log n}$. Using the Taylor expansion on the same interval for the log-function, we also have $-\log(1-\varepsilon) \leq \varepsilon + \frac{1}{2}\varepsilon^2 + \frac{1}{3}\varepsilon^3 + \frac{2}{3}\varepsilon^4$. Combining the two inequalities, we get

$$u(\varepsilon) \leq \frac{1}{2}\varepsilon + \frac{1}{4}\varepsilon^2 + \frac{1}{6}\varepsilon^3 + \frac{1}{3}\varepsilon^4 + \frac{c}{n^8}, \quad -1 \leq \varepsilon \leq \frac{1}{2}. \qquad (10.6)$$

In order to involve the remaining interval $\frac{1}{2} \leq \varepsilon \leq 1$ in the inequality of a similar type, recall that, by (10.5), $|u(\varepsilon)| \leq 4\log T$ for all $|\varepsilon| \leq 1$. Hence, the inequality (10.6) will hold automatically for this interval, if we increase the coefficient in front of ε^4 to a suitable multiple of $\log T$. As a result, we obtain the desired inequality on the whole segment, that is,

$$u(\varepsilon) \leq \frac{1}{2}\varepsilon + \frac{1}{4}\varepsilon^2 + \frac{1}{6}\varepsilon^3 + (c\log T)\,\varepsilon^4 + \frac{c}{n^8}. \quad -1 \leq \varepsilon \leq 1.$$

In particular,

$$\psi\left(\sqrt{1-\xi}\right) \leq \frac{1}{2}\xi + \frac{1}{4}\xi^2 + \frac{1}{6}\xi^3 + (c\log T)\,\xi^4 + \frac{c}{n^8},$$

and taking the expectation, we get

$$\mathbb{E}\,\psi\left(\sqrt{1-\xi}\right) \leq \frac{1}{4n} + \frac{1}{6}\,\mathbb{E}\xi^3 + (c\log T)\,\mathbb{E}\xi^4, \qquad (10.7)$$

where the term cn^{-8} was absorbed by the 4-th moment of ξ.

Now, let us turn to the integral

$$I_n = \int_0^T \frac{J_n(t\sqrt{2n}) - (J_n(t\sqrt{n}))^2}{t} \, dt,$$

appearing in (10.4), and recall the asymptotic formulas (10.1) and (10.2). After integration, the remainder term $\varepsilon_n(t) = O\left(n^{-2} \min(1, t^4)\right)$ will create an error of order at most $n^{-2} \log T$, up to which I_n is equal to

$$-\int_0^T \frac{t^4}{2n} e^{-t^2} \frac{dt}{t} = -\frac{1}{4n} \left(1 - (T^2 + 1) e^{-T^2}\right) = -\frac{1}{4n} + o(n^{-15}).$$

Thus,

$$I_n = -\frac{1}{4n} + O(n^{-2} \log T).$$

Applying this expansion together with (10.7) in (10.4), we therefore obtain that

$$\int_0^T \frac{\Delta_n(t)}{t} \, dt \leq \frac{1}{6} \mathbb{E}\xi^3 + c \log T \, \mathbb{E}\xi^4.$$

One can now apply this estimate in (10.3), and then we eventually arrive at

$$\mathbb{E}_\theta \, \rho^2(F_\theta, F) \leq c_1 (\log n) \, \mathbb{E}\xi^3 + c_2 (\log n)^2 \, \mathbb{E}\xi^4.$$

By (2.3) with $p = \infty$, a similar inequality remains to hold for the standard normal distribution function Φ in place of F. This proves the inequality (1.4). □

11 Relations Between L^1, L^2 and Kolmogorov Distances

Given a random vector X in $\mathbb{R}^n$, let us now compare the L^2 and L^∞ distances on average, between the distributions F_θ of the weighted sums $\langle X, \theta \rangle$ and the typical distribution $F = \mathbb{E}_\theta F_\theta$. Such information will be needed to derive appropriate lower bounds on $\mathbb{E}_\theta \, \rho(F_\theta, F)$.

Proposition 11.1 *If $|X| \leq b\sqrt{n}$ a.s., then, for any $\alpha \in [1, 2]$,*

$$b^{-\alpha/2} \, \mathbb{E}_\theta \, \omega^\alpha(F_\theta, F) \leq 14 \, (\log n)^{\alpha/4} \, \mathbb{E}_\theta \, \rho^\alpha(F_\theta, F) + \frac{8}{n^4}. \tag{11.1}$$

As will be clear from the proof, at the expense of a larger coefficient in front of $\log n$, the last term n^{-4} can be replaced by $n^{-\beta}$ for any prescribed value of β.

A relation similar to (11.1) is also true for the Kantorovich or L^1-distance

$$W(F_\theta, F) = \int_{-\infty}^{\infty} |F_\theta(x) - F(x)| \, dx$$

in place of L^2. We state it for the case $\alpha = 1$.

Proposition 11.2 *If $|X| \leq b\sqrt{n}$ a.s., then*

$$\mathbb{E}_\theta \, W(F_\theta, F) \leq 14 \, b\sqrt{\log n} \; \mathbb{E}_\theta \, \rho(F_\theta, F) + \frac{8b}{n^4}. \tag{11.2}$$

Proof Put $R_\theta(x) = F_\theta(-x) + (1 - F_\theta(x))$ for $x > 0$ and define similarly R on the basis of F. Using

$$(F_\theta(-x) - F(-x))^2 \leq F_\theta(-x)^2 + F(-x)^2,$$
$$(F_\theta(x) - F(x))^2 \leq (1 - F_\theta(x))^2 + (1 - F(x))^2,$$

we have

$$(F_\theta(-x) - F(-x))^2 + (F_\theta(x) - F(x))^2 \leq R_\theta(x)^2 + R(x)^2.$$

Hence, given $T > 0$ (to be specified later on), we have

$$\omega^2(F_\theta, F) = \int_{-T}^{T} (F_\theta(x) - F(x))^2 \, dx + \int_{|x| \geq T} (F_\theta(x) - F(x))^2 \, dx$$

$$\leq 2T\rho^2(F_\theta, F) + \int_{T}^{\infty} R_\theta(x)^2 \, dx + \int_{T}^{\infty} R(x)^2 \, dx.$$

It follows that, for any $\alpha \in [1, 2]$,

$$\omega^\alpha(F_\theta, F) \leq (2T)^{\frac{\alpha}{2}} \rho^\alpha(F_\theta, F) + \left(\int_{T}^{\infty} R_\theta(x)^2 \, dx \right)^{\frac{\alpha}{2}} + \left(\int_{T}^{\infty} R(x)^2 \, dx \right)^{\frac{\alpha}{2}}$$

and therefore, by Jensen's inequality,

$$\mathbb{E}_\theta \, \omega^\alpha(F_\theta, F) \leq (2T)^{\frac{\alpha}{2}} \, \mathbb{E}_\theta \, \rho^\alpha(F_\theta, F)$$

$$+ \left(\int_{T}^{\infty} \mathbb{E}_\theta \, R_\theta(x)^2 \, dx \right)^{\frac{\alpha}{2}} + \left(\int_{T}^{\infty} R(x)^2 \, dx \right)^{\frac{\alpha}{2}}.$$

Next, by Markov's inequality, for any $x > 0$ and $p \geq 1$,

$$R_\theta(x)^2 \leq \left(\frac{\mathbb{E} \, |\langle X, \theta \rangle|^p}{x^p} \right)^2 \leq \frac{\mathbb{E} \, |\langle X, \theta \rangle|^{2p}}{x^{2p}}$$

and

$$\mathbb{E}_\theta R_\theta(x)^2 \le \left(\frac{\mathbb{E}\,|\,\langle X,\theta\rangle\,|^p}{x^p}\right)^2 \le \frac{\mathbb{E}_\theta\,\mathbb{E}\,|\,\langle X,\theta\rangle\,|^{2p}}{x^{2p}}.$$

Since $R = \mathbb{E}_\theta R_\theta$, a similar inequality holds true for R as well (by Cauchy's inequality). Hence

$$\mathbb{E}_\theta\,\omega^\alpha(F_\theta, F) \le (2T)^{\frac{\alpha}{2}}\,\mathbb{E}_\theta\,\rho^\alpha(F_\theta, F) + 2\left(\mathbb{E}_\theta\,\mathbb{E}\,|\,\langle X,\theta\rangle\,|^{2p}\int_T^\infty \frac{1}{x^{2p}}\,dx\right)^{\frac{\alpha}{2}}.$$

When $\theta = (\theta_1,\dots,\theta_n)$ is treated as a random vector with distribution $\mathfrak{s}_{n-1}$, which is independent of X, the inner product $\langle X,\theta\rangle$ has the same distribution as the random variable $|X|\,\theta_1$. Therefore, recalling Lemma 2.5 and using the assumption $|X| \le b\sqrt{n}$ a.e., we have

$$\mathbb{E}_\theta\,\mathbb{E}\,|\,\langle X,\theta\rangle\,|^{2p} = \mathbb{E}\,|X|^{2p}\,\mathbb{E}_\theta\,|\theta_1|^{2p} \le 2\,(2b^2 p)^p,$$

so that

$$2\left(\mathbb{E}_\theta\int_T^\infty \frac{\mathbb{E}\,|\,\langle X,\theta\rangle\,|^{2p}}{x^{2p}}\,dx\right)^{\frac{\alpha}{2}} \le \frac{2^{\frac{\alpha}{2}+1}}{(2p-1)^{\frac{\alpha}{2}}}\,\frac{(2b^2 p)^{\frac{\alpha p}{2}}}{T^{\frac{\alpha(2p-1)}{2}}}.$$

Thus,

$$\mathbb{E}_\theta\,\omega^\alpha(F_\theta, F) \le (2T)^{\frac{\alpha}{2}}\,\mathbb{E}_\theta\,\rho^\alpha(F_\theta, F) + \frac{2^{\frac{\alpha}{2}+1}}{(2p-1)^{\frac{\alpha}{2}}}\,T^{\frac{\alpha}{2}}\left(\frac{2b^2 p}{T^2}\right)^{\frac{\alpha p}{2}}.$$

Let us choose $T = 2b\sqrt{p}$ in which case the above inequality becomes

$$\mathbb{E}_\theta\,\omega^\alpha(F_\theta, F) \le (4b\sqrt{p})^{\frac{\alpha}{2}}\,\mathbb{E}_\theta\,\rho^\alpha(F_\theta, F) + \frac{2^{\alpha+1}}{(2p-1)^{\frac{\alpha}{2}}}\,(b\sqrt{p})^{\frac{\alpha}{2}}\,2^{-\frac{\alpha p}{2}}.$$

To simplify, one can use $\sqrt{p} \le 2p - 1$ for $p \ge 1$ together with $2^{\alpha+1} \le 8$ and $2^{-\frac{\alpha p}{2}} \le 2^{-\frac{p}{2}}$ (since $1 \le \alpha \le 2$), which leads to

$$\mathbb{E}_\theta\,\omega^\alpha(F_\theta, F) \le (4b\sqrt{p})^{\frac{\alpha}{2}}\,\mathbb{E}_\theta\,\rho^\alpha(F_\theta, F) + 8\,b^{\frac{\alpha}{2}}\,2^{-p/2}.$$

Finally, choosing $p = p_n = (8\log n)/\log 2$, we arrive at (11.1).

Now, turning to (11.2), we use the same functions R_θ and R as before and write

$$W(F_\theta, F) = \int_{-T}^{T} |F_\theta(x) - F(x)|\, dx + \int_{|x| \geq T} |F_\theta(x) - F(x)|\, dx$$

$$\leq 2T\rho(F_\theta, F) + \int_{T}^{\infty} R_\theta(x)\, dx + \int_{T}^{\infty} R(x)\, dx,$$

which gives

$$\mathbb{E}_\theta\, W(F_\theta, F) \leq 2T\, \mathbb{E}_\theta\, \rho(F_\theta, F) + 2 \int_{T}^{\infty} R(x)\, dx.$$

By Markov's inequality, for any $x > 0$ and $p > 1$,

$$R_\theta(x) \leq \frac{\mathbb{E}\, |\langle X, \theta \rangle|^p}{x^p}, \quad R(x) = \mathbb{E}_\theta\, R_\theta(x) \leq \frac{\mathbb{E}_\theta\, \mathbb{E}\, |\langle X, \theta \rangle|^p}{x^p}.$$

Hence

$$\mathbb{E}_\theta\, W(F_\theta, F) \leq 2T\, \mathbb{E}_\theta\, \rho(F_\theta, F) + 2\, \mathbb{E}_\theta\, \mathbb{E}\, |\langle X, \theta \rangle|^p \int_{T}^{\infty} \frac{1}{x^p}\, dx.$$

Here, one may use once more the bound (2.10), which yields

$$\mathbb{E}_\theta\, \mathbb{E}\, |\langle X, \theta \rangle|^p = \mathbb{E}\, |X|^p\, \mathbb{E}_\theta\, |\theta_1|^p \leq 2\, (b^2 p)^{p/2}$$

and

$$\mathbb{E}_\theta\, W(F_\theta, F) \leq 2T\, \mathbb{E}_\theta\, \rho(F_\theta, F) + \frac{4}{p-1}\, \frac{(b^2 p)^{p/2}}{T^{p-1}}.$$

Let us take $T = 2b\sqrt{p}$ in which case the above inequality becomes

$$\mathbb{E}_\theta\, W(F_\theta, F) \leq 4b\sqrt{p}\, \mathbb{E}_\theta\, \rho(F_\theta, F) + 8b\, \frac{\sqrt{p}}{p-1}\, 2^{-p}.$$

Here we arrive at (11.2), by choosing again $p = p_n$ and using $\sqrt{p_n} < p_n - 1$. □

12   Lower Bounds: Proof of Theorem 1.3

A lower bound on $\mathbb{E}_\theta\, \rho^2(F_\theta, \Phi)$ which would be close to the upper bound (1.4) may be given with the help of the lower bound on $\mathbb{E}_\theta\, \omega^2(F_\theta, \Phi)$. More precisely, this can be done in the case where the quantity $\frac{1}{n^{3/2}} m_3^3 + \frac{1}{n^2} m_4^4$ asymptotically dominates

n^{-2} (in particular, when m_4 is essentially larger than 1). Combining the asymptotic expansion (1.3) of Theorem 1.1 with the bound (11.1) of Proposition 11.1 for $\alpha = 2$ and $b = 1$, and recalling the second relation in (2.3) on the normal approximation for the typical distribution F, we therefore obtain:

Proposition 12.1 *If X is an isotropic random vector in $\mathbb{R}^n$ with mean zero and such that $|X| = \sqrt{n}$ a.s., then*

$$\sqrt{\log n}\,\mathbb{E}_\theta\,\rho^2(F_\theta, \Phi) \geq \frac{c_1}{n^{3/2}}\,m_3^3 + \frac{c_2}{n^2}\,m_4^4 - \frac{c_3}{n^2}. \tag{12.1}$$

The relation (11.2) for the Kantorovich distance W may be used to answer the following question: Is it possible to sharpen the lower bound (12.1) by replacing $\mathbb{E}_\theta\,\rho^2(F_\theta, \Phi)$ with $\mathbb{E}_\theta\,\rho(F_\theta, \Phi)$? To this aim, we will need an additional information about moments of $\omega(F_\theta, F)$ of order higher than 2.

Lemma 12.2 *If X is isotropic and satisfies $|X| \leq b\sqrt{n}$, then*

$$c\left(\mathbb{E}_\theta\,\omega^3(F_\theta, F)\right)^{1/3} \leq (1+\sigma_4)\sqrt{b}\,\frac{(\log n)^{5/4}}{\sqrt{n}}. \tag{12.2}$$

Proof For any distribution function G with finite first absolute moment, the function on the unit sphere $\mathbb{S}^{n-1}$ of the form $g(\theta) = W(F_\theta, G)$ has a Lipschitz semi-norm $\|g\|_{\mathrm{Lip}} \leq 1$. Therefore, it admits a subgaussian large deviation bound

$$\mathfrak{s}_{n-1}\{W(F_\theta, G) \geq m + r\} \leq e^{-(n-1)r^2/2}, \qquad r \geq 0, \tag{12.3}$$

where $m = \mathbb{E}_\theta\,W(F_\theta, G)$. Indeed, consider the elementary representation

$$W(F_\theta, G) \equiv \int_{-\infty}^{\infty} |F_\theta(x) - G(x)|\,dx$$

$$= \sup_u\left[\int_{-\infty}^{\infty} u\,dF_\theta - \int_{-\infty}^{\infty} u\,dG\right],$$

where the supremum is running over all functions u on $\mathbb{R}$ with $\|u\|_{\mathrm{Lip}} \leq 1$. For any such u,

$$H_u(\theta) = \int_{-\infty}^{\infty} u\,dF_\theta = \mathbb{E}\,u(\langle X, \theta\rangle)$$

is Lipschitz on $\mathbb{R}^n$ and therefore on $\mathbb{S}^{n-1}$. Moreover, $\|g\|_{\mathrm{Lip}} \leq \sup_u \|H_u\|_{\mathrm{Lip}} \leq 1$.

Hence, (12.3) is fulfilled as a consequence of fact that the logarithmic Sobolev constant for the uniform distribution on the unit sphere is equal to $n - 1$ (cf. [21]). In particular, for any $r \geq 0$,

$$\mathfrak{s}_{n-1}\{W(F_\theta, F) \geq m + r\} \leq e^{-(n-1)r^2/2}$$

with $m = \mathbb{E}_\theta \, W(F_\theta, F)$. In turn, the latter ensures that, for any $p \geq 2$,

$$\left(\mathbb{E}_\theta \, W(F_\theta, F)^p\right)^{1/p} \leq m + \frac{\sqrt{p}}{\sqrt{n-1}}. \tag{12.4}$$

For the proof, put $\xi = (W(F_\theta, F) - m)^+$. Using $\Gamma(x+1) \leq x^x$ with $x = p/2 \geq 1$, we have

$$\mathbb{E}_\theta \, \xi^p = \int_0^\infty \mathfrak{s}_{n-1}\{\xi \geq r\} \, dr^p \leq \int_0^\infty e^{-(n-1)r^2/2} \, dr^p$$

$$= \left(\frac{\sqrt{2}}{\sqrt{n-1}}\right)^p \Gamma\left(\frac{p}{2}+1\right) \leq \left(\frac{\sqrt{p}}{\sqrt{n-1}}\right)^p \equiv A^p \quad (A \geq 0).$$

Thus, $\|\xi\|_p = (\mathbb{E}_\theta \, \xi^p)^{1/p} \leq A$. Since $W(F_\theta, F) \leq \xi + m$, we conclude, by the triangle inequality, that

$$\|W(F_\theta, F)\|_p \leq \|\xi\|_p + m \leq A + m,$$

that is, (12.4) holds.

Let us proceed with one elementary general inequality, connecting the three distances,

$$\omega^2(F_\theta, F) = \int_{-\infty}^\infty (F_\theta(x) - F(x))^2 \, dx$$

$$\leq \int_{-\infty}^\infty \sup_x |F_\theta(x) - F(x)| \, |F_\theta(x) - F(x)| \, dx = \rho(F_\theta, F) \, W(F_\theta, F).$$

Putting $\omega = \omega(F_\theta, F)$, $W = W(F_\theta, F)$, $\rho = \rho(F_\theta, F)$, we thus have $\omega^3 \leq W^{3/2} \rho^{3/2}$ and, by Hölder's inequality with exponents $p = 4$ and $q = 4/3$,

$$\|\omega\|_3 = \left(\mathbb{E}_\theta \, \omega^3\right)^{1/3} \leq \left(\mathbb{E}_\theta \, W^6\right)^{1/12} \left(\mathbb{E}_\theta \, \rho^2\right)^{1/4}.$$

By (12.4) with $p = 6$, we have

$$\left(\mathbb{E}_\theta \, W^6\right)^{1/6} \leq \mathbb{E}_\theta \, W + \frac{4}{\sqrt{n}},$$

so that

$$\|\omega\|_3 \leq \left(\mathbb{E}_\theta \, W + \frac{4}{\sqrt{n}}\right)^{1/2} \left(\mathbb{E}_\theta \, \rho^2\right)^{1/4}.$$

Applying Proposition 11.2 and noting that necessarily $b \geq 1$ in the isotrpic case, we get

$$\|\omega\|_3 \leq 4\sqrt{b} \left(\sqrt{\log n} \, \mathbb{E}_\theta \, \rho + \frac{1}{\sqrt{n}} \right)^{1/2} \left(\mathbb{E}_\theta \, \rho^2 \right)^{1/4}.$$

Here we employ the inequality (9.3) with F in place of Φ, i.e.

$$\mathbb{E}_\theta \, \rho(F_\theta, F) \leq \left(\mathbb{E}_\theta \, \rho^2(F_\theta, F) \right)^{1/2} \leq c \, (1 + \sigma_4) \frac{\log n}{\sqrt{n}}.$$

Since the last expression dominates the term $\frac{1}{\sqrt{n}}$, it follows that

$$\|\omega\|_3 \leq c\sqrt{b} \left(\sqrt{\log n} \, (1 + \sigma_4) \frac{\log n}{\sqrt{n}} \right)^{1/2} \left((1 + \sigma_4) \frac{\log n}{\sqrt{n}} \right)^{1/2},$$

and we arrive at the upper bound (12.2). $\square$

Let us now explain how this bound can be used to refine the lower bound (12.1). The argument is based on the following general elementary observation. Given a random variable ξ, introduce the L^p-norms $\|\xi\|_p = (\mathbb{E} \, |\xi|^p)^{1/p}$.

Lemma 12.3 *If* $\xi \geq 0$ *with* $0 < \|\xi\|_3 < \infty$, *then*

$$\mathbb{E} \, \xi \geq \frac{1}{\mathbb{E} \, \xi^3} \, (\mathbb{E} \, \xi^2)^2. \tag{12.5}$$

Moreover,

$$\mathbb{P}\left\{ \xi \geq \frac{1}{\sqrt{2}} \, \|\xi\|_2 \right\} \geq \frac{1}{8} \left(\frac{\|\xi\|_2}{\|\xi\|_3} \right)^6. \tag{12.6}$$

Thus, in the case where $\|\xi\|_2$ and $\|\xi\|_3$ are equivalent within not too large factors, $\|\xi\|_1$ will be of a similar order. Moreover, ξ cannot be much smaller than its mean $\mathbb{E}\xi$ on a large part of the probability space (where it was defined).

Proof Let ξ be defined on the probability space $(\Omega, \mathfrak{F}, \mathbb{P})$. By homogeneity with respect to ξ, we may assume that $\mathbb{E}\xi = 1$, so that $dQ = \xi \, d\mathbb{P}$ is a probability measure. Then, (12.5) follows from the Cauchy inequality $(\mathbb{E}_Q \xi)^2 \leq \mathbb{E}_Q \xi^2$ on the space $(\Omega, \mathfrak{F}, Q)$.

To prove (12.6), given $r > 0$, let $p = \mathbb{P}\{\xi \geq r\}$. By Hölder's inequality with exponents $3/2$ and 3,

$$\mathbb{E} \, \xi^2 \, 1_{\{\xi \geq r\}} \leq \left(\mathbb{E} \, \xi^3 \right)^{2/3} p^{1/3}.$$

Hence, choosing $r = \frac{1}{\sqrt{2}} \|\xi\|_2$, we get

$$\mathbb{E}\,\xi^2 = \mathbb{E}\,\xi^2\,1_{\{\xi \geq r\}} + \mathbb{E}\,\xi^2\,1_{\{\xi < r\}}$$

$$\leq (\mathbb{E}\,\xi^3)^{2/3}\,p^{1/3} + r^2 = (\mathbb{E}\,\xi^3)^{2/3}\,p^{1/3} + \frac{1}{2}\,\mathbb{E}\,\xi^2.$$

Hence $p^{1/3} \geq \frac{1}{2\,(\mathbb{E}\,\xi^3)^{2/3}}\,\mathbb{E}\,\xi^2$ which is the desired bound (12.6). $\square$

We now combine Lemma 12.2 with Lemma 12.3 which is applied on the unit sphere to $\xi(\theta) = \omega(F_\theta, F)$ viewed as a random variable on the probability space $(\mathbb{S}^{n-1}, \mathfrak{s}_{n-1})$. Recall that $b \geq 1$ in the isotropic case.

Proposition 12.4 *Let X be an isotropic random vector in $\mathbb{R}^n$ such that $|X| \leq b\sqrt{n}$ a.s. Assume that*

$$\mathbb{E}_\theta\,\omega^2(F_\theta, F) \geq \frac{D}{n}$$

with some $D > 0$. Then

$$\mathbb{E}_\theta\,\omega(F_\theta, F) \geq \frac{c}{(1+\sigma_4)^3\,b^{\frac{3}{2}}}\,\frac{D^2}{(\log n)^{\frac{15}{4}}\,\sqrt{n}}. \tag{12.7}$$

Moreover,

$$\mathfrak{s}_{n-1}\left\{\omega(F_\theta, F) \geq \frac{1}{\sqrt{2n}}\sqrt{D}\right\} \geq \frac{c}{(1+\sigma_4)^6\,b^3}\,\frac{D^3}{(\log n)^{\frac{15}{2}}}.$$

Proof of Theorem 1.3 The lower bound (12.7) implies a similar assertion about the Kolmogorov distance. Indeed, by Proposition 11.1 with $\alpha = 1$, we have

$$\frac{1}{\sqrt{b}}\,\mathbb{E}_\theta\,\omega(F_\theta, F) \leq 14\,(\log n)^{1/4}\,\mathbb{E}_\theta\,\rho(F_\theta, F) + \frac{8}{n^4}.$$

Using $\frac{8}{n^4} < \frac{1}{n^3} \cdot 14\,(\log n)^{1/4}$, we therefore obtain that

$$\mathbb{E}_\theta\,\rho(F_\theta, F) \geq \frac{1}{14\sqrt{b}\,(\log n)^{1/4}}\,\mathbb{E}_\theta\,\omega(F_\theta, F) - \frac{1}{n^3}$$

$$\geq \frac{c}{(1+\sigma_4)^3\,b^2}\,\frac{D^2}{(\log n)^4\,\sqrt{n}} - \frac{1}{n^3}.$$

To replace F with Φ, it remains to recall the bound $\rho(F, \Phi) \leq \frac{c}{n}\,(1+\sigma_4^2)$, cf. (2.3). $\square$

In the isotropic case with $|X|^2 = n$ a.s., the above lower bound is further simplified to

$$\mathbb{E}_\theta\, \rho(F_\theta, F) \geq \frac{cD^2}{(\log n)^4 \sqrt{n}} - \frac{1}{n^3}.$$

On the other hand, let us note that the rates for the normal approximation of F_θ that are better than $1/n$ (on average) cannot be obtained under the support assumption as above. That is, if $|X| = \sqrt{n}$ a.s., then

$$\mathbb{E}_\theta\, \rho(F_\theta, \Phi) \geq \frac{c}{n}.$$

Indeed, using the convexity of the distance function $G \to \rho(G, \Phi)$ and applying Jensen's inequality, we have that $\mathbb{E}_\theta\, \rho(F_\theta, \Phi) \geq \rho(F, \Phi)$. It remains to appeal to Proposition 2.6.

13 Functional Examples

13.1. For the trigonometric system as in item (i) of the Introduction (with n even), the linear forms

$$\langle X, \theta \rangle = \sqrt{2} \sum_{k=1}^{\frac{n}{2}} \big(\theta_{2k-1} \cos(kt) + \theta_{2k} \sin(kt)\big), \quad \theta = (\theta_1, \ldots, \theta_n) \in \mathbb{S}^{n-1},$$

represent trigonometric polynomials of degree at most $\frac{n}{2}$. The normalization $\sqrt{2}$ is chosen in order to meet the requirement that the random vector X is isotropic with respect to the normalized Lebesgue measure $\mathbb{P}$ on $\Omega = (-\pi, \pi)$. Moreover, in this case $|X| = \sqrt{n}$, so that $\sigma_4 = 0$. Hence, by Theorem 1.1, we have the upper bounds (1.6). On the other hand, since for all $k \leq \frac{n}{2}$

$$|X_k(t) - X_k(s)| \leq k\sqrt{2}\,|t - s| \leq \frac{n}{\sqrt{2}}\,|t - s|, \quad t, s \in \Omega,$$

the Lipschitz condition (7.1) is fulfilled with $L(t) = \frac{t}{\sqrt{2}}$. Hence, Proposition 7.1 is applicable and yields the lower bound

$$\mathbb{E}_\theta\, \omega^2(F_\theta, \Phi) \geq \frac{c_1}{n} - \frac{c_2}{n^2} \geq \frac{c_3}{n},$$

where in the last inequality we assume that $n \geq n_0$ for some universal integer n_0. This restriction may be dropped, since the distances $\omega^2(F_\theta, \Phi)$

are bounded away from zero for $n < n_0$ uniformly over all $\theta \in \mathbb{S}^{n-1}$, just due to the property that the distributions F_θ are supported on the bounded interval $[-\sqrt{n_0}, \sqrt{n_0}]$. Note that the above lower estimate (may also be obtained by applying Theorem 1.1. Thus, for all $n \geq 2$,

$$\frac{c_0}{n} \leq \mathbb{E}_\theta \, \omega^2(F_\theta, \Phi) \leq \frac{c_1}{n}. \tag{13.1}$$

Applying Proposition 12.4, we obtain similar bounds for the L^1-norm (modulo logarithmic factors). Namely, it gives

$$\frac{c_0}{(\log n)^{\frac{15}{4}} \sqrt{n}} \leq \mathbb{E}_\theta \, \omega(F_\theta, \Phi) \leq \frac{c_1}{\sqrt{n}}. \tag{13.2}$$

We also get an analogous pointwise lower bound on the "essential" part of the unit sphere.

A similar statement is also true for the Kolmogorov distance. Here, the upper bound is provided in Proposition 9.1, while the lower bound is obtained when combining Theorem 1.3 with the left inequality in (13.1). That is,

$$\frac{c_0}{(\log n)^4 \sqrt{n}} \leq \mathbb{E}_\theta \, \rho(F_\theta, \Phi) \leq \left(\mathbb{E}_\theta \, \rho^2(F_\theta, \Phi) \right)^{1/2} \leq \frac{c_1 \log n}{\sqrt{n}}. \tag{13.3}$$

13.2. Analogous results remain true for the cosine trigonometric system $X = (X_1, \ldots, X_n)$ as in item (ii). Due to the normalization $\sqrt{2}$, the distribution of X is isotropic in $\mathbb{R}^n$. The property $|X| = \sqrt{n}$ is not true anymore; however, there is a pointwise bound $|X| \leq \sqrt{2n}$. In addition, the variance functional σ_4^2 does not depend on n. Indeed, write

$$X_k^2 = 2\cos^2(kt) = 1 + \cos(2kt) = 1 + \frac{e^{2ikt} + e^{-2ikt}}{2},$$

so that

$$2\left(|X|^2 - n\right) = \sum_{0 < |k| \leq n} e^{2ikt}, \qquad 4\left(|X|^2 - n\right)^2 = \sum_{0 < |k|, |l| \leq n} e^{2i(k+l)t}.$$

It follows that

$$4 \operatorname{Var}(|X|^2) = \sum_{0 < |k|, |l| \leq n} \mathbb{E} \, e^{2i(k+l)t} = \sum_{0 < |k| \leq n, \, l = -k} 1 = 2n.$$

Hence

$$\sigma_4^2 = \frac{1}{n} \operatorname{Var}(|X|^2) = \frac{1}{2}.$$

As before, the Lipschitz condition is fulfilled with the function $L(t) = t\sqrt{2}$. Therefore, with similar arguments we obtain all the bounds (13.1)–(13.3).

Let us also note that the sums $\sum_{k=1}^{n} \cos(kt)$ remain bounded for growing n (for any fixed $0 < t < \pi$). Hence the normalized sums

$$S_n = \frac{1}{\sqrt{n}} \sum_{k=1}^{n} X_k = \frac{\sqrt{2}}{\sqrt{n}} \sum_{k=1}^{n} \cos(kt),$$

which correspond to $\langle X, \theta \rangle$ with equal coefficients, are convergent to zero pointwise on Ω as $n \to \infty$. In particular, they fail to satisfy the central limit theorem.

13.3. An example closely related to the cosine trigonometric system is represented by the normalized Chebyshev's polynomials X_k as in item (iii), which we consider for $k = 1, 2, \ldots, n$. These polynomials are orthonormal on the interval $\Omega = (-1, 1)$ with respect to the probability measure

$$\frac{d\mathbb{P}(t)}{dt} = \frac{1}{\pi \sqrt{1 - t^2}}, \quad -1 < t < 1,$$

cf. e.g. [17]. Similarly to 13.2, for the random vector $X = (X_1, \ldots, X_n)$ we find that

$$4\left(|X|^2 - n\right)^2 = \sum_{0 < |k|, |l| \le n} \exp\{2i(k + l) \arccos t\}.$$

It follows that

$$4 \operatorname{Var}(|X|^2) = \sum_{0 < |k|, |l| \le n} \mathbb{E} \exp\{2i(k + l) \arccos t\} = \sum_{0 < |k| \le n} 1 = 2n,$$

so that $\sigma_4^2 = \frac{1}{n} \operatorname{Var}(|X|^2) = \frac{1}{2}$. In addition, for all $k \le n$,

$$|X_k(t) - X_k(s)| \le k\sqrt{2} \, |\arccos t - \arccos s|, \quad t, s \in \Omega,$$

which implies that the Lipschitz condition is fulfilled with the function $L(t) = \sqrt{2} \arccos t$. As a result, we obtain the bounds (13.1)–(13.3) as well.

13.4. Turning to item (iv), consider the functions of the form

$$X_k(t, s) = \Psi(kt + s),$$

assuming that Ψ is a 1-periodic measurable function on the real line such that

$$\int_0^1 \Psi(x)\,dx = 0 \quad \text{and} \quad \int_0^1 \Psi(x)^2\,dx = 1.$$

These conditions ensure that the random vector $X = (X_1, \ldots, X_n)$ is isotropic in $\mathbb{R}^n$ with respect to the Lebesgue measure $\mathbb{P}$ on the square $\Omega = (0, 1) \times (0, 1)$, with $\mathbb{E}X_k = 0$. In fact, as was emphasized in [5], $\{X_k\}_{k=1}^\infty$ represents a strictly stationary sequence of pairwise independent random variables on Ω. The latter implies in particular that, if Ψ has finite 4-th moment on $(0, 1)$, the variance functional

$$\sigma_4^2 = \frac{1}{n}\,\text{Var}(|X|^2) = \int_0^1 \Psi(x)^4\,dx - 1$$

is finite and does not dependent on n. Hence, by Theorem 1.1, cf. (1.6), the upper bounds in (13.1)–(13.3) hold true with a constant c_1 depending on the 4-th moment of Ψ on $(0, 1)$.

In addition, if the function Ψ has finite Lipschitz constant $\|\Psi\|_{\text{Lip}}$, then for all (t_1, t_2) and (s_1, s_2) in Ω,

$$|X_k(t_1, t_2) - X_k(s_1, s_2)| \le \|\Psi\|_{\text{Lip}}\left(k\,|t_1 - s_1| + |t_2 - s_2|\right).$$

This means that the Lipschitz condition (7.5) is fulfilled with linear functions L_1 and L_2. Hence, one may apply Proposition 7.5 giving the lower bound

$$\mathbb{E}_\theta\,\omega^2(F_\theta, F) \ge \frac{c_\Psi}{n} - \frac{c\,(1 + \sigma_4^4)}{n^2}$$

in full analogy with item (i). Hence $\mathbb{E}_\theta\,\omega^2(F_\theta, \Phi) \ge \frac{c_\Psi'}{n}$ for all $n \ge n_0$, where the positive constants c_Ψ, c_Ψ', and an integer $n_0 \ge 1$ depend on the distribution of Ψ only. Since the collection $\{F_\theta\}$ is separated from Φ in the weak sense for $n < n_0$ (by the uniform boundedness of X_k's), the latter bound holds true for all $n \ge 2$. Also, as Lipschitz functions on $(0, 1)$ are bounded, we have $|X| \le b\sqrt{n}$ with $b = \sup_x |f(x)|$, and one may apply Theorem 1.3.

Let us summarize: *The upper bounds in (13.1)–(13.3) hold true, if Ψ has finite 4-th moment under the uniform distribution on $(0, 1)$. The lower bounds hold under an additional assumption that Ψ has a finite Lipschitz semi-norm (with constants depending on Ψ only).*

Choosing, for example, $\Psi(t) = \cos t$, we obtain the system $X_k(t, s) = \cos(kt + s)$, which is closely related to the cosine trigonometric system. The main difference is however the property that X_k's are now pairwise independent. Nevertheless, the normalized sums $\frac{1}{\sqrt{n}}\sum_{k=1}^n \cos(kt + s)$ fail to satisfy the central limit theorem.

14 The Walsh System; Empirical Measures

14.1. The Walsh system on the discrete cube $\Omega = \{-1, 1\}^d$ with the uniform counting measure $\mathbb{P}$ as in item (v) in Introduction forms a complete orthonormal system in $L^2(\Omega, \mathbb{P})$. Note that each X_τ with $\tau \neq \emptyset$ is a symmetric Bernoulli random variable taking the values -1 and 1 with probability $\frac{1}{2}$. For simplicity, we exclude from this family the constant $X_\emptyset = 1$ and consider $X = \{X_\tau\}_{\tau \neq \emptyset}$ as a random vector in $\mathbb{R}^n$ of dimension $n = 2^d - 1$. As before, F_θ denotes the distribution function of the linear form

$$\langle X, \theta \rangle = \sum_{\tau \neq \emptyset} \theta_\tau X_\tau, \quad \theta = \{\theta_\tau\}_{\tau \neq \emptyset} \in \mathbb{S}^{n-1}.$$

Since $|X_\tau| = 1$ and thus $|X| = \sqrt{n}$, for the study of the asymptotic behavior of the L^2-distance $\omega(F_\theta, \Phi)$ on average, one may apply Theorem 1.1. Let Y be an independent copy of X, which we realize on the product space $\Omega^2 = \Omega \times \Omega$ with product measure $\mathbb{P}^2 = \mathbb{P} \times \mathbb{P}$ by

$$X_\tau(t, s) = \prod_{k \in \tau} t_k, \ Y_\tau(t, s) = \prod_{k \in \tau} s_k \quad t = (t_1, \ldots, t_d), \ s = (s_1, \ldots, s_d) \in \Omega.$$

Then the inner product

$$\langle X, Y \rangle = \sum_{\tau \neq \emptyset} X_\tau(t, s) Y_\tau(t, s) = -1 + \prod_{k=1}^{d} (1 + t_k s_k)$$

takes only two values, namely $2^d - 1$ in the case $t = s$, and -1 if $t \neq s$. Hence

$$\mathbb{E} \langle X, Y \rangle^3 = (2^d - 1)^3 \, 2^{-d} + (1 - 2^{-d}) = \frac{n^3}{n+1} + \left(1 - \frac{1}{n+1}\right) \sim n^2$$

and

$$\mathbb{E} \langle X, Y \rangle^4 = (2^d - 1)^4 \, 2^{-d} + (1 - 2^{-d}) = \frac{n^4}{n+1} + \left(1 - \frac{1}{n+1}\right) \sim n^3.$$

In other words, $m_3^3 \sim \sqrt{n}$ and $m_4^4 \sim n$ as $n \to \infty$. As a result, we may conclude that all inequalities in (13.1)–(13.3) are fulfilled for this system as well.

14.2. Here is another interesting example leading to the similar rate of normal approximation. Let $e_1, \ldots, e_n$ denote the canonical basis in $\mathbb{R}^n$. Assuming that the random vector $X = (X_1, \ldots, X_n)$ takes only n values, $\sqrt{n} e_1, \ldots, \sqrt{n} e_n$, each with probability $1/n$, the linear form $\langle X, \theta \rangle$ also takes n values, namely, $\sqrt{n}\,\theta_1, \ldots, \sqrt{n}\,\theta_n$, each with probability $1/n$, for any $\theta = (\theta_1, \ldots, \theta_n) \in \mathbb{S}^{n-1}$. That is, as a measure, the distribution of $\langle X, \theta \rangle$ is described as

$$F_\theta = \frac{1}{n} \sum_{k=1}^{n} \delta_{\sqrt{n}\,\theta_k},$$

which may be viewed as an empirical measure based on the observations $Z_k = \sqrt{n}\,\theta_k$, $k = 1, \ldots, n$. Each Z_k is almost standard normal, while jointly they are nearly independent (we have already considered in detail its characteristic functions $J_n(t\sqrt{n})$).

Just taking a short break, let us recall that when Z_k are indeed standard normal and independent, it is well-known that the empirical measures $G_n = \frac{1}{n}\sum_{k=1}^{n}\delta_{Z_k}$ approximate the standard normal law Φ with rate $1/\sqrt{n}$ with respect to the Kolmogorov distance. More precisely, $\mathbb{E}\,G_n = \Phi$ and there is a subgaussian deviation bound (cf. [22])

$$\mathbb{P}\big\{\sqrt{n}\,\rho(G_n, \Phi) \geq r\big\} \leq 2e^{-2r^2}, \qquad r \geq 0.$$

In particular, $\mathbb{E}\,\rho(G_n, \Phi) \leq \frac{c}{\sqrt{n}}$. Note that the characteristic function $g_n(t) = \frac{1}{n}\sum_{k=1}^{n}e^{itZ_k}$ of the measure G_n has mean $g(t) = e^{-t^2/2}$ and variance

$$\mathbb{E}\,|g_n(t) - g(t)|^2 = \frac{1}{n}\,\mathrm{Var}(e^{itZ_1}) = \frac{1}{n}\left(1 - |\mathbb{E}\,e^{itZ_1}|^2\right) = \frac{1}{n}\left(1 - e^{-t^2}\right).$$

Hence, applying Plancherel's theorem and using the identity (4.7) for the functions $\psi_r(\alpha)$ with $r = \alpha = 0$, we also have

$$\mathbb{E}\,\omega^2(G_n, \Phi) = \frac{1}{2\pi}\int_{-\infty}^{\infty}\mathbb{E}\left|\frac{g_n(t) - g(t)}{t}\right|^2 dt$$

$$= \frac{1}{2\pi n}\int_{-\infty}^{\infty}\frac{1 - e^{-t^2}}{t^2}\,dt = \frac{1}{n\sqrt{\pi}}.$$

Thus, on average the L^2-distance $\omega(G_n, \Phi)$ is of order $1/\sqrt{n}$ as well.

Similar properties may be expected for the random variables $Z_k = \sqrt{n}\,\theta_k$ and hence for the random vector X. Note that $|X| = \sqrt{n}$, while

$$\mathbb{E}\,\langle X, \theta \rangle^2 = \frac{1}{n}\sum_{k=1}^{n}(\sqrt{n}\,\theta_k)^2 = 1, \qquad \theta \in \mathbb{S}^{n-1},$$

so that X is isotropic. We now involve an asymptotic formula of Corollary 5.1 which yields

$$\mathbb{E}_\theta \, \omega^2(F_\theta, \Phi) = \frac{1}{\sqrt{\pi}} \left(1 + \frac{1}{4n}\right) \mathbb{E}\left(1 - (1-\xi)^{1/2}\right) - \frac{1}{8n\sqrt{\pi}} + O\left(\frac{1}{n^2}\right),$$

where $\xi = \frac{\langle X,Y \rangle}{n}$ with Y being an independent copy of X. By the definition, ξ takes only two values, 1 with probability $\frac{1}{n}$ and 0 with probability $1 - \frac{1}{n}$. Hence, the last expectation is equal to $\frac{1}{n}$, and we get

$$\mathbb{E}_\theta \, \omega^2(F_\theta, \Phi) = \frac{7/8}{n\sqrt{\pi}} + O\left(\frac{1}{n^2}\right).$$

As for the Kolmogorov distance, one may apply again Theorem 1.3, which leads to the two-sided bound (13.3). Apparently, both logarithmic terms can be removed. Their appearance here is explained by the use of the Fourier tools (in the form of the Berry-Esseen bounds), while the proof of the Dvoretzky-Kiefer-Wolfowitz inequality on $\rho(G_n, \Phi)$ in [13] is based on the entirely different arguments.

15   Improved Rates for Lacunary Systems

An orthonormal sequence of random variables $\{X_k\}_{k=1}^\infty$ in $L^2(\Omega, \mathfrak{F}, \mathbb{P})$ is called a lacunary system of order $p > 2$, if for any sequence (a_k) in ℓ^2, the series $\sum_{k=1}^\infty a_k X_k$ converges in L^p-norm to an element of $L^p(\Omega, \mathfrak{F}, \mathbb{P})$. This property is equivalent to the validity of the Khinchine-type inequality

$$\left(\mathbb{E} \, |a_1 X_1 + \cdots + a_n X_n|^p\right)^{1/p} \leq M_p \, (a_1^2 + \cdots + a_n^2)^{1/2} \tag{15.1}$$

for arbitrary $a_k \in \mathbb{R}$ with some constant M_p independent of n and the choice of the coefficients a_k. For basic properties of such systems we refer an interested reader to the books [16, 17].

Starting from an orthonormal lacunary system of order $p = 4$, consider the random vector $X = (X_1, \ldots, X_n)$. According to Theorem 1.1, if $|X|^2 = n$ a.s. and $\mathbb{E}X = 0$, then

$$c \, \mathbb{E}_\theta \, \omega^2(F_\theta, \Phi) \leq \frac{1}{n^3} \mathbb{E}\langle X, Y\rangle^3 + \frac{1}{n^4} \mathbb{E}\langle X, Y\rangle^4, \tag{15.2}$$

where Y is an independent copy of X. A similar bound

$$c \, \mathbb{E}_\theta \, \rho^2(F_\theta, \Phi) \leq \frac{\log n}{n^3} \, \mathbb{E} \, \langle X, Y \rangle^3 + \frac{(\log n)^2}{n^4} \, \mathbb{E} \, \langle X, Y \rangle^4 \tag{15.3}$$

also holds for the Kolmogorov distance. As easily follows from (15.1),

$$\mathbb{E} \, | \, \langle X, Y \rangle \, |^p \leq M_p^{2p} n^{p/2}.$$

In particular,

$$\mathbb{E} \, | \, \langle X, Y \rangle \, |^3 \leq M_3^6 \, n^{3/2}, \quad \mathbb{E} \, \langle X, Y \rangle^4 \leq M_4^8 \, n^2.$$

Hence, the bounds (15.2) and (15.3) lead to the estimates

$$c \, \mathbb{E}_\theta \, \omega^2(F_\theta, \Phi) \leq \frac{1}{n^{3/2}} \, M_3^6 + \frac{1}{n^2} \, M_4^8,$$

$$c \, \mathbb{E}_\theta \, \rho^2(F_\theta, \Phi) \leq \frac{\log n}{n^{3/2}} \, M_3^6 + \frac{(\log n)^2}{n^2} \, M_4^8.$$

Thus, if M_4 is bounded, both distances are at most of order $n^{-3/4}$ on average (modulo a logarithmic factor). Moreover, if

$$\Sigma_3(n) \equiv \mathbb{E} \, \langle X, Y \rangle^3 = \sum_{1 \leq i_1, i_2, i_3 \leq n} \left(\mathbb{E} X_{i_1} X_{i_2} X_{i_3} \right)^2 \tag{15.4}$$

is bounded by a multiple of n, then these distances are on average at most $1/n$ (modulo a logarithmic factor in the case of ρ).

For an illustration, on the interval $\Omega = (-\pi, \pi)$ with the uniform measure $d\mathbb{P}(t) = \frac{1}{2\pi} \, dt$, consider a finite trigonometric system $X = (X_1, \ldots, X_n)$ with components

$$X_{2k-1}(t) = \sqrt{2} \, \cos(m_k t),$$

$$X_{2k}(t) = \sqrt{2} \, \sin(m_k t), \qquad k = 1, \ldots, n/2,$$

where m_k are positive integers such that $\frac{m_{k+1}}{m_k} \geq q > 1$ (assuming that n is even). Then X is an isotropic random vector satisfying $|X|^2 = n$ and $\mathbb{E} X = 0$, and with M_4 bounded by a function of q only. For evaluation of the moment $\Sigma_3(n)$, one may use the identities

$$\cos t = \mathbb{E}_\varepsilon \, e^{i\varepsilon t}, \quad \sin t = \frac{1}{i} \, \mathbb{E}_\varepsilon \, \varepsilon \, e^{i\varepsilon t},$$

where ε is a Bernoulli random variable taking the values ± 1 with probability $\frac{1}{2}$. Let $\varepsilon_1, \varepsilon_2, \varepsilon_3$ be independent copies of ε. Using the property that $\varepsilon_1\varepsilon_3$ and $\varepsilon_2\varepsilon_3$ are independent, the first identity implies that, for all integers $1 \leq n_1 \leq n_2 \leq n_3$,

$$\mathbb{E} \cos(n_1 t) \cos(n_2 t) \cos(n_3 t) = \mathbb{E}_\varepsilon \mathbb{E} \exp\{i(\varepsilon_1 n_1 + \varepsilon_2 n_2 + \varepsilon_3 n_3) t\}$$
$$= \mathbb{E}_\varepsilon I\{\varepsilon_1 n_1 + \varepsilon_2 n_2 + \varepsilon_3 n_3 = 0\}$$
$$= \mathbb{E}_\varepsilon I\{\varepsilon_1 n_1 + \varepsilon_2 n_2 = n_3\} = \frac{1}{4} I\{n_1 + n_2 = n_3\},$$

where $\mathbb{E}_\varepsilon$ means the expectation over $(\varepsilon_1, \varepsilon_2, \varepsilon_3)$, and where $I\{A\}$ denotes the indicator of the event A. Similarly, involving also the identity for the sine function, we have

$$\mathbb{E} \sin(n_1 t) \sin(n_2 t) \cos(n_3 t) = -\mathbb{E}_\varepsilon \mathbb{E} \varepsilon_1\varepsilon_2 \exp\{i(\varepsilon_1 n_1 + \varepsilon_2 n_2 + \varepsilon_3 n_3) t\}$$
$$= -\mathbb{E}_\varepsilon \varepsilon_1\varepsilon_2 I\{\varepsilon_1 n_1 + \varepsilon_2 n_2 + \varepsilon_3 n_3 = 0\}$$
$$= -\mathbb{E}_\varepsilon \varepsilon_1\varepsilon_2 I\{\varepsilon_1 n_1 + \varepsilon_2 n_2 = n_3\}$$
$$= -\frac{1}{4} I\{n_1 + n_2 = n_3\},$$

$$\mathbb{E} \sin(n_1 t) \cos(n_2 t) \sin(n_3 t) = -\mathbb{E}_\varepsilon \mathbb{E} \varepsilon_1\varepsilon_3 \exp\{i(\varepsilon_1 n_1 + \varepsilon_2 n_2 + \varepsilon_3 n_3) t\}$$
$$= -\mathbb{E}_\varepsilon \varepsilon_1\varepsilon_3 I\{\varepsilon_1 n_1 + \varepsilon_2 n_2 + \varepsilon_3 n_3 = 0\}$$
$$= -\mathbb{E}_\varepsilon \varepsilon_1 I\{\varepsilon_1 n_1 + \varepsilon_2 n_2 = n_3\}$$
$$= -\frac{1}{4} I\{n_1 + n_2 = n_3\},$$

$$\mathbb{E} \cos(n_1 t) \sin(n_2 t) \sin(n_3 t) = -\mathbb{E}_\varepsilon \mathbb{E} \varepsilon_2\varepsilon_3 \exp\{i(\varepsilon_1 n_1 + \varepsilon_2 n_2 + \varepsilon_3 n_3) t\}$$
$$= -\mathbb{E}_\varepsilon \varepsilon_2\varepsilon_3 I\{\varepsilon_1 n_1 + \varepsilon_2 n_2 + \varepsilon_3 n_3 = 0\}$$
$$= -\mathbb{E}_\varepsilon \varepsilon_2 I\{\varepsilon_1 n_1 + \varepsilon_2 n_2 = n_3\}$$
$$= -\frac{1}{4} I\{n_1 + n_2 = n_3\}.$$

On the other hand, if the sine function appears in the product once or three times, such expectations will be vanishing. They are thus vanishing in all cases where $n_1 + n_2 \neq n_3$, and do not exceed $\frac{1}{4}$ in absolute value for any combination of sine and cosine terms in all cases with $n_1 + n_2 = n_3$. Therefore, the moment $\Sigma_3(n)$ in (15.4) is bounded by a multiple of

$$T_3(n) = \text{card}\{(i_1, i_2, i_3) : 1 \leq i_1 \leq i_2 < i_3 \leq n, \ m_{i_1} + m_{i_2} = m_{i_3}\}.$$

One can now involve the lacunary assumption. If $q \geq 2$, the property $i_1 \leq i_2 < i_3$ implies $m_{i_1} + m_{i_2} < m_{i_3}$, so that $T_3(n) = \Sigma_3(n) = 0$. In the case $1 < q < 2$, define A_q to be the (finite) collection of all couples (k_1, k_2) of positive integers such that

$$q^{-k_1} + q^{-k_2} \geq 1.$$

By the lacunary assumption, if $1 \leq i_1 \leq i_2 < i_3 \leq n$, we have

$$m_{i_1} + m_{i_2} \leq \left(q^{-(i_3 - i_1)} + q^{-(i_3 - i_2)}\right) m_{i_3} < m_{i_3},$$

as long as the couple $(i_3 - i_1, i_2 - i_1)$ is not in A_q. Hence,

$$T_3(n) \leq \text{card}\left\{(i_1, i_2, i_3) : 1 \leq i_1 \leq i_2 < i_3 \leq n, \ (i_3 - i_1, i_2 - i_1) \in A_q\right\}$$
$$\leq n \, \text{card}(A_q) \leq c_q n$$

with constant depending on q only. Returning to (15.2) and (15.3), we then obtain:

Proposition 15.1 *For the lacunary trigonometric system X of an even length n and with parameter $q > 1$, we have*

$$\mathbb{E}_\theta \, \omega^2(F_\theta, \Phi) \leq \frac{c_q}{n^2}, \quad \mathbb{E}_\theta \, \rho^2(F_\theta, \Phi) \leq \frac{c_q \, (\log n)^2}{n^2},$$

where the constants c_q depend q only.

In this connection one should mention a classical result of Salem and Zygmund concerning distributions of the lacunary sums

$$S_n = \sum_{k=1}^{n} (a_k \cos(m_k t) + b_k \sin(m_k t))$$

with an arbitrary prescribed sequence of the coefficients $(a_k)_{k \geq 1}$ and $(b_k)_{k \geq 1}$. Assume that $\frac{m_{k+1}}{m_k} \geq q > 1$ for all k and put

$$v_n^2 = \frac{1}{2} \sum_{k=1}^{n} (a_k^2 + b_k^2) \qquad (v_n \geq 0),$$

so that the normalized sums $Z_n = S_n / v_n$ have mean zero and variance one under the measure $\mathbb{P}$. It was shown in [25] that Z_n are weakly convergent to the standard normal law, i.e., their distributions F_n under $\mathbb{P}$ satisfy $\rho(F_n, \Phi) \to 0$ as $n \to \infty$, if and only if $\frac{a_n^2 + b_n^2}{v_n^2} \to 0$ (in fact, the weak convergence was established on every subset of Ω of positive measure).

Restricting to the coefficients $\theta_{2k-1} = a_k / v_n$, $\theta_{2k} = b_k / v_n$, Salem-Zygmund's theorem may be stated as the assertion that $\rho(F_\theta, \Phi)$ is small, if and only if $\|\theta\|_\infty =$

$\max_{1\le k\le n}|\theta_k|$ is small. The latter condition naturally appears in the central limit theorem for weighted sums of independent identically distributed random variables. Thus, Proposition 15.1 complements this result in terms of the rate of convergence in the mean on the unit sphere. It would be interesting to describe explicit coefficients θ_k, for which we get a standard rate of normal approximation (perhaps, using other approaches such as the Stein method, cf. e.g. [14]).

The result of [25] was generalized in [26]; it turns out there is no need to assume that all m_k are integers, and the asymptotic normality is preserved for real m_k such that $\inf_k \frac{m_{k+1}}{m_k} > 1$. However, in this more general situation, the rate $1/n$ as in Proposition 15.1 is no longer true (although the rate $1/\sqrt{n}$ is valid). The main reason is that the means

$$\mathbb{E}X_{2k-1} = \sqrt{2}\,\mathbb{E}\cos(m_k t) = \sqrt{2}\,\frac{\sin(\pi m_k)}{\pi m_k}$$

may be non-zero. For example, choosing $m_k = 2^k + \frac{1}{2}$, we obtain an orthonormal system with $\mathbb{E}X_{2k} = 0$, while

$$\mathbb{E}X_{2k-1} = \frac{2\sqrt{2}}{\pi\,(2^{k+1}+1)}.$$

Hence

$$\mathbb{E}\,\langle X, Y\rangle = |\mathbb{E}X|^2 = \frac{8}{\pi^2}\sum_{k=1}^{n}\frac{1}{(2^{k+1}+1)^2} \to c \quad (n\to\infty)$$

for some absolute constant $c > 0$ (where Y is an independent copy of X). In this situation, as was already mentioned in (5.3), cf. Remark 5.3, we have a lower bound

$$\mathbb{E}_\theta\,\omega^2(F_\theta, F) \ge \frac{c}{2\sqrt{\pi}\,n} + O\Big(\frac{1}{n^2}\Big).$$

Since $\mathbb{E}\,\langle X, Y\rangle^3 = O(n)$ and $\mathbb{E}\,\langle X, Y\rangle^4 = O(n^2)$, this inequality may actually be replaced with equality, according to (5.2). A similar asymptotic holds as well when F is replaced with Φ.

16 Improved Rates for Independent and Log-Concave Summands

Let $X = (X_1,\ldots,X_n)$ be an isotropic random vector in $\mathbb{R}^n$ with mean zero. If the components X_k are independent, the normal approximation for the distributions F_θ of the weighted sums

$$S_\theta = \theta_1 X_1 + \cdots + \theta_n X_n, \quad \theta \in \mathbb{S}^{n-1},$$

may be controlled by virtue of the Berry-Esseen theorem under the 3-rd moment assumption. Namely, this theorem provides an upper bound

$$\rho(F_\theta, \Phi) \leq c \sum_{i=1}^{n} |\theta_i|^3 \, \mathbb{E} \, |X_i|^3 \qquad (16.1)$$

(cf. e.g. [23, 24]). Since $\mathbb{E} \, |X_i|^3 \geq 1$, the sum in (16.1) is at least $\frac{1}{\sqrt{n}}$. On the other hand, (16.1) yields an upper estimate on average

$$\mathbb{E}_\theta \, \rho(F_\theta, \Phi) \leq \frac{c\beta_3}{\sqrt{n}}, \quad \beta_3 = \max_{1 \leq i \leq n} \mathbb{E} \, |X_i|^3, \qquad (16.2)$$

which is consistent with the standard rate.

As it turns out, the relations (16.1) and (16.2) are far from being optimal for most of θ, as the following statement due to Klartag and Sodin shows.

Theorem 16.1 ([20]) *If the random variables $X_1, \ldots, X_n$ are independent, have mean zero, variance one, and finite 4-th moments, then*

$$\mathbb{E}_\theta \, \rho(F_\theta, \Phi) \leq \frac{c\beta_4}{n}, \quad \beta_4 = \frac{1}{n} \sum_{i=1}^{n} \mathbb{E}X_i^4. \qquad (16.3)$$

Moreover, for any $r \geq 0$,

$$\mathfrak{s}_{n-1}\{n\rho(F_\theta, \Phi) \geq c\beta_4 r\} \leq 2\,e^{-\sqrt{r}}.$$

In the i.i.d. case, $\beta_4 = \mathbb{E}X_1^4$, and we obtain an upper bound of order at most $1/n$.

In fact, in the i.i.d. case, the relation (16.3) may be further sharpened under the 5-th moment assumption, if $\mathbb{E}X_1^3 = 0$, and if $\Phi(x)$ is slightly modified to

$$G(x) = \Phi(x) - \frac{\beta_4 - 3}{8(n+2)} (x^3 - 3x) \, \varphi(x), \quad x \in \mathbb{R},$$

where $\varphi(x) = \frac{1}{\sqrt{2\pi}} e^{-x^2/2}$ is the standard normal density.

Theorem 16.2 *If the random variables $X_1, \ldots, X_n$ are independent, identically distributed, and have moments $\mathbb{E}X_1 = 0$, $\mathbb{E}X_1^2 = 1$, $\mathbb{E}X_1^3 = 0$, $\mathbb{E}X_1^4 = \beta_4$, $\mathbb{E} \, |X_1|^5 = \beta_5 < \infty$, then*

$$\mathbb{E}_\theta \, \rho(F_\theta, G) \leq \frac{c\beta_5}{n^{3/2}}. \qquad (16.4)$$

Moreover, for any $r \geq 0$,

$$\mathfrak{s}_{n-1}\left\{ n^{3/2}\rho(F_\theta, G) \geq c\beta_4 r \right\} \leq 2 \exp\{-r^{2/5}\}.$$

We refer an interested reader to [4] and [11]. In the i.i.d. case, both inequalities (16.3) and (16.4) are sharp in the following sense. If $\alpha_3 = \mathbb{E}X_1^3 \neq 0$ and $\beta_4 < \infty$, then, for any function G of bounded total variation, such that $G(-\infty) = 0$ and $G(\infty) = 1$, we have

$$\mathbb{E}_\theta \, \rho(F_\theta, G) \geq \frac{c}{n}$$

with a constant $c > 0$ depending on α_3 and β_4. Similarly, if $\alpha_3 = 0$, $\beta_4 \neq 3$, $\beta_5 < \infty$, then

$$\mathbb{E}_\theta \, \rho(F_\theta, G) \geq \frac{c}{n^{3/2}},$$

where the constant $c > 0$ depends on β_4 and β_5 only.

In the upper bounds such as (16.3), the independence assumption may be replaced with closely related hypotheses. The random vector X is said to have a log-concave distribution, when it has a density of the form $p(x) = e^{-V(x)}$ where $V : \mathbb{R}^n \to (-\infty, \infty]$ is a convex function. Recall that the distribution of X is coordinatewise symmetric, if

$$p(\varepsilon_1 x_1, \ldots, \varepsilon_n x_n) = p(x_1, \ldots, x_n), \quad x_i \in \mathbb{R},$$

for any choice of signs $\varepsilon_i = \pm 1$. The following theorem sharpening (16.1) is due to Klartag.

Theorem 16.3 ([18]) *Suppose that the isotropic random vector* $X = (X_1, \ldots, X_n)$ *in* $\mathbb{R}^n$ *has a coordinatewise symmetric log-concave distribution. For all* $\theta = (\theta_1, \ldots, \theta_n) \in \mathbb{S}^{n-1}$,

$$\|F_\theta - \Phi\|_{\mathrm{TV}} \leq c \sum_{i=1}^{n} \theta_i^4. \tag{16.5}$$

Here, the total variation distance is understood in the usual sense as

$$\|F_\theta - \Phi\|_{\mathrm{TV}} = \int_{-\infty}^{\infty} |p_\theta(x) - \varphi(x)| \, dx,$$

where p_θ denotes the density of S_θ. By the assumptions, p_θ is symmetric about the origin and is log-concave for any $\theta \in \mathbb{S}^{n-1}$. Note that, by the coordinatewise symmetry, the isotropy assumption is reduced to the moment condition $\mathbb{E}X_i^2 = 1$ ($1 \leq i \leq n$).

In particular, it follows from (16.5) that

$$\mathbb{E}_\theta \, \rho(F_\theta, \Phi) \le \mathbb{E}_\theta \, \|F_\theta - \Phi\|_{\mathrm{TV}} \le \frac{c}{n}. \tag{16.6}$$

17 Improved Rates Under Correlation-Type Conditions

Up to a logarithmically growing term, the improved rate as in the upper bound (16.3) can be achieved under more flexible correlation-type conditions (in comparison with independence). For example, one may consider an optimal value $\Lambda = \Lambda(X)$ in the relation

$$\mathrm{Var}\left(\sum_{i,j=1}^n a_{ij} X_i X_j \right) \le \Lambda \sum_{i,j=1}^n a_{ij}^2 \quad (a_{ij} \in \mathbb{R}), \tag{17.1}$$

which we call that the random vector $X = (X_1, \ldots, X_n)$ satisfies a second order correlation condition with constant Λ. This quantity is finite as long as the moment $\mathbb{E}\,|X|^4$ is finite.

To relate Λ to the moment-type characteristics which we discussed before, one may apply (17.1) with $a_{ij} = \delta_{ij}$ or (as another option) with $a_{ij} = \theta_i \theta_j$, $\theta = (\theta_1, \ldots, \theta_n) \in \mathbb{S}^{n-1}$. This gives that

$$\sigma_4^2 \le \Lambda, \quad m_4^2 \le \sup_{\theta \in \mathbb{S}^{n-1}} \mathbb{E} S_\theta^4 \le 1 + \Lambda,$$

where in the last inequality we should assume that $\mathbb{E} S_\theta^2 = 1$ for all θ (i.e. X is isotropic). In the latter case, necessarily $\Lambda \ge \frac{n-1}{n}$, so that Λ is bounded away from zero.

If the distribution of X is "regular" in some sense, one may also bound Λ from above. For example, this is the case when it shares a Poincaré-type inequality

$$\lambda_1 \mathrm{Var}(u(X)) \le \mathbb{E}\,|\nabla u(X)|^2, \tag{17.2}$$

which is required to hold in the class of all bounded, smooth functions u on $\mathbb{R}^n$ with a constant $\lambda_1 > 0$ independent of u (called the spectral gap). We then have

$$\Lambda \le \frac{4}{\lambda_1^2}, \quad \Lambda \le \frac{4}{\lambda_1}, \tag{17.3}$$

where in the second inequality we assume that X is isotropic.

The following relation is established in [9].

Theorem 17.1 *If the distribution of X is isotropic and symmetric about the origin, then*

$$\mathbb{E}_\theta \, \rho(F_\theta, \Phi) \leq c\Lambda \, \frac{\log n}{n}. \tag{17.4}$$

The proof is based on the second order spherical concentration phenomenon which was developed in [6] with the aim of applications to randomized central limit theorems. It indicates that the deviations of any smooth function $u(\theta)$ on $\mathbb{S}^{n-1}$ from the mean $\mathbb{E}_\theta u(\theta)$ are at most of the order $1/n$, provided that u is orthogonal in $L^2(\mathbb{R}^n, \mathfrak{s}_{n-1})$ to all linear functions and has a "bounded" Hessian (the matrix of second order partial derivatives). Being applied to the characteristic functions $u(\theta) = f_\theta(t)$, this property yields an upper bound

$$\mathbb{E}_\theta \, |f_\theta(t) - f(t)|^2 \leq \frac{c\Lambda t^4}{n^2}$$

on every interval $|t| \leq An^{1/5}$ with constants $c > 0$ depending on the parameter $A \geq 1$ only. This estimate can be used to bound the integrals in (8.4) to get a similar variant of (17.4).

The symmetry hypothesis in Theorem 17.1 may be dropped, if Λ is replaced by λ_1^{-1} which is a larger quantity according to (17.3). In addition, one can control large deviations of the distance $\rho(F_\theta, \Phi)$ for most of the directions θ (rather than on average). The corresponding assertions are obtained in [10].

Theorem 17.2 *Let X be an isotropic random vector in $\mathbb{R}^n$ with mean zero and a positive Poincaré constant λ_1. Then*

$$\mathbb{E}_\theta \, \rho(F_\theta, \Phi) \leq c\lambda_1^{-1} \, \frac{\log n}{n}. \tag{17.5}$$

Moreover, for all $r > 0$,

$$\mathfrak{s}_{n-1}\left\{\rho(F_\theta, \Phi) \geq c\lambda_1^{-1}\frac{\log n}{n} r\right\} \leq 2\,e^{-\sqrt{r}}.$$

The logarithmic term in (17.5) may be removed using the less sensitive L^2-distance:

$$\mathbb{E}_\theta \, \omega^2(F_\theta, \Phi) \leq \frac{c}{\lambda_1^2 n^2}.$$

There is an extensive literature devoted to bounding the spectral gap λ_1 from below. In particular, it is positive for any log-concave probability distribution on $\mathbb{R}^n$. A well-known conjecture raised by Kannan, Lovász and Simonovits asserts that λ_1 is actually bounded away from zero, as long as the random vector X has an isotropic log-concave distribution (cf. [15]). The best known dimensional lower bound up to date is due to Klartag and Lehec [19] who showed that

$$\lambda_1 \geq \frac{c}{(\log n)^\alpha}$$

for some absolute positive constants c and α (one may take $\alpha = 10$). Applying this bound in Theorem 17.2, we therefore obtain:

Corollary 17.3 *Let X be an isotropic random vector in $\mathbb{R}^n$ with mean zero and a log-concave probability distribution. Then with some absolute positive constants c and α*

$$\mathbb{E}_\theta \, \rho(F_\theta, \Phi) \leq \frac{c(\log n)^\alpha}{n}. \tag{17.6}$$

Thus, there is a certain extension of Klartag's bound (16.6) at the expense of a logarithmic factor to the entire class of isotropic log-concave probability distributions on $\mathbb{R}^n$.

One may also argue in the opposite direction: upper bounds of the form

$$\mathbb{E}_\theta \, \rho(F_\theta, \Phi) \leq \frac{c(\log n)^\beta}{n}, \quad \beta > 0,$$

in the class of log-concave probability distributions on $\mathbb{R}^n$ imply lower bounds $\lambda_1 \geq c \, (\log n)^{-\beta'}$ with some $\beta' > 0$, cf. [9].

Acknowledgments Research was supported by SFB 1283, BSF grant 2016050, and NSF grant DMS-2154001.

References

1. M. Anttila, K. Ball, I. Perissinaki, The central limit problem for convex bodies. Trans. Am. Math. Soc. **355**(12), 4723–4735 (2003)
2. S.G. Bobkov, On concentration of distributions of random weighted sums. Ann. Probab. **31**(1), 195–215 (2003)
3. S.G. Bobkov, Closeness of probability distributions in terms of Fourier-Stieltjes transforms. Russ. Math. Surv. **71**(6), 1021–1079 (2016). Translated from: Uspekhi Matem. Nauk, vol. 71, issue 6 (432), (2016), 37–98
4. S.G. Bobkov, Edgeworth corrections in randomized central limit theorems. Geom. Aspects Funct. Anal. **2256**, 71–97 (2020)
5. S.G. Bobkov, F. Götze, Concentration inequalities and limit theorems for randomized sums. Probab. Theory Relat. Fields **137**(1–2), 49–81 (2007)
6. S.G. Bobkov, G.P. Chistyakov, F. Götze, Second-order concentration on the sphere. Commun. Contemp. Math. **19**(5), 1650058, 20pp. (2017)
7. S.G. Bobkov, G.P. Chistyakov, F. Götze, Gaussian mixtures and normal approximation for V. N. Sudakov's typical distributions. Zap. Nauchn. Sem. S.-Peterburg. Otdel. Mat. Inst. Steklov.

(POMI) **457** (2017). Veroyatnost i Statistika. 25, 37–52; reprinted in J. Math. Sci. (N.Y.) 238 (2019), no. 4, 366–376

8. S.G. Bobkov, G.P. Chistyakov, F. Götze, Berry-Esseen bounds for typical weighted sums. J. Electron. Probab. **23**(92), 1–22 (2018)

9. S.G. Bobkov, G.P. Chistyakov, F. Götze, Normal approximation for weighted sums under a second order correlation condition. Ann. Probab. **48**(3), 1202–1219 (2020)

10. S.G. Bobkov, G.P. Chistyakov, F. Götze, Poincaré-type inequalities and normal approximation for weighted sums. Electron. J. Probab. **25**, 155, 31 pp. (2020)

11. S.G. Bobkov, G.P. Chistyakov, F. Götze, *Concentration and Gaussian Approximation for Randomized Sums*. Probability Theory and Stochastic Modelling, vol. 104 (Springer Cham, 2023), 434 pp.

12. S. Brazitikos, A. Giannopoulos, P. Valettas, B.-H. Vritsiou, *Geometry of Isotropic Convex Bodies*. Mathematical Surveys and Monographs, vol. 196 (American Mathematical Society, Providence, 2014). xx+594pp.

13. A. Dvoretzky, J. Kiefer, J. Wolfowitz, Asymptotic minimax character of the sample distribution function and of the classical multinomial estimator. Ann. Math. Stat. **27**, 642–669 (1956)

14. L. Goldstein, G. Reinert, Stein's method and the zero bias transformation with application to simple random sampling. Ann. Appl. Probab. **7**(4), 935–952 (1997)

15. R. Kannan, L. Lovász, M. Simonovits, Isoperimetric problems for convex bodies and a localization lemma. Discrete Comput. Geom. **13**, 541–559 (1995)

16. S. Kaczmarz, G. Steinhaus, *Theory of Orthogonal Series* (Warszawa, Lwow, 1935); Russian ed.: Izdat. Fiz.-Mat. Lit., Moscow, 1958, 507pp.

17. B.S. Kashin, A.A Saakyan, *Orthogonal Series*. Translated from the Russian by Ralph P. Boas. Translation edited by Ben Silver. Translations of Mathematical Monographs, vol. 75 (American Mathematical Society, Providence, 1989), xii+451pp.

18. B. Klartag, A Berry-Esseen type inequality for convex bodies with an unconditional basis. Probab. Theory Relat. Fields **145**(1–2), 1–33 (2009),

19. B. Klartag, J. Lehec, Bourgain's slicing problem and KLS isoperimetry up to polylog (2022). arXiv:2203.15551v2

20. B. Klartag, S. Sodin, Variations on the Berry-Esseen theorem. Teor. Veroyatn. Primen. **56**(3), 514–533 (2011); Reprinted in: Theory Probab. Appl. 56 (2012), no. 3, 403–419

21. M. Ledoux, Concentration of measure and logarithmic Sobolev inequalities, in *Séminaire de Probabilités XXXIII*. Lecture Notes in Mathematics, vol. 1709 (Springer, Berlin, 1999), pp. 120–216

22. P. Massart, The tight constant in the Dvoretzky-Kiefer-Wolfowitz inequality. Ann. Probab. **18**(3), 1269–1283 (1990)

23. V.V. Petrov, *Sums of Independent Random Variables*. Translated from the Russian by A. A. Brown. Ergebnisse der Mathematik und ihrer Grenzgebiete, Band 82 (Springer, New York, 1975), x+346pp.

24. V.V. Petrov, *Limit Theorems for Sums of Independent Random Variables* (in Russian) (Nauka, Moscow, 1987), 318pp.

25. R. Salem, A. Zygmund, On lacunary trigonometric systems. Proc. Nat. Acad. Sci. USA **33**, 333–338 (1947)

26. R. Salem, A. Zygmund, On lacunary trigonometric series. II. Proc. Nat. Acad. Sci. USA **34**, 54–62 (1948)

27. V.N. Sudakov, Typical distributions of linear functionals in finite-dimensional spaces of high dimension (in Russian). Soviet Math. Dokl. **19**, 1578–1582 (1978); translation in: Dokl. Akad. Nauk SSSR, 243 (1978), no. 6, 1402–1405

The Case of Equality in Geometric Instances of Barthe's Reverse Brascamp-Lieb Inequality

Karoly J. Boroczky, Pavlos Kalantzopoulos, and Dongmeng Xi

Abstract The works of Bennett, Carbery, Christ, Tao and of Valdimarsson have clarified when equality holds in the Brascamp-Lieb inequality. Here we characterize the case of equality in the Geometric case of Barthe's reverse Brascamp-Lieb inequality.

1 Introduction

For a proper linear subspace E of $\mathbb{R}^n$ ($E \neq \mathbb{R}^n$ and $E \neq \{0\}$), let P_E denote the orthogonal projection into E. We say that the subspaces $E_1, \ldots, E_k$ of $\mathbb{R}^n$ and $c_1, \ldots, c_k > 0$ form a Geometric Brascamp-Lieb data if they satisfy

$$\sum_{i=1}^{k} c_i P_{E_i} = I_n. \tag{1}$$

The name "Geometric Brascamp-Lieb data" coined by Bennett et al. [15] comes from the following theorem, originating in the work of Brascamp and Lieb [21] and Ball [3, 4] in the rank one case ($\dim E_i = 1$ for $i = 1, \ldots, k$), and Lieb [51] and Barthe [8] in the general case. In the rank one case, the Geometric Brascamp-Lieb data is known as Parseval frame in coding theory and computer science (see for example Casazza et al. [32]).

K. J. Boroczky (✉)
Alfred Renyi Institute of Mathematics, Budapest, Hungary
e-mail: BoroczkyK@ceu.hu

P. Kalantzopoulos
Central European University, Budapest, Hungary

D. Xi
Department of Mathematics, Shanghai University, Shanghai, China

© The Author(s), under exclusive license to Springer Nature Switzerland AG 2023   129
R. Eldan et al. (eds.), *Geometric Aspects of Functional Analysis*, Lecture Notes
in Mathematics 2327, https://doi.org/10.1007/978-3-031-26300-2_4

Theorem 1 (Brascamp-Lieb, Ball, Barthe) *For the linear subspaces $E_1, \ldots, E_k$ of $\mathbb{R}^n$ and $c_1, \ldots, c_k > 0$ satisfying (1), and for non-negative $f_i \in L_1(E_i)$, we have*

$$\int_{\mathbb{R}^n} \prod_{i=1}^k f_i(P_{E_i} x)^{c_i} \, dx \leq \prod_{i=1}^k \left(\int_{E_i} f_i \right)^{c_i} \tag{2}$$

Remark This is Hölder's inequality if $E_1 = \ldots = E_k = \mathbb{R}^n$ and $B_i = I_n$, and hence $\sum_{i=1}^k c_i = 1$.

We note that equality holds in Theorem 1 if $f_i(x) = e^{-\pi \|x\|^2}$ for $i = 1, \ldots, k$; and hence, each f_i is a Gaussian density. Actually, Theorem 1 is an important special case discovered by Ball [4, 5] in the rank one case and by Barthe [8] in the general case of the general Brascamp-Lieb inequality Theorem 5.

After partial results by Barthe [8], Carlen et al. [31] and Bennett et al. [15], it was Valdimarsson [67] who characterized equality in the Geometric Brascamp-Lieb inequality. In order to state his result, we need some notation. Let $E_1, \ldots, E_k$ the proper linear subspaces of $\mathbb{R}^n$ and $c_1, \ldots, c_k > 0$ satisfy (1). In order to understand extremizers in (5), following Carlen et al. [31] and Bennett et al. [15], we say that a non-zero linear subspace V is a critical subspace if

$$\sum_{i=1}^k c_i \dim(E_i \cap V) = \dim V,$$

which is turn equivalent saying that

$$E_i = (E_i \cap V) + (E_i \cap V^\perp) \text{ for } i = 1, \ldots, k$$

according to [15] (see also Lemma 7). We say that a critical subspace V is indecomposable if V has no proper critical linear subspace.

Valdimarsson [67] introduced the so called independent subspaces and the dependent space. We write J to denote the set of 2^k functions $\{1, \ldots, k\} \to \{0, 1\}$. If $\varepsilon \in J$, then let $F_{(\varepsilon)} = \cap_{i=1}^k E_i^{(\varepsilon(i))}$ where $E_i^{(0)} = E_i$ and $E_i^{(1)} = E_i^\perp$ for $i = 1, \ldots, k$. We write J_0 to denote the subset of $\varepsilon \in J$ such that $\dim F_{(\varepsilon)} \geq 1$, and such an $F_{(\varepsilon)}$ is called independent following Valdimarsson [67]. Readily $F_{(\varepsilon)}$ and $F_{(\tilde{\varepsilon})}$ are orthogonal if $\varepsilon \neq \tilde{\varepsilon}$ for $\varepsilon, \tilde{\varepsilon} \in J_0$. In addition, we write F_{dep} to denote the orthogonal component of $\oplus_{\varepsilon \in J_0} F_{(\varepsilon)}$. In particular, $\mathbb{R}^n$ can be written as a direct sum of pairwise orthogonal linear subspaces in the form

$$\mathbb{R}^n = \left(\oplus_{\varepsilon \in J_0} F_{(\varepsilon)} \right) \oplus F_{\text{dep}}. \tag{3}$$

Here it is possible that $J_0 = \emptyset$, and hence $\mathbb{R}^n = F_{\text{dep}}$, or $F_{\text{dep}} = \{0\}$, and hence $\mathbb{R}^n = \oplus_{\varepsilon \in J_0} F_{(\varepsilon)}$ in that case.

For a non-zero linear subspace $L \subset \mathbb{R}^n$, we say that a linear transformation $A : L \to L$ is positive definite if $\langle Ax, y \rangle = \langle x, Ay \rangle$ and $\langle x, Ax \rangle > 0$ for any $x, y \in L \backslash \{0\}$.

Theorem 2 (Valdimarsson) *For the proper linear subspaces* $E_1, \ldots, E_k$ *of* $\mathbb{R}^n$ *and* $c_1, \ldots, c_k > 0$ *satisfying* (1), *let us assume that equality holds in the Brascamp-Lieb inequality* (2) *for non-negative* $f_i \in L_1(E_i)$, $i = 1, \ldots, k$. *If* $F_{\text{dep}} \neq \mathbb{R}^n$, *then let* $F_1, \ldots, F_\ell$ *be the independent subspaces, and if* $F_{\text{dep}} = \mathbb{R}^n$, *then let* $\ell = 1$ *and* $F_1 = \{0\}$. *There exist* $b \in F_{\text{dep}}$ *and* $\theta_i > 0$ *for* $i = 1, \ldots, k$, *integrable non-negative* $h_j : F_j \to [0, \infty)$ *for* $j = 1, \ldots, \ell$, *and a positive definite matrix* $A : F_{\text{dep}} \to F_{\text{dep}}$ *such that the eigenspaces of* A *are critical subspaces and*

$$f_i(x) = \theta_i e^{-\langle A P_{F_{\text{dep}}} x, P_{F_{\text{dep}}} x - b \rangle} \prod_{F_j \subset E_i} h_j(P_{F_j}(x)) \quad \text{for Lebesgue a.e. } x \in E_i.$$

$$(4)$$

On the other hand, if for any $i = 1, \ldots, k$, f_i *is of the form as in* (4), *then equality holds in* (2) *for* $f_1, \ldots, f_k$.

Theorem 2 explains the term "independent subspaces" because the functions h_j on F_j are chosen freely and independently from each other.

A reverse form of the Geometric Brascamp-Lieb inequality was proved by Barthe [8]. We write $\int_{\mathbb{R}^n}^* \varphi$ to denote the outer integral for a possibly non-integrable function $\varphi : \mathbb{R}^n \to [0, \infty)$; namely, the infimum (actually minimum) of $\int_{\mathbb{R}^n} \psi$ where $\psi \geq \varphi$ is Lebesgue measurable.

Theorem 3 (Barthe) *For the non-trivial linear subspaces* $E_1, \ldots, E_k$ *of* $\mathbb{R}^n$ *and* $c_1, \ldots, c_k > 0$ *satisfying* (1), *and for non-negative* $f_i \in L_1(E_i)$, *we have*

$$\int_{\mathbb{R}^n}^* \sup_{x = \sum_{i=1}^k c_i x_i, \, x_i \in E_i} \prod_{i=1}^k f_i(x_i)^{c_i} \, dx \geq \prod_{i=1}^k \left(\int_{E_i} f_i \right)^{c_i}. \quad (5)$$

Remark This is the Prékopa-Leindler inequality Theorem 16 if $E_1 = \ldots = E_k = \mathbb{R}^n$ and $B_i = I_n$, and hence $\sum_{i=1}^k c_i = 1$.

We say that a function $h : \mathbb{R}^n \to [0, \infty)$ is log-concave if $h((1 - \lambda)x + \lambda y) \geq h(x)^{1-\lambda} h(y)^\lambda$ for any $x, y \in \mathbb{R}^n$ and $\lambda \in (0, 1)$; or in other words, $h = e^{-W}$ for a convex function $W : \mathbb{R}^n \to (-\infty, \infty]$. Our main result is the following characterization of equality in the Geometric Barthe's inequality (5).

Theorem 4 *For linear subspaces* $E_1, \ldots, E_k$ *of* $\mathbb{R}^n$ *and* $c_1, \ldots, c_k > 0$ *satisfying* (1), *if* $F_{\text{dep}} \neq \mathbb{R}^n$, *then let* $F_1, \ldots, F_\ell$ *be the independent subspaces, and if* $F_{\text{dep}} = \mathbb{R}^n$, *then let* $\ell = 1$ *and* $F_1 = \{0\}$.

If equality holds in the Geometric Barthe's inequality (5) *for non-negative* $f_i \in$
$L_1(E_i)$ *with* $\int_{E_i} f_i > 0$, $i = 1, \ldots, k$, *then*

$$f_i(x) = \theta_i e^{-\langle A P_{F_{\text{dep}}} x, P_{F_{\text{dep}}} x - b_i \rangle} \prod_{F_j \subset E_i} h_j(P_{F_j}(x - w_i)) \quad \text{for Lebesgue a.e. } x \in E_i$$

(6)

where

- $\theta_i > 0$, $b_i \in E_i \cap F_{\text{dep}}$ *and* $w_i \in E_i$ *for* $i = 1, \ldots, k$,
- $h_j \in L_1(F_j)$ *is non-negative for* $j = 1, \ldots, \ell$, *and in addition,* h_j *is log-concave if there exist* $\alpha \neq \beta$ *with* $F_j \subset E_\alpha \cap E_\beta$,
- $A : F_{\text{dep}} \rightarrow F_{\text{dep}}$ *is a positive definite matrix such that the eigenspaces of* A *are critical subspaces.*

On the other hand, if for any $i = 1, \ldots, k$, f_i *is of the form as in* (6) *and equality holds for all* $x \in E_i$ *in* (6), *then equality holds in* (5) *for* $f_1, \ldots, f_k$.

In particular, if for any $\alpha = 1, \ldots, k$, $\{E_i\}_{i \neq \alpha}$ spans $\mathbb{R}^n$ in Theorem 4, then any extremizer of the Geometric Barthe's inequality is log-concave.

The explanation for the phenomenon concerning the log-concavity of h_j in Theorem 4 is as follows (see the proof of Proposition 17). Let $\ell \geq 1$ and $j \in \{1, \ldots, \ell\}$, and hence $\sum_{E_i \supset F_j} c_i = 1$. If $f_1, \ldots, f_k$ are of the form (6), then equality in Barthe's inequality (5) yields

$$\int_{F_j}^{*} \sup_{\substack{x = \sum_{E_i \supset F_j} c_i x_i \\ x_i \in F_j}} h_j \left(x_i - P_{F_j} w_i \right)^{c_i} dx$$

$$= \prod_{E_i \supset F_j} \left(\int_{F_j} h_j \left(x - P_{F_j} w_i \right) dx \right)^{c_i} \left(= \int_{F_j} h_j(x) \, dx \right).$$

Therefore, if there exist $\alpha \neq \beta$ with $F_j \subset E_\alpha \cap E_\beta$, then the equality conditions in the Prékopa-Leindler inequality Proposition 16 imply that h_j is log-concave. On the other hand, if there exists $\alpha \in \{1, \ldots, k\}$ such that $F_j \subset E_\beta^{\perp}$ for $\beta \neq \alpha$, then we do not have any condition on h_j, and $c_\alpha = 1$.

For completeness, let us state and discuss the general Brascamp-Lieb inequality and its reverse form due to Barthe. The following was proved by Brascamp and Lieb [21] in the rank one case and Lieb [51] in general.

Theorem 5 (Brascamp-Lieb Inequality) *Let* $B_i : \mathbb{R}^n \rightarrow H_i$ *be surjective linear maps where* H_i *is* n_i-*dimensional Euclidean space,* $n_i \geq 1$, *for* $i = 1, \ldots, k$, *and let* $c_1, \ldots, c_k > 0$ *satisfy* $\sum_{i=1}^{k} c_i n_i = n$. *For non-negative* $f_i \in L_1(H_i)$, *we have*

$$\int_{\mathbb{R}^n} \prod_{i=1}^{k} f_i(B_i x)^{c_i} \, dx \leq C \prod_{i=1}^{k} \left(\int_{H_i} f_i \right)^{c_i}$$

(7)

where C is determined by choosing centered Gaussians $f_i(x) = e^{-\langle A_i x, x \rangle}$, A_i positive definite.

Remark The Geometric Brascamp-Lieb Inequality is readily a special case of (7). We note that (7) is Hölder's inequality if $H_1 = \ldots = H_k = \mathbb{R}^n$ and each $B_i = I_n$, and hence $C = 1$ and $\sum_{i=1}^{k} c_i = 1$ in that case.

We say that two Brascamp-Lieb data $\{(B_i, c_i)\}_{i=1,\ldots,k}$ and $\{(B_i', c_i')\}_{i=1,\ldots,k'}$ as in Theorem 5 are called equivalent if $k' = k$, $c_i' = c_i$, and there exists linear isomorphism $\Phi_i : H_i \to H_i'$ for $i = 1, \ldots, k$ such that $B_i' = \Phi_i \circ B_i$. It was proved by Carlen et al. [31] in the rank one case, and by Bennett et al. [15] in general that there exists a set of extremizers $f_1, \ldots, f_k$ for (7) if and only if the Brascamp-Lieb data $\{(B_i, c_i)\}_{i=1,\ldots,k}$ is equivalent to some Geometric Brascamp-Lieb data. Therefore, Valdimarsson's Theorem 2 provides a full characterization of the equality case in Theorem 5, as well.

The following reverse version of the Brascamp-Lieb inequality was proved by Barthe in [7] in the rank one case, and in [8] in general.

Theorem 6 (Barthe's Inequality) *Let $B_i : \mathbb{R}^n \to H_i$ be surjective linear maps where H_i is n_i-dimensional Euclidean space, $n_i \geq 1$, for $i = 1, \ldots, k$, and let $c_1, \ldots, c_k > 0$ satisfy $\sum_{i=1}^{k} c_i n_i = n$. For non-negative $f_i \in L_1(H_i)$, we have*

$$\int_{\mathbb{R}^n}^{*} \sup_{x = \sum_{i=1}^{k} c_i B_i^* x_i, \, x_i \in H_i} \prod_{i=1}^{k} f_i(x_i)^{c_i} \, dx \geq D \prod_{i=1}^{k} \left(\int_{H_i} f_i \right)^{c_i} \qquad (8)$$

where D is determined by choosing centered Gaussians $f_i(x) = e^{-\langle A_i x, x \rangle}$, A_i positive definite.

Remark The Geometric Barthe's Inequality is readily a special case of (8). We note that (8) is the Prékopa-Leindler inequality if $H_1 = \ldots = H_k = \mathbb{R}^n$ and each $B_i = I_n$, and hence $D = 1$ and $\sum_{i=1}^{k} c_i = 1$ in that case.

Concerning extremals in Theorem 6, Lehec [48] proved that if there exists some Gaussian extremizers for Barthe's Inequality (8), then the corresponding Brascamp-Lieb data $\{(B_i, c_i)\}_{i=1,\ldots,k}$ is equivalent to some Geometric Brascamp-Lieb data; therefore, the equality case of (8) can be understood via Theorem 4 in that case.

However, it is still not known whether having any extremizers in Barthe's Inequality (8) yields the existence of Gaussian extremizers. One possible approach is to use iterated convolutions and renormalizations as in Bennett et al. [15] in the case of Brascamp-Lieb inequality.

There are three main methods of proofs that work for proving both the Brascamp-Lieb Inequality and its reverse form due to Barthe. The paper Barthe [8] used optimal transportation to prove Barthe's Inequality ("the Reverse Brascamp-Lieb inequality") and reprove the Brascamp-Lieb Inequality simultaneously. A heat equation argument was provided in the rank one case by Carlen et al. [31] for the Brascamp-Lieb Inequality and by Barthe and Cordero-Erausquin [10] for Barthe's inequality. The general versions of both inequalities are proved via the heat equation approach by Barthe and Huet [11]. Finally, simultaneous probabilistic arguments for the two inequalities are due to Lehec [48].

We note that Chen et al. [33] and Courtade and Liu [35], as well, deal systematically with finiteness conditions in Brascamp-Lieb and Barthe's inequalities. The importance of the Brascamp-Lieb inequality is shown by the fact that besides harmonic analysis, probability and convex geometry, it has been also even applied in number theory, see eg. Guo and Zhang [43]. Various versions of the Brascamp-Lieb inequality and its reverse form have been obtained by Balogh and Kristaly [6] Barthe [9], Barthe and Cordero-Erausquin [10], Barthe et al. [14], Barthe and Wolff [12, 13], Bennett et al. [16], Bennett et al. [17], Bobkov et al. [18], Bueno and Pivarov [24], Carlen, Cordero-Erausquin [30], Chen et al. [33], Courtade and Liu [35], Duncan [38], Ghilli and Salani [41], Kolesnikov and Milman [47], Livshyts [52, 53], Lutwak et al. [55, 56], Maldague [57], Marsiglietti [58], Rossi and Salani [65, 66].

Concerning the proof of Theorem 4, we discuss the structure theory of a Brascamp-Lieb data, Barthe's crucial determinantal inequality (*cf.* Proposition 11) and the extremality of Gaussians (*cf.* Proposition 13) in Sect. 2. Section 3 explains how Barthe's proof of his inequality using optimal transportation in [8] yields the splitting along independent and dependent subspaces in the case of equality in Barthe's inequality for positive C^1 probality densities $f_1, \ldots, f_k$, and how the equality case of the Prékopa-Leindler inequality leads to the log-concavity of certain functions involved. However, one still needs to produce suitably smooth extremizers given any extremizers of Barthe's inequality. In order to achieve this, we discuss that convolution and suitable products of extremizers are also extremizers in Sect. 4. To show that extremizers are Gaussians on the dependent subspace, we use a version of Caffarelli's Contraction Principle in Sect. 5. Finally, all ingredients are pieced together to prove Theorem 4 in Sect. 6.

As an application of the understanding the equality case of Barthe's inequality, we discuss the equality case of Liakopoulos' dual Bollobas-Thomason inequality in Sect. 7.

2 The Structure Theory of the Geometric Brascamp-Lieb Data and Barthe's Determinantal Inequality

We review the structural theory for a Geometric Brascamp-Lieb data based on Barthe [8], Bennett et al. [15] and Valdimarsson [67]. All these statements but Proposition 13 can be found or indicated in Valdimarsson [67], and Proposition 13 is due to Barthe [8]. Let $E_1, \ldots, E_k$ be non-zero linear subspaces of $\mathbb{R}^n$, and let $c_1, \ldots, c_k > 0$ satisfying the Geometric Brascamp-Lieb condition

$$\sum_{i=1}^{k} c_i P_{E_i} = I_n. \tag{9}$$

Lemma 7 *For linear subspaces $E_1, \ldots, E_k$ of $\mathbb{R}^n$ and $c_1, \ldots, c_k > 0$ satisfying (9),*

(i) if $x \in \mathbb{R}^n$, then $\sum_{i=1}^{k} c_i \| P_{E_i} x \|^2 = \|x\|^2$;
(ii) if $V \subset \mathbb{R}^n$ is a proper linear subspace, then

$$\sum_{i=1}^{k} c_i \dim(E_i \cap V) \le \dim V \tag{10}$$

where equality holds if and only if $E_i = (E_i \cap V) + (E_i \cap V^{\perp})$ for $i = 1, \ldots, k$; or equivalently, when $V = (E_i \cap V) + (E_i^{\perp} \cap V)$ for $i = 1, \ldots, k$.

We say that a non-zero linear subspace V is a critical subspace with respect to the proper linear subspaces $E_1, \ldots, E_k$ of $\mathbb{R}^n$ and $c_1, \ldots, c_k > 0$ satisfying (9) if

$$\sum_{i=1}^{k} c_i \dim(E_i \cap V) = \dim V.$$

In particular, $\mathbb{R}^n$ is a critical subspace by calculating traces of both sides of (9). For a proper linear subspace $V \subset \mathbb{R}^n$, Lemma 7 yields that V is critical if and only if $V^{\perp}$ is critical, which is turn equivalent saying that

$$E_i = (E_i \cap V) + (E_i \cap V^{\perp}) \text{ for } i = 1, \ldots, k; \tag{11}$$

or in other words,

$$V = (E_i \cap V) + (E_i^{\perp} \cap V) \text{ for } i = 1, \ldots, k. \tag{12}$$

We observe that (11) has the following consequence: If V_1 and V_2 are orthogonal critical subspaces, then

$$E_i \cap (V_1 + V_2) = (E_i \cap V_1) + (E_i \cap V_2) \text{ for } i = 1, \ldots, k. \tag{13}$$

We recall that a critical subspace V is indecomposable if V has no proper critical linear subspace.

Lemma 8 *If $E_1, \ldots, E_k$ are linear subspaces of $\mathbb{R}^n$ and $c_1, \ldots, c_k > 0$ satisfying (9), and V, W are proper critical subspaces, then $V^{\perp}$ and $V + W$ are critical subspaces, and even $V \cap W$ is critical provided that $V \cap W \ne \{0\}$.*

We deduce from Lemma 8 that any critical subspace can be decomposed into indecomposable ones.

Corollary 9 *If $E_1, \ldots, E_k$ are proper linear subspaces of $\mathbb{R}^n$ and $c_1, \ldots, c_k > 0$ satisfy (9), and W is a critical subspace or $W = \mathbb{R}^n$, then there exist pairwise* *orthogonal indecomposable critical subspaces* $V_1, \ldots, V_m$, $m \geq 1$, *such that* $W = V_1 + \ldots + V_m$ *(possibly $m = 1$ and $W = V_1$).*

We note that the decomposition of $\mathbb{R}^n$ into indecomposable critical subspaces is not unique in general for a Geometric Brascamp-Lieb data. Valdimarsson [67] provides some examples, and in addition, we provide an example where we have a continuous family of indecomposable critical subspaces.

Example 10 (Continuous family of indecomposable critical subspaces) In $\mathbb{R}^4$, let us consider the following six unit vectors: $u_1(1, 0, 0, 0)$, $u_2(\frac{1}{2}, \frac{\sqrt{3}}{2}, 0, 0)$, $u_3(\frac{-1}{2}, \frac{\sqrt{3}}{2}, 0, 0)$, $v_1(0, 0, 1, 0)$, $v_2(0, 0, \frac{1}{2}, \frac{\sqrt{3}}{2})$, $v_3(0, 0, \frac{-1}{2}, \frac{\sqrt{3}}{2})$, which satisfy $u_2 = u_1 + u_3$ and $v_2 = v_1 + v_3$.

For any $x \in \mathbb{R}^4$, we have

$$\|x\|^2 = \sum_{i=1}^{3} \frac{2}{3} \cdot (\langle x, u_i \rangle^2 + \langle x, v_i \rangle^2)$$

Therefore, we define the Geometric Brascamp-Lieb Data $E_i = \mathrm{lin}\{u_i, v_i\}$ and $c_i = \frac{2}{3}$ for $i = 1, 2, 3$ satisfying (1). In this case, $F_{\mathrm{dep}} = \mathbb{R}^4$.

For any angle $t \in \mathbb{R}$, we have a two-dimensional indecomposable critical subspace

$$V_t = \mathrm{lin}\{(\cos t)u_1 + (\sin t)v_1, (\cos t)u_2 + (\sin t)v_2, (\cos t)u_3 + (\sin t)v_3\}.$$

Next we state the crucial determinantal inequality Proposition 11 from Barthe [8]. While Proposition 11 has a crucial role in proving both the Brascamp-Lieb inequality (2) and the Barthe's inequality (5) and their equality cases, Proposition 11 can be actually derived from the Brascamp-Lieb inequality (2) and the characterization of the equality cases by Valdimarsson [67] (*cf.* Theorem 2), only we needs to choose $f_i(z) = e^{-\pi \langle A_i z, z \rangle}$ for $z \in E_i$ and $i = 1, \ldots, k$ in the Brascamp-Lieb inequality.

Proposition 11 *For linear subspaces $E_1, \ldots, E_k$ of $\mathbb{R}^n$, $n \geq 1$ and $c_1, \ldots, c_k > 0$ satisfying (9), if $A_i : E_i \to E_i$ is a positive definite linear transformation for $i = 1, \ldots, k$, then*

$$\det\left(\sum_{i=1}^{k} c_i A_i P_{E_i}\right) \geq \prod_{i=1}^{k} (\det A_i)^{c_i}. \tag{14}$$

Equality holds in (14) if and only if there exist linear subspaces $V_1, \ldots, V_m$ where $V_1 = \mathbb{R}^n$ if $m = 1$ and $V_1, \ldots, V_m$ are pairwise orthogonal indecomposable critical subspaces spanning $\mathbb{R}^n$ if $m \geq 2$, and a positive definite $n \times n$ matrix Φ such that $V_1, \ldots, V_m$ are eigenspaces of Φ and $\Phi|_{E_i} = A_i$ for $i = 1, \ldots, k$. In addition, $\Phi = \sum_{i=1}^{k} c_i A_i P_{E_i}$ in the case of equality.

The next lemma shows how useful is the indecomposability of the critical subspaces in Proposition 11.

Lemma 12 *Let the linear subspaces $E_1, \ldots, E_k$ of $\mathbb{R}^n$ and $c_1, \ldots, c_k > 0$ satisfy (9), let $F_{\text{dep}} \neq \mathbb{R}^n$, and let $F_1, \ldots, F_l$ be the independent subspaces, $l \geq 1$. If V is an indecomposable critical subspace, then either $V \subset F_{\text{dep}}$, or there exists an independent subspace F_j, $j \in \{1, \ldots, l\}$ such that $V \subset F_j$.*

Finally, we exhibit the basic type of Gaussian exemizers of Barthe's inequality.

Proposition 13 *For linear subspaces $E_1, \ldots, E_k$ of $\mathbb{R}^n$, $n \geq 1$ and $c_1, \ldots, c_k > 0$ satisfying (9), if Φ is a positive definite linear transform whose eigenspaces are critical subspaces, then*

$$\int_{\mathbb{R}^n}^{*} \left(\sup_{\substack{x = \sum_{i=1}^{k} c_i x_i \\ x_i \in E_i}} \prod_{i=1}^{k} e^{-c_i \|\Phi x_i\|^2} \right) dx = \prod_{i=1}^{k} \left(\int_{E_i} e^{-\|\Phi x_i\|^2} dx_i \right)^{c_i}.$$

3 Splitting Smooth Extremizers Along Independent and Dependent Subspaces

Optimal transportation as a tool proving geometric inequalities was introduced by Gromov in his Appendix to [62] in the case of the Brunn-Minkowski inequality. Actually, Barthe's inequality in [8] was one of the first inequalities in probability, analysis or geometry that was obtained via optimal transportation.

We write $\nabla \Theta$ to denote the first derivative of a C^1 vector valued function Θ defined on an open subset of $\mathbb{R}^n$, and $\nabla^2 \varphi$ to denote the Hessian of a real C^2 function φ. We recall that a vector valued function Θ on an open set $U \subset \mathbb{R}^n$ is C^α for $\alpha \in (0, 1)$ if for any $x_0 \in U$ there exist an open neighbourhood U_0 of x_0 and a $c_0 > 0$ such that $\|\Theta(x) - \Theta(y)\| \leq c_0 \|x - y\|^\alpha$ for $x, y \in U_0$. In addition, a real function φ is $C^{2,\alpha}$ if φ is C^2 and $\nabla^2 \varphi$ is C^α.

Combining Corollary 2.30, Corollary 2.32, Theorem 4.10 and Theorem 4.13 in Villani [68] on the Brenier map based on McCann [59, 60] for the first two, and on Caffarelli [25–27] for the last two theorems, we deduce the following:

Theorem 14 (Brenier, McCann, Caffarelli) *If f and g are positive C^α probability density functions on $\mathbb{R}^n$, $n \geq 1$, for $\alpha \in (0, 1)$, then there exists a $C^{2,\alpha}$ convex function φ on $\mathbb{R}^n$ (unique up to additive constant) such that $T = \nabla \varphi : \mathbb{R}^n \to \mathbb{R}^n$ is bijective and*

$$g(x) = f(T(x)) \cdot \det \nabla T(x) \text{ for } x \in \mathbb{R}^n. \tag{15}$$

Remarks The derivative $T = \nabla\varphi$ is the Brenier (transportation) map pushing forward the measure on $\mathbb{R}^n$ induced by g to the measure associated to f; namely, $\int_{T(X)} f = \int_X g$ for any measurable $X \subset \mathbb{R}^n$.

In addition, $\nabla T = \nabla^2\varphi$ is a positive definite symmetrix matrix in Theorem 14, and if f and g are C^k for $k \geq 1$, then T is C^{k+1}.

Sometimes it is practical to consider the case $n = 0$, when we set $T : \{0\} \rightarrow \{0\}$ to be the trivial map.

Proof of Theorem 3 Based on Barthe [8] First we assume that each f_i is a C^1 positive probability density function on $\mathbb{R}^n$, and let us consider the Gaussian densiy $g_i(x) = e^{-\pi\|x\|^2}$ for $x \in E_i$. According to Theorem 14, if $i = 1, \ldots, k$, then there exists a C^3 convex function φ_i on E_i such that for the C^2 Brenier map $T_i = \nabla\varphi_i$, we have

$$g_i(x) = \det \nabla T_i(x) \cdot f_i(T_i(x)) \text{ for all } x \in E_i. \tag{16}$$

It follows from the Remark after Theorem 14 that $\nabla T_i = \nabla^2\varphi_i(x)$ is positive definite symmetric matrix for all $x \in E_i$. For the C^2 transformation $\Theta : \mathbb{R}^n \rightarrow \mathbb{R}^n$ given by

$$\Theta(y) = \sum_{i=1}^{k} c_i T_i \left(P_{E_i} y \right), \qquad y \in \mathbb{R}^n, \tag{17}$$

its differential

$$\nabla\Theta(y) = \sum_{i=1}^{k} c_i \nabla T_i \left(P_{E_i} y \right)$$

is positive definite by Proposition 11. It follows that $\Theta : \mathbb{R}^n \rightarrow \mathbb{R}^n$ is injective (see [8]), and actually a diffeomorphism. Therefore Proposition 11, (16) and Lemma 7 (i) imply

$$\int_{\mathbb{R}^n}^{*} \sup_{x=\sum_{i=1}^{k} c_i x_i, \, x_i \in E_i} \prod_{i=1}^{k} f_i(x_i)^{c_i} \, dx$$

$$\geq \int_{\mathbb{R}^n}^{*} \left(\sup_{\Theta(y)=\sum_{i=1}^{k} c_i x_i, \, x_i \in E_i} \prod_{i=1}^{k} f_i(x_i)^{c_i} \right) \det\left(\nabla\Theta(y)\right) dy$$

$$\geq \int_{\mathbb{R}^n} \left(\prod_{i=1}^{k} f_i \left(T_i \left(P_{E_i} y \right) \right)^{c_i} \right) \det\left(\sum_{i=1}^{k} c_i \nabla T_i \left(P_{E_i} y \right) \right) dy$$

$$\geq \int_{\mathbb{R}^n} \left(\prod_{i=1}^{k} f_i \left(T_i \left(P_{E_i} y \right) \right)^{c_i} \right) \prod_{i=1}^{k} \left(\det \nabla T_i \left(P_{E_i} y \right) \right)^{c_i} dy \qquad (18)$$

$$= \int_{\mathbb{R}^n} \left(\prod_{i=1}^{k} g_i \left(P_{E_i} y \right)^{c_i} \right) dy = \int_{\mathbb{R}^n} e^{-\pi \|y\|^2} dy = 1.$$

Finally, Barthe's inequality (5) for arbitrary non-negative integrable functions f_i follows by scaling and approximation (see Barthe [8]). □

We now prove that if equality holds in Barthe's inequality (5), then the diffeomorphism Θ in (17) in the proof of Barthe's inequality splits along the independent subspaces and the dependent subspace. First we explain how Barthe's inequality behaves under the shifts of the functions involved. Given proper linear subspaces $E_1, \ldots, E_k$ of $\mathbb{R}^n$ and $c_1, \ldots, c_k > 0$ satisfying (9), first we discuss in what sense Barthe's inequality is translation invariant. For non-negative integrable function f_i on E_i, $i = 1, \ldots, k$, let us define

$$F(x) = \sup_{x = \sum_{i=1}^{k} c_i x_i, \, x_i \in E_i} \prod_{i=1}^{k} f_i(x_i)^{c_i}.$$

We observe that for any $e_i \in E_i$, defining $\tilde{f}_i(x) = f_i(x + e_i)$ for $x \in E_i$, $i = 1, \ldots, k$, we have

$$\tilde{F}(x) = \sup_{x = \sum_{i=1}^{k} c_i x_i, \, x_i \in E_i} \prod_{i=1}^{k} \tilde{f}_i(x_i)^{c_i} = F \left(x + \sum_{i=1}^{k} c_i e_i \right). \qquad (19)$$

Proposition 15 For non-trivial linear subspaces $E_1, \ldots, E_k$ of $\mathbb{R}^n$ and $c_1, \ldots, c_k > 0$ satisfying (1), we write $F_1, \ldots, F_l$ to denote the independent subspaces (if exist), and F_0 to denote the dependent subspace (possibly $F_0 = \{0\}$). Let us assume that equality holds in (5) for positive C^1 probability densities f_i on E_i, $i = 1, \ldots, k$, let $g_i(x) = e^{-\pi \|x\|^2}$ for $x \in E_i$, let $T_i : E_i \to E_i$ be the C^2 Brenier map satisfying

$$g_i(x) = \det \nabla T_i(x) \cdot f_i(T_i(x)) \text{ for all } x \in E_i, \qquad (20)$$

and let

$$\Theta(y) = \sum_{i=1}^{k} c_i T_i \left(P_{E_i} y \right), \qquad y \in \mathbb{R}^n.$$

(i) *For any* $i \in \{1, \ldots, k\}$ *there exists positive* C^1 *integrable* $h_{i0} : F_0 \cap E_i \to [0, \infty)$ *(where* $h_{i0}(0) = 1$ *if* $F_0 \cap E_i = \{0\}$*), and for any* $i \in \{1, \ldots, k\}$ *and* $j \in \{1, \ldots, l\}$ *with* $F_j \subset E_i$*, there exists positive* C^1 *integrable* $h_{ij} : F_j \to [0, \infty)$ *such that*

$$f_i(x) = h_{i0}(P_{F_0}x) \cdot \prod_{\substack{F_j \subset E_i \\ j \geq 1}} h_{ij}(P_{F_j}x) \quad \text{for } x \in E_i.$$

(ii) *For* $i = 1, \ldots, k$, $T_i(E_i \cap F_p) = E_i \cap F_p$ *whenever* $E_i \cap F_p \neq \{0\}$ *for* $p\{0, \ldots, l\}$, *and if* $x \in E_i$, *then*

$$T_i(x) = \bigoplus_{\substack{E_i \cap F_p \neq \{0\} \\ p \geq 0}} T_i(P_{F_p}x).$$

(iii) *For* $i = 1, \ldots, k$, *there exist* C^2 *functions* $\Omega_i : E_i \to E_i$ *and* $\Gamma_i : E_i^\perp \to E_i^\perp$ *such that*

$$\Theta(y) = \Omega_i(P_{E_i}y) + \Gamma_i(P_{E_i^\perp}y) \quad \text{for } y \in \mathbb{R}^n.$$

(iv) *If* $y \in \mathbb{R}^n$, *then the eigenspaces of the positive definite matrix* $\nabla\Theta(y)$ *are critical subspaces, and* $\nabla T_i(P_{E_i}y) = \nabla\Theta(y)|_{E_i}$ *for* $i = 1, \ldots, k$.

Proof According to (19), we may assume that

$$T_i(0) = 0 \quad \text{for } i = 1, \ldots, k, \tag{21}$$

If equality holds in (5), then equality holds in the determinantal inequality in (18) in the proof of Barthe's inequality; therefore, we apply the equality case of Proposition 11. In particular, for any $x \in \mathbb{R}^n$, there exist $m_x \geq 1$ and linear subspaces $V_{1,x}, \ldots, V_{m_x,x}$ where $V_{1,x} = \mathbb{R}^n$ if $m_x = 1$, and $V_{1,x}, \ldots, V_{m_x,x}$ are pairwise orthogonal indecomposable critical subspaces spanning $\mathbb{R}^n$ if $m_x \geq 2$, and there exist $\lambda_{1,x}, \ldots, \lambda_{m_x,x} > 0$ such that if $E_i \cap V_{j,x} \neq \{0\}$, then

$$\nabla T_i(P_{E_i}x)|_{E_i \cap V_{j,x}} = \lambda_{j,x} I_{E_i \cap V_{j,x}}; \tag{22}$$

and in addition, each E_i satisfies (*cf.* (13))

$$E_i = \bigoplus_{E_i \cap V_{j,x} \neq \{0\}} E_i \cap V_{j,x}. \tag{23}$$

Let us consider a fixed E_i, $i \in \{1, \ldots, k\}$. First we claim that if $y \in E_i$, then

$$\nabla T_i(y)(F_p) = F_p \qquad \text{if } p \geq 1 \text{ and } E_i \cap F_p \neq \{0\}$$

$$\nabla T_i(y)(F_0 \cap E_i) = F_0 \cap E_i. \tag{24}$$

To prove (24), we take $y = x$ in (22). If $p \geq 1$ and $E_i \cap F_p \neq \{0\}$, then $F_p \subset E_i$, and Lemma 12 yields that

$$\oplus_{F_p \cap V_{j,y} \neq \{0\}} V_{j,y} \subset F_p$$

$$\oplus_{F_p \cap V_{j,y} = \{0\}} V_{j,y} \subset F_p^{\perp}.$$

Since the subspaces $V_{j,y}$ span $\mathbb{R}^n$, we have

$$F_p = \oplus_{\substack{E_i \cap V_{j,y} \neq \{0\} \\ V_{j,y} \subset F_p}} V_{j,y};$$

therefore, (22) implies (24) if $p \geq 1$.

For the case of F_0 in (24), it follows from (23) and Lemma 12 that if $E_i \cap F_0 \neq \{0\}$, then

$$E_i \cap F_0 = \oplus_{\substack{E_i \cap V_{j,y} \neq \{0\} \\ V_{j,y} \subset F_0}} E_i \cap V_{j,y}. \tag{25}$$

Therefore, (22) completes the proof of (24).

It follows from (24) that if $E_i \cap F_p \neq \{0\}$, $y \in E_i$, $v \in E_i \cap F_p \cap S^{n-1}$ and $w \in E_i \cap F_p^{\perp} \cap S^{n-1}$, then

$$\left\langle v, \frac{\partial}{\partial t} T_i(y + tw) \Big|_{t=0} \right\rangle = 0. \tag{26}$$

In turn, (24), (26) and $T_i(0) = 0$ (cf. (21)) imply that if $y \in E_i$, then

$$T_i(E_i \cap F_p) = E_i \cap F_p \quad \text{whenever } E_i \cap F_p \neq \{0\} \text{ for } p \geq 0, \tag{27}$$

$$T_i(y) = \bigoplus_{\substack{E_i \cap F_p \neq \{0\} \\ p \geq 0}} T_i(P_{F_p} y). \tag{28}$$

We deduce from (28) that if $y \in E_i$, then

$$\det \nabla T_i(y) = \prod_{\substack{E_i \cap F_p \neq \{0\} \\ p \geq 0}} \det \left(\nabla T_i(P_{F_p} y)|_{F_p} \right). \tag{29}$$

We conclude (i) from (26), (27), (28), and (29) as (20) yields that if $y \in E_i$, then

$$f_i(T_i(y)) = \prod_{\substack{E_i \cap F_p \neq \{0\} \\ p \geq 0}} \frac{e^{-\pi \|P_{F_p} y\|^2}}{\det \left(\nabla T_i(P_{F_p} y)|_{F_p} \right)}.$$

We deduce (ii) from (27) and (28).

For (iii), it follows from Proposition 11 that for any $x \in \mathbb{R}^n$, the spaces $V_{j,x}$ are eigenspaces for $\nabla\Theta(x)$ and span $\mathbb{R}^n$; therefore, (12) implies that if $x \in \mathbb{R}^n$ and $i \in \{1, \ldots, k\}$, then

$$\nabla\Theta(x) = \nabla\Theta(x)|_{E_i} \oplus \nabla\Theta(x)|_{E_i^\perp}.$$

Since $\Theta(0) = 0$ by (21), for fixed $i \in \{1, \ldots, k\}$, we conclude

$$\Theta(E_i) = E_i;$$
$$\Theta(x) = \Theta\left(P_{E_i}x\right)\big|_{E_i} \oplus \Theta\left(P_{E_i^\perp}x\right)\big|_{E_i^\perp} \quad \text{if } x \in \mathbb{R}^n.$$

Finally, (iv) directly follows from Proposition 11, completing the proof of Proposition 15. □

Next we show that if the extremizers $f_1, \ldots, f_k$ in Proposition 15 are of the form as in (i), then for any given $F_j \neq \{0\}$, the functions h_{ij} on F_j for all i with $E_i \cap F_j \neq \{0\}$ are also extremizers. We also need the Prékopa-Leindler inequality Theorem 16 (proved in various forms by Prékopa [63, 64], Leindler [49] and Borell [20]) whose equality case was clarified by Dubuc [37] (see the survey Gardner [40]). In turn, the Prékopa-Leindler inequality (30) is of the very similar structure like Barthe's inequality (5).

Theorem 16 (Prékopa, Leindler, Dubuc) *For $m \geq 2$, $\lambda_1, \ldots, \lambda_m \in (0,1)$ with $\lambda_1 + \ldots + \lambda_m = 1$ and integrable $\varphi_1, \ldots, \varphi_m : \mathbb{R}^n \to [0, \infty)$, we have*

$$\int_{\mathbb{R}^n}^* \sup_{x=\sum_{i=1}^m \lambda_i x_i, \, x_i \in \mathbb{R}^n} \prod_{i=1}^m \varphi_i(x_i)^{\lambda_i} \, dx \geq \prod_{i=1}^m \left(\int_{\mathbb{R}^n} \varphi_i\right)^{\lambda_i}, \tag{30}$$

and if equality holds and the left hand side is positive and finite, then there exist a log-concave function φ and $a_i > 0$ and $b_i \in \mathbb{R}^n$ for $i = 1, \ldots, m$ such that

$$\varphi_i(x) = a_i \, \varphi(x - b_i)$$

for Lebesgue a.e. $x \in \mathbb{R}^n$, $i = 1, \ldots, m$.

For linear subspaces $E_1, \ldots, E_k$ of $\mathbb{R}^n$ and $c_1, \ldots, c_k > 0$ satisfying (1), we assume that $F_{\text{dep}} \neq \mathbb{R}^n$, and write $F_1, \ldots, F_l$ to denote the independent subspaces. We verify that if $j \in \{1, \ldots, l\}$, then

$$\sum_{E_i \supset F_j} c_i = 1. \tag{31}$$

For this, let $x \in F_j \backslash \{0\}$. We observe that for any E_i, either $F_j \subset E_i$, and hence $P_{E_i} x = x$, or $F_j \subset E_i^\perp$, and hence $P_{E_i} x = o$. We deduce from (1) that

$$x = \sum_{i=1}^{k} c_i P_{E_i} x = \left(\sum_{F_j \subset E_i} c_i \right) \cdot x,$$

which formula in turn implies (31).

Proposition 17 *For linear subspaces $E_1, \ldots, E_k$ of $\mathbb{R}^n$ and $c_1, \ldots, c_k > 0$ satisfying (1), we write $F_1, \ldots, F_l$ to denote the independent subspaces (if exist), and F_0 denote the dependent subspace (possibly $F_0 = \{0\}$). Let us assume that equality holds in Barthe's inequality (5) for probability densities f_i on E_i, $i = 1, \ldots, k$, and for any $i \in \{1, \ldots, k\}$ there exists integrable $h_{i0} : F_0 \cap E_i \to [0, \infty)$ (where $h_{i0}(0) = 1$ if $F_0 \cap E_i = \{0\}$), and for any $i \in \{1, \ldots, k\}$ and $j \in \{1, \ldots, l\}$ with $F_j \subset E_i$, there exists non-negative integrable $h_{ij} : F_j \to [0, \infty)$ such that*

$$f_i(x) = h_{i0}(P_{F_0} x) \cdot \prod_{\substack{F_j \subset E_i \\ j \geq 1}} h_{ij}(P_{F_j} x) \quad \text{for } x \in E_i. \tag{32}$$

(i) If $F_0 \neq \{0\}$, then $\sum_{E_i \cap F_0 \neq \{0\}} c_i P_{E_i \cap F_0} = \mathrm{Id}_{F_0}$ and

$$\int_{F_0}^{*} \sup_{x = \sum \{c_i x_i : x_i \in E_i \cap F_0 \ \& \ E_i \cap F_0 \neq \{0\}\}} \prod_{E_i \cap F_0 \neq \{0\}} h_{i0}(x_i)^{c_i} \, dx$$

$$= \prod_{E_i \cap F_0 \neq \{0\}} \left(\int_{E_i \cap F_0} h_{i0} \right)^{c_i}.$$

(ii) If $F_0 \neq \mathbb{R}^n$, then there exist integrable $\psi_j : F_j \to [0, \infty)$ for $j = 1, \ldots, l$ where ψ_j is log-concave whenever $F_j \subset E_\alpha \cap E_\beta$ for $\alpha \neq \beta$, and there exist $a_{ij} > 0$ and $b_{ij} \in F_j$ for any $i \in \{1, \ldots, k\}$ and $j \in \{1, \ldots, l\}$ with $F_j \subset E_i$ such that $h_{ij}(x) = a_{ij} \cdot \psi_j(x - b_{ij})$ for $i \in \{1, \ldots, k\}$ and $j \in \{1, \ldots, l\}$ with $F_j \subset E_i$.

Proof We only present the argument in the case $F_0 \neq \mathbb{R}^n$ and $F_0 \neq \{0\}$. If $F_0 = \mathbb{R}^n$, then the same argument works ignoring the parts involving $F_1, \ldots, F_l$, and if $F_0 = \{0\}$, then the same argument works ignoring the parts involving F_0.

Since $F_0 \oplus F_1 \oplus \ldots \oplus F_l = \mathbb{R}^n$ and $F_0, \ldots, F_l$ are critical subspaces, (13) yields for $i = 1, \ldots, k$ that

$$E_i = (E_i \cap F_0) \oplus \bigoplus_{\substack{F_j \subset E_i \\ j \geq 1}} F_j; \tag{33}$$

therefore, the Fubini theorem and (32) imply that

$$\int_{E_i} f_i = \left(\int_{E_i \cap F_0} h_{i0} \right) \cdot \prod_{\substack{F_j \subset E_i \\ j \geq 1}} \int_{F_j} h_{ij}. \tag{34}$$

On the other hand, using again $F_0 \oplus F_1 \oplus \ldots \oplus F_l = \mathbb{R}^n$, we deduce that if $x = \sum_{j=0}^{l} z_j$ where $z_j \in F_j$ for $j \geq 0$, then $z_j = P_{F_j} x$. It follows from (33) that for any $x \in \mathbb{R}^n$, we have

$$\sup_{\substack{x=\sum_{i=1}^{k} c_i x_i, \\ x_i \in E_i}} \prod_{i=1}^{k} f_i(x_i)^{c_i} = \left(\sup_{\substack{P_{F_0} x=\sum_{i=1}^{k} c_i x_{0i}, \\ x_{0i} \in E_i \cap F_0}} \prod_{i=1}^{k} h_{i0}(x_{i0}) \right) \times$$

$$\times \prod_{j=1}^{l} \left(\sup_{\substack{P_{F_j} x=\sum_{F_j \subset E_i} c_i x_{ji}, \\ x_{ji} \in F_j}} \prod_{F_j \subset E_i} h_{ij}(x_{ji})^{c_i} \right),$$

and hence

$$\int_{\mathbb{R}^n}^{*} \sup_{\substack{x=\sum_{i=1}^{k} c_i x_i, \\ x_i \in E_i}} \prod_{i=1}^{k} f_i(x_i)^{c_i} \, dx = \left(\int_{F_0}^{*} \sup_{\substack{x=\sum_{i=1}^{k} c_i x_i, \\ x_i \in E_i \cap F_0}} \prod_{i=1}^{k} h_{i0}(x_i) \, dx \right) \times \tag{35}$$

$$\times \prod_{j=1}^{l} \left(\int_{F_j}^{*} \sup_{\substack{x=\sum_{F_j \subset E_i} c_i x_i, \\ x_i \in F_j}} \prod_{F_j \subset E_i} h_{ij}(x_i)^{c_i} \, dx \right).$$

As F_0 is a critical subspace, we have

$$\sum_{i=1}^{k} c_i P_{E_i \cap F_0} = \mathrm{Id}_{F_0},$$

and hence Barthe's inequality (5) yields

$$\int_{F_0}^{*} \sup_{\substack{x=\sum_{i=1}^{k} c_i x_i, \\ x_i \in E_i \cap F_0}} \prod_{i=1}^{k} h_{i0}(x_i) \, dx \geq \prod_{i=1}^{k} \left(\int_{E_i \cap F_0} h_{i0} \right)^{c_i}. \tag{36}$$

We deduce from (31) and the Prékopa-Leindler inequality (30) that if $j = 1, \ldots, l$, then

$$\int_{F_j}^{*} \sup_{\substack{x=\sum_{F_j \subset E_i} c_i x_i, \\ x_i \in F_j}} \prod_{F_j \subset E_i} h_{ij}(x_i)^{c_i} \, dx \geq \prod_{E_i \supset F_j} \left(\int_{F_j} h_{ij} \right)^{c_i}. \tag{37}$$

Combining (34), (35), (36) and (37) with the fact that $f_1, \ldots, f_k$ are extremizers for Barthe's inequality (5) implies that if $j = 1, \ldots, l$, then

$$\int_{F_0}^{*} \sup_{\substack{x=\sum_{i=1}^{k} c_i x_i, \\ x_i \in E_i \cap F_0}} \prod_{i=1}^{k} h_{i0}(x_i) \, dx = \prod_{i=1}^{k} \left(\int_{E_i \cap F_0} h_{i0} \right)^{c_i} \tag{38}$$

$$\int_{F_j}^{*} \sup_{\substack{x=\sum_{F_j \subset E_i} c_i x_i, \\ x_i \in F_j}} \prod_{F_j \subset E_i} h_{ij}(x_i)^{c_i} \, dx = \prod_{E_i \supset F_j} \left(\int_{F_j} h_{ij} \right)^{c_i}. \tag{39}$$

We observe that (38) is just (i). In addition, (ii) follows from the equality conditions in the Prékopa-Leindler inequality (see Theorem 16). □

4 Convolution and Product of Extremizers

Given proper linear subspaces $E_1, \ldots, E_k$ of $\mathbb{R}^n$ and $c_1, \ldots, c_k > 0$ satisfying (9), we say that the non-negative integrable functions $f_1, \ldots, f_k$ with positive integrals are extremizers if equality holds in (5). In order to deal with positive smooth functions, we use convolutions. More precisely, Lemma 2 in Barthe [8] states the following.

Lemma 18 *Given proper linear subspaces $E_1, \ldots, E_k$ of $\mathbb{R}^n$ and $c_1, \ldots, c_k > 0$ satisfying (9), if $f_1, \ldots, f_k$ and $g_1, \ldots, g_k$ are extremizers in Barthe's inequality (5), then $f_1 * g_1, \ldots, f_k * g_k$ are also are extremizers.*

Since in a certain case we want to work with Lebesgue integral instead of outer integrals, we use the following statement that can be proved via compactness argument.

Lemma 19 *Given proper linear subspaces $E_1, \ldots, E_k$ of $\mathbb{R}^n$ and $c_1, \ldots, c_k > 0$ satisfying (9), if h_i is a positive continuous functions satisfying $\lim_{x \to \infty} h_i(x) = 0$*

for $i = 1, \ldots, k$, then the function

$$h(x) = \sup_{\substack{x = \sum_{i=1}^{k} c_i x_i, \\ x_i \in E_i}} \prod_{i=1}^{k} h_i(x_i)^{c_i}$$

of $x \in \mathbb{R}^n$ is continuous.

Next we show that the product of a shift of a smooth extremizer and a Gaussian is also an extremizer for Barthe's inequality.

Lemma 20 *Given proper linear subspaces $E_1, \ldots, E_k$ of $\mathbb{R}^n$ and $c_1, \ldots, c_k > 0$ satisfying (9), if $f_1, \ldots, f_k$ are positive bounded C^1 are extremizers in Barthe's inequality (5), and $g_i(x) = e^{-\pi \|x\|^2}$ for $x \in E_i$, then there exist $z_i \in E_i$, $i = 1, \ldots, k$, such that the functions $y \mapsto f_i(y - z_i) g_i(y)$ of $y \in E_i$, $i = 1, \ldots, k$, are also extremizers for (5).*

Proof We may assume that $f_1, \ldots, f_k$ are probability densities.

Readily the functions $\tilde{f}_1, \ldots, \tilde{f}_k$ defined by $\tilde{f}_i(y) = f_i(-y)$ for $y \in E_i$ and $i = 1, \ldots, k$ are also extremizers. We deduce from Lemma 18 that the functions $\tilde{f}_i * g_i$ for $i = 1, \ldots, k$ are also extremizers where each $\tilde{f}_i * g_i$ is a probability density on E_i. According to Theorem 14, if $i = 1, \ldots, k$, then there exists a C^2 Brenier map $S_i : E_i \to E_i$ such that

$$g_i(x) = \det \nabla S_i(x) \cdot (\tilde{f}_i * g_i)(S_i(x)) \quad \text{for all } x \in E_i,$$

and $\nabla S_i(x)$ is a positive definite symmetric matrix for all $x \in E_i$. As in the proof of Theorem 3 above, we consider the C^2 diffeomorphism $\Theta : \mathbb{R}^n \to \mathbb{R}^n$ given by

$$\Theta(y) = \sum_{i=1}^{k} c_i S_i \left(P_{E_i} y \right), \qquad y \in \mathbb{R}^n.$$

whose positive definite differential is

$$\nabla \Theta(y) = \sum_{i=1}^{k} c_i \nabla S_i \left(P_{E_i} y \right).$$

On the one hand, we note that if $x = \sum_{i=1}^{k} c_i x_i$ for $x_i \in E_i$, then

$$\|x\|^2 \le \sum_{i=1}^{k} c_i \|x_i\|^2$$

holds according to Barthe [8]; or equivalently,

$$\prod_{i=1}^{k} g_i(x_i)^{c_i} \le e^{-\pi \|x\|^2}.$$

Since f_i is positive, bounded, continuous and in $L_1(E_i)$ for $i = 1, \ldots, k$, we observe that the function

$$z \mapsto \int_{\mathbb{R}^n} \sup_{\substack{x = \sum_{i=1}^{k} c_i x_i, \\ x_i \in E_i}} \prod_{i=1}^{k} f_i \left(x_i - S_i(P_{E_i} \Theta^{-1} z)\right)^{c_i} g_i(x_i)^{c_i} dx \qquad (40)$$

of $z \in \mathbb{R}^n$ is continuous.

Using also that $\tilde{f}_1, \ldots, \tilde{f}_k$ are extremizers and probability density functions, we have

$$\int_{\mathbb{R}^n}^* \int_{\mathbb{R}^n}^* \sup_{\substack{z = \sum_{i=1}^{k} c_i z_i, \\ z_i \in E_i}} \sup_{\substack{x = \sum_{i=1}^{k} c_i x_i, \\ x_i \in E_i}} \prod_{i=1}^{k} f_i(x_i - z_i)^{c_i} g_i(x_i)^{c_i} dx \, dz$$

$$= \int_{\mathbb{R}^n}^* \int_{\mathbb{R}^n}^* \sup_{\substack{x = \sum_{i=1}^{k} c_i x_i, \\ x_i \in E_i}} \left(\prod_{i=1}^{k} g_i(x_i)^{c_i}\right) \sup_{\substack{z = \sum_{i=1}^{k} c_i z_i, \\ z_i \in E_i}} \prod_{i=1}^{k} f_i(x_i - z_i)^{c_i} dz \, dx$$

$$\le \int_{\mathbb{R}^n}^* e^{-\pi \|x\|^2} \int_{\mathbb{R}^n}^* \sup_{\substack{x = \sum_{i=1}^{k} c_i x_i, \\ x_i \in E_i}} \sup_{\substack{z = \sum_{i=1}^{k} c_i z_i, \\ z_i \in E_i}} \prod_{i=1}^{k} f_i(x_i - z_i)^{c_i} dz \, dx$$

$$= \int_{\mathbb{R}^n}^* e^{-\pi \|x\|^2} \int_{\mathbb{R}^n}^* \sup_{\substack{w = \sum_{i=1}^{k} c_i y_i, \\ y_i \in E_i}} \prod_{i=1}^{k} \tilde{f}_i(y_i)^{c_i} dw \, dx$$

$$= \int_{\mathbb{R}^n} e^{-\pi \|x\|^2} dx = 1.$$

Using Lemma 19 and (40) in (41), Barthe's inequality (5) in (42) and Proposition 11 in (43), we deduce that

$$1 \ge \int_{\mathbb{R}^n}^* \int_{\mathbb{R}^n}^* \sup_{\substack{z = \sum_{i=1}^{k} c_i z_i \\ z_i \in E_i}} \sup_{\substack{x = \sum_{i=1}^{k} c_i x_i \\ x_i \in E_i}} \prod_{i=1}^{k} f_i(x_i - z_i)^{c_i} g_i(x_i)^{c_i} dx \, dz$$

$$\geq \int_{\mathbb{R}^n}^* \int_{\mathbb{R}^n}^* \sup_{\substack{x=\sum_{i=1}^k c_i x_i \\ x_i \in E_i}} \prod_{i=1}^k f_i \left(x_i - S_i(P_{E_i}\Theta^{-1}z)\right)^{c_i} g_i(x_i)^{c_i} \, dx \, dz \qquad (41)$$

$$= \int_{\mathbb{R}^n} \int_{\mathbb{R}^n} \sup_{\substack{x=\sum_{i=1}^k c_i x_i, \\ x_i \in E_i}} \prod_{i=1}^k f_i \left(x_i - S_i(P_{E_i}\Theta^{-1}z)\right)^{c_i} g_i(x_i)^{c_i} \, dx \, dz \qquad (42)$$

$$\geq \int_{\mathbb{R}^n} \prod_{i=1}^k \left(\int_{E_i} f_i \left(x_i - S_i(P_{E_i}\Theta^{-1}z)\right) g_i(x_i) \, dx_i \right)^{c_i} \, dz$$

$$= \int_{\mathbb{R}^n} \left(\prod_{i=1}^k (\tilde{f}_i * g_i) \left(S_i \left(P_{E_i}y\right)\right)^{c_i}\right) \det \left(\sum_{i=1}^k c_i \nabla S_i \left(P_{E_i}y\right)\right) dy \qquad (43)$$

$$\geq \int_{\mathbb{R}^n} \left(\prod_{i=1}^k (\tilde{f}_i * g_i) \left(S_i \left(P_{E_i}y\right)\right)^{c_i}\right) \prod_{i=1}^k \left(\det \nabla S_i \left(P_{E_i}y\right)\right)^{c_i} \, dy$$

$$= \int_{\mathbb{R}^n} \left(\prod_{i=1}^k g_i \left(P_{E_i}y\right)^{c_i}\right) dy = \int_{\mathbb{R}^n} e^{-\pi \|y\|^2} \, dy = 1.$$

In particular, we conclude that

$$1 \geq \int_{\mathbb{R}^n} \int_{\mathbb{R}^n} \sup_{\substack{x=\sum_{i=1}^k c_i x_i, \\ x_i \in E_i}} \prod_{i=1}^k f_i \left(x_i - S_i(P_{E_i}\Theta^{-1}z)\right)^{c_i} g_i(x_i)^{c_i} \, dx \, dz$$

$$\geq \int_{\mathbb{R}^n} \prod_{i=1}^k \left(\int_{E_i} f_i \left(x_i - S_i(P_{E_i}\Theta^{-1}z)\right) g_i(x_i) \, dx_i \right)^{c_i} \, dz \geq 1.$$

Because of Barthe's inequality (5), it follows from (40) that

$$\int_{\mathbb{R}^n} \sup_{\substack{x=\sum_{i=1}^k c_i x_i, \\ x_i \in E_i}} \prod_{i=1}^k f_i \left(x_i - S_i(P_{E_i}\Theta^{-1}z)\right)^{c_i} g_i(x_i)^{c_i} \, dx$$

$$= \prod_{i=1}^k \left(\int_{E_i} f_i \left(x_i - S_i(P_{E_i}\Theta^{-1}z)\right) g_i(x_i) \, dx_i \right)^{c_i}$$

for any $z \in \mathbb{R}^n$; therefore, we may choose $z_i = S_i(0)$ for $i = 1, \ldots, k$ in Lemma 20.
$\square$

5 h_{i0} is Gaussian in Proposition 15

For positive C^α probability density functions f and g on $\mathbb{R}^n$ for $\alpha \in (0, 1)$, the C^1 Brenier map $T : \mathbb{R}^n \to \mathbb{R}^n$ in Theorem 14 pushing forward the the measure on $\mathbb{R}^n$ induced by g to the measure associated to f satisfies that ∇T is positive definite. We deduce that

$$\langle T(y) - T(x), y - x \rangle$$
$$= \int_0^1 \langle \nabla T(x + t(y - x)) \cdot (y - x), y - x \rangle \, dt \geq 0 \quad \text{for any } x, y \in \mathbb{R}^n.$$
(44)

We say that a continuous function $T : \mathbb{R}^n \to \mathbb{R}^m$ has linear growth if there exists a positive constant $c > 0$ such that

$$\|T(x)\| \leq c\sqrt{1 + \|x\|^2}$$

for $x \in \mathbb{R}^n$. It is equivalent saying that

$$\limsup_{\|x\| \to \infty} \frac{\|T(x)\|}{\|x\|} < \infty.$$
(45)

In general, T has polynomial growth, if there exists $k \geq 1$ such that

$$\limsup_{\|x\| \to \infty} \frac{\|T(x)\|}{\|x\|^k} < \infty.$$

Proposition 21 related to Caffarelli Contraction Principle in Caffarelli [28] was proved by Emanuel Milman, see for example Colombo and Fathi [34], De Philippis and Figalli [36], Fathi et al. [39], Kim and Milman [44], Klartag and Putterman [45], Kolesnikov [46], Livshyts [52] for relevant results.

Proposition 21 (Emanuel Milman) *If a Gaussian probability density g and a positive C^α, $\alpha \in (0, 1)$, probability density f on $\mathbb{R}^n$ satisfy $f \leq c \cdot g$ for some positive constant $c > 0$, then the Brenier map $T : \mathbb{R}^n \to \mathbb{R}^n$ pushing forward the measure on $\mathbb{R}^n$ induced by g to the measure associated to f has linear growth.*

Proof We may assume that $g(x) = e^{-\pi \|x\|^2}$.

We observe that $T : \mathbb{R}^n \to \mathbb{R}^n$ is bijective as both f and g are positive. Let S be the inverse of T; namely, $S : \mathbb{R}^n \to \mathbb{R}^n$ is the bijective Brenier map pushing forward the measure on $\mathbb{R}^n$ induced by f to the measure associated to g. In particular, any Borel $X \subset \mathbb{R}^n$ satisfies

$$\int_{S(X)} g = \int_X f.$$
(46)

We note that (45), and hence Proposition 21 is equivalent saying that

$$\liminf_{x \to \infty} \frac{\|S(x)\|}{\|x\|} > 0. \tag{47}$$

The main idea of the argument is the following observation. For any unit vector u and $\theta \in (0, \pi)$, we consider

$$\Xi(u, \theta) = \{y : \langle y, u \rangle \geq \|y\| \cdot \cos \theta\}.$$

Since S is surjective, and $\langle S(z) - S(w), z - w \rangle \geq 0$ for any $z, w \in \mathbb{R}^n$ according to (44), we deduce that

$$S(w) + \Xi(u, \theta) \subset S\left(w + \Xi\left(u, \theta + \frac{\pi}{2}\right)\right) \tag{48}$$

for any $u \in S^{n-1}$ and $\theta \in (0, \frac{\pi}{2})$.

We suppose that T does not have linear growth, and seek a contradiction. According to (47), there exists a sequence $\{x_k\}$ of points of $\mathbb{R}^n \backslash \{0\}$ tending to infinity such that

$$\lim_{k \to \infty} \|x_k\| = \infty \quad \text{and} \quad \lim_{k \to \infty} \frac{\|S(x_k)\|}{\|x_k\|} = 0.$$

In particular, we may assume that

$$\|S(x_k)\| < \frac{\|x_k\|}{8}. \tag{49}$$

For any k, we consider the unit vector $e_k = x_k/\|x_k\|$. We observe that $X_k = x_k + \Xi(e_k, \frac{3\pi}{4})$ avoids the interior of the ball $\frac{\|x_k\|}{\sqrt{2}} B^n$; therefore, if k is large, then

$$\int_{X_k} f \leq c \cdot n\kappa_n \int_{\|x_k\|/\sqrt{2}}^{\infty} r^{n-1} e^{-\pi r^2} dr < \int_{\|x_k\|/\sqrt{2}}^{\infty} e^{-2r^2} \sqrt{2} r \, dr = e^{-\|x_k\|^2} \tag{50}$$

On the other hand, $S(x_k) + \Xi(e_k, \frac{\pi}{4})$ contains the ball

$$\widetilde{B} = S(x_k) + \frac{x_k}{8} + \frac{\|x_k\|}{8\sqrt{2}} B^n \subset \frac{\|x_k\|}{2} B^n$$

where we have used (49). It follows form (46) and (48) that if k is large, then

$$\int_{X_k} f = \int_{S(X_k)} g \geq \int_{\widetilde{B}} g \geq \kappa_n \left(\frac{\|x_k\|}{8\sqrt{2}}\right)^n e^{-\pi(\|x_k\|/2)^2} > e^{-\|x_k\|^2}.$$

This inequality contradicts (50), and in turn proves (47). □

Proposition 24 shows that if the whole space is the dependent subspace and the Brenier maps corresponding to the extremizers $f_1, \ldots, f_k$ in Proposition 15 have at most linear growth, then each f_i is actually Gaussian. The proof of Proposition 24 uses classical Fourier analysis, and we refer to Grafakos [42] for the main properties. For our purposes, we need only the action of a tempered distribution on the space of $C_0^\infty(\mathbb{R}^m)$ of C^∞ functions with compact support, do not need to consider the space of Schwarz functions in general. We recall that if u is a tempered distribution on Schwarz functions on $\mathbb{R}^n$, then the support $\operatorname{supp} u$ is the intersection of all closed sets K such that if $\varphi \in C_0^\infty(\mathbb{R}^n)$ with $\operatorname{supp} \varphi \subset \mathbb{R}^n \backslash K$, then $\langle u, \varphi \rangle = 0$. We write $\hat{u}$ to denote the Fourier transform of a u. In particular, if θ is a function of polynomial growth and $\varphi \in C_0^\infty(\mathbb{R}^n)$, then

$$\langle \hat{\theta}, \varphi \rangle = \int_{\mathbb{R}^n} \int_{\mathbb{R}^n} \theta(x) \varphi(y) e^{-2\pi i \langle x, y \rangle} \, dx \, dy. \tag{51}$$

We consider the two well-known statements Lemma 22 and Lemma 23 about the support of a Fourier transform to prepare the proof of Proposition 24.

Lemma 22 *If θ is a measurable function of polynomial growth on $\mathbb{R}^n$, and there exist linear subspace E with $1 \leq \dim E \leq n - 1$ and function ω on E such that $\theta(x) = \omega(P_E x)$, then $\operatorname{supp} \hat{\theta} \subset E$.*

Proof We write a $z \in \mathbb{R}^n$ in the form $z = (z_1, z_2)$ with $z_1 \in E$ and $z_2 \in E^\perp$. Let $\varphi \in C_0^\infty(\mathbb{R}^n)$ satisfy that $\operatorname{supp} \varphi \subset \mathbb{R}^n \backslash E$, and hence $\varphi(x_1, o) = 0$ for $x_1 \in E$, and the Fourier Integral Theorem in $E^\perp$ implies

$$\varphi(x_1, z) = \int_{E^\perp} \int_{E^\perp} \varphi(x_1, x_2) e^{2\pi i \langle z - x_2, y_2 \rangle} \, dx_2 \, dy_2$$

for $x_1 \in E$ and $z \in E^\perp$. It follows from (51) that

$$\langle \hat{\theta}, \varphi \rangle = \int_{E^\perp} \int_E \int_{E^\perp} \int_E \omega(x_1) \varphi(x_1, x_2) e^{-2\pi i \langle x_1, y_1 \rangle} e^{-2\pi i \langle x_2, y_2 \rangle} \, dx_1 \, dx_2 \, dy_1 \, dy_2$$

$$= \int_E \int_E \omega(x_1) e^{-2\pi i \langle x_1, y_1 \rangle} \left(\int_{E^\perp} \int_{E^\perp} \varphi(x_1, x_2) e^{2\pi i \langle -x_2, y_2 \rangle} \, dx_2 \, dy_2 \right) dy_1 \, dx_1$$

$$= \int_E \int_E \omega(x_1) e^{-2\pi i \langle x_1, y_1 \rangle} \varphi(x_1, 0) \, dy_1 \, dx_1 = 0.$$

□

Next, Lemma 23 directly follows from Proposition 2.4.1 in Grafakos [42].

Lemma 23 *If θ is a continuous function of polynomial growth on $\mathbb{R}^n$ and $\operatorname{supp} \hat{\theta} \subset \{0\}$, then θ is a polynomial.*

Proposition 24 *For linear subspaces $E_1, \ldots, E_k$ of $\mathbb{R}^m$ and $c_1, \ldots, c_k > 0$ satisfying* (1), *we assume that*

$$\cap_{i=1}^{k} (E_i \cup E_i^{\perp}) = \{0\}. \tag{52}$$

Let $g_i(x) = e^{-\pi \|x\|^2}$ for $i = 1, \ldots, k$ and $x \in E_i$, let equality hold in (5) *for positive C^1 probability densities f_i on E_i, $i = 1, \ldots, k$, and let $T_i : E_i \to E_i$ be the C^2 Brenier map satisfying*

$$g_i(x) = \det \nabla T_i(x) \cdot f_i(T_i(x)) \ \text{ for all } x \in E_i. \tag{53}$$

If each T_i, $i = 1, \ldots, k$, has linear growth, then there exist a positive definite matrix $A : \mathbb{R}^m \to \mathbb{R}^m$ whose eigenspaces are critical subspaces, and $a_i > 0$ and $b_i \in E_i$, $i = 1, \ldots, k$, such that

$$f_i(x) = a_i e^{-\langle Ax, x + b_i \rangle} \ \text{ for } x \in E_i.$$

Proof We may assume that each linear subspace is non-zero.

We note that the condition (52) is equivalent saying that $\mathbb{R}^m$ itself is the dependent subspace with respect to the Brascamp-Lieb data. We may assume that for some $1 \le l \le k$, we have $1 \le \dim E_i \le m - 1$ if $i = 1, \ldots, l$, and still

$$\cap_{i=1}^{l} (E_i \cup E_i^{\perp}) = \{0\}. \tag{54}$$

We use the diffeomorphism $\Theta : \mathbb{R}^m \to \mathbb{R}^m$ of Proposition 15 defined by

$$\Theta(y) = \sum_{i=1}^{k} c_i T_i \left(P_{E_i} y \right), \qquad y \in \mathbb{R}^m.$$

It follows from (19) that we may asssume

$$T_i(0) = 0 \ \text{ for } i = 1, \ldots, k, \text{ and hence } \Theta(0) = 0. \tag{55}$$

We claim that there exists a positive definite matrix $B : \mathbb{R}^m \to \mathbb{R}^m$ whose eigenspaces are critical subspaces, and

$$\nabla \Theta(y) = B \ \text{ for } y \in \mathbb{R}^m. \tag{56}$$

Let $\Theta(y) = (\theta_1(y), \ldots, \theta_m(y))$ for $y \in \mathbb{R}^m$ and $\theta_j \in C^2(\mathbb{R}^m)$, $j = 1, \ldots, m$. Since each T_i, $i = 1, \ldots, k$ has linear growth, it follows that Θ has linear growth, and in turn each θ_j, $j = 1, \ldots, m$, has linear growth.

According to Proposition 15 (iii), there exist C^2 functions $\Omega_i : E_i \to E_i$ and $\Gamma_i : E_i^{\perp} \to E_i^{\perp}$ such that

$$\Theta(y) = \Omega_i(P_{E_i} y) + \Gamma_i(P_{E_i^{\perp}} y)$$

for $i = 1, \ldots, k$ and $y \in \mathbb{R}^n$. We write $\Omega_i(x) = (\omega_{i1}(x), \ldots, \omega_{im}(x))$ and $\Gamma_i(x) = (\gamma_{i1}(x), \ldots, \gamma_{im}(x))$; therefore,

$$\theta_j(y) = \omega_{ij}(P_{E_i} y) + \gamma_{ij}(P_{E_i^\perp} y) \tag{57}$$

for $j = 1, \ldots, m$ and $i = 1, \ldots, k$.

Fix a $j \in \{1, \ldots, m\}$. It follows from Lemma 22 and (57) that

$$\mathrm{supp}\, \hat{\theta}_j \subset E_i \cup E_i^\perp$$

for $i = 1, \ldots, l$. Thus (54) yields that

$$\mathrm{supp}\, \hat{\theta}_j \subset \{0\},$$

and in turn we deduce from Lemma 23 that θ_j is a polynomial. Given that θ_j has linear growth, it follows that there exist $w_j \in \mathbb{R}^m$ and $\alpha_j \in \mathbb{R}$ such that $\theta_j(y) = \langle w_j, y \rangle + \alpha_j$. We deduce from $\theta_j(0) = 0$ (cf. (55)) that $\alpha_j = 0$.

The argument so far yields that there exists an $m \times m$ matrix B such that $\Theta(y) = By$ for $y \in \mathbb{R}^m$. As $\nabla\Theta(y) = B$ is positive definite and its eigenspaces are critical subspaces, we conclude the claim (56).

Since $\nabla T_i(P_{E_i} y) = \nabla\Theta(y)|_{E_i}$ for $i = 1, \ldots, k$ and $y \in \mathbb{R}^m$ by Proposition 15 (iv), we deduce that $T_i^{-1} = B^{-1}|_{E_i}$ for $i = 1, \ldots, k$. It follows from (53) that

$$f_i(x) = e^{-\pi \|B^{-1}x\|^2} \cdot \det\left(B^{-1}|_{E_i}\right) \quad \text{for } x \in E_i$$

for $i = 1, \ldots, k$. Therefore, we can choose $A = \pi B^{-2}$. □

6   Proof of Theorem 4

We may assume that each linear subspace E_i is non-zero in Theorem 4. Let f_i be a probability density on E_i in a way such that equality holds for $f_1, \ldots, f_k$ in (5). For $i = 1, \ldots, k$ and $x \in E_i$, let $g_i(x) = e^{-\pi \|x\|^2}$, and hence g_i is a probability distribution on E_i, and $g_1, \ldots, g_k$ are extremizers in Barthe's inequality (5).

It follows from Lemma 18 that the convolutions $f_1 * g_1, \ldots, f_k * g_k$ are also extremizers for (5). We observe that for $i = 1, \ldots, k$, $f_i * g_i$ is a bounded positive C^∞ probability density on E_i. Next we deduce from Lemma 20 that there exist $z_i \in E_i$ and $\gamma_i > 0$ for $i = 1, \ldots, k$ such that defining

$$\tilde{f}_i(x) = \gamma_i \cdot g_i(x) \cdot (f_i * g_i)(x - z_i) \quad \text{for } x \in E_i,$$

$\tilde{f}_1, \ldots, \tilde{f}_k$ are probability densities that are extremizers for (5). We note that if $i = 1, \ldots, k$, then $\tilde{f}_i$ is positive and C^∞, and there exists $c > 1$ satisfying

$$\tilde{f}_i \le c \cdot g_i. \tag{58}$$

Let $\tilde{T}_i : E_i \to E_i$ be the C^∞ Brenier map satisfying

$$g_i(x) = \det \nabla \tilde{T}_i(x) \cdot \tilde{f}_i(\tilde{T}_i(x)) \quad \text{for all } x \in E_i, \tag{59}$$

We deduce from (58) and Proposition 21 that $\tilde{T}_i$ has linear growth.

For $i = 1, \ldots, k$ and $x \in F_0 \cap E_i$, let $g_{i0}(x) = e^{-\pi \|x\|^2}$. It follows from Proposition 15 (i) that for $i \in \{1, \ldots, k\}$, there exists positive C^1 integrable $h_{i0} : F_0 \cap E_i \to [0, \infty)$ (where $h_{i0}(0) = 1$ if $F_0 \cap E_i = \{0\}$), and for any $i \in \{1, \ldots, k\}$ and $j \in \{1, \ldots, l\}$ with $F_j \subset E_i$, there exists positive C^1 integrable $\tilde{h}_{ij} : F_j \to [0, \infty)$ such that

$$\tilde{f}_i(x) = \tilde{h}_{i0}(P_{F_0}x) \cdot \prod_{\substack{F_j \subset E_i \\ j \ge 1}} \tilde{h}_{ij}(P_{F_j}x) \quad \text{for } x \in E_i.$$

We deduce from Proposition 15 (ii) that $\tilde{T}_{i0} = \tilde{T}_i|_{F_0 \cap E_i}$ is the Brenier map pushing forward the measure on $F_0 \cap E_i$ determined g_{i0} onto the measure determined by $\tilde{h}_{i0}$. Since $\tilde{T}_i$ has linear growth, $\tilde{T}_{i0}$ has linear growth, as well, for $i = 1, \ldots, k$.

We deduce from Proposition 17 (i) that $\sum_{i=1}^k c_i P_{E_i \cap F_0} = \mathrm{Id}_{F_0}$, the Geometric Brascamp Lieb data $E_1 \cap F_0, \ldots, E_k \cap F_0$ in F_0 has no independent subspaces, and $\tilde{h}_{10}, \ldots, \tilde{h}_{k0}$ are extremizers in Barthe's inequality for this data in F_0.

As $\tilde{T}_{i0}$ has linear growth for $i = 1, \ldots, k$, Proposition 24 yields the existence of a positive definite matrix $\tilde{A} : F_0 \to F_0$ whose eigenspaces are critical subspaces, and $\tilde{a}_i > 0$ and $\tilde{b}_i \in F_0 \cap E_i$ for $i = 1, \ldots, k$, such that

$$\tilde{f}_i(x) = \tilde{a}_i e^{-\langle \tilde{A}x, x + \tilde{b}_i \rangle} \cdot \prod_{\substack{F_j \subset E_i \\ j \ge 1}} \tilde{h}_{ij}(P_{F_j}x) \quad \text{for } x \in E_i.$$

Dividing by g_i and shifting, we deduce that there exist a symmetric matrix $\bar{A} : F_0 \to F_0$ whose eigenspaces are critical subspaces, and $\bar{a}_i > 0$ and $\bar{b}_i \in F_0 \cap E_i$ for $i = 1, \ldots, k$, and for any $i \in \{1, \ldots, k\}$ and $j \in \{1, \ldots, l\}$ with $F_j \subset E_i$, there exists positive C^1 $\bar{h}_{ij} : F_j \to [0, \infty)$ such that

$$f_i * g_i(x) = \bar{a}_i e^{-\langle \bar{A}x, x + \bar{b}_i \rangle} \cdot \prod_{\substack{F_j \subset E_i \\ j \ge 1}} \bar{h}_{ij}(P_{F_j}x) \quad \text{for } x \in E_i.$$

Since $f_i * g_i$ is a probability density on E_i, it follows that $\bar{A}$ is positive definite and $\bar{h}_{ij} \in L_1(E_i \cap F_j)$ for $i \in \{1, \ldots, k\}$ and $j \in \{1, \ldots, l\}$ with $F_j \subset E_i$.

For any $i = 1, \ldots, k$, we write $\hat{\varrho}$ for the Fourier transform of a function $\varrho \in L_1(E_i)$, thus we can take the inverse Fourier transform in the sense that ϱ is a.e. the L_1 limit of

$$ x \mapsto \int_{\mathbb{R}^n} \hat{\varrho}(\xi) e^{-a|\xi|^2} e^{2\pi i \langle \xi, x \rangle} \, d\xi $$

as $a > 0$ tends to zero. For $i = 1, \ldots, k$, using that $\widehat{f_i * g_i} = \hat{f_i} \cdot \hat{g_i}$, we deduce that the restriction of $\hat{f_i}$ to $F_0 \cap E_i$ is the quotient of two Gaussian densities. Since $\hat{f_i}$ is bounded and zero at infinity, we deduce that the restriction of $\hat{f_i}$ to $F_0 \cap E_i$ is a Gaussian density for $i = 1, \ldots, k$, as well, with the symmetric matrix involved being positive definite. We conclude using the inverse Fourier transform above and the fact that the linear subspaces F_j, $j = 0, \ldots, l$, are pairwise orthogonal that there exist a symmetric matrix $A : F_0 \to F_0$ whose eigenspaces are critical subspaces, and $a_i > 0$ and $b_i \in F_0 \cap E_i$ for $i = 1, \ldots, k$, and for any $i \in \{1, \ldots, k\}$ and $j \in \{1, \ldots, l\}$ with $F_j \subset E_i$, there exists $h_{ij} : F_j \to [0, \infty)$ such that

$$ f_i(x) = a_i e^{-\langle Ax, x + b_i \rangle} \cdot \prod_{\substack{F_j \subset E_i \\ j \geq 1}} h_{ij}(P_{F_j} x) \quad \text{for } x \in E_i. $$

Since f_i is a probability density on E_i, it follows that A is positive definite and each h_{ij} is non-negative and integrable. Finally, Proposition 17 (ii) yields that there exist integrable $\psi_j : F_j \to [0, \infty)$ for $j = 1, \ldots, l$ where ψ_j is log-concave whenever $F_j \subset E_\alpha \cap E_\beta$ for $\alpha \neq \beta$, and there exist $a_{ij} > 0$ and $b_{ij} \in F_j$ for any $i \in \{1, \ldots, k\}$ and $j \in \{1, \ldots, l\}$ with $F_j \subset E_i$ such that $h_{ij}(x) = a_{ij} \cdot \psi_j(x - b_{ij})$ for $i \in \{1, \ldots, k\}$ and $j \in \{1, \ldots, l\}$ with $F_j \subset E_i$.

Finally, we assume that $f_1, \ldots, f_k$ are of the form as described in (6) and equality holds for all $x \in E_i$ in (6). According to (19), we may assume that there exist a positive definite matrix $\Phi : F_0 \to F_0$ whose proper eigenspaces are critical subspaces and a $\tilde{\theta}_i > 0$ for $i = 1, \ldots, k$ such that

$$ f_i(x) = \tilde{\theta}_i e^{-\|\Phi P_{F_0} x\|^2} \prod_{F_j \subset E_i} h_j(P_{F_j}(x)) \quad \text{for } x \in E_i. \tag{60} $$

We recall that according to (31), if $j \in \{1, \ldots, l\}$, then

$$ \sum_{E_i \supset F_j} c_i = 1. \tag{61} $$

We set $\theta = \prod_{i=1}^{k} \tilde{\theta}_i^{c_i}$ and $h_0(x) = e^{-\|\Phi x\|^2}$ for $x \in F_0$. On the left hand side of Barthe's inequality (5), we use first (61) and the log-concavity of h_j whenever

$j \geq 1$ and $F_j \subset E_\alpha \cap E_\beta$ for $\alpha \neq \beta$, secondly Proposition 13, thirdly (61), fourth the Fubini Theorem, and finally (61) again to prove that

$$
\int_{\mathbb{R}^n}^* \sup_{\substack{x=\sum_{i=1}^k c_i x_i \\ x_i \in E_i}} \prod_{i=1}^k f_i(x_i)^{c_i} \, dx = \theta \int_{\mathbb{R}^n}^* \sup_{\substack{x=\sum_{i=1}^k \sum_{j=0}^l c_i x_{ij} \\ x_{ij} \in E_i \cap F_j}} \prod_{j=0}^l \prod_{i=1}^k h_j(x_{ij})^{c_i} \, dx
$$

$$
= \theta \int_{\mathbb{R}^n}^* \prod_{j=0}^l \sup_{\substack{P_{F_j} x = \sum_{i=1}^k c_i x_{ij} \\ x_{ij} \in E_i \cap F_j}} \prod_{i=1}^k h_j(x_{ij})^{c_i} \, dx
$$

$$
= \theta \int_{\mathbb{R}^n}^* \left(\sup_{\substack{P_{F_0} x = \sum_{i=1}^k c_i x_{i0} \\ x_{i0} \in E_i \cap F_0}} \prod_{i=1}^k e^{-c_i \|\Phi x_{i0}\|^2} \right)
$$

$$
\times \prod_{j=1}^l h_j(P_{F_j} x) \, dx
$$

$$
= \theta \left(\prod_{i=1}^k \left(\int_{F_0 \cap E_i} e^{-\|\Phi y\|^2} \, dy \right)^{c_i} \right) \times \prod_{j=1}^l \int_{F_j} h_j
$$

$$
= \prod_{i=1}^k \left(\int_{E_i} f_i \right)^{c_i},
$$

completing the proof of Theorem 4.   □

7 An Application: Equality in Liakopoulos' Dual Bollobas-Thomason Inequality

We write $e_1, \ldots, e_n$ to denote an orthonomal basis of $\mathbb{R}^n$. For a compact set $K \subset \mathbb{R}^n$ with dim aff $K = m$, we write $|K|$ to denote the m-dimensional Lebesgue measure of K.

The starting point of this section is the classical Loomis-Whitney inequality [54].

Theorem 25 (Loomis, Whitney) *If $K \subset \mathbb{R}^n$ is compact and affinely spans $\mathbb{R}^n$, then*

$$
|K|^{n-1} \leq \prod_{i=1}^k |P_{e_i^\perp} K|, \tag{62}
$$

with equality if and only if $K = \oplus_{i=1}^{n} K_i$ where affK_i *is a line parallel to* e_i.

Meyer [61] provided a dual form of the Loomis-Whitney inequality where equality holds for affine crosspolytopes.

Theorem 26 (Meyer) *If $K \subset \mathbb{R}^n$ is compact convex with $o \in \text{int} K$, then*

$$|K|^{n-1} \geq \frac{n!}{n^n} \prod_{i=1}^{k} |K \cap e_i^\perp|, \qquad (63)$$

with equality if and only if $K = \text{conv}\{\pm \lambda_i e_i\}_{i=1}^{n}$ for $\lambda_i > 0$, $i = 1, \ldots, n$.

We note that various Reverse and dual Loomis-Whitney type inequalities are proved by Campi et al. [29], Brazitikos et al. [22, 23], Alonso-Gutiérrez et al. [1, 2].

To consider a genarization of the Loomis-Whitney inequality and its dual form, we set $[n] := \{1, \ldots, n\}$, and for a non-empty proper subset $\sigma \subset [n]$, we define $E_\sigma = \text{lin}\{e_i\}_{i \in \sigma}$. For $s \geq 1$, we say that the not necessarily distinct proper non-empty subsets $\sigma_1, \ldots, \sigma_k \subset [n]$ form an s-uniform cover of $[n]$ if each $j \in [n]$ is contained in exactly s of $\sigma_1, \ldots, \sigma_k$.

The Bollobas-Thomason inequality [19] reads as follows.

Theorem 27 (Bollobas, Thomason) *If $K \subset \mathbb{R}^n$ is compact and affinely spans $\mathbb{R}^n$, and $\sigma_1, \ldots, \sigma_k \subset [n]$ form an s-uniform cover of $[n]$ for $s \geq 1$, then*

$$|K|^s \leq \prod_{i=1}^{k} |P_{E_{\sigma_i}} K|. \qquad (64)$$

We note that additional the case when $k = n$, $s = n - 1$, and hence when we may assume that $\sigma_i = [n] \backslash e_i$, is the Loomis-Whitney inequality Therem 25. •

Liakopoulos [50] managed to prove a dual form of the Bollobas-Thomason inequality. For a finite set σ, we write $|\sigma|$ to denote its cardinality.

Theorem 28 (Liakopoulos) *If $K \subset \mathbb{R}^n$ is compact convex with $o \in \text{int} K$, and $\sigma_1, \ldots, \sigma_k \subset [n]$ form an s-uniform cover of $[n]$ for $s \geq 1$, then*

$$|K|^s \geq \frac{\prod_{i=1}^{k} |\sigma_i|!}{(n!)^s} \cdot \prod_{i=1}^{k} |K \cap E_{\sigma_i}|. \qquad (65)$$

The equality case of the Bollobas-Thomason inequality Theorem 27 based on Valdimarsson [67] has been known to the experts. Let $s \geq 1$, and let $\sigma_1, \ldots, \sigma_k \subset [n]$ be an s-uniform cover of $[n]$. We say that $\tilde{\sigma}_1, \ldots, \tilde{\sigma}_l \subset [n]$ form a 1-uniform cover of $[n]$ induced by the s-uniform cover $\sigma_1, \ldots, \sigma_k$ if $\{\tilde{\sigma}_1, \ldots, \tilde{\sigma}_l\}$ consists of all non-empty distinct subsets of $[n]$ of the form $\cap_{i=1}^{k} \sigma_i^{\varepsilon(i)}$ where $\varepsilon(i) \in \{0, 1\}$ and $\sigma_i^0 = \sigma_i$ and $\sigma_i^1 = [n] \setminus \sigma_i$. We observe that $\tilde{\sigma}_1, \ldots, \tilde{\sigma}_l \subset [n]$ actually form a 1-uniform cover of $[n]$; namely, $\tilde{\sigma}_1, \ldots, \tilde{\sigma}_l$ is a partition of $[n]$.

Theorem 29 (Folklore) *Let $K \subset \mathbb{R}^n$ be compact and affinely span $\mathbb{R}^n$, and let $\sigma_1, \ldots, \sigma_k \subset [n]$ form an s-uniform cover of $[n]$ for $s \geq 1$. Then equality holds in (64) if and only if $K = \oplus_{i=1}^{l} P_{E_{\tilde{\sigma}_i}} K$ where $\tilde{\sigma}_1, \ldots, \tilde{\sigma}_l$ is the 1-uniform cover of $[n]$ induced by $\sigma_1, \ldots, \sigma_k$.*

Our main result in this section is the characterization of the equality case of the dual Bollobas-Thomason inequality Theorem 28 relating it to the Geometric Barthe's inequality.

Theorem 30 *Let $K \subset \mathbb{R}^n$ be compact convex with $o \in \text{int} K$, and let $\sigma_1, \ldots, \sigma_k \subset [n]$ form an s-uniform cover of $[n]$ for $s \geq 1$. Then equality holds in (65) if and only if $K = \text{conv}\{K \cap F_{\tilde{\sigma}_i}\}_{i=1}^{l}$ where $\tilde{\sigma}_1, \ldots, \tilde{\sigma}_l$ is the 1-uniform cover of $[n]$ induced by $\sigma_1, \ldots, \sigma_k$.*

We set $\sigma_i^0 = \sigma_i$ and $\sigma_i^1 = [n] \setminus \sigma_i$. When we write $\tilde{\sigma}_1, \ldots, \tilde{\sigma}_l$ for the induced cover from $\sigma_1, \ldots, \sigma_k$, we assume that the sets $\tilde{\sigma}_1, \ldots, \tilde{\sigma}_l$ are pairwise disjoint.

Lemma 31 *For $s \geq 1$, let $\sigma_1, \ldots, \sigma_k \subset [n]$ form an s-uniform cover of $[n]$, and let $\tilde{\sigma}_1, \ldots, \tilde{\sigma}_\ell$ be the 1-uniform cover of $[n]$ induced by $\sigma_1, \ldots, \sigma_k$. Then*

(i) *the subspaces $E_{\sigma_i} := \text{lin}\{e_j : i \in \sigma_i\}$ satisfy*

$$\sum_{i=1}^{k} \frac{1}{s} P_{E_{\sigma_i}} = I_n \tag{66}$$

i.e. form a Geometric Brascamp Lieb data;
(ii) *For $r \in \tilde{\sigma}_j$, $j = 1, \ldots, \ell$, we have*

$$\tilde{\sigma}_j := \bigcap_{r \in \sigma_i} \sigma_i^0 \cap \bigcap_{r \notin \sigma_i} \sigma_i^1; \tag{67}$$

(iii) *the subspaces $F_{\tilde{\sigma}_j} := \text{lin}\{e_r : r \in \tilde{\sigma}_j\}$ are the independent subspaces of the Geometric Brascamp Lieb data (66) and $F_{\text{dep}} = \{0\}$.*

Proof Since $\sigma_1, \ldots, \sigma_k$ form a s-uniform cover, every $e_i \in \mathbb{R}^n$ is contained in exactly s of $E_{\sigma_1}, \ldots, E_{\sigma_k}$, yielding (i).

For (ii), the definition of $\tilde{\sigma}_j$ directly implies (67).

For (iii), the linear subspaces $F_{\tilde{\sigma}_1}, \ldots, F_{\tilde{\sigma}_\ell}$ are pairwise orthogonal because $\sigma_i^0 \cap \sigma_i^1 = \emptyset$ for $i = 1, \ldots, k$. On the other hand, for any $r \in [n]$, $r \in \cap_{i=1}^{n} \sigma_i^{\varepsilon(i)}$ where $\varepsilon(i) = 0$ if $r \in \sigma_i$, and $\varepsilon(i) = 1$ if $r \notin \sigma_i$; therefore, $F_{\tilde{\sigma}_1}, \ldots, F_{\tilde{\sigma}_\ell}$ span $\mathbb{R}^n$. In particular, $F_{\text{dep}} = \{0\}$. □

Proof of Theorem 30 Let us introduce the notation that we use in the proof of Theorem 30. Let $\sigma_1, \ldots, \sigma_k$ be the s cover of $[n]$ occuring in Theorem 30, and hence $E_i = E_{\sigma_i}, i = 1, \ldots, k$, satisfies

$$\frac{1}{s} \sum_{i=1}^{k} P_{E_{\sigma_i}} = I_n. \tag{68}$$

Let $\tilde{\sigma}_1, \ldots, \tilde{\sigma}_l$ be the 1-uniform cover of $[n]$ induced by $\sigma_1, \ldots, \sigma_k$. It follows that

$$F_j = E_{\tilde{\sigma}_j} \quad \text{for } j = 1, \ldots, l \text{ are the independent subspaces}, \tag{69}$$

$$F_{\text{dep}} = \{0\}. \tag{70}$$

For any $i \in \{1, \ldots, k\}$, we set

$$I_i = \{j \in \{1, \ldots, l\} : F_j \subset E_i\},$$

and for any $j \in \{1, \ldots, l\}$, we set

$$J_j = \{i \in \{1, \ldots, k\} : F_j \subset E_i\}.$$

To prove Theorem 30, we use two additional observations. First if M is any convex body with $o \in \text{int } M$, then

$$\int_{\mathbb{R}^n} e^{-\|x\|_M} \, dx = \int_0^\infty e^{-r} n r^{n-1} |M| \, dr = n! |M|. \tag{71}$$

Secondly, if F_j are pairwise orthogonal subspaces and $M = \text{conv}\{M_1, \ldots, M_l\}$ where $M_j \subset F_j$ is a $\dim F_j$-dimensional compact convex set with $o \in \text{relint } M_j$, then for any $x \in \mathbb{R}^n$

$$\|x\|_M = \sum_{i=1}^l \|P_{F_j} x\|_{M_j}. \tag{72}$$

In addition, we often use the fact, for a subspace F of $\mathbb{R}^n$ and $x \in F$, then $\|x\|_K = \|x\|_{K \cap F}$.

We define

$$f(x) = e^{-\|x\|_K}, \tag{73}$$

which is a log-concave function with $f(0) = 1$, and satisfying (cf (71))

$$\int_{\mathbb{R}^n} f(y)^n \, dy = \int_{R^n} e^{-n\|y\|_K} \, dy = \int_{R^n} e^{-\|y\|_{\frac{1}{n}K}} = n! \left|\frac{1}{n} K\right| = \frac{n!}{n^n} \cdot |K|. \tag{74}$$

We claim that

$$n^n \int_{\mathbb{R}^n} f(y)^n \, dy \geq \prod_{i=1}^k \left(\int_{E_i} f(x_i) \, dx_i\right)^{1/s}. \tag{75}$$

Equating the traces of the two sides of (66), we deduce that, $d_i := |\sigma_i| = \dim E_i$

$$\sum_{i=1}^{k} \frac{d_i}{sn} = 1. \tag{76}$$

For $z = \sum_{i=1}^{k} \frac{1}{s} x_i$ with $x_i \in E_i$, the log-concavity of f and its definition (73), imply

$$f(z/n) \geq \prod_{i=1}^{k} f(x_i/d_i)^{\frac{d_i}{ns}} = \prod_{i=1}^{k} f(x_i)^{\frac{1}{ns}}. \tag{77}$$

Now, the monotonicity of the integral and Barthe's inequality yield

$$\int_{\mathbb{R}^n} f(z/n)^n \, dz \geq \int_{\mathbb{R}^n}^{*} \sup_{z=\sum_{i=1}^{k} \frac{1}{s} x_i, \, x_i \in E_i} \prod_{i=1}^{k} f(x_i)^{1/s} \, dz \geq \prod_{i=1}^{k} \left(\int_{E_i} f(x_i) \, dx_i \right)^{1/s}. \tag{78}$$

Making the change of variable $y = z/n$ we conclude to (75). Computing the right hand side of (75), we have

$$\int_{E_i} f(x_i) \, dx_i = \int_{E_i} e^{-\|x_i\|_K} \, dx_i = \int_{E_i} e^{-\|x_i\|_{K \cap E_i}} \, dx_i = d_i! \, |K \cap E_i|. \tag{79}$$

Therefore, (74), (75) and (79) yield (65).

Let us assume that equality holds in (65), and hence we have two equalities in (78). We set

$$M = \mathrm{conv}\{K \cap F_j\}_{1 \leq j \leq l}.$$

Clearly, $K \supseteq M$. For the other inclusion, we start with $z \in \mathrm{int} K$, namely $\|z\|_K < 1$. Equality in the first inequality in (78) means,

$$\left(e^{-\|z/n\|_K} \right)^n = \sup_{z=\sum_{i=1}^{k} \frac{1}{s} x_i, \, x_i \in E_i} \prod_{i=1}^{k} e^{-\|x_i\|_K 1/s},$$

or in other words,

$$\|z\|_K = \frac{1}{s} \cdot \inf_{z=\sum_{i=1}^{k} \frac{1}{s} x_i, \, x_i \in E_i} \sum_{i=1}^{k} \|x_i\|_K = \inf_{z=\sum_{i=1}^{k} y_i, \, y_i \in E_i} \sum_{i=1}^{k} \|y_i\|_K. \tag{80}$$

We deduce that there exist $y_i \in E_i, i = 1, \ldots, k$ such that

$$z = \sum_{i=1}^{k} y_i \text{ and } \sum_{i=1}^{k} \|y_i\|_K < 1, \tag{81}$$

Therefore, from (81), then (72) and after the triangle inequality for $\|\cdot\|_{K \cap F_j}$, we have

$$\|z\|_M = \left\| \sum_{i=1}^{k} \sum_{j \in I_i} P_{F_j} y_i \right\|_M = \sum_{i=1}^{k} \left\| \sum_{i \in I_i} P_{F_j} y_i \right\|_{K \cap F_j} \leq \sum_{i=1}^{k} \sum_{i \in I_i} \|P_{F_j} y_i\|_{K \cap F_j}. \tag{82}$$

It suffices to show that

$$K \cap E_i = \text{conv}\{K \cap F_j\}_{j \in I_i} \tag{83}$$

because then, from (82), applying (72) and (81), we have

$$\|z\|_M \leq \sum_{j=1}^{l} \sum_{i \in J_j} \|P_{F_j} y_i\|_{K \cap F_j} = \sum_{i=1}^{k} \|y_i\|_{K \cap E_i} < 1,$$

which means $z \in M$. Now, to show (83), we start with the equality case of Barthe's inequality which has been applied in (78). From Theorem 4, there exist $\theta_i > 0$ and $w_i \in E_i$ and log-concave $h_j : F_j \to [0, \infty)$, namely $h_j = e^{-\varphi_j}$ for a convex functon φ_j, such that

$$e^{-\|x_i\|_{K \cap E_i}} = \theta_i \prod_{j \in I_i} h_j(P_{F_j}(x_i - w_i)). \tag{84}$$

for Lebesgue a.e. $x_i \in E_i$. For $i \in [k]$ and $j \in I_i$ we set, $\psi_{ij} : F_j \to \mathbb{R}$ by

$$\psi_{ij}(x) = \varphi_j\left(x - P_{F_j} w_i\right) - \varphi_j\left(-P_{F_j} w_i\right) + \frac{\ln \theta_i}{|I_i|}.$$

We see

$$\psi_{ij}(0) = 0 \text{ and } \psi_{ij} \text{ is convex on } F_j. \tag{85}$$

and also (84) yields, for $x \in E_i$

$$e^{-\|x\|_{K \cap E_i}} = \exp\left(-\sum_{j \in I_i} \psi_{ij}(P_{F_j} x)\right). \tag{86}$$

For $x \in F_j$, we apply λx to (86) with $\lambda > 0$, and we have from $\psi_{im}(0) = 0$ for $m \in I_i \backslash \{j\}$ that

$$\psi_{ij}(\lambda x) = \lambda \psi_{ij}(x) \text{ and } \psi_{ij}(x) > 0. \tag{87}$$

We deduce from (85) and (87) that ψ_{ij} is a norm. Therefore, $\psi_{ij}(x) = \|x\|_{C_{ij}}$ for some (dim F_j)-dimensional compact convex set $C_{ij} \subset F_j$ with $o \in \text{relint} \, C_{ij}$. Now (86) becomes,

$$\|x\|_{K \cap E_i} = \sum_{j \in I_i} \|P_{F_j} x\|_{C_{ij}}$$

and hence by (72) we conclude to

$$K \cap E_i = \text{conv} \, \{C_{ij}\}_{j \in I_i}.$$

In particular, if $i \in [k]$ and $j \in I_i$, then $C_{ij} = (K \cap E_i) \cap F_j = K \cap F_j$, completing the proof of (83), and in turn yielding Theorem 30. □

Acknowledgments We thank Alessio Figalli, Greg Kuperberg and Christos Saroglou for helpful discussions. We are especially grateful to Emanuel Milman for providing the proof of Proposition 21, and for Franck Barthe for providing the proof of Proposition 11 and insight on the history of the subject, and for further ideas and extremely helpful discussions. We thank the referee for correcting a mistake in Theorem 4 and signicantly improving the presentation of the whole paper.

The first named author is also grateful for the hospitality and excellent working environment provided by University of California, Davis and by ETH Zürich during various parts of this project.

Karoly J. Boroczky was Supported by NKFIH 132002. Dongmeng Xi was Supported by National Natural Science Foundation of China (12071277).

References

1. D. Alonso-Gutiérrez, S. Brazitikos, Reverse Loomis-Whitney inequalities via isotropicity. arXiv:2001.11876
2. D. Alonso-Gutiérrez, J. Bernués, S. Brazitikos, A. Carbery, On affine invariant and local Loomis-Whitney type inequalities. arXiv:2002.05794
3. K.M. Ball, Volumes of sections of cubes and related problems, in *Israel Seminar on Geometric Aspects of Functional Analysis*, ed. by J. Lindenstrauss, V.D. Milman. Lectures Notes in Mathematics, vol. 1376 (Springer, Berlin, 1989)
4. K.M. Ball, Volume ratios and a reverse isoperimetric inequality. J. Lond. Math. Soc. **44**, 351–359 (1991)
5. K.M. Ball, Convex geometry and functional analysis, in *Handbook of the Geometry of Banach Spaces*, ed. by W.B. Johnson, L. Lindenstrauss, vol. 1 (2003), pp. 161–194
6. Z. Balogh, A. Kristaly, Equality in Borell-Brascamp-Lieb inequalities on curved spaces. Adv. Math. **339**, 453–494 (2018)
7. F. Barthe, Inégalités de Brascamp-Lieb et convexité. C. R. Acad. Sci. Paris **324**, 885–888 (1997)

8. F. Barthe, On a reverse form of the Brascamp-Lieb inequality. Invent. Math. **134**, 335–361 (1998)
9. F. Barthe, A continuous version of the Brascamp-Lieb inequalities, in *Geometric Aspects of Functional Analysis*. Lecture Notes in Mathematics, vol. 1850 (2004), pp. 53–63
10. F. Barthe, D. Cordero-Erausquin, Inverse Brascamp-Lieb inequalities along the heat equation, in *Geometric Aspects of Functional Analysis*. Lecture Notes in Mathematics, vol. 1850 (Springer, Berlin, 2004), pp. 65–71
11. F. Barthe, N. Huet, On Gaussian Brunn-Minkowski inequalities. Stud. Math. **191**, 283–304 (2009)
12. F. Barthe, P. Wolff, Positivity improvement and Gaussian kernels. C. R. Math. Acad. Sci. Paris **352**, 1017–1021 (2014)
13. F. Barthe, P. Wolff, Positive Gaussian kernels also have Gaussian minimizers. Mem. Am. Math. Soc. **276**(1359), iii+90pp. (2022)
14. F. Barthe, D. Cordero-Erausquin, M. Ledoux, B. Maurey, Correlation and Brascamp-Lieb inequalities for Markov semigroups. Int. Math. Res. Not. **10**, 2177–2216 (2011)
15. J. Bennett, T. Carbery, M. Christ, T. Tao, The Brascamp–Lieb inequalities: finiteness, structure and extremals. Geom. Funct. Anal. **17**, 1343–1415 (2008)
16. J. Bennett, N. Bez, T.C. Flock, S. Lee, Stability of the Brascamp–Lieb constant and applications. Am. J. Math. **140**(2), 543–569 (2018)
17. J. Bennett, N. Bez, S. Buschenhenke, M.G. Cowling, T.C. Flock, On the nonlinear Brascamp-Lieb inequality. Duke Math. J. **169**(17), 3291–3338 (2020)
18. S.G. Bobkov, A. Colesanti, I. Fragalà, Quermassintegrals of quasi-concave functions and generalized Prékopa-Leindler inequalities. Manuscripta Math. **143**, 131–169 (2014)
19. B. Bollobas, A. Thomason, Projections of bodies and hereditary properties of hypergraphs. Bull. Lond. Math. Soc. **27**, 417–424 (1995)
20. C. Borell, The Brunn-Minkowski inequality in Gauss spaces. Invent. Math. **30**, 207–216 (1975)
21. H.J. Brascamp, E.H. Lieb, Best constants in Young's inequality, its converse, and its generalization to more than three functions. Adv. Math. **20**, 151–173 (1976)
22. S. Brazitikos, S. Dann, A. Giannopoulos, A. Koldobsky, On the average volume of sections of convex bodies. Isr. J. Math. **222**, 921–947 (2017)
23. S. Brazitikos, A. Giannopoulos, D.-M. Liakopoulos, Uniform cover inequalities for the volume of coordinate sections and projections of convex bodies. Adv. Geom. **18**, 345–354 (2018)
24. J.R. Bueno, P. Pivarov, A stochastic Prékopa-Leindler inequality for log-concave functions. Commun. Contemp. Math. **23**(2), 2050019, 17pp. (2021)
25. L.A. Caffarelli, A localization property of viscosity solutions to the Monge-Ampère equation and their strict convexity. Ann. Math. **131**, 129–134 (1990)
26. L.A. Caffarelli, Interior $W^{2,p}$ estimates for solutions of the Monge-Ampère equation. Ann. Math. **131**, 135–150 (1990)
27. L.A. Caffarelli, The regularity of mappings with a convex potential. J. Am. Math. Soc. **5**, 99–104 (1992)
28. L.A. Caffarelli, Monotonicity properties of optimal transportation and the FKG and related inequalities. Commun. Math. Phys. **214**(3), 547–563 (2000)
29. S. Campi, R. Gardner, P. Gronchi, Reverse and dual Loomis-Whitney-type inequalities. Trans. Am. Math. Soc. **368**, 5093–5124 (2016)
30. E. Carlen, D. Cordero-Erausquin, Subadditivity of the entropy and its relation to Brascamp-Lieb type inequalities. Geom. Funct. Anal. **19**, 373-405 (2009)
31. E. Carlen, E.H. Lieb, M. Loss, A sharp analog of Young's inequality on S^N and related entropy inequalities. J. Geom. Anal. **14**, 487–520 (2004)
32. P.G. Casazza, T.T. Tran, J.C. Tremain, Regular two-distance sets. J. Fourier Anal. Appl. **26**(3), 49, 32pp. (2020)
33. W.-K. Chen, N. Dafnis, G. Paouris, Improved Hölder and reverse Hölder inequalities for Gaussian random vectors. Adv. Math. **280**, 643–689 (2015)
34. M. Colombo, M. Fathi, Bounds on optimal transport maps onto log-concave measures. J. Differ. Equ. **271**, 1007–1022 (2021)

35. T.A. Courtade, J. Liu, Euclidean forward-reverse Brascamp-Lieb inequalities: finiteness, structure, and extremals. J. Geom. Anal. **31**, 3300–3350 (2021)
36. G. De Philippis, A. Figalli, Rigidity and stability of Caffarelli's log-concave perturbation theorem. Nonlinear Anal. **154**, 59–70 (2017)
37. S. Dubuc, Critères de convexité et inégalités intégrales. Ann. Inst. Fourier Grenoble **27**(1), 135–165 (1977)
38. J. Duncan, An algebraic Brascamp-Lieb inequality. J. Geom. Anal. **31**, 10136–10163 (2021)
39. M. Fathi, N. Gozlan, M. Prod'hommem, A proof of the Caffarelli contraction theorem via entropic regularization. Calc. Var. Partial Differ. Equ. **59**(3), 96, 18pp. (2020)
40. R. Gardner, The Brunn-Minkowski inequality. Bull. Am. Math. Soc. **39**, 355–405 (2002)
41. D. Ghilli, P. Salani, Quantitative Borell-Brascamp-Lieb inequalities for power concave functions. J. Convex Anal. **24**, 857–888 (2017)
42. L. Grafakos, *Classical Fourier Analysis*. Graduate Texts in Mathematics, vol. 249 (Springer, Berlin, 2014)
43. S. Guo, R. Zhang, On integer solutions of Parsell-Vinogradov systems. Invent. Math. **218**, 1–81 (2019)
44. Y.-H. Kim, E. Milman, A generalization of Caffarelli's contraction theorem via (reverse) heat flow. Math. Ann. **354**(3), 827–862 (2012)
45. B. Klartag, E. Putterman, Spectral monotonicity under Gaussian convolution. arXiv:2107.09496
46. A.V. Kolesnikov, On Sobolev regularity of mass transport and transportation inequalities. Theory Probab. Appl. **57**(2), 243–264 (2013)
47. A.V. Kolesnikov, E. Milman, Local L_p-Brunn-Minkowski inequalities for $p < 1$. Mem. Am. Math. Soc. (accepted). arXiv:1711.01089
48. J. Lehec, Short probabilistic proof of the Brascamp-Lieb and Barthe theorems. Can. Math. Bull. **57**, 585–597 (2014)
49. L. Leindler, On a certain converse of Hölder's inequality. II. Acta Sci. Math. **33**, 217–223 (1972)
50. D.-M. Liakopoulos, Reverse Brascamp-Lieb inequality and the dual Bollobás-Thomason inequality. Arch. Math. **112**, 293–304 (2019)
51. E.H. Lieb, Gaussian kernels have only Gaussian maximizers. Invent. Math. **102**, 179–208 (1990)
52. G.V. Livshyts, Some remarks about the maximal perimeter of convex sets with respect to probability measures. Commun. Contemp. Math. **23**(5), 2050037, 19pp. (2021)
53. G.V. Livshyts, On a conjectural symmetric version of Ehrhard's inequality. arXiv:2103.11433
54. L.H. Loomis, H. Whitney, An inequality related to the isoperimetric inequality. Bull. Am. Math. Soc. **55**, 961–962 (1949)
55. E. Lutwak, D. Yang, G. Zhang, Volume inequalities for subspaces of L_p. J. Differ. Geom. **68**, 159–184 (2004)
56. E. Lutwak, D. Yang, G. Zhang, Volume inequalities for isotropic measures. Am. J. Math. **129**, 1711–1723 (2007)
57. D. Maldague, Regularized brascamp–lieb inequalities and an application. Quart. J. Math. **73**, 311–331 (2022). https://doi.org/10.1093/qmath/haab032
58. A. Marsiglietti, Borell's generalized Prékopa-Leindler inequality: a simple proof. J. Convex Anal. **24**, 807–817 (2017)
59. R.J. McCann, Existence and uniqueness of monotone measure-preserving maps. Duke Math. J. **80**, 309–323 (1995)
60. R.J. McCann, A convexity principle for interacting gases. Adv. Math. **128**, 153–179 (1997)
61. M. Meyer, A volume inequality concerning sections of convex sets. Bull. Lond. Math. Soc. **20**, 15–155 (1988)
62. V.D. Milman, G. Schechtman, *Asymptotic Theory of Finite-Dimensional Normed Spaces*. With an appendix by M. Gromov (Springer, Berlin, 1986)
63. A. Prékopa, Logarithmic concave measures with application to stochastic programming. Acta Sci. Math. **32**, 301–316 (1971)

64. A. Prékopa, On logarithmic concave measures and functions. Acta Sci. Math. **34**, 335–343 (1973)
65. A. Rossi, P. Salani, Stability for Borell-Brascamp-Lieb inequalities, in *Geometric Aspects of Functional Analysis*. Lecture Notes in Mathematics, vol. 2169 (Springer, Cham, 2017), pp. 339–363
66. A. Rossi, P. Salani, Stability for a strengthened Borell-Brascamp-Lieb inequality. Appl. Anal. **98**, 1773–1784 (2019)
67. S.I. Valdimarsson, Optimisers for the Brascamp-Lieb inequality. Israel J. Math. **168**, 253–274 (2008)
68. C. Villani, *Topics in Optimal Transportation* (AMS, Providence, 2003)

A Journey with the Integrated Γ2 Criterion and its Weak Forms

Patrick Cattiaux and Arnaud Guillin

Abstract As the title indicates this paper will describe several extensions and applications of the Γ_2 integrated criterion introduced by M. Ledoux following ideas of B. Hellffer. We introduce general weak versions and show that they are equivalent to the weak Poincaré inequalities introduced by M. Röckner and F. Y. Wang. We also discuss special weak versions appropriate to the study of log-concave measures and log-concave perturbations of product measures.

Keywords Poincaré inequality · Γ_2 operator · Log-concave measures

1 Introduction, Framework and Presentation of the Results

Introduced in [2] the Γ_2 criterion (also called $CD(\rho, \infty)$ curvature condition) is the best known sufficient condition for Poincaré and log-Sobolev inequalities to hold for some probability measure μ. It reads as

$$\Gamma_2(f) \geq \rho \, \Gamma(f)$$

for some $\rho > 0$ (see the definitions in the next subsection), i.e. is a pointwise condition. In [28], M. Ledoux introduced an integrated version

$$\mu(\Gamma_2(f)) \geq \rho \, \mu(\Gamma(f))$$

P. Cattiaux (✉)
Institut de Mathématiques de Toulouse, CNRS UMR 5219, Université Paul Sabatier, Toulouse Cedex 09, France
e-mail: patrick.cattiaux@math.univ-toulouse.fr

A. Guillin
Université Clermont Auvergne, CNRS, LMBP, Clermont-Ferrand, France
e-mail: arnaud.guillin@uca.fr

© The Author(s), under exclusive license to Springer Nature Switzerland AG 2023
R. Eldan et al. (eds.), *Geometric Aspects of Functional Analysis*, Lecture Notes in Mathematics 2327, https://doi.org/10.1007/978-3-031-26300-2_5

and proved that this integrated version for some $\rho > 0$ is equivalent to a Poincaré inequality (see Theorem 1.3 below). The Poincaré inequality is thus a mean curvature condition.

As it is well known, Poincaré inequality is related to the "exponential" concentration of measure, to the $\mathbb{L}^2(\mu)$ contraction of some associated Markov semi-group (implying exponential stabilization) and to some isoperimetric questions.

During the last years weaker (and also stronger) forms of the Poincaré inequality have been discussed. They allow us to describe weaker concentration properties (polynomial for instance) and slower rates of convergence to equilibrium (see Sect. 1.2). It is natural to ask whether these weak Poincaré inequalities are equivalent to some weak integrated Γ_2 criteria. This was the starting point of this work.

We then describe some applications of weak integrated Γ_2 criteria to log-concave measures, perturbation of product measures or of radial measures.

1.1 Framework (The Heart of Darkness Following [4])

We will first introduce the objects we are dealing with. The aficionados of [4] will (almost) recognize what is called a full Markov triple therein. Nevertheless in order to understand some of our approaches, one has to understand why this framework is the good one.

Let $\mu(dx) = Z_V^{-1} e^{-V(x)} dx$ be a probability measure defined on an open domain $D \subseteq \mathbb{R}^n$. When needed, we will require some regularity for V and assume that it takes finite values. We denote by $\mu(f)$ the integral of f w.r.t. μ.

If V is in $C^2(D)$, we may introduce the operator

$$A = \Delta - \nabla V . \nabla$$

and the diffusion process

$$X_t^x = x + \sqrt{2}\, B_t - \int_0^t \nabla V(X_s^x) ds$$

living in D up to an explosion time T_∂^x since ∇V is local Lipschitz. Of course here B is a standard Brownian motion.

When $D = \mathbb{R}^n$, $T_\partial^x = \sup_{k \in \mathbb{N}_*} T_k^x$ where T_k^x denotes the exit time of the euclidean ball of radius k, while if D is a bounded open subset, T_∂^x denotes the hitting time of the boundary ∂D, i.e.

$$T_\partial^x = \sup_k T_k^x \quad \text{where} \quad T_k^x = \inf\{t, d(X_t^x, D^c) \le 1/k\},$$

where $d(.,.)$ denotes the euclidean distance. In the sequel we will assume that

$$T_{\partial}^x = +\infty \quad a.s. \text{ for all } x \in D. \tag{1.1}$$

In other words the process X is conservative (in D) and we define $P_t f(x) = \mathbb{E}(f(X_t^x))$ for bounded f's, so that P_t is a markovian semi-group of contractions in $\mathbb{L}^{\infty}(D)$.

Definition 1.1 We shall say that Assumption (H) is satisfied if (1.1) holds true and if in addition

$$\mu \text{ is a reversible (symmetric) measure for the process.} \tag{1.2}$$

We will denote $\langle u, v \rangle$ the usual scalar product and introduce

$$\Gamma(f, g) = \langle \nabla f, \nabla g \rangle \quad , \quad \mathcal{E}(f, g) = \mu(\Gamma(f, g))$$

the associated Dirichlet form, with domain $\mathcal{D}(\mathcal{E})$. We will write $\Gamma(f)$ for $\Gamma(f, f)$.

The next result is the key of the construction

Proposition 1.2 *Assume that (H) is satisfied. In the following two cases*

(1) $D = \mathbb{R}^n$,
(2) D is an open bounded domain and $V \in C^{\infty}(D)$,

then P_t extends to a μ-symmetric continuous Markov semi-group on $\mathbb{L}^2(\mu)$ with generator $\tilde{A}$ and domain $\mathcal{D}(\tilde{A})$.

In addition the generator $\tilde{A}$ is essentially self-adjoint on $C_0^{\infty}(D)$ (C^{∞} functions with compact support). We shall call ESA this property. In particular $C_0^{\infty}(D)$ is a core for $\mathcal{D}(\tilde{A})$. The latter is exactly the set of $f \in H_{loc}^2(D)$ such that f and Af are in $\mathbb{L}^2(\mu)$.

We shall give a proof of this Proposition in Sect. 7, where sufficient conditions for (H) are discussed as well as examples. For simplicity we will only use the notation A in the sequel both for A and $\tilde{A}$.

If $g \in \mathcal{D}(A)$ it holds

$$\mathcal{E}(f, g) = -\mu(f \, Ag). \tag{1.3}$$

If $f \in \mathbb{L}^2(\mu)$ it is well known that $P_t f \in \mathcal{D}(A)$ for $t > 0$ and

$$\partial_t P_t f = A P_t f. \tag{1.4}$$

If in addition $f \in \mathcal{D}(A)$,

$$\partial_t P_t f = A P_t f = P_t A f. \tag{1.5}$$

In particular if f is in $\mathcal{D}(A)$, for $t > 0$,

$$\partial_t \, AP_t f = \partial_t \, P_t Af = A \, P_t Af \, . \tag{1.6}$$

1.2   Presentation of the Main Results

We define the Poincaré constant $C_P(\mu)$ as the smallest constant C satisfying

$$\mathrm{Var}_\mu(f) := \mu(f^2) - \mu^2(f) \leq C \, \mu(|\nabla f|^2) \, , \tag{1.7}$$

for all $f \in C_b^1(D)$ the set of C^1 functions which are bounded with a bounded derivative. For simplicity we will say that μ satisfies a Poincaré inequality provided $C_P(\mu)$ is finite.

As it is well known, the Poincaré constant is linked to the exponential stabilization of the Markov semi-group P_t.

For a Diffusion Markov Triple, the following is well known (see chapter 4 in [4]), it extends to our situation

Theorem 1.3 *If (H) is satisfied, the following three statements are equivalent*

(1) μ satisfies a Poincaré inequality,
(2) there exists C such that for every $f \in C_0^\infty(D)$ (or $C_b^\infty(D)$ the set of smooth functions with bounded derivatives of any order), it holds

$$\mu(|\nabla f|^2) \leq C \, \mu((Af)^2) \, , \tag{1.8}$$

(3) there exists $C > 0$ such that for every $f \in \mathbb{L}^2(\mu)$,

$$Var_\mu(P_t f) \leq e^{-2t/C} \, Var_\mu(f) \, . \tag{1.9}$$

In addition the optimal constants in (1.8) and (1.9) are equal to $C_P(\mu)$.

It is important to check that the previous theorem only requires the properties we have recalled before. Actually the proof of (1) $\Leftrightarrow$ (3) ([4] Theorem 4.2.5) only requires (1.4) so that it is always satisfied. The one of (2) $\Leftrightarrow$ (1) ([4] Proposition 4.8.3) requires to use ESA. In addition one has to check that the semi-group is ergodic, i.e. that the only invariant functions ($P_t f = f$ for all t) are the constants. A proof is provided in the Appendix.

Following D. Bakry we may define (provided V is C^2) the Γ_2 operator

$$\Gamma_2(f, g) = \frac{1}{2} \, [A\Gamma(f, g) - \Gamma(f, Ag) - \Gamma(Af, g)] \, . \tag{1.10}$$

for f, g in $C_b^\infty(D)$. A simple calculation yields in this case

$$\Gamma_2(f) := \Gamma_2(f, f) = \| \, Hess(f) \, \|_{HS}^2 + \langle \nabla f, Hess(V) \nabla f \rangle, \qquad (1.11)$$

where $Hess(f)$ denotes the Hessian matrix of f, and $\| \, Hess(f) \, \|_{HS}^2 = \sum_{i,j} |\partial_{i,j}^2 f|^2$.

Using symmetry we get

$$\mu(\Gamma_2(f, g)) = \mu((Af)(Ag)). \qquad (1.12)$$

still for C_b^∞ functions since if (H) is satisfied, they belong to $\mathcal{D}(A)$. The latter extends to f, g in $\mathcal{D}(A)$ thanks to ESA.

It is important to see that without (H) this result is wrong in general. To justify (1.12) it is at least necessary to know that $\Gamma(f, g) \in \mathcal{D}(A)$ which is not always the case even for C_b^∞ functions if they are not all in $\mathcal{D}(A)$, as in the case of reflected diffusions for instance. Fortunately if (H) is satisfied it suffices to verify it for C_b^∞ functions.

Assume from now on that (1.12) is satisfied for f and g in the domain of A. It immediately follows that, if the curvature-dimension condition $CD(\rho, N)$ i.e.

$$\Gamma_2(f) \geq \rho \, |\nabla f|^2 + \frac{1}{N} (Af)^2$$

is satisfied, then

$$C_P(\mu) \leq \frac{N-1}{\rho N}$$

the result being true for $N \in]1, +\infty]$. This is the famous Bakry-Emery criterion for the Poincaré inequality. For $N = +\infty$ the criterion is satisfied provided V is strictly convex in which case it is also a consequence of Brascamp-Lieb inequality.

The second statement in Theorem 1.3 is thus sometimes called "the integrated Γ2 criterion". This statement appears in Proposition 1.3 of M. Ledoux's paper [28] as "*a simple instance of the Witten Laplacian approach of Sjöstrand and Helffer*", but part of the argument goes back to Hörmander (see e.g. [1, p. 14]). It is worth noticing that, if the semi-group does not appear in the statement, it is an essential tool of Ledoux's proof.

The integrated Γ2 criterion is used in M. Ledoux's work [28] on Gibbs measures. Under the denomination of "Bochner's method" it appeared more or less at the same time in the statistical mechanics world. More recently it was used in the context of convex geometry in [5, 27] under the denomination of $\mathbb{L}^2$ method. Lemma 1 in [5] contains another proof (without using the semi-group) of $(2) \Rightarrow (1)$ in the previous Theorem.

The third statement in Theorem 1.3 can be improved in the following way

Proposition 1.4 *The third statement (hence the first two too) of Theorem 1.3 is equivalent to the following one: there exists $C > 0$, such that for every f in a dense subset $\mathcal{C}$ of $\mathbb{L}^2(\mu)$ one can find a constant $c(f)$ such that*

$$Var_\mu(P_t f) \leq c(f) e^{-2t/C}$$

and the optimal C is again $C_P(\mu)$.

The proof of this proposition lies on the log-convexity of $t \mapsto \mu(P_t^2 f)$ for which several proofs are available (see the simplest one in [22] lemma 2.11 or in [4]).

A natural subset $\mathcal{C}$ is furnished by $\mathbb{L}^\infty(\mu)$. An exponential decay to 0 of the variance controlled by the initial uniform norm thus implies that the same holds for the $\mathbb{L}^2$ norm and is equivalent to the Poincaré inequality.

The semi-group property shows that $\mathbb{L}^2$ decay to 0 cannot be faster than exponential and the previous result that any uniform decay i.e. $Var_\mu(P_T f) \leq c\, Var_\mu(f)$ for some $T > 0$, $c < 1$ and $f \in \mathbb{L}^2(\mu)$ implies exponential decay. A natural question is then to describe what happens for slower decays. After a pioneering work by T. Liggett ([32]), this question was tackled by M. Röckner and F. Y. Wang in [38]. These authors introduced the notion of weak Poincaré inequalities and relate them to all possible decays of the variance along the semi-group. Let us recall the main result in this direction

Theorem 1.5 *Consider the following two statements*

(1) There exists a non-increasing function $\beta_{WP} : (0, +\infty) \to \mathbb{R}^+$, such that for all $s > 0$ and any bounded and Lipschitz function f,

$$Var_\mu(f) \leq \beta_{WP}(s)\, \mu(|\nabla f|^2) + s\, Osc^2(f), \tag{1.13}$$

where $Osc(f) = \sup f - \inf f$ denotes the Oscillation of f. (1.13) is called a weak Poincaré inequality (WPI) and it is clear that we may always choose $\beta_{WP}(s) = 1$ for $s \geq 1$.

(2) There exists a non-increasing function ξ going to 0 at infinity such that

$$Var_\mu(P_t f) \leq \xi(t)\, Osc^2(f).$$

The weak Poincaré inequality (1) implies statement (2) with

$$\xi(t) = 2 \inf\{s > 0, \beta_{WP}(s) \ln(1/s) \leq 2t\} = \inf_{s>0}\left(s + e^{-2t/\beta_{WP}(s)}\right).$$

Conversely statement (2) implies statement (1) with

$$\beta_{WP}(s) = 2s \inf_{r>0}\left(\frac{1}{r}\xi^{-1}(r \exp(1 - \frac{r}{s}))\right)$$

where ξ^{-1} denotes the converse of ξ, i.e. $\xi^{-1}(r) = \inf\{s > 0, \xi(s) \leq r\}$.

Of course if $\beta_{WP}(0) < +\infty$ one recovers the usual Poincaré inequality.

Remark 1.6 Röckner and Wang (see [38] Corollary 2.4 (2)) introduce a trick that allows to improve ξ in the previous result. The basic idea is to use repeatedly (1.13). We will choose four sequences:

(1) a decreasing sequence of positive numbers $(\theta_i)_{i \in \mathbb{N}}$ such that $\theta_0 = 1$ and $\theta_i \to 0$ as $i \to +\infty$,
(2) for $i \geq 1, \alpha_i = \theta_{i-1} - \theta_i$ so that $\sum_i \alpha_i = 1$,
(3) a sequence $(\gamma_i)_{i \geq 0}$ of positive numbers such that $\gamma_0 = 1$ and $\prod_i \gamma_i = 0$,
(4) for $i \geq 1$, $s_i(t)$ is defined by $e^{-2t \alpha_i / \beta_{WP}(s_i(t))} = \gamma_i$, hence $s_i(t) = \beta_{WP}^{-1}(2t\alpha_i / \ln(1/\gamma_i))$.

Applying (1.13) between $t\theta_i$ and $t\theta_{i-1}$ we thus have

$$\text{Var}_\mu(P_{\theta_{i-1}t}f) \leq e^{-2\alpha_i t / \beta_{WP}(s_i(t))} \text{Var}_\mu(P_{\theta_i t}f) + s_i(t) \text{Osc}^2(f)$$

$$= \gamma_i \text{Var}_\mu(P_{\theta_i t}f) + s_i(t) \text{Osc}^2(f),$$

which yields

$$\text{Var}_\mu(P_t f) \leq \sum_{i \geq 0}(\gamma_i \, s_{i+1}(t)) \text{Osc}^2(f). \tag{1.14}$$

So that we may choose $\xi(t) = \sum_{i \geq 0}(\gamma_i \, s_{i+1}(t))$.   ◇

Remark 1.7 In order to prove that statement (2) implies statement (1) we may follow another route. Using

$$\text{Var}_\mu(f) - \text{Var}_\mu(P_t f) = 2 \int_0^t \mu(|\nabla P_u f|^2) \, du \tag{1.15}$$

and the fact that $t \mapsto \mu(|\nabla P_t f|^2)$ is non-increasing (we shall recall a proof in the next section), we have

$$\text{Var}_\mu(f) \leq 2t \, \mu(|\nabla f|^2) + \text{Var}_\mu(P_t f) \leq 2t \, \mu(|\nabla f|^2) + \xi(t) \text{Osc}^2(f)$$

from which we deduce that

$$\beta_{WP}(s) \leq 2\xi^{-1}(s).$$

This expression is simpler than the one in [38] we recalled in Theorem 1.5, but can be slightly worse.   ◇

Example 1.8 Let us give some examples of (non optimal) pairs for (β_{WP}, ξ)

(1) If for $p > 0$, $\xi(t) = c' t^{-p}$ one can take $\beta_{WP}(s) = c s^{-1/p}$. Conversely if $\beta_{WP}(s) = c s^{-1/p}$ the previous Theorem yields $\xi(t) = c' t^{-p} \ln^p(t)$.

Using the trick in remark 1.6, when $\beta_{WP}(s) = c s^{-1/p}$, and choosing $\gamma_i = 2^{-i}$ and $\alpha_i = \frac{6}{\pi^2} i^{-2}$ we get that $\xi(t) \sim t^{-p}$, for large t's, i.e. the logarithmic term disappeared as expected.

Examples of measures satisfying such a weak Poincaré inequality are for $V(x) = (n+q) \ln(1+|x|)$ for some $q > 0$, i.e. measures with polynomial tails. For the explicit link between p and q see [38] example 1.4.a).

(2) For $p > 0$, $\xi(t) = c' \ln^{-p}(1+t)$ and $\beta_{WP}(s) = c e^{\delta/s^{1/p}}$.

This time it corresponds to $\mu(dx) = C \frac{1}{|x|^n \ln^q(1+|x|)} dx$ for some $q > 1$, see [38] example 1.4.b).

(3) For $0 < p \le 1$, $\xi(t) = c e^{-c't^p}$ and $\beta_{WP}(s) = d' + d \ln^{(1-p)/p}(1 + 1/s)$.

This case covers the Subbotin distributions $\mu(dx) = Z^{-1} e^{-|x|^\delta} dx$ for $\delta \le 1$ with $p = \delta/(4 - 3\delta)$, see [38] example 1.4.c). In particular for $p = 1$ one recovers the radial exponential distribution which satisfies an usual Poincaré inequality.

All the constants depend on p. $\diamond$

A natural question is thus to understand whether there is an integrated Γ_2 version of these weak inequalities or not. This will be done in the next section where we introduce a first weak version: for some decreasing β for any bounded $g \in D(A)$ and any $s > 0$,

$$\textbf{(WI}\Gamma_2\textbf{Osc)} \quad \mu(|\nabla g|^2) \le \beta(s) \mu((Ag)^2) + s \operatorname{Osc}^2(g). \tag{1.16}$$

We shall see that (WIΓ_2Osc) can be compared with the weak Poincaré inequality.

In Sect. 3 we introduce another, perhaps more natural, weak version

$$\textbf{(WI}\Gamma_2\textbf{grad)} \quad \mu(|\nabla g|^2) \le \beta(s) \mu((Ag)^2) + s \, |||\nabla g|^2||_\infty, \tag{1.17}$$

which is useful in the log-concave situation, i.e. provided V is convex (not necessarily strictly convex). It is known since S. Bobkov's work [8], that a log-concave probability measure always satisfies some Poincaré inequality (see [3] for a direct proof using Lyapunov functions). Recent results by E. Milman [35] combined with Brascamp-Lieb inequality allow us to get the following result: if μ is log-concave,

$$C_P(\mu) \le C_{univ} \, \mu(|||Hess^{-1}V|||)$$

for some universal C_{univ}. Here for a real and non-negative symmetric matrix, we denote by

$$|||M||| := \sup_{|u|=1} \langle u, Mu \rangle = \lambda_{max}(M)$$

the operator norm of M, $\lambda_{max}(M)$ being the largest eigenvalue of M.

We recover this result in corollary 3.2 as a consequence of (WIΓ$_2$ grad) (and not Brascamp-Lieb) and obtain new explicit bounds in corollary 3.5 involving

$$\mu(\ln^{1+\varepsilon}(1 + |||Hess^{-1}V|||)$$

only.

The next Sect. 4 deals with log-concave perturbations of either log-concave product measures or log-concave radially symmetric measures. Actually M. Ledoux introduced the integrated Γ$_2$ criterion in order to study the Poincaré inequality of perturbations (non necessarily log-concave but wit a potential whose curvature is bounded from below) of product measures and to obtain results for Gibbs measures on continuous spin systems [28]. In the same paper he extended his approach to the log-Sobolev constant (see [28] Proposition 1.5 and the comments immediately after its statement). This approach was then developed in [36] and several works.

In their subsection 3.4, Barthe and Klartag [6] indicate that this method should be used in order to get some results on log-concave perturbations of product measures that are uniformly log-concave in the large, but not for heavy tailed product measures. In Sect. 4 we show that the weak integrated Γ$_2$ criterion allows us to (partly) recover similar but slightly worse results as in [6]. Other results in this direction are shown in [17]. We then extend the method and replace product measures by radial distributions.

In all the paper, unless explicitly stated, we assume for simplicity that assumption (H) is in force.

Dedication *A tribute to Michel Ledoux.*

The origin of this work was an attempt to convince M. Ledoux of the interest of weak inequalities of Poincaré type. After reading the beautiful wink to Michel's heroes [31], we understood that the only way to succeed was to introduce some "curvature condition" inside. It was thus natural to weaken the integrated Γ$_2$ criterion introduced in [28] and to see what happens. The byproduct results in the paper were a nice surprise.

2 Weak Integrated Γ$_2$

Let us start with an obvious remark: since ∇f and Af are unchanged when replacing f by $f - a$ for any constant a, we have

$$\mu(|\nabla f|^2) = \mu(|\nabla(f - a)|^2) = -\mu((f - a) A(f - a))$$

$$= -\mu((f - a) Af) \leq \frac{1}{2} \text{Osc}(f) \mu(|Af|),$$

by choosing $a = (\sup(f) + \inf(f))/2$. Using for $s > 0$, $2uv \leq \frac{1}{s}u^2 + s\,v^2$ we thus deduce, using Cauchy-Schwarz inequality, that for all $s > 0$,

$$\mu(|\nabla f|^2) \leq \frac{1}{16s}\mu((Af)^2) + s\,\mathrm{Osc}^2(f). \tag{2.1}$$

This is a special instance of (1.16) we recall here: for some decreasing β for any bounded $g \in D(A)$ and any $s > 0$,

$$\textbf{(WI}\Gamma_2\textbf{Osc)} \quad \mu(|\nabla g|^2) \leq \beta(s)\,\mu((Ag)^2) + s\,\mathrm{Osc}^2(g). \tag{2.2}$$

Hence some (very) weak form of the integrated Γ_2 is always satisfied. The previous inequality is thus certainly insufficient in order to get interesting consequences.

Remark 2.1 Contrary to (WPI), (2.2) is not always satisfied for $s \geq 1$, so that, apriori, β does not necessarily goes to 0 at infinity. However if (2.2) is satisfied with two functions β_1 and β_2, it is also satisfied with $\beta = \min(\beta_1, \beta_2)$. According to (2.1), it is thus always satisfied for $s \mapsto \min(\beta(s), 1/16s)$, so that we may always assume without loss of generality that β goes to 0 at infinity. Again in all what follows we denote $\beta^{-1}(t) = \inf\{s > 0, \ \beta(s) \leq t\}$. $\qquad \diamond$

To see how to reinforce (2.1) it is enough to look at the proof of (2) implies (1) in Theorem 1.3. We follow the proof in [28].

The starting point is again (1.15),

$$\mathrm{Var}_\mu(f) - \mathrm{Var}_\mu(P_t f) = 2\int_0^t \mu(|\nabla P_u f|^2)\,du$$

yielding the equality (1.7) in [28],

$$\mathrm{Var}_\mu(f) = 2\int_0^{+\infty} \mu(|\nabla P_t f|^2)\,d\mu$$

as soon as

$$\mathrm{Var}_\mu(P_t f) \to 0 \quad \text{as } t \to +\infty.$$

Since μ is symmetric, the latter is satisfied as soon as the semi-group is ergodic, i.e. the eigenspace of A associated to the eigenvalue 0 is reduced to the constants. Actually this property is ensured by our assumptions: as shown for instance in [38] Theorem 3.1 and the remark following this theorem, if $\mu(dx) = e^{-V}dx$ is a probability measure with V of C^1 class, hence locally bounded, μ satisfies some weak Poincaré inequality so that the above convergence holds true.

Now defining $F(t) = \mu(|\nabla P_t f|^2)$, one can check (using (1.3) and (1.6)) that

$$F'(t) = -2\mu((AP_t f)^2). \tag{2.3}$$

Notice that this equality shows that F is non increasing.

Using this property in (1.15) we get

$$\text{Var}_\mu(f) \geq \text{Var}_\mu(P_t f) + 2t\,\mu(|\nabla P_t f|^2) \geq 2t\,\mu(|\nabla P_t f|^2)$$

so that

$$\mu(|\nabla P_t f|^2) \leq \frac{1}{2t}\,\text{Var}_\mu(f) \leq \frac{1}{2t}\,\text{Osc}^2(f). \tag{2.4}$$

Assuming that a weak integrated Γ2 inequality (2.2) is satisfied we get, using that $\text{Osc}(P_t f) \leq \text{Osc}(f)$,

$$F'(t) \leq -\frac{2}{\beta(s)}\,F(t) + \frac{2s}{\beta(s)}\,\text{Osc}^2(f).$$

This immediately yields

$$\mu(|\nabla P_t f|^2) = F(t) \leq e^{-2t/\beta(s)}\,\mu(|\nabla f|^2) + s\left(1 - e^{-2t/\beta(s)}\right)\text{Osc}^2(f). \tag{2.5}$$

We may apply the previous inequality replacing f by $P_a f$, next t by $t - a$ and use again $\text{Osc}(P_a f) \leq \text{Osc}(f)$. Using (2.4) we thus have for $t > a > 0$,

$$\mu(|\nabla P_t f|^2) \leq \inf_{s>0}\left(s + \frac{1}{2a}e^{-2(t-a)/\beta(s)}\right)\text{Osc}^2(f) = \eta(t)\,\text{Osc}^2(f). \tag{2.6}$$

We have thus obtained

Proposition 2.2 *Assume that μ satisfies a weak integrated Γ2 inequality (WI Γ2Osc) (2.2). Define for $t > a > 0$,*

$$\eta(t) = \inf_{s>0}\left(s + \frac{1}{2a}e^{-2(t-a)/\beta(s)}\right) = 2\inf\{s > 0;\; \beta(s)\ln(1/as) \leq 2(t-a)\}.$$

If η is integrable at infinity, then for $t > a$,

$$\text{Var}_\mu(P_t f) \leq 2\left(\int_t^{+\infty}\eta(u)du\right)\text{Osc}^2(f).$$

In particular μ satisfies a (WPI) where β_{WP} is given in Theorem 1.5 with

$$\xi(t) = 2\left(\int_t^{+\infty}\eta(u)du\right).$$

Remark 2.3 Notice that if μ satisfies a Poincaré inequality we recover the correct exponential decay thanks to proposition 1.4.

If we come back to (2.1) we may always use $\beta(s) \sim c/s$. Using proposition 2.2 with $a = t/2$ the best possible $\eta(t)$ is of order $1/t$ (for large t's) and thus is not integrable, in accordance with the fact that (2.1) cannot furnish the rate of decay to 0 since it is satisfied for all measures μ.

Notice that, as for the (WPI), if $\beta(s) = cs^{-1/(p+1)}$ we obtain $\eta(t) = c'(\ln(t)/t)^{p+1}$ and finally $\xi(t) \sim c'(\ln(t)/t)^p$. But here again we may apply the trick of remark 1.6, simply replacing (1.13) by (2.5), yielding

$$\mu(|\nabla P_t f|^2) \le \sum_{i \ge 0} (\gamma_i \, s_{i+1}(t)) \, \mathrm{Osc}^2(f) , \tag{2.7}$$

with

$$s_i(t) = \beta^{-1}(2t\alpha_i / \ln(1/\gamma_i)) .$$

As for (WPI) this remark allows us to skip the logarithmic term.   ◇

Remark 2.4 Taking $a = \mu(f)$ we may replace (2.1) by

$$\mu(|\nabla f|^2) \le \frac{1}{4s} \mu((Af)^2) + s \, \mathrm{Var}_\mu(f) ,$$

so that we could also consider weak inequalities of the form

$$\mu(|\nabla f|^2) \le \beta(s) \, \mu((Af)^2) + s \, \mathrm{Var}_\mu(f) . \tag{2.8}$$

It is immediately seen that the previous derivation is unchanged if we replace $\mathrm{Osc}^2(f)$ by $\mathrm{Var}_\mu(f)$ so that if η is integrable we get

$$\mathrm{Var}_\mu(P_t f) \le 2 \left(\int_t^{+\infty} \eta(u) du \right) \mathrm{Var}_\mu(f) .$$

But according to what we already said, such a decay implies that μ satisfies a Poincaré inequality, hence thanks to Theorem 1.3 that β is constant equal to $C_P(\mu)$ (or if one prefers that $\beta(0) < +\infty$). Thus, in the other cases, (2.8) furnishes a non-integrable η.   ◇

Let us look at the converse statement. According to Theorem 1.5 we may associate some (WPI) inequality to any decay controlled by the Oscillation. Thus for $a = \mu(f)$,

$$\mu^2(|\nabla f|^2) = -\mu^2((f-a)Af) \le \mu((Af)^2) \, \mathrm{Var}_\mu(f)$$

$$\le \mu((Af)^2) \left(\beta_{WP}(s) \, \mu(|\nabla f|^2) + s \, \mathrm{Osc}^2(f) \right) . \tag{2.9}$$

Since $u^2 \leq Cu + B$ implies that

$$u \leq \frac{1}{2}\left(C + (C^2 + 4B)^{\frac{1}{2}}\right) \leq C + B^{\frac{1}{2}},$$

we obtain

$$\mu(|\nabla f|^2) \leq \beta_{WP}(s)\,\mu((Af)^2)) + s^{\frac{1}{2}}\,\mu^{\frac{1}{2}}((Af)^2)\,\mathrm{Osc}(f),$$

$$\leq (\beta_{WP}(s) + \frac{1}{2})\,\mu((Af)^2)) + \frac{1}{2}\,s\,\mathrm{Osc}^2(f).$$

We have thus obtained (since we know that μ always satisfies some (WPI) inequality) and according to remark 2.1

Proposition 2.5 μ *always satisfies a weak integrated* Γ_2 *inequality (WI* Γ_2 *Osc)* (2.2), *with*

$$\beta(s) = \min\left(1/2 + \beta_{WP}(2s), \frac{1}{16s}\right).$$

The previous results need some comments.

In first place, if we cannot assume that $\beta(s) = 1$ for $s \geq 1$ in the weak integrated Γ_2 inequality (2.2), the interesting behaviour of this function is nevertheless as $s \to 0$ for proposition 2.2 to have some interest.

In second place proposition 2.5 is certainly non sharp. In particular we do not recover the same β when β_{WP} is constant, i.e. when μ satisfies a Poincaré inequality, while using (2.9) with $s = 0$ yields the correct value.

A still worse remark is that the previous proposition cannot be always used in conjunction with proposition 2.2. Indeed if $\beta_{WP}(s) \geq c/s$ as it is the case in the second case of example 1.8 the η obtained in proposition 2.2 is not integrable.

Let us look at some other example.

Example 2.6 Assume that for some $p > 0$, $\beta_{WP}(s) = cs^{-1/p}$. In this case one can improve upon the result of proposition 2.5. Indeed we may replace the weak Poincaré inequality by its equivalent Nash type inequality

$$\mathrm{Var}_\mu(f) \leq c\,(p + (1/p)^p)^{\frac{1}{p+1}}\,\mu^{\frac{p}{p+1}}(|\nabla f|^2)\,\mathrm{Osc}^{\frac{2}{p+1}}(f).$$

We thus deduce

$$\mu^2(|\nabla f|^2) \leq \mu((Af)^2)\,\mathrm{Var}_\mu(f) \leq c(p)\,\mu((Af)^2)\,\mu^{\frac{p}{p+1}}(|\nabla f|^2)\,\mathrm{Osc}^{\frac{2}{p+1}}(f)$$

for some $c(p)$ that may change from line to line, so that

$$\mu^2(|\nabla f|^2) \leq c(p)\,\mu^{\frac{p+1}{p+2}}((Af)^2)\,\mathrm{Osc}^{\frac{2}{p+2}}(f)$$

and finally that μ satisfies a weak integrated Γ_2 inequality with

$$\beta(s) = c(p)\, s^{-1/(p+1)}.$$

This result is of course better than the $s^{-1/p}$ obtained by directly using proposition 2.5 and according to remark 1.6 allows to recover the correct decay for $\xi(t)$.
$\diamond$

3   The Log-Concave Case

If one wants to mimic (WPI) it seems more natural to consider another type of weak integrated Γ_2 inequalities, namely

$$\textbf{(WI } \Gamma_2 \textbf{ grad)} \quad \mu(|\nabla g|^2) \leq \beta(s)\, \mu((Ag)^2) + s\, |||\nabla g|^2||_\infty. \tag{3.1}$$

But contrary to the previous derivation it is no more true that $|||\nabla P_t f|^2||_\infty \leq |||\nabla f|^2||_\infty$ so that the analogue of (2.4) will involve $\sup_{u \leq t} |||\nabla P_u f|^2||_\infty$ which is not really tractable.

If we want to guarantee $|||\nabla P_t f|^2||_\infty \leq |||\nabla f|^2||_\infty$ a sufficient condition is that μ is log-concave, i.e. V is convex. Indeed in this case on can show (see a stochastic immediate proof in [15]) that

$$|\nabla P_t f|^2 \leq P_t^2(|\nabla f|) \leq P_t(|\nabla f|^2) \leq |||\nabla f|^2||_\infty. \tag{3.2}$$

In this case we will thus obtain the analogue of (2.5)

$$\mu(|\nabla P_t f|^2) \leq e^{-2t/\beta(s)}\, \mu(|\nabla f|^2) + s\, |||\nabla f|^2||_\infty. \tag{3.3}$$

The difference with the previous section is that (3.1) is satisfied by $\beta(s) = 0$ for $s \geq 1$. The converse function β^{-1} is thus bounded by 1, hence integrable at the origin.

Now we can use the trick described in Remark 1.6 which yields,

$$\mu(|\nabla P_t f|^2) \leq \left(\sum_{i=0}^{+\infty} \gamma_i\, \beta^{-1}(2\alpha_{i+1}\, t/\ln(1/\gamma_{i+1})) \right) |||\nabla f|^2||_\infty$$

$$= \eta(t)\, |||\nabla f|^2||_\infty. \tag{3.4}$$

Since β^{-1} is integrable at 0, we have thus obtained after a simple change of variable, provided η is integrable at infinity

$$\text{Var}_\mu(f) \leq 2 \left(\int_0^{+\infty} \eta(u)du \right) |||\nabla f|^2||_\infty$$

$$\leq \left(\sum_{i=0}^{+\infty} \frac{\gamma_i \ \ln(1/\gamma_{i+1})}{\alpha_{i+1}} \right) \left(\int_0^{+\infty} \beta^{-1}(t)dt \right) |||\nabla f|^2||_\infty . \quad (3.5)$$

As before we may choose $\gamma_i = 2^{-i}$ and $\alpha_i = \frac{6}{\pi^2}i^{-2}$ so that

$$\sum_{i=0}^{+\infty} \frac{\gamma_i \ \ln(1/\gamma_{i+1})}{\alpha_{i+1}} = \kappa$$

where κ is thus a universal constant. Hence if β^{-1} is integrable with integral equal to M_β we have obtained

$$\text{Var}_\mu(f) \leq \kappa M_\beta |||\nabla f|^2||_\infty . \quad (3.6)$$

As first shown by E. Milman in [35], for log-concave measures (3.6) implies a Poincaré inequality. A semi-group proof of E. Milman's result was then given by M. Ledoux in [30]. Another semi-group proof and various improvements were recently shown in [16]. We shall follow the latter to give a precise result.

Starting with

$$\mu(|f - \mu(f)|) \leq \text{Var}_\mu^{1/2}(f) \leq \kappa^{1/2} M_\beta^{\frac{1}{2}} |||\nabla f|||_\infty ,$$

we deduce from [16] Theorem 2.7 that the $\mathbb{L}^1$ Poincaré constant $C'_C(\mu)$ is less than $16 \sqrt{\kappa M_\beta / \pi}$. Using Cheeger's inequality $C_P(\mu) \leq 4 (C'_C(\mu))^2$ we have thus obtained

Proposition 3.1 *Assume that μ is log-concave and satisfies a weak integrated Γ_2 inequality (WI Γ_2 grad) (3.1). Then*

$$C_P(\mu) \leq \frac{1024}{\pi^2} \kappa M_\beta ,$$

where κ is some (explicit) universal constant and $M_\beta = \int_0^{+\infty} \beta^{-1}(t)dt$.

It turns out that there always exists a (non necessarily optimal) function β such that (3.1) is satisfied for a log-concave measure μ

Indeed recall (1.10) and (1.12). We have

$$\mu((Af)^2) = \mu(\|Hessf\|^2_{HS}) + \mu(\langle \nabla f, Hess V \nabla f \rangle)$$

$$\geq \frac{1}{u} \mu \left(|\nabla f|^2 \mathbf{1}_{\lambda_{min}(HessV)\geq\frac{1}{u}} \right)$$

$$\geq \frac{1}{u} \mu(|\nabla f|^2) - \frac{1}{u} \mu \left(\lambda_{min}(HessV) \leq \frac{1}{u} \right) \||\nabla f|^2\|_\infty .$$

It follows

$$\mu(|\nabla f|^2) \leq \beta(s) \mu((Af)^2) + s \||\nabla f|^2\|_\infty$$

where, using $\||Hess^{-1}V\|| = 1/\lambda_{min}(HessV)$ one has

$$\beta^{-1}(s) = \mu(\||Hess^{-1}V\|| \geq s). \tag{3.7}$$

Since

$$\mu(\||Hess^{-1}V\||) = \int_0^{+\infty} \mu(\||Hess^{-1}V\|| \geq s) \, ds$$

we have obtained

Corollary 3.2 *If μ is log-concave and such that $\mu(\||Hess^{-1}V\||) < +\infty$, then*

$$C_P(\mu) \leq C_{univ} \, \mu(\||Hess^{-1}V\||),$$

for some universal constant C_{univ}.

This result is not new and as remarked by E. Milman is an immediate consequence of the fact that (3.6) implies that μ satisfies some Poincaré inequality and of one of the favorite inequality of M. Ledoux, namely the Brascamp-Lieb inequality

$$Var_\mu(f) \leq \mu(\langle \nabla f, Hess^{-1}V \nabla f \rangle) \leq \mu(\||Hess^{-1}V\||) \||\nabla f|^2\|_\infty .$$

Actually this method furnishes a slightly better pre-constant than the one obtained with our method (since our $\kappa \geq 1$).

Still in the log-concave situation, if we assume (2.2) we may derive another control for the Poincaré constant.

Proposition 3.3 *Assume that μ is log-concave and satisfies a weak integrated Γ_2 inequality (WI Γ_2Osc) (2.2). If in addition there exists a function $s(t)$ such that*

$$\int_0^{+\infty} s(t) \, dt = \frac{s_0}{2} < \frac{1}{12} \quad and \quad \int_0^{+\infty} e^{-2t/\beta(s(t))} \, dt = \kappa/2 < +\infty,$$

then

$$C_P(\mu) \leq \frac{64 \ln(2)\,\kappa}{(1 - 6s_0)^2}\,. \tag{3.8}$$

Proof Starting with (2.5) in the simplified form

$$\mu(|\nabla P_t f|^2) = F(t) \leq e^{-2t/\beta(s(t))}\,\mu(|\nabla f|^2) + s(t)\,\mathrm{Osc}^2(f)\,,$$

we get

$$\mathrm{Var}_\mu(f) \leq \kappa\,\mu(|\nabla f|^2) + s_0\,\mathrm{Osc}^2 f$$

so that the conclusion follows from [16] Theorem 9.2.14. □

Still in the log-concave case it was shown by M. Ledoux in [29] that

$$|||\nabla P_t f|||_\infty \leq \frac{1}{\sqrt{2t}}\,||f||_\infty$$

so that replacing f by $f - a$ with $a = \frac{1}{2}(\inf f + \sup f)$ we have

$$|||\nabla P_t f|||_\infty \leq \frac{1}{2\sqrt{2t}}\mathrm{Osc}(f)\,.$$

This bound was improved in [15] replacing $\sqrt{2}$ by $\sqrt{\pi}$ and is one of the key element in the proof of Theorem 2.7 in [16].

We may combine this bound with the (WI $\Gamma_2$grad) inequality in order to improve upon the previous result. If a (WI $\Gamma_2$grad) inequality is satisfied we have

$$\mu(|\nabla P_{2t} f|^2) \leq e^{-2t/\beta(s)}\,\mu(|\nabla P_t f|^2) + s\,|||\nabla P_t f|^2||_\infty$$
$$\leq e^{-2t/\beta(s)}\,\mu(|\nabla f|^2) + \frac{s}{4\pi t}\,\mathrm{Osc}^2(f)\,.$$

We have thus obtained

Proposition 3.4 *Assume that μ is log-concave and satisfies a weak integrated Γ_2 inequality (WI Γ_2 grad) (3.1). If in addition there exists a function $s(t)$ such that*

$$\int_0^{+\infty} \frac{s(t)}{4\pi t}\,dt = \frac{s_0}{4} < \frac{1}{24} \quad and \quad \int_0^{+\infty} e^{-2t/\beta(s(t))}\,dt = \kappa/4 < +\infty\,,$$

then

$$C_P(\mu) \leq \frac{64 \ln(2)\,\kappa}{(1 - 6s_0)^2}\,. \tag{3.9}$$

In the previous proposition we can choose a generic function $s(t)$ given by

$$s(t) = \frac{\theta}{16} \left(t \, \mathbf{1}_{t \leq 2} + \ln^{-(1+\theta)}(t) \, \mathbf{1}_{t > 2} \right) , \tag{3.10}$$

so that

$$\int_0^{+\infty} \frac{s(t)}{4\pi \, t} \, dt = \frac{\theta}{32\pi} + \frac{1}{64\pi \, \ln^\theta(2)} \leq \frac{1}{48}$$

as soon as $0 < \theta \leq 1$. So we may always choose

$$\kappa = 4 \left(2 + \int_2^{+\infty} e^{-2t / \beta((\theta/16) \, \ln^{-(1+\theta)}(t))} dt \right) , \quad s_0 = \frac{1}{12} , \quad C_P(\mu) \leq 256 \ln(2) \, \kappa .$$
$$\tag{3.11}$$

As we previously saw, we may also use the previous proposition with

$$\beta^{-1}(s) = \mu(|||Hess^{-1}V||| \geq s) .$$

This yields

Corollary 3.5 *If* μ *is log-concave and such that* $M_\varepsilon := \mu(\ln^{1+\varepsilon}(1 + |||Hess^{-1}V|||)) < +\infty$ *for some* $\varepsilon > 0$, *then*

$$C_P(\mu) \leq c + 4 \max \left(2, \exp \left(\left[\frac{2^\varepsilon \, 64 \, M_\varepsilon}{\theta} \right]^{\frac{1}{\varepsilon - \theta}} \right) \right) ,$$

with $\theta = 1$ *if* $\varepsilon \geq 2$ *and* $\theta = \varepsilon/2$ *if* $\varepsilon \leq 2$, *where* c *is some universal constant.*

Proof Denote by $M_\varepsilon = \mu(\ln^{1+\varepsilon}(1 + |||Hess^{-1}V|||))$. According to Markov inequality

$$\beta^{-1}(s) \leq \frac{M_\varepsilon}{\ln^{1+\varepsilon}(1 + s)} .$$

It follows

$$\beta(t) \leq \exp \left[\left(\frac{M_\varepsilon}{t} \right)^{\frac{1}{1+\varepsilon}} \right]$$

so that for $t \geq 2$,

$$\beta(s(t)) \leq \exp \left[\left(\frac{8 M_\varepsilon}{\theta} \right)^{\frac{1}{1+\varepsilon}} \ln^{\frac{1+\theta}{1+\varepsilon}}(1 + t) \right] .$$

In particular, using $t^2 \geq t + 1$ for $t \geq 2$,

$$\frac{1}{2} \ln(t) \geq \left(\frac{8M_\varepsilon}{\theta}\right)^{\frac{1}{1+\varepsilon}} \ln^{\frac{1+\theta}{1+\varepsilon}}(1+t)$$

as soon as

$$t \geq \max\left(2, \exp\left[\frac{2^\varepsilon \, 64 \, M_\varepsilon}{\theta}\right]^{\frac{1}{\varepsilon-\theta}}\right).$$

For such t's we thus have

$$e^{-2t/\beta(s(t))} = e^{-2\exp(\ln(t)-\ln(\beta(s(t))))} \leq e^{-2\sqrt{t}}$$

so that finally

$$\frac{\kappa}{4} \leq \max\left(2, \exp\left[\frac{2^\varepsilon \, 64 \, M_\varepsilon}{\theta}\right]^{\frac{1}{\varepsilon-\theta}}\right) + \int_2^{+\infty} e^{-2\sqrt{t}}\, dt\,.$$

Hence the result choosing $\theta = 1$ if $\varepsilon \geq 2$ and $\theta = \varepsilon/2$ otherwise. □

Of course our bounds are far from being sharp. Notice that the previous corollary allows to look at Subbotin distributions $\mu(dx) = Z^{-1} e^{-|x|^p}\, dx$ for large p's, while Brascamp-Lieb inequality cannot be used. However other known methods (see e.g. S. Bobkov's results on radial measures in [9]) furnish better bounds in this case. Of course the previous corollary covers non radial cases.

Remark 3.6 If $M_\varepsilon := \mu(\||Hess^{-1}V\||^\varepsilon) < +\infty$ for some $\varepsilon > 0$ we can obtain another explicit bound choosing $\theta = 1$ in (3.10). Using again Markov inequality we have $\beta(s) \leq (M_\varepsilon/s)^{1/\varepsilon}$ so that

$$\int_2^{+\infty} e^{-2t/\beta(s(t))}\, dt \leq \int_2^{+\infty} e^{-2t/\ln^{2/\varepsilon}(M_\varepsilon^{1/\varepsilon} t)}\, dt$$

$$\leq \frac{1}{2} M_\varepsilon^{1/\varepsilon} \ln^{2/\varepsilon}(M_\varepsilon^{2/\varepsilon}) + \int_2^{+\infty} e^{-2t/\ln^{2\varepsilon}(t^2)}\, dt$$

and finally

$$C_P(\mu) \leq c(\varepsilon) + \max\left(2, \frac{1}{2} M_\varepsilon^{1/\varepsilon} \ln^{2/\varepsilon}(M_\varepsilon^{2/\varepsilon})\right).$$

Notice that for $\varepsilon = 1$ we recover a slightly worse result than corollary 3.2 since an extra logarithm appears. Of course choosing $s(t)$ with a slower decay, we may improve upon this result but it seems that in all cases an extra worse term always appears. In addition constants are quite bad. But of course the result is new for $\varepsilon < 1$. ◇

4 Some Applications: Perturbation of Product Measures and Radial Measures

We will first recall how the (usual) integrated Γ_2 criterion can be used in order to relate the Poincaré constant of μ to the ones of its one dimensional conditional distributions, in some special situations. We copy here Proposition (3.1) in [28] and its proof to see how to potentially extend it. In the sequel we denote

$$SG(\mu) = \frac{1}{C_P(\mu)}$$

the spectral gap of μ.

Proposition 4.1 (M. Ledoux) *Let* $\mu(dx) = Z^{-1} e^{-W(x)-\sum_{i=1}^n h_i(x_i)} dx = Z^{-1} e^{-V(x)} dx$ *be a probability measure on* $\mathbb{R}^n$, *W and the h_i's being C^2. Introduce the one dimensional conditional distributions*

$$\eta_{i,x}(dt) = Z_{i,x}^{-1} e^{-W(x_1,...,x_{i-1},t,x_i,...x_n)-h_i(t)} dt.$$

Let

$$S = \inf_{i,x} SG(\eta_{i,x}).$$

Assume that $Hess\,W(x) \geq w$ and $\max_i \partial_{ii}^2 W(x) \leq \bar{w}$ for all $x \in \mathbb{R}^n$. Then

$$SG(\mu) \geq S + w - \bar{w}.$$

Proof It holds

$$\Gamma_2 f = \sum_{i,j}(\partial_{ij}^2 f)^2 + \sum_i h_i''(x_i)(\partial_i f)^2 + \langle \nabla f, Hess\,W \nabla f \rangle$$

$$\geq \sum_i (\partial_{ii}^2 f)^2 + \sum_i h_i''(x_i)(\partial_i f)^2 + w|\nabla f|^2 \qquad (4.1)$$

$$\geq \sum_i (\partial_{ii}^2 f)^2 + \sum_i (h_i''(x_i) + \partial_{ii}^2 W)(\partial_i f)^2 + (w - \bar{w})|\nabla f|^2$$

$$\geq \sum_i \Gamma_{2,i} f + (w - \bar{w})|\nabla f|^2.$$

It follows

$$\mu((Af)^2) = \mu(\Gamma_2 f) \geq \sum_i \mu(SG(\eta_{i,x}) |\partial_i f|^2) + (w - \bar{w}) \mu(|\nabla f|^2) \quad (4.2)$$

$$\geq (S + w - \bar{w}) \mu(|\nabla f|^2),$$

hence the result applying Theorem 1.3. $\qquad\qquad\qquad\qquad\qquad\qquad\qquad$ □

Remark 4.2 Choosing $W = 0$ the previous result contains the renowned tensorization property of Poincaré inequality if $\mu = \otimes_i \mu_i$,

$$C_P(\otimes_i \mu_i) \leq \max_i C_P(\mu_i).$$

Similar results for weak Poincaré inequalities involve a "dimension dependence" (see e.g. [7]). $\qquad\qquad\qquad\qquad\qquad\qquad\qquad\qquad\qquad\qquad\qquad\qquad\qquad\qquad$ ◇

Remark 4.3 For the proof of proposition 4.1 to be rigorous, it is enough to assume that ESA is satisfied for $C_0^\infty(\mathbb{R}^n)$ (which is implicit in M. Ledoux's work). Indeed in this case one only has to consider such test functions. The delicate point in the previous proof is that one has to check

$$\mu(\Gamma_{2,i} f) = \mu((A_i f)^2)$$

where $A_i f = \partial_{ii}^2 f - (h'(x_i) + \partial_i W)\partial_i f$ in order to use the integrated Γ_2 criterion. If f is compactly supported, this is immediate as we already discussed in the introduction. Hence for $D = \mathbb{R}^n$, (H) ensures that the result holds true.

The case of a bounded domain D will be discussed later. $\qquad\qquad\qquad\qquad\qquad$ ◇

In the previous proof, assume that $w = 0$ (W is convex), we thus obtain

$$\Gamma_2 f \geq \sum_i h_i''(x_i) (\partial_i f)^2$$

which is interesting only if the right and side is non-negative for any f, i.e. if $h_i'' \geq 0$. Hence as we did for obtaining (3.7) we have for $u > 0$, since we may integrate w.r.t. μ,

$$\mu(|\nabla f|^2) \leq u \, \mu((Af)^2) + \mu \left(\min_i(h_i''(x_i) \leq 1/u) \right) |||\nabla f|^2||_\infty \quad (4.3)$$

that furnishes a (WI Γ_2 grad) inequality. Of course

$$\mu \left(\min_i(h_i''(x_i) \leq 1/u) \right) \leq n \, \max_i \mu \left(h_i''(x_i) \leq \frac{1}{u} \right).$$

We have seen that such a weak inequality is interesting provided on one hand μ is log-concave and on the other hand $u \mapsto \max_i \mu \left(h_i''(x_i) \le \frac{1}{u} \right)$ which is clearly non-increasing goes to 0 as $u \to +\infty$. We will thus assume that all h_i are convex, yielding thanks to Proposition 3.4 with the choice (3.10) with $\theta = 1$.

Lemma 4.4 *Let* $\mu(dx) = Z^{-1} e^{-W(x)-\sum_{i=1}^{n} h_i(x_i)} dx = Z^{-1} e^{-V(x)} dx$ *be a probability measure on* $\mathbb{R}^n$, *W and the h_i's being convex and C^2. Define*

$$\alpha(v) = \max_i \mu(h_i''(x_i) \le v)$$

and assume that (the non-decreasing) α goes to 0 as $v \to 0$. Then

$$C_P(\mu) \le 256 \ln(2)\,\kappa$$

with

$$\kappa = 4 \left(2 + \int_2^{+\infty} e^{-2t/\alpha^{-1}(1/16n\,\ln^2(t))}\,dt \right).$$

Let us illustrate this situation in the particular case $h_i(u) = |u|^p$ for $p > 1$. We immediately see that the situation is completely different depending on whether $p < 2$ or $p > 2$. Denote by μ_i the probability distribution of x_i under μ. For $p < 2$ we have to control the tails of μ_i while for $p > 2$ we have to control the mass of small intervals centered at the origin.

Remark 4.5 For $p < 2$, $h_p : u \mapsto |u|^p$ is not C^2. But if $p > 1$, the only problem lies at the origin, and using that h_p'' is integrable at the origin it is not difficult to check (regularizing h_p at the origin for instance) that all what was done above is still true. $\qquad\qquad\qquad\qquad\qquad\qquad\qquad\qquad\qquad\qquad\qquad\qquad \diamond$

More generally we may consider h_i's who satisfy similar concentration bounds. Let us state a first result

Proposition 4.6 *Let* $\mu(dx) = Z^{-1} e^{-W(x)-\sum_{i=1}^{n} h_i(x_i)} dx$ *be a probability measure on* $\mathbb{R}^n$. *We assume that the h_i's are even convex functions. In addition we assume that for all $i = 1, \dots, n$,*

$$h_i''(u) \ge \rho(|u|)$$

where ρ is a non-increasing positive function going to 0 at infinity. Then for all even convex function W it holds

$$C_P(\mu) \le 4 \left(2 + \int_2^{+\infty} e^{-2t\,\rho(\sqrt{2\max_i C_P(\eta^i)}\,\ln(n\,\ln^2(t)))}\,dt \right),$$

where $\eta^i(du) = Z_i^{-1} e^{-h_i(u)}\,du$.

Proof According to Prekopa-Leindler theorem we know that the i-th marginal law μ_i of μ, i.e. the μ distribution of x_i, is a one dimensional distribution, that can be written

$$\mu_i(du) = Z_i^{-1} \, \rho_i(u) \, e^{-h_i(u)} \, du \, , \tag{4.4}$$

with an even and log-concave (thus non increasing on $\mathbb{R}^+$) function ρ_i. For such one dimensional distributions we may use a remarkable result due to O. Roustant, F. Barthe and B. Ioos (see [39]) recalled in proposition 6 of [6], namely

Lemma 4.7 (Roustant-Barthe-Ioos) *Let $\eta(du) = e^{-V(u)} \, \mathbf{1}_{(-b,b)}(u) \, du$ be a probability measure on $\mathbb{R}$, with V a continuous and even function. For any even function ρ which is non-increasing on $\mathbb{R}^+$ and such that $v(du) = \rho(u) \, \eta(du)$ is a probability measure, it holds*

$$C_P(v) \leq C_P(\eta) \, .$$

Applying the lemma we get

$$C_P(\mu_i) \leq C_P(Z^{-1} e^{-h_i(u)} \, du) := C_P(\eta^i) \, . \tag{4.5}$$

We can thus use the concentration of measure property obtained via the Poincaré inequality, first shown by Bobkov and Ledoux [10]. Here we use an explicit form we found in [4] (4.4.6). Since $u \mapsto u$ is 1-Lipschitz and centered (again thanks to symmetry), it yields

$$\mu(h''(x_i) \leq 1/u) \leq \mu(|x_i| \geq \rho^{-1}(1/u)) \leq 6 \exp\left(-\frac{\rho^{-1}(1/u)}{\sqrt{C_P(\eta^i)}}\right) \, . \tag{4.6}$$

We thus have for $v > 0$

$$v \, \mu((Af)^2) + 6n \, \max_i \exp\left(-\frac{\rho^{-1}(1/v)}{\sqrt{C_P(\eta^i)}}\right) \, \|\nabla f\|^2\|_\infty \geq \mu(|\nabla f|^2) \, , \tag{4.7}$$

yielding, for $s > 0$ small enough,

$$\beta(s) = \frac{1}{\rho\left(\sqrt{\max_i C_P(\eta^i) \, \ln(6n/s)}\right)} \, . \tag{4.8}$$

It remains to use the lemma 4.4. □

Remark 4.8 When $h_i(u) = |u|^p$ for some $1 < p \le 2$, one knows that $C_P(\eta^i) \le \frac{4}{p^{2(1-1/p)}}$ according to [11] Theorem 2.1. It follows that for some (explicit) constant $c(p)$,

$$C_P(\mu) \le \frac{c(p)}{p(p-1)} \left(1 + \ln^{2-p}(6n)\right).$$

The study of such μ's is not new. A much better result has been recently shown by F. Barthe and B. Klartag (see Theorem 1 in [6]),

Theorem 4.9 *(Barthe-Klartag) Let* $\mu(dx) = Z^{-1} e^{-W(x) - \sum_{i=1}^{n} |x_i|^p} dx$ *be a probability measure. We assume that* $1 \le p \le 2$ *and that* W *is an even convex function. Then*

$$C_P(\mu) \le C \ln^{\frac{2-p}{p}} \left(\max(n, 2)\right),$$

where C is some universal constant.

The key point here is naturally that the result holds true for any even and convex W. The proof by Barthe and Klartag lies on a lot of properties of log-concave measures and uses in particular the extension of the gaussian correlation inequality shown by Royen, to mixtures of gaussian measures. We of course refer the reader to [6]. We do not only loose something on the power of the logarithm, but the constant becomes infinite as p goes to 1, which is natural since the Γ_2 requires some strict convexity except at some point. However our result does not require the full machinery of gaussian mixtures, and shows that the result only depends on the behaviour of the second derivative of the h's at infinity.   ◇

Remark 4.10 In the previous proof we implicitly have used the fact that (H) is satisfied. We know that it is the case when $W \in C^2(\mathbb{R}^n)$. If W is only continuous, we may replace W by $W_\varepsilon = W * \gamma_\varepsilon$ where γ_ε is a tiny centered gaussian density. W_ε is still even and convex, so that the Theorem applies. Since the bound does not depend on ε we may take limits in the corresponding Poincaré inequalities and get the same bound for W.   ◇

Remark 4.11 Let now consider the case $p > 2$. This time we have to control

$$\mu\left(|x_i| \le (p(p-1)u)^{-1/(p-2)}\right),$$

for large u's.

Using (4.4) and since ρ_i is even and log-concave, we see that

$$\mu\left(|x_i| \le u^{-1/(p-2)}\right) \le Z_i^{-1} \rho_i(0) u^{-1/(p-2)}.$$

But

$$Z_i^{-1}\rho_i(0) = \frac{\int e^{-W(x_1,\ldots,x_{i-1},0,x_{i+1},\ldots,x_n)-\sum_{j\neq i}|x_j|^p}\prod_{i\neq j}dx_j}{\int e^{-W(x)-\sum_j|x_j|^p}\prod_j dx_j}.$$

Denote by

$$\alpha = \max_i Z_i^{-1}\rho_i(0). \tag{4.9}$$

Then we get

$$\beta(s) = \frac{1}{p(p-1)}\left(\frac{\alpha n}{s}\right)^{p-2}$$

so that we have to estimate (choosing $\theta = 1$ in (3.10))

$$\int_2^{+\infty} \exp\left(-\frac{2p(p-1)}{(\alpha n)^{p-2}}\frac{t}{\ln^{2(p-2)}(t)}\right) dt.$$

Using that $t/\ln^k(t)$ is bounded below by $c(k,\varepsilon)t^{1-\varepsilon}$ for any $\varepsilon > 0$, we easily obtain

Proposition 4.12 *Let $\mu(dx) = Z^{-1}e^{-W(x)-\sum_{i=1}^n|x_i|^p}dx$. We assume that $p > 2$ and that W is convex and even so that μ is log-concave. Then for all $\varepsilon > 0$, there exists a constant $c(p,\varepsilon)$ such that*

$$C_P(\mu) \leq c(p,\varepsilon)(\alpha n)^{(p-2)(1+\varepsilon)}$$

where

$$\alpha = \max_i \frac{\int e^{-W(x_1,\ldots,x_{i-1},0,x_{i+1},\ldots,x_n)-\sum_{j\neq i}|x_j|^p}\prod_{i\neq j}dx_j}{\int e^{-W(x)-\sum_j|x_j|^p}\prod_j dx_j}.$$

For instance if we assume that $t \mapsto W(x_1,\ldots t,\ldots,x_n)$ is β Hölder continuous, uniformly in x and i, using $|W(x_1,\ldots t,\ldots,x_n) - W(x_1,\ldots 0,\ldots,x_n)| \leq L|t|^\beta$, we get

$$\alpha \leq \frac{1}{\int e^{-L|t|^\beta-|t|^p}dt}.$$

The previous result may have some interest only if $2 < p < 3$. This is also quite natural: for large p's, $|x|^p$ becomes flat near the origin so that one cannot expect to use some convexity approach.

The best general control (thus including the case $p > 2$ for Subbotin distributions) is obtained in Theorem 18 of [6], and says that

$$C_P(\mu) \leq c\, n \, \max_i (C_P(\nu_i)).$$

In addition, in subsection 3.4 of [6], it is shown that the factor n is optimal by considering log concave perturbations of Subbotin distributions ν_i with exponent p for large p's for which $C_P(\mu)$ is at least of order $n^{(p-2)/p}$.

However if W is unconditional (i.e. $W(\sigma x) = W(x)$ for all $\sigma \in \{-1, 1\}^n$), one can deeply reinforce the previous result and show that $C_P(\mu) \leq \max_i(C_P(\nu_i))$ as shown in [6] Theorem 17.   ◇

Remark 4.13 Denote by $Cov(\mu)$ the covariance matrix, i.e. $Cov_{i,j}(\mu) = \mu(x_i x_j) - \mu(x_i)\mu(x_j)$. It is immediate that $\lambda_{max}(Cov(\mu)) \leq C_P(\mu)$. Our proof thus gives an universal bound (that does not depend on W) for the Covariance matrix. The proofs by Barthe and Klartag use first estimates for this covariance matrix.   ◇

Looking at log-concave perturbations of log-concave product measures as above, can be partly motivated by statistical issues. We refer to [17] (in particular the final section) for some of them. Of course looking at product measures is interesting thanks to the tensorization property of Poincaré inequality, furnishing dimension free bounds. For log-concave measures, another case is well understood since S. Bobkov's work [9], namely radial measures. The following version is due to Bonnefont et al. [12, Theorem 1.2]

Theorem 4.14 (Bobkov, Bonnefont-Joulin-Ma) *Let μ be a spherically symmetric (radial) log-concave probability measure on $\mathbb{R}^n$, $n \geq 2$. Then*

$$C_P(\mu) \leq \frac{\mu(|x|^2)}{n-1}.$$

We can obtain a result similar to Proposition 4.6 or Proposition 4.12

Theorem 4.15 *Let $\mu(dx) = Z_\mu^{-1} e^{-W(x)-h(|x|^2)} dx$ be a probability measure on $\mathbb{R}^n$. We assume that W is even and convex and that h is convex and non-decreasing on $\mathbb{R}^+$, so that μ is log-concave. W and h are also normalized so that $W(0) = h(0) = 0$ (and consequently W and h are non-negative). Introduce*

$$\nu_h(dx) = e^{-h(|x|^2)} dx.$$

There exists an universal constant c such that

$$C_P(\mu) \leq c \left(1 + \int_2^{+\infty} e^{-4t\, h'\left(\frac{1}{(c_n(\mu)\,\ln^2(t))^{2/n}} \right)} dt \right),$$

with

$$c_n(\mu) = Z_\mu^{-1} \frac{\pi^{n/2}}{n\Gamma(n/2)} \le \inf_\theta \left\{ \frac{\pi^{n/2}}{n\Gamma(n/2)} \frac{e^{\max_{|x|=\theta} W(x)}}{v_h(|x| \le \theta)} \right\}$$

$$= \inf_\theta \left\{ \frac{e^{\max_{|x|=\theta} W(x)}}{n \int_0^\theta r^{n-1} e^{-h(r^2)} \, dr} \right\}.$$

Remark 4.16 Let $\mu_\lambda(dx) = Z_\mu^{-1} \lambda^{-n} e^{-W(x/\lambda) - h(|x|^2/\lambda^2)} \, dx$ a dilation of μ. Notice that $\lambda^2 c_n^{2/n}(\mu_\lambda) = c_n(\mu)$. Since one has a factor $1/\lambda^2$ in front of h', we partly recover the homogeneity of the Poincaré constant under dilations. ◇

Proof Once again we may assume that W and h are smooth, convolving with a tiny gaussian kernel, that preserves convexity and parity. For simplicity we also assume that h' is (strictly) increasing, so that h' is one to one.

For two vectors x and y we write xy for the vector with coordinates $(xy)_i = x_i y_i$. It holds

$$\Gamma_2 f = \sum_{i,j} (\partial_{ij}^2 f)^2 + \langle \nabla f, \operatorname{Hess} W \nabla f \rangle + 4h''(|x|^2) |x \nabla f|^2 + 2h'(|x|^2)|\nabla f|^2$$

$$\ge 2h'(|x|^2)|\nabla f|^2$$

so that

$$u\,\mu((Af)^2) + \mu\left(2h'(|x|^2) \le \frac{1}{u}\right) |||\nabla f|^2||_\infty \ge \mu(|\nabla f|^2). \tag{4.10}$$

So μ satisfies a (WI Γ_2 grad) inequality, with

$$\beta^{-1}(u) = \mu\left(2h'(|x|^2) \le \frac{1}{u}\right) = \mu_r\left(h'(r^2) \le \frac{1}{2u}\right) = \mu_r\left(r \le \sqrt{(h')^{-1}(1/2u)}\right)$$

where μ_r denotes the probability distribution of the radial part of μ. We have

$$\mu_r(dv) = Z_\mu^{-1} n \omega_n v^{n-1} e^{-h(v)} \left(\int_{S^{n-1}} e^{-W(v\theta)} \sigma_n(d\theta) \right) dv$$

where σ_n denotes the uniform measure on the sphere S^{n-1} and $\omega_n = \frac{\pi^{n/2}}{n\Gamma(n/2)}$ denotes the volume of the unit (euclidean) ball. It follows, since W and h are non-negative,

$$\mu_r\left(r \le \sqrt{(h')^{-1}(1/2u)}\right) \le Z_\mu^{-1} \frac{\pi^{n/2}}{n\,\Gamma(n/2)} ((h')^{-1}(1/2u))^{n/2}$$

from which we deduce that we can choose

$$\beta(t) = \frac{1}{2\,h'((s/c_n)^{2/n})} \quad \text{with } c_n = Z_\mu^{-1}\frac{\pi^{n/2}}{n\Gamma(n/2)} .$$

It remains to apply Proposition 3.1.

The next step is thus to get some tractable bound for c_n, i.e a lower bound for Z_μ. The simplest way to do it is to use the fact that W is non-decreasing on each radial direction so that for all $\theta > 0$

$$Z_\mu \geq \int_{|x|\leq\theta} e^{-W(x)-h(|x|^2)}\,dx \geq e^{-\max_{|x|=\theta} W(x)}\,v_h(|x|\leq\theta) .$$

□

Corollary 4.17 *In particular if $h(u) = u^p$ with $p \geq 1$, we have*

$$C_P(\mu) \leq 12288\,\ln(2)\,\frac{c_n^{\frac{2(p-1)}{n}}(\mu)}{4p}\,(4(p-1))^{\frac{4(p-1)}{n}} .$$

Proof If $h(u) = u^p$ for $p > 1$, the corresponding dilation μ_λ is given by $h_\lambda(u) = \lambda^{-2p}u^p$. Recall that $c_n(\mu_\lambda) = \lambda^{-n}c_n(\mu)$.

We shall use that

$$\ln(t) \leq \frac{1}{\alpha\,2^\alpha}\,t^\alpha + (\ln(2) - (1/\alpha)) \ \text{ for } t \geq 2 \text{ and } \alpha > 0.$$

If $t \geq 2$, we thus have $\ln(t) \leq \frac{1}{\alpha 2^\alpha}\,t^\alpha$ if $\alpha \leq 1$ and $\ln(t) \leq t^\alpha$ if $\alpha \geq 1$.

It follows

$$\ln^{\frac{4(p-1)}{n}}(t) \leq c_\beta\,t^\beta$$

for $t \geq 2$ and $0 < \beta$, with

$$c_\beta = 2^{-\beta}\left(\frac{4(p-1)}{\beta n}\right)^{\frac{4(p-1)}{n}} \ \text{ if } \frac{\beta n}{4(p-1)} \leq 1 \ ; \ \ c_\beta = 1 \text{ if } \frac{\beta n}{4(p-1)} \geq 1 .$$

This yields

$$e^{-4t\,h_\lambda'\left(\frac{1}{(c_n(\mu_\lambda)\,\ln^2(t))^{2/n}}\right)} \leq e^{-\kappa_\beta\,t^{1-\beta}}$$

for

$$\kappa_\beta = \frac{4p}{c_\beta \, \lambda^2 \, c_n^{\frac{2(p-1)}{n}}(\mu)}.$$

A simple change of variables $u = \kappa_\beta \, t^{1-\beta}$, together with the positivity of all constants yields

$$\int_2^{+\infty} e^{-4t\, h'_\lambda\left(\frac{1}{(c_n(\mu_\lambda)\, \ln^2(t))^{2/n}}\right)} dt \le ((1-\beta)\kappa_\beta)^{-1} \int_0^{+\infty} u^{\frac{\beta}{1-\beta}} e^{-u^{1-\beta}} du.$$

Choosing for simplicity $\beta = 1/n$, so that $\beta n/4(p-1) \le 1$, for $n \ge 2$ the final integral is bounded independently of n for instance by $c = \int_0^{+\infty} u\, e^{-\sqrt{u}} du = 12$. It follows

$$C_P(\mu_\lambda) \le 1024 \ln(2) \left(2 + c\, \frac{\lambda^2 \, c_n^{\frac{2(p-1)}{n}}(\mu)}{4p}(4(p-1))^{\frac{4(p-1)}{n}}\right).$$

Using $C_P(\mu) = \lambda^{-2} C_P(\mu_\lambda)$ and letting λ go to infinity furnishes the result.
    For $p = 1$ the result follows from strict convexity.   □

Remark 4.18 If μ_r denotes the radial distribution of μ,

$$\mu_r(dv) = \rho(v)\, v^{n-1}\, e^{-h(v)}.$$

ρ is clearly an even function. Since for a fixed θ, $v \mapsto W(v\theta)$ is even and convex, it is non-decreasing, so that $v \mapsto \rho(v)$ is non-increasing. ρ is non necessarily log-concave, but we can again apply Proposition 6 in [6] furnishing, with $\bar{\nu}_h = Z^{-1} \nu_h$,

$$C_P(\mu_r) \le C_P(\bar{\nu}_h). \tag{4.11}$$

The measure $\bar{\nu}_h$ being log-concave, we know that

$$C_P(\bar{\nu}_h) \le 12\, \mathrm{Var}_{\bar{\nu}_h}(v).$$

What is important here is that the Poincaré constant of the radial measure μ_r can be bounded independently of W.   ◇

Remark 4.19 Is the bound in Corollary 4.17 of the good order ? To see it look at the particular case $W = 0$. In this case $Z_\mu = \frac{n\omega_n}{2p} \Gamma(n/2p)$ so that $c_n(\mu) = \frac{2p}{n\Gamma(n/2p)}$, and our bound furnishes

$$C_P(\mu) \le c\, \frac{2^{10(p-1)/n}\, p^{6(p-1)/n}}{4p\, n^{2(p-1)/n}}\, \frac{1}{\Gamma^{2(p-1)/n}(n/2p)},$$

for some universal c. In this case the following very precise bounds were obtained by Bonnefont, Joulin and Ma in [12],

$$\frac{\mu(|x|^2)}{n} \leq C_P(\mu) \leq \frac{\mu(|x|^2)}{n-1}.$$

Since

$$\mu(|x|^2) = \frac{\Gamma((n+2)/2p)}{\Gamma(n/2p)}$$

for $n/p \gg 1$ (even for large p's) we may use

$$\Gamma(z) \sim_{z \to +\infty} \sqrt{2\pi} z^{z-(1/2)} e^{-z}$$

so that the Bonnefont, Joulin, Ma theorem furnishes

$$C_P(\mu) \sim (2ep)^{-1/p} n^{1-(1/p)}. \tag{4.12}$$

For the same asymptotics our bound furnishes (for some new constant c)

$$C_P(\mu) \leq c \, p^{3(p-1)/n} n^{1-(1/p)}. \tag{4.13}$$

Hence provided $p \ln(p) \leq Cn$, we get the good order (but of course not the good constant). This shows that our bound is not so bad.. $\diamond$

5 The Case of Compactly Supported Measures

Let us come back to the proof of Proposition 4.1 starting with

$$\Gamma_2 f = \sum_{i,j} (\partial_{ij}^2 f)^2 + \sum_i h_i''(x_i)(\partial_i f)^2 + \langle \nabla f, Hess W \nabla f \rangle. \tag{5.1}$$

If W is convex, we thus have

$$\mu(\Gamma_2 f) \geq \sum_i \mu((\partial_{ii}^2 f)^2 + \sum_i h_i''(x_i)(\partial_i f)^2)$$

$$= \sum_i \mu(\eta_{i,x}((\partial_{ii}^2 f)^2 + h_i''(x_i)(\partial_i f)^2)) \tag{5.2}$$

Instead of adding and substracting $\partial_{ii}^2 W (\partial_i f)^2$, consider $\eta_{i,x}$ as a perturbation of

$$\theta_i(dt) = z_i^{-1} e^{-h_i(t)} dt$$

using the notation

$$\eta_{i,x}(dt) = Z_{i,x}^{-1} e^{-W_{i,x}(t)} \theta_i(dt).$$

Since we integrate a non-negative quantity it holds

$$\eta_{i,x}((\partial_{ii}^2 f)^2 + h_i''(x_i)(\partial_i f)^2) \geq e^{-\sup W_{i,x}} \theta_i((\partial_{ii}^2 f)^2 + h_i''(x_i)(\partial_i f)^2)$$

$$\geq e^{-\sup W_{i,x}} SG(\theta_i)\, \theta_i((\partial_i f)^2)$$

$$\geq e^{-\operatorname{Osc} W_{i,x}} SG(\theta_i)\, \eta_{i,x}((\partial_i f)^2), \qquad (5.3)$$

provided

$$\theta_i(\Gamma_2 g) = \theta_i((L_i g)^2)$$

with $L_i g = g'' - h_i' g'$. Notice since $W_{i,x}$ is convex, its Oscillation cannot be bounded on $\mathbb{R}$, unless $W_{i,x}$ is constant. Hence the previous result has no interest on $\mathbb{R}^n$ and we shall only consider the case where the process lives in a bounded domain D.

We have thus obtained some variation of the renowned Holley-Stroock perturbation result namely

Proposition 5.1 *Let* $\mu(dx) = Z^{-1} e^{-W(x) - \sum_{i=1}^n h_i(x_i)} \mathbf{1}_D(x)\, dx$ *be a probability measure on the hypercube* $D = \prod_i]a_i, b_i[$ *. Assume that*

(G1) For all i, the one dimensional diffusion $dy_t^i = \sqrt{2}\, dB_t^i - h_i'(y_t^i) dt$ satisfies (H) on $]a_i, b_i[$ with reversible measure $\theta_i(du) = z_i^{-1} e^{-h_i(u)} \mathbf{1}_{u \in]a_i, b_i[}\, du$.
(G2) $W \in C^\infty(\mathbb{R}^n)$ and is convex.

Introduce the one dimensional conditional log-density

$$W_{i,x}(t) = W(x_1, \ldots, x_{i-1}, t, x_i, ..x_n).$$

Then

$$C_P(\mu) \leq \max_i \sup_x e^{\operatorname{Osc}(W_{i,x})} \max_i C_P(\theta_i).$$

Since $\max_i \sup_x e^{\operatorname{Osc}(W_{i,x})} \leq e^{\operatorname{Osc} W}$ we recover (provided W is convex) Holley-Stroock result for a product reference measure on a hypercube. But what is important here is that we only have to consider the Oscillation of W along lines parallel to the axes.

Proof The only thing remaining to prove is that we can work with $C_b^\infty(D)$ functions f so that $u \mapsto f(x_1, \ldots, x_{i-1}, u, x_{i+1}, \ldots, x_n)$ is also $C_b^\infty(]a_i, b_i[)$ and we may use (G1) to justify the calculations we have done before. It is thus enough to show that (H) is satisfied for the full process i.e. with $V = W + \sum_i h_i$.

Since ∇W and ΔW are bounded on $\bar{D}$, the law of X^x is absolutely continuous w.r.t. to the one of $(y^1, \ldots, y^n)$ thanks to Girsanov theory. It follows that the exit time of D is almost surely infinite since the same holds for $(y^1, \ldots, y^n)$ according to (G1). In addition the Feynman-Kac representation of the density F_T (on $C^0([0, T], D)$) is again given by the formula of Example 7.1, so that, as we have seen, (H) is satisfied.   □

Corollary 5.2 *Let $\mu(dx) = Z^{-1} e^{-W(x) - \sum_{i=1}^n h_i(x_i)} \mathbf{1}_D(x) \, dx$ be a probability measure on the hypercube $D = \prod_i]a_i, b_i[$. Assume that the h_i's and W are convex and $C_b^2(\bar{D})$. Then, with the notations of Proposition 5.1 we have*

$$C_P(\mu) \le 12 \max_i \sup_x e^{Osc(W_i,x)} \max_i C_P(\theta_i).$$

Proof As usual, using smooth approximations, we may assume that $W \in C^\infty(\mathbb{R}^n)$. We shall perturb μ in order to apply the previous proposition. To this end, on the interval $]a_i, b_i[$ define

$$h_\varepsilon^i(u) = \varepsilon \left(\frac{1}{u - a_i} + \frac{1}{b_i - u} \right).$$

Consider

$$\mu_\varepsilon(dx) = Z^{-1} e^{-W(x) - \sum_i (h_i(x_i) + h_\varepsilon^i(x_i))} \mathbf{1}_D(x) \, dx.$$

Denote $g_\varepsilon^i = h_i + h_\varepsilon^i$.

Assumptions (G1) and (G2) of Proposition 5.1 are satisfied. We already assumed (G2). In order to show (G1) it is first enough to use Feller test of non explosion for a one dimensional diffusion, i.e. to check that for $c_i = \frac{1}{2}(a_i + b_i)$,

$$\int_{c_i}^{a_i} \exp \left(\int_{c_i}^y (g_\varepsilon^i)'(u) du \right) dy = -\infty$$

(replacing a_i by b_i we similarly get $+\infty$) according for instance to [26] Chapter VI, Theorem 3.1, which is immediate. It follows that (1.1) is satisfied. In addition $(g_\varepsilon^i)' \in \mathbb{L}^2(\theta_\varepsilon^i(du))$ where $\theta_\varepsilon^i(du) = z_\varepsilon^{-1} e^{-g_\varepsilon^i(u)} \mathbf{1}_{]a_i, b_i[}(u) \, du$, so that we are in the situation of Example 7.2 ensuring that the one dimensional y in (G1) satisfies (H).

We have thus obtained

$$C_P(\mu_\varepsilon) \le \max_i \sup_x e^{Osc(W_i,x)} \max_i C_P(\theta_\varepsilon^i(dt)).$$

Using Lebesgue's bounded convergence Theorem, for all $f \in C_b^0(D)$ it holds

$$\lim_{\varepsilon \to 0} \int_D f(x) \, e^{-W(x) - \sum_{i=1}^n g_\varepsilon^i(x_i)} \, dx = \int_D f(x) \, e^{-W(x) - \sum_{i=1}^n h_i(x_i)} \, dx$$

so that using this result for f and 1, μ_ε weakly converges to μ. It follows

$$C_P(\mu) \leq \liminf_{\varepsilon \to 0} C_P(\mu_\varepsilon) \leq \max_i \sup_x e^{\mathrm{Osc}(W_{i,x})} \liminf_{\varepsilon \to 0} \max_i C_P(\theta_\varepsilon^i).$$

We may now use the fact that θ_ε^i is log-concave since both h_i and h_ε^i are convex. We thus have

$$C_P(\theta_\varepsilon^i) \leq 12 \, \mathrm{Var}_{\theta_\varepsilon^i}(x_i).$$

Once again θ_ε^i weakly converges to θ_i and since $x_i \mapsto x_i^2$ is continuous and bounded on $[a_i, b_i]$,

$$\mathrm{Var}_{\theta_\varepsilon^i}(x_i) \to \mathrm{Var}_{\theta_i}(x_i)$$

so that the conclusion follows from the immediate $Var_{\theta_i}(x_i) \leq C_P(\theta_i)$. $\qquad \square$

6 Super Γ_2 Condition

As there are weak Poincaré inequalities, Super Poincaré inequalities (SPI) have also been introduced by Wang [42] as a concise description of functional inequalities strictly stronger than Poincaré inequalities, in particular logarithmic Sobolev (or more generally F-Sobolev) inequalities.

(SPI) is often written in the following form: $\forall s > 0$, there exists a non-increasing $\beta :]0, \infty[\mapsto [1, +\infty[$ such that

$$\mu(f^2) \leq s\mu(|\nabla f|^2) + \beta(s)\mu(|f|)^2. \tag{6.1}$$

Applying (6.1) to constant functions one sees that $\beta(s) \geq 1$ for all s. Since 1 is assumed to belong to the range of β, the (SPI) inequality implies a Poincaré inequality with $C_P(\mu) \leq \beta^{-1}(1)$, and one has $\beta(s) = 1$ for $s \geq C_P(\mu)$. When $\beta(s) = ae^{b/s}$ for positive a and b, then the Super Poincaré inequality is equivalent to a logarithmic Sobolev inequality (see [23] lemma 2.5 and lemma 2.6 for a precise statement).

It is also possible to consider SPI with a L^p norm rather than the L^1 norm, so that we will introduce general (p-SPI) for $1 \leq p < 2$ and all $s > 0$,

$$\mu(f^2) \leq s\mu(|\nabla f|^2) + \beta(s)\mu(|f|^p)^{2/p}. \tag{6.2}$$

This time, (6.2) does not imply a Poincaré inequality, so that it is natural to assume in addition that $C_P(\mu) < +\infty$. In this case we have the following

Lemma 6.1 *Assume that $C_P(\mu) < +\infty$ and that the following centered (cp-SPI) inequality is satisfied for all $s > 0$,*

$$Var_\mu(f) \le s\mu(|\nabla f|^2) + \beta_c(s)\mu(|f - \mu(f)|^p)^{2/p}, \qquad (6.3)$$

where β is non increasing. Then (p-SPI) holds with $\beta(s) = 1 + 4\beta_c(s)$.

Proof Since $C_P(\mu) < +\infty$ we may choose $\beta_c(s) = 0$ for $s > C_P(\mu)$. Let f be given. It holds

$$\mu(|f - \mu(f)|^p) \le 2^{p-1}\left(\mu(|f|^p) + \mu^p(|f|)\right) \le 2^p\,\mu(|f|^p)$$

yielding

$$\mu(|f|^2) = Var_\mu(f) + \mu^2(f) \le s\mu(|\nabla f|^2) + \beta_c(s)\mu^{2/p}(|f - \mu(f)|^p) + \mu^2(|f|)$$

$$\le s\mu(|\nabla f|^2) + (4\beta_c(s) + 1)\,\mu^{2/p}(|f|^p).$$

$\square$

It is then natural to introduce an integrated super Γ_2 condition: for some $1 \le p < 2$, there exists a positive non-increasing function β such that $\forall s > 0$

$$(pSI - \Gamma_2) \qquad \mu(|\nabla f|^2) \le s\,\mu((Af)^2) + \beta(s)\mu(|f|^p)^{2/p}.$$

In the sequel we assume that $C_P(\mu) < +\infty$, so that for all $s \ge C_P(\mu)$ one may take $\beta(s) = 0$.

Let us begin by this simple proposition

Proposition 6.2 *We have the following*

(1) A $(p - SPI)$ inequality is equivalent to

$$\mu((P_t f)^2 \le e^{-2t/s}\mu(f^2) + \beta(s)\mu(|f|^p)^{2/p}(1 - e^{-2t/s}), \qquad (6.4)$$

for all $s > 0$ and all $t \ge 0$.
(2) A $(pSI - \Gamma_2)$ condition is equivalent to

$$\mu(|\nabla P_t f|^2 \le e^{-2t/s}\mu(|\nabla f|^2) + \beta(s)\mu(|f|^p)^{2/p}(1 - e^{-2t/s}). \qquad (6.5)$$

for all $s > 0$ and all $t \ge 0$.

Proof The first part is well known and is included in Wang's work [42]. The second point will follow the same line of proof. As already emphasized in the previous sections, denoting

$$F(t) = \mu(|\nabla P_t f|^2)$$

one has

$$F'(t) = -2\mu((AP_t f)^2)$$

so that the $(pSI - \Gamma_2)$ gives directly

$$F'(t) \leq -\frac{2}{s}F(t) + \frac{2\beta(s)}{s}\mu(|P_t f|^p)^{2/p}$$

and since $\mu(|P_t f|^p)^{2/p} \leq \mu(|f|^p)^{2/p}$ we conclude thanks to Gronwall's lemma. The other implication comes from differentiating with respect to time at time 0. $\quad\square$

6.1 From (p-SPI) to (pSI-Γ_2)

We follow the same proof as in Sect. 2, assuming that a $(p - SPI)$ holds, i.e. we use Cauchy-Schwartz inequality in order to get

$$\mu(|\nabla f|^2) = \mu(-fAf)$$

$$\leq \sqrt{\mu(f^2)\,\mu((Af)^2)}$$

$$\leq \left(s\,\mu(|\nabla f|^2)\,\mu((Af)^2) + \beta(s)\mu(|f|^p)^{2/p}\mu((Af)^2)\right)^{1/2}.$$

Recall now the already used following fact: if $0 \leq u \leq \sqrt{Au + B}$ then $u \leq A + B^{1/2}$. It yields

$$\mu(|\nabla f|^2) \leq s\,\mu((Af)^2) + \sqrt{\beta(s)\mu(|f|^p)^{2/p}\mu((Af)^2)}$$

$$\leq \frac{3}{2}s\,\mu((Af)^2) + \frac{\beta(s)}{2s}\mu(|f|^p)^{2/p}.$$

We thus see that we have "lost" a factor $1/s$ but if we think to the logarithmic Sobolev inequality, it roughly means the loss of a constant.

6.2   From (pSI-Γ_2) to (p-SPI)

Starting with

$$\mu(|\nabla P_t f|^2 \leq e^{-2t/s}\mu(|\nabla f|^2) + \beta(s)\mu(|P_t f|^p)^{2/p}(1 - e^{-2t/s})$$

and using

$$\mathrm{Var}_\mu(f) = 2\int_0^\infty \mu(|\nabla P_u f|^2)du$$

we get

$$\mathrm{Var}_\mu(f) \leq 2\int_0^\infty e^{-2u/s}\mu(|\nabla f|^2)du + 2\beta(s)\int_0^\infty \mu(|P_u f|^p)^{2/p}(1 - e^{-2u/s})du.$$

Assume first that f is centered. If $p > 1$ then Poincaré inequality implies back an exponential convergence in L^p norm (see [21] Theorem 1.3) so that for all centered f we get

$$\mu(f^2) \leq s\mu(|\nabla f|^2) + K_p\beta(s)\mu(|f|^p)^{2/p}$$

where K_p depends on p and is going to infinity as p goes to 1. Applying Lemma 6.1 we thus obtain

$$\mu(|f|^2) \leq s\mu(|\nabla f|^2) + (1 + 4K_p\beta(s))\mu(|f|^p)^{2/p}.$$

7   Appendix: About the Heart of Darkness

Let us come back to the framework of this work especially Proposition 1.2.

First of all, if (H) is satisfied, according to Theorem 2.2.25 and its proof in Royer's book [40] (also see the english version [41]), the following holds

(A1) P_t extends to a μ-symmetric continuous Markov semi-group $e^{-t\tilde{A}}$ on $\mathbb{L}^2(\mu)$. We denote by $\mathcal{D}(\tilde{A})$ the domain of the generator $\tilde{A}$ of this $\mathbb{L}^2(\mu)$ semi-group.
(A2) Any $f \in C^2(D)$ such that $|\nabla f|$ is bounded and $Af \in \mathbb{L}^2(\mu)$ belongs to $\mathcal{D}(\tilde{A})$, and $\tilde{A}f = Af$.
(A3) If $f \in \mathcal{D}(\tilde{A})$ the set of Schwartz distributions on D, then $f \in \mathcal{D}'(D)$ and satisfies $\tilde{A}f = Af$ in $\mathcal{D}'(D)$.

Actually Royer only considers the case $D = \mathbb{R}^n$, but the key point in the proof is that one can apply Ito's formula for such an f up to time t (without any stopping time) which is ensured by the conservativeness in D.

In the case $D = \mathbb{R}^n$ the proof of ESA for C_0^∞ the set of smooth compactly supported functions is contained in [43] using an elliptic regularity result Theorem 2.1 in [24] (actually the latest result certainly appeared in other places). The proof is explained in Theorem 2.2.7 of [40] (also see Proposition 3.2.1 in [4]) when V is C^∞. The structure of $\mathcal{D}(\tilde{A})$ is also proved in the same Theorem.

We shall explain the proof when D is bounded, still assuming for simplicity that $V \in C^\infty(D)$. The same elliptic regularity should be used to extend the result to $V \in C^2(D)$, but will introduce too much intricacies to be explained here.

Proof First consider the Dirichlet form $\mathcal{E}(f, g) = \mu(\langle \nabla f, \nabla g \rangle)$ whose domain is the closure of $C_0^\infty(D)$ denoted by $H_0^1(\mu, D)$. Since $\mathcal{E}$ is regular, Fukushima's theory (see [25]) allows us to build a symmetric Hunt process associated to $(\mathcal{E}, H_0^1(\mu, D))$. This process is then a solution to the martingale problem associated to A and $C_0^\infty(D)$. Since T_∂^x is almost surely infinite, this martingale problem has an unique solution given by the (distribution) of the stochastic process X^x.

In order to prove ESA it is enough to show that if $g \in \mathbb{L}^2(\mu)$ satisfies $\mu(g (A\varphi - \varphi)) = 0$ for all $\varphi \in C_0^\infty(D)$ then g vanishes (see the beginning of the proof in [40, p. 31]). According to the proof in [40, p. 31], it implies in particular that $g \in \mathcal{D}'(D)$ and satisfies $Ag = g$. Using that A is hypoelliptic since $V \in C^\infty(D)$, we deduce that $g \in C^\infty(D)$.

Using Ito's formula (since the process is conservative) we have

$$\sqrt{2} \int_0^t \langle \nabla g(X_s), dB_s \rangle = g(X_t) - g(X_0) - \int_0^t g(X_s)\, ds \tag{7.1}$$

almost surely. If X_0 is distributed according to μ, the right hand side belongs to $\mathbb{L}^2(\mathbb{P})$ ($\mathbb{P}$ being the underlying probability measure on the path space), so that the left hand side also belongs to $\mathbb{L}^2(\mathbb{P})$. The $\mathbb{L}^2$ norm of this left hand side is equal to $2t\, \mu(|\nabla g|^2)$ so that $\nabla g \in \mathbb{L}^2(\mu)$.

As a consequence

$$t \mapsto \int_0^t \langle \nabla g(X_s), dB_s \rangle$$

is a $\mathbb{P}$ martingale so that for all bounded h,

$$\mathbb{E}(h(X_0)g(X_t)) = \mu(gh) + \int_0^t \mathbb{E}(h(X_0)g(X_s))\, ds.$$

Since a regular disintegration of $\mathbb{P}$ is furnished by the distribution of the X^x's, it follows

$$P_t g = g + \int_0^t P_s g\, ds$$

μ almost surely, so that $g \in \mathcal{D}(\tilde{A})$ and satisfies $\tilde{A}g = g$. Hence

$$\mu(g^2) = \mu(gAg) = -\mu(|\nabla g|^2)$$

so that $g = 0$.

The proof of the remaining part of the Theorem is the same as in [40, p. 42]. □

Finally we will indicate how to show that the semi-group is ergodic when D is bounded (we already mention a possible way for $D = \mathbb{R}^n$ in Sect. 2). If $P_t f = f$ for all $t > 0$ it follows that $f \in \mathcal{D}(A)$ and satisfies $Af = 0$ so that f is smooth thanks to hypoellipticity. Applying Ito's formula we deduce that $f(X_t^x) = f(x)$ a.s. for all $t > 0$. Thanks to the Support Theorem ([26] chapter 6 section 8) we know that the distribution of X_t^x admits a positive density w.r.t. Lebesgue measure, so that if $f(y) \neq f(x)$ for some y, hence all z in a neighborhood N of y by continuity, $\mathbb{P}(X_t^x \in N) > 0$ and thus $f(X_t^x) \neq f(x)$ with positive probability, which is a contradiction.

Let us give now some of the most important examples. In these examples we assume that $V \in C^3$.

Example 7.1 If either

(H1) $V(x) \to +\infty$ as $x \to \partial D$ (i.e. $|x| \to +\infty$ if $D = \mathbb{R}^n$), and $\frac{1}{2}|\nabla V|^2 - \Delta V$
 is bounded from below, or

(H2) $D = \mathbb{R}^n$ and $\langle x, \nabla V(x) \rangle \geq -a|x|^2 - b$ for some a, b in $\mathbb{R}$,

then (H) is satisfied. If V is convex (H2) is satisfied with $a = b = 0$.

If $D = \mathbb{R}^n$ these two cases are detailed in [40] subsection 2.2.2 (conservativeness is shown in Theorem 2.2.19 therein). In the (H1) case for a bounded domain the only thing to do is to replace the exit times of large balls by the T_k's in Lemme 2.2.21 of [40].

In all cases the law of $(X_t^x)_{t \leq T}$ is given by $dQ = F_T dP$ where P is the law of a Wiener process starting from 0 and

$$F_T = \exp\left(\frac{1}{2}V(x) - \frac{1}{2}V(x+\sqrt{2}\,W_T) + \frac{1}{2}\int_0^T (\frac{1}{2}|\nabla V|^2 - \Delta V)(x+\sqrt{2}\,W_s)ds\right). \tag{7.2}$$

◇

Example 7.2 Assume now

$$\mu(|\nabla V|^2) < +\infty. \tag{7.3}$$

(7.3) is an entropy condition related to the stationary Nelson processes (see [18–20, 33, 34]). The stationary (symmetric) conservative diffusion process is built in these papers. Conservative means here that

$$T_{\partial D} = +\infty \quad \mathbb{P}_\mu \text{ a.s} \tag{7.4}$$

i.e. if X_0 is distributed according to μ.

The proof in the bounded case is a simple modification of the one in [18]. The modification is as follows (we refer to the notations therein):

(1) First the flow v_t is stationary with $v_t = \mu$.
(2) Next the drift $B = -\nabla V$. Assuming in addition that D has a smooth boundary, one can approximate in $\mathbb{L}^2(\mu)$, B by B_k's which are $C_b^\infty(\mathbb{R}^n)$ and coincide with B on $\bar{D}_k = \{x ; , d(x, \partial D) \geq 1/k\}$ (for this we need ∂D_k to be smooth).

One can then follow the "Outline of proof" (4.9 bis) in [18] replacing the T_n therein by the T_k we have introduced before (the exit times of D_k) so that (4.14) in [18] is trivially satisfied. (4.16) is then justified when (1.1) is satisfied. The only remaining thing to prove is thus (4.10) in [18]. For $f \in C_0^\infty(\mathbb{R}^n)$ whose support contains $\bar{D}$ we may then proceed as in the proof of Theorem (4.18) in [18] in order to prove it.

In order to show the strong existence of the diffusion process starting from x (and not the stationary measure) it is enough to show (1.1) is satisfied (the strong existence of the diffusion process up to $T_{\partial D}^x$ is ensured since V is local-Lipschitz). Since the stationary process is conservative, so is X^x for μ, hence Lebesgue, almost all x. Standard results in Dirichlet forms theory show that this result extends to all x outside some polar set. Actually it is true for all x using the following (itself more or less standard 40 years ago): choose a small ball $B(x, \varepsilon)$ with $\varepsilon < d(x, \partial D)/2$ and introduce S^x the exit time of this ball. For $t > 0$ the distribution of $X_t \mathbf{1}_{t < S^x}$ has a density w.r.t. Lebesgue's measure restricted to the ball (using e.g. Malliavin calculus), hence w.r.t μ. It follows from the Markov property and (7.4)

$$\mathbb{P}(T_{\partial D}^x < +\infty) = \lim_{t \to 0} \mathbb{P}(t < S^x ; T_{\partial D}^x < +\infty) = \lim_{t \to 0} \mathbb{E}(\mathbf{1}_{t < S^x} \mathbb{E}(\mathbf{1}_{T_{\partial D}^{X_t^x} < +\infty})) = 0$$

since the second expectation in the last formula vanishes identically.

In all cases the Feynman-Kac representation of F_T in (7.2) is obtained by using Ito's formula with V which is allowed since $V \in C^3(D)$ and (1.1) again. $\diamond$

Example 7.3 If we do no more assume that the hitting time of the boundary is infinite, assumption (H) is not satisfied. The space of interest should be $H_b^1(\mu)$ the closure of $C_b^\infty(D)$ for the Dirichlet form. The corresponding process is the symmetric reflected diffusion process. A good reference is [37] where this normally reflected diffusion process is built (under much more general conditions). Assume that the boundary is smooth.

A little bit more is needed. First if $f \in H_b^1(\mu)$ and $g \in \mathcal{D}(A)$, one has, according to Fukushima's theory (see [25] (1.3.10)),

$$\mu(\langle \nabla f, \nabla g \rangle) = -\mu(f \, Ag). \tag{7.5}$$

If f is smooth $(C^2(\bar{D}))$ and belongs to $\mathcal{D}(A)$ it also holds

$$- \mu(f\,Ag) = \mu(\langle \nabla f, \nabla g \rangle) + \int_{\partial D} g \langle n_D, \nabla f \rangle e^{-V} d\sigma_D$$

according to Green's identity. Here n_D denotes the normalized inward normal vector on ∂D and σ_D denotes the surface measure on ∂D. Since the set of the traces on ∂D of bounded functions in $H^1(\mu)$ is dense in $\mathbb{L}^\infty(\sigma_D)$, we deduce that

$$\langle n_D, \nabla f \rangle_{|\partial D} = 0.$$

It is however not clear in general that $P_t f$ is smooth (even if V is). If one assumes that ∂D is C^∞ and $V \in C^\infty(\bar{D})$, $P_t f \in C^\infty(\bar{D})$ is shown in [14] Theorem 2.9 by using the method of [13] (see the proof of Theorem 2.11 therein). Notice that the proof of regularity is using Sobolev imbedding theorem, so that one should relax the C^∞ assumption but with dimension dependent regularity assumptions. Other more important difficulties will be pointed out later.

The other major difficulty is that ESA is not satisfied in general.

In comparison with the previous example, the boundary term will disappear if $e^{-V} = 0$ on ∂D. It is what happens if (H) is satisfied, but here again we do not need such a proof which is not useful in the present work. $\diamondsuit$

References

1. D. Alonso-Gutierrez, J. Bastero, *Approaching the Kannan-Lovasz-Simonovits and Variance Conjectures*. Lecture Notes in Mathematics, vol. 2131 (Springer, Berlin, 2015)
2. D. Bakry, M. Émery, Diffusions hypercontractives, in *Séminaire de Probabilités, XIX, 1983/84*. Lecture Notes in Mathematics, vol. 1123 (Springer, Berlin, 1985), pp. 177–206
3. D. Bakry, F. Barthe, P. Cattiaux, A. Guillin, A simple proof of the Poincaré inequality for a large class of probability measures. Electron. Commun. Probab **13**, 60–66 (2008)
4. D. Bakry, I. Gentil, M. Ledoux, *Analysis and Geometry of Markov Diffusion Operators*. Grundlehren der Mathematischen Wissenchaften, vol. 348 (Springer, Berlin, 2014)
5. F. Barthe, D. Cordero-Erausquin, Invariances in variance estimates. Proc. Lond. Math. Soc. **106**(1), 33–64 (2013)
6. F. Barthe, B. Klartag, Spectral gaps, symmetries and log-concave perturbations. Bull. Hellenic Math. Soc. **64**, 1–31 (2020)
7. F. Barthe, P. Cattiaux, C. Roberto, Concentration for independent random variables with heavy tails. AMRX **2005**(2), 39–60 (2005)
8. S.G. Bobkov, Isoperimetric and analytic inequalities for log-concave probability measures. Ann. Probab. **27**(4), 1903–1921 (1999)
9. S.G. Bobkov, Spectral gap and concentration for some spherically symmetric probability measures, in *Geometric Aspects of Functional Analysis, Israel Seminar 2000–2001*. Lecture Notes in Mathematics, vol. 1807 (Springer, Berlin, 2003), pp. 37–43
10. S. Bobkov, M. Ledoux, Poincaré's inequalities and Talagrand's concentration phenomenon for the exponential distribution. Probab. Theory Relat. Fields **107**(3), 383–400 (1997)
11. M. Bonnefont, A. Joulin, Y. Ma, A note on spectral gap and weighted Poincaré inequalities for some one-dimensional diffusions. ESAIM Probab. Stat. **20**, 18–29 (2016)

12. M. Bonnefont, A. Joulin, Y. Ma, Spectral gap for spherically symmetric log-concave probability measures, and beyond. J. Funct. Anal. **270**(7), 2456–2482 (2016)
13. P. Cattiaux, Regularité au bord pour les densités et les densités conditionnelles d'une diffusion réfléchie hypoelliptique. Stochastics **20**(4), 309–340 (1987)
14. P. Cattiaux, Stochastic calculus and degenerate boundary value problems. Ann. Inst. Fourier **42**(3), 541–624 (1992)
15. P. Cattiaux, A. Guillin, Semi log-concave Markov diffusions, in *Séminaire de Probabilités XLVI*. Lecture Notes in Mathematics, vol. 2123 (Springer, Cham, 2014), pp. 231–292
16. P. Cattiaux, A. Guillin, On the Poincaré constant of log-concave measures, in *Geometric Aspects of Functional Analysis, Israel Seminar 2017–2019, Vol. 1*. Lecture Notes in Mathematics, vol. 2256 (Springer, Berlin, 2020), pp. 171–217
17. P. Cattiaux, A. Guillin, Functional inequalities for perturbed measures with applications to log-concave measures and to some Bayesian problems. Bernoulli. ArXiv 2101.11257 [math PR] (2021)
18. P. Cattiaux, C. Léonard, Minimization of the Kullback information of diffusion processes. Ann. Inst. H. Poincaré Probab. Stat. **30**(1), 83–132 (1994)
19. P. Cattiaux, C. Léonard, Correction to: "Minimization of the Kullback information of diffusion processes" Ann. Inst. H. Poincaré Probab. Stat. **30**(1), 83–132 (1994); MR1262893 (95d:60056). Ann. Inst. H. Poincaré Probab. Stat. 31(4):705–707 (1995)
20. P. Cattiaux, C. Léonard, Minimization of the Kullback information for some Markov processes, in *Séminaire de Probabilités, XXX*. Lecture Notes in Mathematics, vol. 1626 (Springer, Berlin, 1996), pp. 288–311
21. P. Cattiaux, A. Guillin, C. Roberto, Poincaré inequality and the L^p convergence of semigroups. Electron. Commun. Probab. **15**, 270–280 (2010)
22. P. Cattiaux, A. Guillin, P.A. Zitt, Poincaré inequalities and hitting times. Ann. Inst. Henri Poincaré. Prob. Stat. **49**(1), 95–118 (2013)
23. P. Cattiaux, A. Guillin, L. Wu, Poincaré and Logarithmic Sobolev inequalities for nearly radial measures. Preliminary version available on Math. ArXiv 1912.10825 [math PR]. The revised one to appear in Acta Math. Sin. on the homepage of the first named author (2021)
24. J. Frehse, Essential selfadjointness of singular elliptic operators. Bol. Soc. Brasil. Mat. **8**(2), 87–107 (1977)
25. M. Fukushima, *Dirichlet Forms and Markov Processes*. North-Holland Mathematical Library, vol. 23 (North-Holland, Amsterdam/Kodansha, Tokyo, 1980)
26. N. Ikeda, S. Watanabe, *Stochastic Differential Equations and Diffusion Processes*. North-Holland Mathematical Library, vol. 24 (North-Holland, Amsterdam/Kodansha, Tokyo, 1981)
27. B. Klartag, A Berry-Esseen type inequality for convex bodies with an unconditional basis. Probab. Theory Relat. Fields **145**(1–2), 1–33 (2009)
28. M. Ledoux, Logarithmic Sobolev inequalities for unbounded spin systems revisited, in *Séminaire de Probabilités, XXXV*. Lecture Notes in Mathematics, vol. 1755, pp. 167–194 (Springer, Berlin, 2001)
29. M. Ledoux, Spectral gap, logarithmic Sobolev constant, and geometric bounds, in *Surveys in Differential Geometry*, vol. IX (International Press, Somerville, 2004), pp. 219–240
30. M. Ledoux, From concentration to isoperimetry: semigroup proofs, in *Concentration, Functional Inequalities and Isoperimetry*. Contemporary Mathematics, vol. 545 (American Mathematical Society, Providence, 2011), pp. 155–166
31. M. Ledoux, γ_2 and Γ_2. In honour of D. Bakry and M. Talagrand (2015). https://perso.math.univ-toulouse.fr/ledoux/publications-3/
32. T.M. Liggett, L^2 rates of convergence for attractive reversible nearest particle systems. Ann. Probab. **19**, 935–959 (1991)
33. P.-A. Meyer, W.A. Zheng, Tightness criteria for laws of semimartingales. Ann. Inst. H. Poincaré Probab. Stat. **20**(4), 353–372 (1984)
34. P.-A. Meyer, W.A. Zheng, Construction de processus de Nelson réversibles, in *Séminaire de Probabilités, XIX, 1983/84*. Lecture Notes in Mathematics, vol. 1123 (Springer, Berlin, 1985), pp. 12–26

35. E. Milman, On the role of convexity in isoperimetry, spectral-gap and concentration. Invent. Math. **177**, 1–43 (2009)
36. F. Otto, M.G. Reznikoff, A new criterion for the logarithmic Sobolev inequality and two applications. J. Funct. Anal. **243**(1), 121–157 (2007)
37. É. Pardoux, R.J. Williams, Symmetric reflected diffusions. Ann. Inst. H. Poincaré Probab. Stat. **30**(1), 13–62 (1994)
38. M. Röckner, F.Y. Wang, Weak Poincaré inequalities and L^2-convergence rates of Markov semigroups. J. Funct. Anal. **185**(2), 564–603 (2001)
39. O. Roustant, F. Barthe, B. Iooss, Poincaré inequalities on intervals - application to sensitivity analysis. Electron. J. Stat. **11**(2), 3081–3119 (2017)
40. G. Royer, *Une Initiation aux inégalités de Sobolev Logarithmiques*. Cours Spécialisés [Specialized Courses], vol. 5 (Société Mathématique de France, Paris, 1999)
41. G. Royer, *An Initiation to Logarithmic Sobolev Inequalities*. SMF/AMS Texts and Monographs, vol. 14 (American Mathematical Society, Providence/Société Mathématique de France, Paris, 2007). Translated from the 1999 French original by Donald Babbitt
42. F.Y. Wang, Functional inequalities for empty essential spectrum. J. Funct. Anal. **170**(1), 219–245 (2000)
43. N. Wielens, The essential selfadjointness of generalized Schrödinger operators. J. Funct. Anal. **61**(1), 98–115 (1985)

The Entropic Barrier Is n-Self-Concordant

Sinho Chewi

Abstract For any convex body $K \subseteq \mathbb{R}^n$, S. Bubeck and R. Eldan introduced the entropic barrier on K and showed that it is a $(1 + o(1))\, n$-self-concordant barrier. In this note, we observe that the optimal bound of n on the self-concordance parameter holds as a consequence of the dimensional Brascamp–Lieb inequality.

1 Introduction

Let $K \subseteq \mathbb{R}^n$ be a convex body. In [9], S. Bubeck and R. Eldan introduced the *entropic barrier* $f^\star : \operatorname{int} K \to \mathbb{R}$, defined as follows. First, let $f : \mathbb{R}^n \to \mathbb{R}$ denote the logarithmic Laplace transform of the uniform measure on K,

$$f(\theta) := \ln \int_K \exp\langle \theta, x \rangle \, \mathrm{d}x . \tag{1}$$

Then, define $f^\star$ to be the Fenchel conjugate of f,

$$f^\star(x) := \sup_{\theta \in \mathbb{R}^n} \{\langle \theta, x \rangle - f(\theta)\} .$$

They proved the following result.

Theorem 1 ([9, Theorem 1]) *The function $f^\star$ is strictly convex on $\operatorname{int} K$. Also, the following statements hold.*

1. $f^\star$ is self-concordant, i.e.,

$$\nabla^3 f^\star(x)[h, h, h] \le 2 \, |\langle h, \nabla^2 f^\star(x)\, h \rangle|^{3/2} , \qquad \textit{for all } x \in \operatorname{int} K , \ h \in \mathbb{R}^n .$$

S. Chewi (✉)
Massachusetts Institute of Technology (MIT), Cambridge, MA, USA
e-mail: schewi@mit.edu

2. $f^\star$ is a ν-self-concordant barrier, i.e.,

$$\nabla^2 f^\star(x) \succeq \frac{1}{\nu} \nabla f^\star(x) \nabla f^\star(x)^\mathsf{T}, \qquad \text{for all } x \in \text{int } K,$$

with $\nu = (1 + o(1)) \, n$.

Self-concordant barriers are most well-known for their prominent role in the theory of interior-point methods for optimization [17], but they also find applications to numerous other problems such as online linear optimization with bandit feedback [1] (indeed, the latter was a motivating example for the introduction of the entropic barrier in [9]).

A central theoretical question in the study of self-concordant barriers is: for any convex domain $K \subseteq \mathbb{R}^n$, does there exist a ν-self-concordant barrier for K, and if so, what the optimal value of the parameter ν? In their seminal work [17], Y. Nesterov and A. Nemirovskii constructed for each K a *universal barrier* with $\nu = O(n)$. On the other hand, explicit examples (e.g., the simplex and the cube) show that the best possible self-concordance parameter is $\nu = n$ [17, Proposition 2.3.6]. The situation was better understood for convex cones, on which the *canonical barrier* was shown to be n-self-concordant independently by Hildebrand and Fox [13, 15]. Then, in [9], S. Bubeck and R. Eldan introduced the entropic barrier and showed that it is $(1 + o(1)) \, n$-self-concordant on general convex bodies, and n-self-concordant on convex cones; further, they showed that the universal barrier is also n-self-concordant on convex cones. Subsequently, Y. Lee and M. Yue settled the question of obtaining optimal self-concordant barriers for general convex bodies by proving that the universal barrier is always n-self-concordant [16].

The purpose of this note is to describe the following observation.

Theorem 2 *The entropic barrier on any convex body $K \subseteq \mathbb{R}^n$ is an n-self-concordant barrier.*

Besides improving the result of [9], the theorem shows that the entropic barrier provides a second example of an optimal self-concordant barrier for general convex bodies; to the best of the author's knowledge, no other optimal self-concordant barriers are known.

We will provide two distinct proofs of Theorem 2. First, we will observe that Theorem 2 is an immediate consequence of the following theorem, which was obtained independently in [18, 21]; see also [14].

Theorem 3 *Let $\mu \propto \exp(-V)$ be a log-concave density on $\mathbb{R}^n$. Then,*

$$\text{Var}_\mu V \le n.$$

In turn, as discussed in [7, 18], Theorem 3 is related to certain dimensional improvements of the *Brascamp–Lieb inequality*. We state a version of this inequality which is convenient for the present discussion.

Theorem 4 ([7, Proposition 4.1]) *Let $\mu \propto \exp(-V)$ be a log-concave density on $\mathbb{R}^n$, where V is of class C^2 and $\nabla^2 V \succ 0$. Then, for all C^1 compactly supported $g : \mathbb{R}^n \to \mathbb{R}$, it holds that*

$$\mathrm{Var}_\mu \, g \leq \mathbb{E}_\mu \langle \nabla g, (\nabla^2 V)^{-1} \nabla g \rangle - \frac{\mathrm{cov}_\mu(g, V)^2}{n - \mathrm{Var}_\mu \, V} \, .$$

It is straightforward to see that Theorem 4 implies Theorem 3. Indeed, via a routine approximation argument, we may assume that μ satisfies the hypothesis of Theorem 4. Taking $g = V$ (which is justified via another approximation argument) and rearranging the inequality of Theorem 4 yields

$$\mathrm{Var}_\mu \, V \leq \frac{n \mathbb{E}_\mu \langle \nabla V, (\nabla^2 V)^{-1} \nabla V \rangle}{n + \mathbb{E}_\mu \langle \nabla V, (\nabla^2 V)^{-1} \nabla V \rangle} \leq n \, .$$

Next, in our second approach to Theorem 2, we observe that a key step in the proof of Theorem 3 given by [21] is a tensorization principle. It is then natural to wonder whether such a principle can be applied directly to deduce Theorem 2. Indeed, we have the following elementary lemma.

Lemma 1 *Suppose that for each $n \in \mathbb{N}^+$ and each convex body $K \subseteq \mathbb{R}^n$, we have a function $\phi_{n,K} : \mathrm{int}\, K \to \mathbb{R}$ such that $\phi_{n,K}$ is a $\nu(n)$-self-concordant barrier for K. Also, suppose that the following consistency condition holds:*

$$\phi_{m+n, K \times K'}(x, x') = \phi_{m,K}(x) + \phi_{n,K'}(x') \, , \tag{2}$$

for all $m, n \in \mathbb{N}^+$, all convex bodies $K \subseteq \mathbb{R}^m$, $K' \subseteq \mathbb{R}^n$, and all $x \in K$, $x' \in K'$. Then, $\phi_{n,K}$ is a $\inf_{k \in \mathbb{N}^+} \nu(kn)/k$-self-concordant barrier for K.

We will check that the entropic barrier satisfies the consistency condition described in the previous lemma in Sect. 4. Combined with the second statement in Theorem 1, it yields another proof of Theorem 2.

The remainder of this note is organized as follows. In Sect. 2, we will explain the connection between Theorems 2 and 3, thereby deducing the former from the latter. Then, so as to make this note more self-contained, in Sect. 3 we will provide two proofs of the dimensional Brascamp–Lieb inequality (Theorem 4). The first proof follows [7] and proceeds via a dimensional improvement of Hörmander's L^2 method. The second "proof", which is only sketched, shows how the dimensional Brascamp–Lieb inequality may be obtained from a convexity principle: the entropy functional is convex along generalized Wasserstein geodesics which arise from Bregman divergence couplings [2]. The second argument appears to be new. Finally, in Sect. 4, we present the tensorization argument as encapsulated in Theorem 1.

2 From the Entropic Barrier to the Dimensional Brascamp–Lieb Inequality

In this section, we follow [9]. The entropic barrier has a fruitful interpretation in terms of an exponential family of probability distributions defined over the convex body $K \subseteq \mathbb{R}^n$. For each $\theta \in \mathbb{R}^n$, we define the density p_θ on K via

$$p_\theta(x) := \frac{\exp \langle \theta, x \rangle}{\int_K \exp \langle \theta, x' \rangle \, dx'} \, \mathbb{1}\{x \in K\}. \tag{3}$$

Since f (defined in (1)) is essentially the logarithmic moment-generating function of p_θ, then the derivatives of f yield cumulants of p_θ. In particular,

$$\nabla f(\theta) = \mathbb{E}_{p_\theta} X, \qquad \nabla^2 f(\theta) = \operatorname{cov}_{p_\theta} X.$$

By convex duality, the mappings $\nabla f : \mathbb{R}^n \to \operatorname{int} K$ and $\nabla f^\star : \operatorname{int} K \to \mathbb{R}^n$ are inverses of each other. From the classical duality between the logarithmic moment-generating function and entropy, we can also deduce that

$$f^\star(x) = \mathcal{H}(p_{\nabla f^\star(x)}),$$

where $\mathcal{H}$ denotes the entropy functional[1]

$$\mathcal{H}(p) := \int p \ln p. \tag{4}$$

The self-concordance parameter of $f^\star$ is the least $\nu \geq 0$ such that

$$\langle \nabla f^\star(x), [\nabla^2 f^\star(x)]^{-1} \nabla f^\star(x) \rangle \leq \nu, \qquad \text{for all } x \in \operatorname{int} K.$$

Taking $x = \nabla f(\theta)$, equivalently we require

$$\langle \theta, \nabla^2 f(\theta) \theta \rangle \leq \nu, \qquad \text{for all } \theta \in \mathbb{R}^n,$$

which has the probabilistic interpretation

$$\operatorname{Var}_{p_\theta} \langle \theta, X \rangle \leq \nu, \qquad \text{for all } \theta \in \mathbb{R}^n. \tag{5}$$

From the definition (3), we see that the density $p_\theta \propto \exp(-V)$ is log-concave, where $V(x) = \langle \theta, x \rangle$ for $x \in \operatorname{int} K$. By applying Theorem 3 to p_θ, we immediately deduce that (5) holds with $\nu = n$.

---

[1] Note the sign convention, which is opposite the usual one in information theory. We use this convention as it is convenient for $\mathcal{H}$ to be convex.

3 Proof of the Dimensional Brascamp–Lieb Inequality

Next, we wish to give some proofs of the dimensional Brascamp–Lieb inequality (Theorem 4). Classically, the Brascamp–Lieb inequality reads as follows.

Theorem 5 ([8]) *Let $\mu \propto \exp(-V)$ be a density on $\mathbb{R}^n$, where V is a convex function of class C^2. Then, for every locally Lipschitz $g : \mathbb{R}^n \to \mathbb{R}$,*

$$\operatorname{Var}_\mu g \leq \mathbb{E}_\mu \langle \nabla g, (\nabla^2 V)^{-1} \nabla g \rangle. \tag{6}$$

The Brascamp–Lieb inequality is a Poincaré inequality for the measure μ corresponding to the Newton–Langevin diffusion [10]. When V is strongly convex, $\nabla^2 V \succeq \alpha I_n$, it recovers the usual Poincaré inequality

$$\operatorname{Var}_\mu g \leq \frac{1}{\alpha} \mathbb{E}_\mu [\|\nabla g\|^2].$$

See [4, 6, 11] for various proofs of Theorem 5.

Since the inequality (6) makes no explicit reference to the dimension, it actually holds in infinite-dimensional space. In contrast, Theorem 4 asserts that (6) can be improved by subtracting an additional non-negative term from the right-hand side in any finite dimension. This is referred to as a *dimensional improvement* of the Brascamp–Lieb inequality.

3.1 Proof by Hörmander's L^2 Method

We now present the proof of Theorem 4 given in [7]. The starting point for Hörmander's L^2 method is to first dualize the Poincaré inequality.

Proposition 1 ([5, Lemma 1]) *Let $\mu \propto \exp(-V)$ be a probability density on $\mathbb{R}^n$, where V is of class C^1. Define the corresponding generator $\mathscr{L}$ on smooth functions $g : \mathbb{R}^n \to \mathbb{R}$ via*

$$\mathscr{L}g := \Delta g - \langle \nabla V, \nabla g \rangle.$$

Suppose $A : \mathbb{R}^n \to \mathrm{PD}(n)$ is a matrix-valued function mapping into the space of symmetric positive definite matrices such that for all smooth $u : \mathbb{R}^n \to \mathbb{R}$,

$$\mathbb{E}_\mu [(\mathscr{L}u)^2] \geq \mathbb{E}_\mu \langle \nabla u, A \nabla u \rangle. \tag{7}$$

Then, for all $g \in L^2(\mu)$, it holds that

$$\operatorname{Var}_\mu g \leq \mathbb{E}_\mu \langle \nabla g, A^{-1} \nabla g \rangle.$$

Proof We may assume $\mathbb{E}_\mu g = 0$. This condition is certainly necessary for the Poisson equation $-\mathscr{L}u = g$ to be solvable; in order to streamline the proof, we will assume that a solution u exists. (This assumption can be avoided by invoking [12] and using a density argument; see [5] for details.)

Using the integration by parts formula for the generator,

$$- \mathbb{E}_\mu[g \, \mathscr{L}u] = \mathbb{E}_\mu \langle \nabla g, \nabla u \rangle \,,$$

we obtain

$$\mathrm{Var}_\mu \, g = \mathbb{E}_\mu[g^2] = -2\mathbb{E}_\mu[g \, \mathscr{L}u] - \mathbb{E}_\mu[(\mathscr{L}u)^2]$$
$$\leq 2\mathbb{E}_\mu \langle \nabla g, \nabla u \rangle - \mathbb{E}_\mu \langle \nabla u, A \, \nabla u \rangle \,.$$

Next, since $2 \langle x, y \rangle \leq \langle x, A x \rangle + \langle y, A^{-1} y \rangle$ for all $x, y \in \mathbb{R}^n$, it implies

$$\mathrm{Var}_\mu \, g \leq \mathbb{E}_\mu \langle \nabla g, A^{-1} \nabla g \rangle \,.$$

$\square$

The key idea now is that the condition (7) can be verified with the help of the *curvature* of the potential V. Indeed, assume now that V is of class $\mathcal{C}^2$ and that $\nabla^2 V \succ 0$. By direct calculation, one verifies the commutation relation

$$\nabla \mathscr{L}u = (\mathscr{L} - \nabla^2 V) \nabla u \,. \tag{8}$$

Hence,

$$\mathbb{E}_\mu[(\mathscr{L}u)^2] = -\mathbb{E}_\mu \langle \nabla u, \nabla \mathscr{L}u \rangle = -\mathbb{E}_\mu \langle \nabla u, (\mathscr{L} - \nabla^2 V) \nabla u \rangle$$
$$= \mathbb{E}_\mu \langle \nabla u, \nabla^2 V \nabla u \rangle + \mathbb{E}_\mu[\|\nabla^2 u\|_{\mathrm{HS}}^2] \,, \tag{9}$$

where the last equality follows from the integration by parts formula for the generator applied to each coordinate separately: $-\mathbb{E}_\mu \langle \nabla u, \mathscr{L}\nabla u \rangle = \mathbb{E}_\mu[\|\nabla^2 u\|_{\mathrm{HS}}^2]$. Since the second term is non-negative, Proposition 1 now implies the Brascamp–Lieb inequality (Theorem 5).

In order to obtain the dimensional improvement of the Brascamp–Lieb inequality (Theorem 4), we will imitate the proof of Proposition 1, only now we will use the additional term $\mathbb{E}_\mu[\|\nabla^2 u\|_{\mathrm{HS}}^2]$ in the above identity.

Proof of Theorem 4 As before, let $\mathbb{E}_\mu g = 0$. However, we introduce an additional trick and consider u not necessarily satisfying $-\mathscr{L}u = g$; this will help to optimize the bound at the end of the argument. Following the computations in Proposition 1 and using the key identity (9), we obtain

$$
\begin{aligned}
\operatorname{Var}_\mu g &= \mathbb{E}_\mu[g^2] = \mathbb{E}_\mu[(g + \mathscr{L}u)^2] - 2\mathbb{E}_\mu[g\,\mathscr{L}u] - \mathbb{E}_\mu[(\mathscr{L}u)^2] \\
&= \mathbb{E}_\mu[(g + \mathscr{L}u)^2] + 2\mathbb{E}_\mu\langle \nabla g, \nabla u \rangle - \mathbb{E}_\mu\langle \nabla u, \nabla^2 V\, \nabla u \rangle - \mathbb{E}_\mu[\|\nabla u\|_{\mathrm{HS}}^2] \\
&\le \mathbb{E}_\mu[(g + \mathscr{L}u)^2] + \mathbb{E}_\mu\langle \nabla g, (\nabla^2 V)^{-1}\, \nabla g \rangle - \mathbb{E}_\mu[\|\nabla u\|_{\mathrm{HS}}^2].
\end{aligned}
$$

For the second term, we use the inequality

$$
\mathbb{E}_\mu[\|\nabla u\|_{\mathrm{HS}}^2] \ge \frac{1}{n}\,(\mathbb{E}_\mu \Delta u)^2.
$$

From integration by parts,

$$
\mathbb{E}_\mu \Delta u = \mathbb{E}_\mu\langle \nabla V, \nabla u \rangle = -\mathbb{E}_\mu[V\,\mathscr{L}u] = \operatorname{cov}_\mu(g, V) - \mathbb{E}_\mu[V\,(\mathscr{L}u + g)].
$$

We now choose $-\mathscr{L}u = g + a\,(V - \mathbb{E}_\mu V)$ for some $a \ge 0$ to be chosen later. For brevity of notation, write $\mathbf{C} := \operatorname{cov}_\mu(g, V)$ and $\mathbf{V} := \operatorname{Var}_\mu V$. Then,

$$
\begin{aligned}
\operatorname{Var}_\mu g - \mathbb{E}_\mu\langle \nabla g, (\nabla^2 V)^{-1}\, \nabla g \rangle &\le a^2 \mathbf{V} - \frac{1}{n}\,(\mathbf{C} + a\mathbf{V})^2 \\
&= -\frac{\mathbf{V}\,(n - \mathbf{V})}{n}\left(a - \frac{\mathbf{C}}{n - \mathbf{V}}\right)^2 - \frac{\mathbf{C}^2 \mathbf{V}}{n\,(n - \mathbf{V})} - \frac{\mathbf{C}^2}{n}.
\end{aligned}
$$

Observe that this inequality entails $\mathbf{V} \le n$, or else we could send $a \to \infty$ and arrive at a contradiction. Optimizing over a, we obtain

$$
\operatorname{Var}_\mu g \le \mathbb{E}_\mu\langle \nabla g, (\nabla^2 V)^{-1}\, \nabla g \rangle - \frac{\mathbf{C}^2}{n - \mathbf{V}}.
$$

$\square$

3.2 Proof by Convexity of the Entropy Along Bregman Divergence Couplings

It is well-known that Poincaré inequalities are obtained from linearizing transportation inequalities. In [11], D. Cordero-Erausquin obtained the Brascamp–Lieb inequality (Theorem 5) by linearizing the following inequality:

$$
\mathcal{D}_V(\rho \,\|\, \mu) \le \mathsf{KL}(\rho \,\|\, \mu), \qquad \text{for all } \rho \in \mathcal{P}(\mathbb{R}^n). \tag{10}
$$

Here, $\mu \propto \exp(-V)$ on $\mathbb{R}^n$; $\mathcal{P}(\mathbb{R}^n)$ denotes the space of probability measures on $\mathbb{R}^n$; $\mathrm{KL}(\cdot \| \cdot)$ is the Kullback–Leibler (KL) divergence; and $\mathcal{D}_V(\cdot \| \cdot)$ is the Bregman divergence coupling cost, defined as

$$\mathcal{D}_V(\rho \| \mu) = \inf_{\gamma \in \text{couplings}(\rho, \mu)} \int D_V(x, y)\, d\gamma(x, y),$$

with

$$D_V(x, y) := V(x) - V(y) - \langle \nabla V(y), x - y \rangle.$$

On the other hand, together with K. Ahn in [2], the author obtained the transportation inequality (10) as a consequence of a convexity principle in optimal transport. It is therefore natural to ask whether the dimensional Brascamp–Lieb inequality (Theorem 4) can be obtained directly from (a strengthening of) this principle. This is indeed the case, and it is the goal of the present section to describe this argument.

Making the argument fully rigorous, however, would entail substantial technical complications which would detract from the focus of this note. In any case, a complete proof of the dimensional Brascamp–Lieb inequality is already present in [7]. Hence, we will work on a purely formal level and assume that everything is smooth, bounded, etc. Also, the computations are rather similar to the proof of Theorem 4 given in the previous section. Nevertheless, the argument seems interesting enough to warrant presenting it here.

The main difference with the preceding proof is that the Bochner formula (implicit in the commutation relation (8)) is replaced by the convexity principle.

Proof Sketch of Theorem 4 Throughout the proof, let $\varepsilon > 0$ be small. Let h be bounded and satisfy $\mathbb{E}_\mu h = 0$, so that $\mu_\varepsilon := (1 + \varepsilon h)\,\mu$ defines a valid probability density on $\mathbb{R}^n$. Our aim is to first strengthen the transportation inequality (10), at least infinitesimally, and then to linearize it.

Let (X_ε, X) be an optimal coupling for the Bregman divergence coupling cost $\mathcal{D}_V(\mu_\varepsilon \| \mu)$. In [2], the following facts were proven:

1. There is a function $u_\varepsilon : \mathbb{R}^n \to \mathbb{R}$ such that $\nabla V(X) = \nabla V(X_\varepsilon) - \nabla u_\varepsilon(X_\varepsilon)$, and $V - u_\varepsilon$ is convex.
2. The entropy functional (defined in (4)) is convex in the sense that

$$\mathcal{H}(\mu_\varepsilon) \geq \mathcal{H}(\mu) + \mathbb{E}\langle [\nabla_{W_2}\mathcal{H}(\mu)](X), X_\varepsilon - X \rangle. \tag{11}$$

Here, $\nabla_{W_2}\mathcal{H}(\mu) = \nabla \ln \mu$ is the Wasserstein gradient of the entropy functional, c.f. [3, 19, 20].

Write $T_\varepsilon(x) := (\nabla V - \nabla u_\varepsilon)^{-1}(\nabla V(x))$. Since $(T_\varepsilon)_\# \mu = \mu_\varepsilon$, the change of variables formula implies

$$\frac{\mu(x)}{\mu_\varepsilon(T_\varepsilon(x))} = \frac{\mu(x)}{\mu(T_\varepsilon(x))\,(1 + \varepsilon h(T_\varepsilon(x)))} = \det \nabla T_\varepsilon(x). \tag{12}$$

To linearize this equation, write $u_\varepsilon = \varepsilon u + o(\varepsilon)$ and $T_\varepsilon(x) = x + \varepsilon T(x) + o(\varepsilon)$. Then, the definition of T_ε yields

$$\nabla V(x) = (\nabla V - \nabla u_\varepsilon)\big(x + \varepsilon T(x) + o(\varepsilon)\big)$$
$$= \nabla V(x) + \varepsilon \nabla^2 V(x)\, T(x) - \varepsilon \nabla u(x) + o(\varepsilon)$$

which implies

$$T_\varepsilon(x) = x + \varepsilon\, [\nabla^2 V(x)]^{-1} \nabla u(x) + o(\varepsilon).$$

Taking logarithms and expanding to first order in ε,

$$\ln \mu(x) - \ln \mu(T_\varepsilon(x)) - \ln(1 + \varepsilon h(T_\varepsilon(x)))$$
$$= -\varepsilon\, \langle \nabla \ln \mu(x), [\nabla^2 V(x)]^{-1} \nabla u(x)\rangle - \varepsilon h(x) + o(\varepsilon)$$
$$= \varepsilon\, \langle \nabla V(x), [\nabla^2 V(x)]^{-1} \nabla u(x)\rangle - \varepsilon h(x) + o(\varepsilon)$$

and

$$\ln \det \nabla T_\varepsilon(x) = \ln \det \nabla\big(\mathrm{Id} + \varepsilon\, [\nabla^2 V]^{-1} \nabla u + o(\varepsilon)\big)(x)$$
$$= \ln \det\big(I_n + \varepsilon \nabla([\nabla^2 V]^{-1} \nabla u)(x) + o(\varepsilon)\big)$$
$$= \varepsilon \operatorname{div}([\nabla^2 V]^{-1} \nabla u)(x) + o(\varepsilon).$$

To interpret this, we introduce a new generator, denoted $\hat{\mathscr{L}}$ to avoid confusion with the previous section, defined by

$$\hat{\mathscr{L}} u := \operatorname{div}([\nabla^2 V]^{-1} \nabla u) - \langle \nabla V, [\nabla^2 V]^{-1} \nabla u\rangle.$$

This new generator satisfies the integration by parts formula

$$\mathbb{E}_\mu[u\, \hat{\mathscr{L}} v] = \mathbb{E}_\mu \langle \nabla u, [\nabla^2 V]^{-1} \nabla v\rangle.$$

In this notation, the preceding computations yield

$$\hat{\mathscr{L}} u = -h + o(1).$$

Next, to strengthen (11), we repeat the proof. From (12),

$$\mathcal{H}(\mu_\varepsilon) = \int \mu_\varepsilon \ln \mu_\varepsilon = \int \mu \ln(\mu_\varepsilon \circ T_\varepsilon) = \int \mu \ln \frac{\mu}{\det \nabla T_\varepsilon}$$

$$= \mathcal{H}(\mu) - \int \mu \ln \det \nabla T_\varepsilon .$$

From the second-order expansion of $-\ln \det$ around I_n,

$$-\int \mu \ln \det \nabla T_\varepsilon$$

$$\geq -\int \mu \ln \det I_n - \int \mu \langle I_n, \nabla T_\varepsilon - I_n \rangle + \frac{1}{2} \int \mu \|\nabla T_\varepsilon - I_n\|_{\mathrm{HS}}^2 + o(\varepsilon^2)$$

$$\geq -\int \mu \operatorname{tr}(\nabla T_\varepsilon - I_n) + \frac{1}{2n} \left(\int \mu \operatorname{tr}(\nabla T_\varepsilon - I_d) \right)^2 + o(\varepsilon^2)$$

$$= -\int \mu \operatorname{div}(T_\varepsilon - \mathrm{Id}) + \frac{1}{2n} \left(\int \mu \operatorname{div}(T_\varepsilon - \mathrm{Id}) \right)^2 + o(\varepsilon^2)$$

$$= \int \mu \langle \nabla \ln \mu, T_\varepsilon - \mathrm{Id} \rangle + \frac{1}{2n} \left(\int \mu \langle \nabla \ln \mu, T_\varepsilon - \mathrm{Id} \rangle \right)^2 + o(\varepsilon^2) .$$

Recalling that $\nabla_{W_2} \mathcal{H}(\mu) = \nabla \ln \mu$, we have established

$$\mathcal{H}(\mu_\varepsilon) - \mathcal{H}(\mu) - \mathbb{E}\langle [\nabla_{W_2} \mathcal{H}(\mu)](X), X_\varepsilon - X \rangle$$

$$\geq \frac{1}{2n} \left(\int \mu \langle \nabla V, T_\varepsilon - \mathrm{Id} \rangle \right)^2 + o(\varepsilon^2)$$

$$= \frac{\varepsilon^2}{2n} \left(\int \mu \langle \nabla V, [\nabla^2 V]^{-1} \nabla u \rangle \right)^2 + o(\varepsilon^2)$$

$$= \frac{\varepsilon^2}{2n} \{\mathbb{E}_\mu[V \, \mathscr{L} u]\}^2 + o(\varepsilon^2) .$$

The next step is to write down the strengthened transportation inequality. Indeed, if we add a suitable additive constant to V so that $\mu = \exp(-V)$, then

$$\mathsf{KL}(\mu_\varepsilon \parallel \mu) = \mathbb{E}_{\mu_\varepsilon} V + \mathcal{H}(\mu_\varepsilon)$$

$$\geq \underbrace{\mathbb{E} V(X) + \mathcal{H}(\mu)}_{=\mathsf{KL}(\mu\|\mu)=0} + \underbrace{\mathbb{E}\langle [\nabla V + \nabla_{W_2} \mathcal{H}(\mu)](X), X_\varepsilon - X \rangle}_{=[\nabla_{W_2} \mathsf{KL}(\cdot\|\mu)](\mu)=0}$$

$$+ \underbrace{\mathbb{E}[V(X_\varepsilon) - V(X) - \langle \nabla V(X), X_\varepsilon - X \rangle]}_{=\mathcal{D}_V(\mu_\varepsilon\|\mu)}$$

$$+ \frac{\varepsilon^2}{2n} \{\mathbb{E}_\mu[h V]\}^2 + o(\varepsilon^2)$$

$$\geq \mathcal{D}_V(\mu_\varepsilon \parallel \mu) + \frac{\varepsilon^2}{2n} \{\mathbb{E}_\mu[h V]\}^2 + o(\varepsilon^2) .$$

Finally, it remains to linearize the transportation inequality. On one hand, it is classical that

$$\mathsf{KL}(\mu_\varepsilon \parallel \mu) = \frac{\varepsilon^2}{2}\mathbb{E}_\mu[h^2] + o(\varepsilon^2).$$

On the other hand, we can guess that

$$\mathcal{D}_V(\mu_\varepsilon \parallel \mu) = \frac{1}{2}\mathbb{E}\langle X_\varepsilon - X, \nabla^2 V(X)(X_\varepsilon - X)\rangle + o(\varepsilon^2)$$

$$= \frac{\varepsilon^2}{2}\mathbb{E}_\mu\langle \nabla u, (\nabla^2 V)^{-1}\nabla u\rangle + o(\varepsilon^2)$$

$$\geq \frac{\varepsilon^2}{2}\frac{\{\mathbb{E}_\mu\langle \nabla g, (\nabla^2 V)^{-1}\nabla u\rangle\}^2}{\mathbb{E}_\mu\langle \nabla g, (\nabla^2 V)^{-1}\nabla g\rangle} + o(\varepsilon^2)$$

$$= \frac{\varepsilon^2}{2}\frac{\{\mathbb{E}_\mu[g\,\mathscr{L}u]\}^2}{\mathbb{E}_\mu\langle \nabla g, (\nabla^2 V)^{-1}\nabla g\rangle} + o(\varepsilon^2)$$

$$= \frac{\varepsilon^2}{2}\frac{\{\mathbb{E}_\mu[gh]\}^2}{\mathbb{E}_\mu\langle \nabla g, (\nabla^2 V)^{-1}\nabla g\rangle} + o(\varepsilon^2).$$

A rigorous proof of this inequality is given as [11, Lemma 3.1].

Thus, we obtain

$$\frac{1}{2}\frac{\{\mathbb{E}_\mu[gh]\}^2}{\mathbb{E}_\mu\langle \nabla g, (\nabla^2 V)^{-1}\nabla g\rangle} + \frac{1}{2n}\{\mathbb{E}_\mu[hV]\}^2 \leq \frac{1}{2}\mathbb{E}_\mu[h^2] + o(1).$$

Now we let $\varepsilon \searrow 0$ and choose $h = g + a(V - \mathbb{E}_\mu V)$ for some $a \in \mathbb{R}$. Writing $\mathbf{C} := \mathrm{cov}_\mu(g, V)$ and $\mathbf{V} := \mathrm{Var}_\mu V$, it yields

$$\frac{(\mathrm{Var}_\mu g + a\mathbf{C})^2}{\mathbb{E}_\mu\langle \nabla g, (\nabla^2 V)^{-1}\nabla g\rangle} + \frac{1}{n}(\mathbf{C} + a\mathbf{V})^2 \leq \mathrm{Var}_\mu g + 2a\mathbf{C} + a^2\mathbf{V}.$$

Actually, choosing a to optimize this inequality and simplifying the resulting expression may be cumbersome, so with our foresight from the earlier proof of Theorem 4, we now take $a = \mathbf{C}/(n - \mathbf{V})$. After some algebra,

$$\frac{(\mathrm{Var}_\mu g + \mathbf{C}^2/(n - \mathbf{V}))^2}{\mathbb{E}_\mu\langle \nabla g, (\nabla^2 V)^{-1}\nabla g\rangle} \leq \mathrm{Var}_\mu g + \frac{\mathbf{C}^2}{n - \mathbf{V}},$$

which of course yields

$$\mathrm{Var}_\mu g \leq \mathbb{E}_\mu\langle \nabla g, (\nabla^2 V)^{-1}\nabla g\rangle - \frac{\mathbf{C}^2}{n - \mathbf{V}}.$$

$\square$

4 A Tensorization Trick

We begin by verifying that the entropic barrier has the consistency property (2). Let f_K denote the function (1), where we now explicitly denote the dependence on the convex body K. Also, let $f_K^\star$ denote the corresponding entropic barrier. Then, we see that

$$f_{K \times K'}(\theta, \theta') = \ln \int_{K \times K'} \exp(\langle \theta, x \rangle + \langle \theta', x' \rangle) \, dx \, dx'$$

$$= \ln \int_K \exp \langle \theta, x \rangle \, dx + \ln \int_{K'} \exp \langle \theta', x' \rangle \, dx' = f_K(\theta) + f_{K'}(\theta').$$

Hence,

$$f_{K \times K'}^\star(x, x') = \sup_{\theta, \theta' \in \mathbb{R}^n} \{\langle \theta, x \rangle + \langle \theta', x' \rangle - f_K(\theta) - f_{K'}(\theta')\} = f_K^\star(x) + f_{K'}^\star(x').$$

Finally, we check that the tensorization property automatically improves the bound on the self-concordance parameter of $f_K^\star$ obtained in [9].

Proof of Lemma 1 Let $x := (x_1, \ldots, x_k) \in (\mathbb{R}^n)^k$. By assumption, the self-concordant barrier ϕ_{kn, K^k} on K^k satisfies $\phi_{kn, K^k}(x) = \sum_{j=1}^k \phi_{n, K}(x_j)$. Also, we are given that

$$\nabla^2 \phi_{kn, K^k}(x) \succeq \frac{1}{\nu(kn)} \nabla \phi_{kn, K^k}(x) \, \nabla \phi_{kn, K^k}(x)^\mathsf{T}. \tag{13}$$

Via elementary calculations,

$$\nabla \phi_{kn, K^k}(x) = \big(\nabla \phi_{n, K}(x_1), \ldots, \nabla \phi_{n, K}(x_k) \big)$$

and

$$\nabla^2 \phi_{kn, K^k}(x) = \begin{bmatrix} \nabla^2 \phi_{n, K}(x_1) & & \\ & \ddots & \\ & & \nabla^2 \phi_{n, K}(x_k) \end{bmatrix}.$$

Let $v \in \mathbb{R}^n$ and let $\boldsymbol{v} := (v, \ldots, v) \in (\mathbb{R}^n)^k$. Also, take $x_1 = \cdots = x_k = x$. By (13), we know that

$$k \langle v, \nabla^2 \phi_{n, K}(x) \, v \rangle = \langle \boldsymbol{v}, \nabla^2 \phi_{kn, K^k}(x) \, \boldsymbol{v} \rangle \geq \frac{1}{\nu(kn)} \langle \boldsymbol{v}, \nabla \phi_{kn, K^k}(x) \rangle^2$$

$$= \frac{k^2}{\nu(kn)} \langle v, \nabla \phi_{n, K}(x) \rangle^2$$

which proves

$$\nabla^2 \phi_{n,K}(x) \succeq \frac{k}{v(kn)} \nabla \phi_{n,K}(x) \nabla \phi_{n,K}(x)^{\mathsf{T}}$$

and gives the claim.   □

Proof of Theorem 2 According to Theorem 1, we know that the entropic barrier in n dimensions is $(1 + \varepsilon_n) n$-self-concordant, with $\varepsilon_n \to 0$ as $n \to \infty$. By Lemma 1, it is actually $(1 + \varepsilon_{kn}) n$-self-concordant, for any $k \in \mathbb{N}^+$. Let $k \to \infty$ to deduce that it is in fact n-self-concordant.   □

Acknowledgments The author thanks Sébastien Bubeck for helpful comments. The author was supported by the Department of Defense (DoD) through the National Defense Science and Engineering Graduate Fellowship (NDSEG) Program.

References

1. J. Abernethy, E. Hazan, A. Rakhlin, Competing in the dark: an efficient algorithm for bandit linear optimization English (US), in *21st Annual Conference on Learning Theory (COLT 2008)* (2008), pp. 263–273
2. K. Ahn, S. Chewi, Efficient constrained sampling via the mirror-Langevin algorithm (2021). arXiv:2010.16212
3. L. Ambrosio, N. Gigli, G. Savaré, *Gradient Flows in Metric Spaces and in the Space of Probability Measures*. Second Lectures in Mathematics ETH Zürich (Birkhäuser Verlag, Basel, 2008), pp. x+334
4. D. Bakry, I. Gentil, M. Ledoux, *Analysis and Geometry of Markov Diffusion Operators*. Grundlehren der Mathematischen Wissenschaften [Fundamental Principles of Mathematical Sciences], vol. 348 (Springer, Cham, 2014), pp. xx+552
5. F. Barthe, D. Cordero-Erausquin, Invariances in variance estimates. Proc. Lond. Math. Soc. **106**(1), 33–64 (2013)
6. S.G. Bobkov, M. Ledoux, From Brunn-Minkowski to Brascamp-Lieb and to logarithmic Sobolev inequalities. Geom. Funct. Anal. **10**(5), 1028–1052 (2000)
7. F. Bolley, I. Gentil, A. Guillin, Dimensional improvements of the logarithmic Sobolev, Talagrand and Brascamp-Lieb inequalities. Ann. Probab. **46**(1), 261–301 (2018)
8. H.J. Brascamp, E.H. Lieb, On extensions of the Brunn- Minkowski and Prékopa-Leindler theorems, including inequalities for log concave functions, and with an application to the diffusion equation. J. Funct. Anal. **22**(4), 366–389 (1976)
9. S. Bubeck, R. Eldan, The entropic barrier: exponential families, log-concave geometry, and self-concordance. Math. Oper. Res. **44**(1), 264–276 (2019)
10. S. Chewi et al., Exponential ergodicity of mirror-Langevin diffusions, in *Advances in Neural Information Processing Systems*, ed. by H. Larochelle et al., vol. 33 (Curran Associates, Red Hook, 2020), pp. 19573–19585
11. D. Cordero-Erausquin, Transport inequalities for log-concave measures, quantitative forms, and applications. Can. J. Math. **69**(3), 481–501 (2017)
12. D. Cordero-Erausquin, M. Fradelizi, B. Maurey, The (B) conjecture for the Gaussian measure of dilates of symmetric convex sets and related problems. J. Funct. Anal. **214**(2), 410–427 (2004)

13. D.J.F. Fox, A Schwarz lemma for Kähler affine metrics and the canonical potential of a proper convex cone. Ann. Mat. Pura Appl. **194**(1), 1–42 (2015)
14. M. Fradelizi, M. Madiman, L. Wang, Optimal concentration of information content for log-concave densities, in *High Dimensional Probability VII.* (vol. 71). Progress in Probability (Springer, Cham, 2016), pp. 45–60
15. R. Hildebrand, Canonical barriers on convex cones. Math. Oper. Res. **39**(3), 841–850 (2014)
16. Y.T. Lee, M.-C. Yue, Universal barrier is n-self-concordant. Math. Oper. Res. **46**(3), 1129–1148 (2021)
17. Y. Nesterov, A. Nemirovskii, *Interior-Point Polynomial Algorithms in Convex Programming*, vol. 13. SIAM Studies in Applied Mathematics (Society for Industrial and Applied Mathematics (SIAM), Philadelphia, PA, 1994), pp. x+405
18. V.H. Nguyen, Dimensional variance inequalities of Brascamp- Lieb type and a local approach to dimensional Prékopa's theorem. J. Funct. Anal. **266**(2), 931–955 (2014)
19. F. Santambrogio, *Optimal Transport for Applied Mathematicians*, vol. 87. Progress in Nonlinear Differential Equations and their Applications. Calculus of variations, PDEs, and Modeling (Birkhäuser/Springer, Cham, 2015), pp. xxvii+353
20. C. Villani, *Optimal Transport*, vol. 338. Grundlehren der Mathematischen Wissenschaften [Fundamental Principles of Mathematical Sciences]. Old and new (Springer, Berlin, 2009), pp. xxii+973
21. L. Wang, *Heat Capacity Bound, Energy Fluctuations and Convexity*. Thesis (Ph.D.)-Yale University (ProQuest LLC, Ann Arbor, MI, 2014), p. 114

Local Tail Bounds for Polynomials on the Discrete Cube

Bo'az Klartag and Sasha Sodin

Abstract Let P be a polynomial of degree d in independent Bernoulli random variables which has zero mean and unit variance. The Bonami hypercontractivity bound implies that the probability that $|P| > t$ decays exponentially in $t^{2/d}$. Confirming a conjecture of Keller and Klein, we prove a local version of this bound, providing an upper bound on the difference between the e^{-r} and the e^{-r-1} quantiles of P.

1 Introduction

This note is concerned with concentration inequalities for polynomials on the discrete cube. Concentration inequalities, i.e. tail bounds on the distribution of functions on high-dimensional spaces belonging to certain classes, were put forth by Vitali Milman in the 1970-s and have since found numerous applications; see e.g. [2, 3] and references therein.

Let $X_1, \ldots, X_n$ be independent, identically distributed symmetric Bernoulli variables, so that $X = (X_1, \ldots, X_n)$ is distributed uniformly on the discrete cube $\{-1, 1\}^n$. The starting point for this work is the concentration inequality for

Supported by a grant from the Israel Science Foundation (ISF).

Supported in part by a Royal Society Wolfson Research Merit Award (WM170012), and a Philip Leverhulme Prize of the Leverhulme Trust (PLP-2020-064).

B. Klartag
Department of Mathematics, Weizmann Institute of Science, Rehovot, Israel
e-mail: boaz.klartag@weizmann.ac.il

S. Sodin (✉)
School of Mathematical Sciences, Queen Mary University of London, London, UK
e-mail: a.sodin@qmul.ac.uk

polynomials in X (see e.g. [3, Theorem 9.23]), which we now recall. Let $d \geq 1$, and consider a polynomial of the form

$$P_d(x) = \sum_{\#(S)=d} a_S \cdot \left(\prod_{i \in S} x_i \right) \tag{1}$$

where the sum runs over all subsets $S \subseteq \{1, \ldots, n\}$ of cardinality d, and the coefficients (a_S) are arbitrary real numbers. In other words, P_d is a d-homogeneous, square-free polynomial in $\mathbb{R}^n$. The Bonami hypercontractivity theorem [3, Chapter 9] tells us that for any $1 < p \leq q$,

$$\|P_d(X)\|_q \leq \left(\frac{q-1}{p-1} \right)^{d/2} \|P_d(X)\|_p. \tag{2}$$

A general polynomial P of degree at most d on $\{-1, 1\}^n$ takes the form

$$P(x) = \sum_{k=0}^{d} P_k(x) \tag{3}$$

where P_k is a k-homogeneous, square-free polynomial. Thanks to orthogonality relations we have

$$\|P(X)\|_2^2 = \sum_{k=0}^{d} \|P_k(X)\|_2^2.$$

Hence, by the Bonami bound (2) and the Cauchy-Schwarz inequality, for any polynomial P of degree at most d and any $q \geq 3$,

$$\|P(X)\|_q \leq \sum_{k=0}^{d} \|P_k(X)\|_q \leq \sum_{k=0}^{d} (q-1)^{k/2} \|P_k(X)\|_2$$

$$\leq \sqrt{\sum_{k=0}^{d} (q-1)^k} \cdot \sqrt{\sum_{k=0}^{d} \|P_k(X)\|_2^2}$$

$$\leq \sqrt{2} \cdot (q-1)^{d/2} \cdot \|P(X)\|_2 \leq \sqrt{2} q^{d/2} \|P(X)\|_2. \tag{4}$$

For $r > 0$ (not necessarily integer), write a_r for a e^{-r}-quantile of $P(X)$, i.e. a number satisfying

$$\mathbb{P}(P(X) \geq a_r) \geq \frac{1}{e^r} \quad \text{and also} \quad \mathbb{P}(P(X) \leq a_r) \geq 1 - \frac{1}{e^r}.$$

Assume the normalization $\|P(X)\|_2 = 1$. It follows from (4) that if $q \geq 3$ then

$$\frac{1}{e^r} \leq \mathbb{P}(P(X) \geq a_r) \leq \frac{\mathbb{E}|P(X)|^q}{a_r^q} \leq \left(\sqrt{2} \cdot \frac{q^{d/2}}{a_r}\right)^q.$$

Substituting $q = 2r/d$ (when $r \geq 3d/2$), we get

$$a_r \leq \sqrt{2} \cdot (2er/d)^{d/2} \leq (Cr/d)^{d/2} \quad (r \geq 3d/2), \tag{5}$$

with a universal constant $C = 4$. Without assuming any normalisation, we obtain

$$a_r - a_1 \leq C^d \left(\frac{r}{d} + 1\right)^{d/2} \|P(X)\|_2 \tag{6}$$

(with a different numerical constant $C > 0$), which is valid for all $r \geq 1$.

The estimate (6) is a a tail bound for the distribution of $P(X)$, i.e. concentration inequality. We refer to [2] and references therein for background on concentration inequalities, particularly, for polynomials, and to [3] for applications of (6) .

In some applications, it is important to have bounds on $a_s - a_r$ when $s \geq r$ are close to one another, e.g. $s = r + 1$. Such bounds are called *local* tail bounds; see [1] and references therein. The following proposition, confirming a conjecture of Nathan Keller and Ohad Klein, provides a local version of (6). In the case $d = 1$, it follows from the results in the aforementioned work [1].

Proposition 1 *Let P be a polynomial of degree at most d on $\{-1, 1\}^n$. Then for all $r \geq 1$,*

$$a_{r+1} - a_r \leq C^d \left(\frac{r}{d} + 1\right)^{\frac{d}{2} - 1} \|P(X)\|_2 , \tag{7}$$

where $C > 0$ is a universal constant.

Clearly, (7) implies (6). The estimate (7) gives the right magnitude of $a_r - a_{r+1}$, say, for

$$P(X) = (X_1 + \cdots + X_n)^d , \quad n \gg 1 . \tag{8}$$

2 Proofs

We now turn to the proof of Proposition 1. Write $\partial_i P$ for the partial derivative of P with respect to the ith variable. Thus

$$\partial_i P(x) = \frac{P(T_i^1 x) - P(T_i^{-1} x)}{2} \qquad \text{for } x \in \{-1, 1\}^n,$$

where T_i^j is the map that sets the ith-coordinate of x to the value j, and keeps the other coordinates intact. Observe that $\partial_i P$ is a polynomial of degree at most $d-1$ if P is of degree d. We denote by ∇P the vector function with coordinates $\partial_i P$. The first step in the proof of Proposition 1 is to sharpen the quantile bound (5).

Lemma 2 *Let P be a polynomial of degree at most d with $\mathbb{E}|P(X)|^2 = 1$. Then for any non-empty subset $A \subseteq \{-1, 1\}^n$ of relative size $\varepsilon = \#(A)/2^n$ we have*

$$\frac{1}{\#(A)} \sum_{x \in A} |P(x)|^2 \le C^d \cdot \max\left\{1, \left(\frac{|\log \varepsilon|}{d}\right)^d\right\}, \tag{9}$$

and

$$\frac{1}{\#(A)} \sum_{x \in A} |\nabla P(x)|^2 \le C^d \cdot \max\left\{1, \left(\frac{|\log \varepsilon|}{d}\right)^{d-1}\right\}, \tag{10}$$

for a universal constant $C > 0$.

Proof Let $q \ge 3$. By Hölder's inequality followed by an application of (4),

$$\sum_{x \in A} |P(x)|^2 \le (\#(A))^{1-2/q} \cdot \left(\sum_{x \in A} |P(x)|^q\right)^{2/q} = (\#(A))^{1-2/q} \cdot 2^{2n/q} \cdot \|P(X)\|_q^2$$

$$\le (\#(A))^{1-2/q} \cdot 2^{2n/q} \cdot 2q^d ,$$

whence

$$\frac{1}{\#(A)} \sum_{x \in A} |P(x)|^2 \le 2\varepsilon^{-2/q} q^d .$$

The estimate (9) clearly holds for $\varepsilon \ge e^{-\frac{3d}{2}}$, therefore we assume that $\varepsilon < e^{-\frac{3d}{2}}$. Set

$$q = 2|\log \varepsilon|/d \ge 3$$

and obtain

$$\frac{1}{\#(A)} \sum_{x \in A} |P(x)|^2 \le \left(\frac{C}{d}\right)^d |\log \varepsilon|^d.$$

This proves (9). Since $\partial_i P$ is a polynomial of degree at most $d-1$, from (9),

$$\frac{1}{\#(A)} \sum_{x \in A} |(\partial_i P)(x)|^2 \le C^d \cdot \max\left\{1, \left(\frac{|\log \varepsilon|}{d}\right)^{d-1}\right\} \cdot \mathbb{E}|(\partial_i P)(X)|^2 ,$$

whence

$$\frac{1}{\#(A)} \sum_{x \in A} |(\nabla P)(x)|^2 \le C^d \cdot \max \left\{ 1, \left(\frac{|\log \varepsilon|}{d} \right)^{d-1} \right\} \cdot \mathbb{E}|(\nabla P)(X)|^2 .$$

We decompose $P(X) = \sum_{k=0}^d P_k(X)$ as in (3), and use the orthogonality relations

$$\mathbb{E}|\nabla P(X)|^2 = \sum_{k=0}^d \mathbb{E}|\nabla P_k(X)|^2 = \sum_{k=0}^d k \cdot \mathbb{E}|P_k(X)|^2 \le d \cdot \mathbb{E}|P(X)|^2 = d.$$

This proves (10). □

Note that for any $f : \{-1, 1\}^n \to \mathbb{R}$,

$$\sum_{x \in \{-1,1\}^n} |\nabla f(x)|^2 \le 2 \cdot \sum_{x \in \{-1,1\}^n} |\nabla f(x)|^2 \cdot 1_{\{f(x) \ne 0\}}. \tag{11}$$

Indeed, the expression on the left-hand side of (11) is the sum over all oriented edges $(x, y) \in E$ in the Hamming cube of the squared difference $|f(x) - f(y)|^2/4$. This is clearly at most twice the sum over all oriented edges $(x, y) \in E$ of the quantity $|f(x) - f(y)|^2 \cdot 1_{\{f(x) \ne 0\}}/4$.

Recall the log-Sobolev inequality (e.g. [3, Chapter 10]) which states that for any function $f : \{-1, 1\}^n \to \mathbb{R}$,

$$\mathbb{E}f^2(X) \log f^2(X) - \mathbb{E}f^2(X) \cdot \log \mathbb{E}f^2(X) \le 2\mathbb{E}|\nabla f(X)|^2. \tag{12}$$

Moreover, let $A \subseteq \{-1, 1\}^n$ be a non-empty set and denote $\varepsilon = \#(A)/2^n$. If the function f is supported in A and is not identicaly zero, then denoting $g = f/\sqrt{\mathbb{E}f^2(X)}$,

$$\mathbb{E}f^2(X) \log f^2(X) - \mathbb{E}f^2(X) \cdot \log \mathbb{E}f^2(X) = \mathbb{E}f^2(X) \cdot \mathbb{E}g^2(X) \log g^2(X)$$
$$\ge \mathbb{E}f^2(X) \cdot |\log \varepsilon|, \tag{13}$$

because g^2 is supported in A, and among all probability distributions supported in A, the maximal entropy is attained for the uniform distribution.

Proof of Proposition 1 Without loss of generality $\|P(X)\|_2 = 1$. We may assume that $a_{r+1} > a_r$, as otherwise there is nothing to prove. Let $U = \{x \in \{-1, 1\}^n ; f(x) > a_r\}$ and set $\varepsilon = \#(U)/2^n$. Then $e^{-(r+1)} \le \varepsilon \le e^{-r}$, by the definition of the quantiles a_r and a_{r+1}. Denote $\chi(t) = \max(t - a_r, 0)$; this is a

1-Lipschitz function on the real line. Applying the log-Sobolev inequality (12) to the function $h = \chi \circ P : \{-1, 1\}^n \to \mathbb{R}$ we get

$$\mathbb{E}h^2(X) \log h^2(X) - \mathbb{E}h^2(X) \cdot \log \mathbb{E}h^2(X) \le 2\mathbb{E}|\nabla h|^2(X). \tag{14}$$

Since h is supported in U, with $\varepsilon = \#(U)/2^n$, by (13) and (14),

$$\mathbb{E}h^2(X) \cdot |\log \varepsilon| \le 2\mathbb{E}|\nabla h|^2(X) \le 4\mathbb{E}|\nabla h(X)|^2 \cdot 1_{\{h(X)>0\}}.$$

The last passage is the content of (11). Since χ is 1-Lipschitz, we know that $|\nabla h|^2 \le |\nabla P|^2$. Hence, by (10),

$$\mathbb{E}|\nabla h(X)|^2 \cdot 1_{\{h(X)>0\}} \le \mathbb{E}|\nabla P(X)|^2 1_{\{X \in U\}} \le \varepsilon \cdot C^d \cdot \max\left\{1, \left(\frac{|\log \varepsilon|}{d}\right)^{d-1}\right\}.$$

To summarize,

$$\mathbb{E}h^2(X) \cdot |\log \varepsilon| \le \varepsilon \cdot C_1^d \cdot \max\left\{1, \left(\frac{|\log \varepsilon|}{d}\right)^{d-1}\right\}, \tag{15}$$

for a universal constant $C_1 > 0$. Recall that $e^{-(r+1)} \le \varepsilon \le e^{-r}$. By the definition of a_{r+1}, we know that $h(X) \ge a_{r+1} - a_r$ with probability at least $e^{-(r+1)}$. Therefore, from (15),

$$e^{-(r+1)} \cdot (a_{r+1} - a_r)^2 \cdot \frac{r}{2} \le e^{-r} \cdot C_1^d \cdot \max\left\{1, \left(\frac{2r}{d}\right)^{d-1}\right\}$$

or

$$a_{r+1} - a_r \le C_2^d \cdot \max\left\{\frac{1}{\sqrt{r}}, \left(\frac{r}{d}\right)^{d/2-1}\right\} \le C_3^d \left(\frac{r}{d} + 1\right)^{\frac{d}{2}}.$$

$\square$

3 Remarks

We remark that Proposition 1 implies the following corollary which holds true without the normalization by $\|P(X)\|_2$.

Corollary 3 *There exists* $C > 0$ *such that the following holds. Let* P *be a polynomial of degree at most d with* $\mathbb{E}P(X) = 0$. *Then for* $r \geq Cd$,

$$a_{r+1} \leq a_r \left[1 + C^d \left(\frac{r}{d} + 1 \right)^{\frac{d}{2}-1} \right]. \tag{16}$$

Remark 4 We conjecture that (16) also holds with the power -1 in place of $\frac{d}{2} - 1$. Such an estimate would give the right order of magnitude for the polynomial (8).

Proof of Corollary 3 Write $\sigma^2 = \mathbb{E}|P(X)|^2$. We shall prove that $\sigma \leq C_1^d a_r$. Once this inequality is established, we deduce from Proposition 1 that

$$\frac{a_{r+1} - a_r}{\sigma} \leq C^d \left(\frac{r}{d} + 1 \right)^{\frac{d}{2}-1} ,$$

whence

$$a_{r+1} \leq a_r + \sigma \cdot C^d \left(\frac{r}{d} + 1 \right)^{\frac{d}{2}-1} \leq a_r \left(1 + (CC_1)^d \left(\frac{r}{d} + 1 \right)^{\frac{d}{2}-1} \right),$$

as claimed.

Let $\sigma_\pm = \sqrt{\mathbb{E}(P(X)_\pm)^2}$. First, we claim that $\sigma_+ \geq C_2^{-d}\sigma$. Indeed, if $\sigma_+ \geq \sigma_-$ then $\sigma_+ \geq \sigma/\sqrt{2}$. If $\sigma_+ < \sigma_-$, then, using (4),

$$\sigma_+ \geq \mathbb{E}P(X)_+ = \mathbb{E}P(X)_- \geq \frac{(\mathbb{E}P(X)_-^2)^{3/2}}{(\mathbb{E}P(X)_-^4)^{1/2}}$$

$$\geq \frac{\sigma_-^3}{2 \cdot 3^d \cdot (\sigma_+^2 + \sigma_-^2)} \geq \frac{1}{4 \cdot 3^d}\sigma_- .$$

Second, another application of (4) yields

$$\mathbb{E}P(X)_+^4 \leq \mathbb{E}P(X)^4 \leq 4 \cdot 3^d \sigma^4 \leq C_3^d \sigma_+^4 ,$$

thus by the Paley–Zygmund inequality

$$e^{-Cd} \geq e^{-r} \geq \mathbb{P}\{P(X) > a_r\} \geq \frac{(1 - a_r^2/\sigma_+^2)_+^2}{C_3^d} ,$$

whence $\sigma_+ \leq 2a_r$ if we ensure that, say, $e^C \geq 2C_3$. This concludes the proof. $\square$

Finally, we remark that both Proposition 1 and Corollary 3 can be generalised in several directions. For example, instead of the Hamming cube, one can consider a general measure which is invariant under a Markov diffusion satisfying the Bakry–Émery CD(R, ∞) condition; in this setting, linear combinations of eigenfunctions of the generator play the rôle of polynomials. The proof requires only notational modifications.

Acknowledgments We are grateful to Nathan Keller for helpful correspondence.

References

1. L. Devroye, G. Lugosi, Local tail bounds for functions of independent random variables. Ann. Probab. **36**(1), 143–159 (2008)
2. A.A. Giannopoulos, V.D. Milman, Concentration property on probability spaces. Adv. Math. **156**(1), 77–106 (2000)
3. R. O'Donnell, *Analysis of Boolean Functions* (Cambridge University Press, New York, 2014)

Stable Recovery and the Coordinate Small-Ball Behaviour of Random Vectors

Shahar Mendelson and Grigoris Paouris

Abstract Recovery procedures in Data Science are often based on *stable point separation*. In its simplest form, stable point separation implies that if f is "far away" from 0, and one is given a random sample $(f(Z_i))_{i=1}^m$ where a proportional number of the sample points may be corrupted by noise—even maliciously, that information is still enough to exhibit that f is far from 0.

Stable point separation is well understood in the context of iid sampling, and to explore it for general sampling methods we introduce a new notion—the *coordinate small-ball* of a random vector X. Roughly put, this feature captures the number of "relatively large coordinates" of $(|\langle TX, u_i \rangle|)_{i=1}^m$, where $T : \mathbb{R}^n \to \mathbb{R}^m$ is an arbitrary linear operator and $(u_i)_{i=1}^m$ is any fixed orthonormal basis of $\mathbb{R}^m$.

We show that under the bare-minimum assumptions on X, and with high probability, many of the values $|\langle TX, u_i \rangle|$ are at least of the order $\|T\|_{S_2}/\sqrt{m}$. As a result, the "coordinate structure" of TX exhibits the typical Euclidean norm of TX and does so in a stable way.

One outcome of our analysis is that random sub-sampled convolutions satisfy stable point separation under minimal assumptions on the generating random vector—a fact that was known previously only in a highly restrictive setup, namely, for random vectors with iid subgaussian coordinates. As an application we address the problem of sparse signal recovery using a circulant matrix when a proportion of the given sample is corrupted by malicious noise.

Supported by NSF grant DMS-1812240 and Simons Fellows in Mathematics Award 823432.

S. Mendelson (✉)
MSI, The Australian National University, Canberra, ACT, Australia

Department of Statistics, University of Warwick, Coventry, UK
e-mail: shahar.mendelson@anu.edu.au

G. Paouris
Department of Mathematics, Texas A&M University, College Station, TX, USA

Department of Mathematics, Princeton University, Princeton, NJ, USA
e-mail: grigoris@math.tamu.edu; gp5172@princeton.edu

1   Introduction

One of the key questions in Data Science is to identify (or at least approximate) an unknown function using partial information. In standard recovery problems the data one receives consists of a finite sample of the unknown function and the sample points are assumed to be independent. The sample is then used to construct a suitable 'guess' of the function and the hope is that the guess is a good approximation in some appropriate sense.

Off-hand, the significance of having sample points that are selected independently is not clear. A closer inspection shows that independence has a strong geometric impact: it leads to *point separation*.

1.1   Point Separation and Stable Point Separation

To explain what we mean by point separation, let us first consider it in its simplest form, separation of a function from 0.

Given a function f on a probability space (Ω, μ), let Z be distributed according to μ and consider a sample $Z_1, \ldots, Z_m$, consisting of independent points distributed as Z. The sample $(Z_1, \ldots, Z_m)$ naturally endows a random vector $X \in \mathbb{R}^m$, whose coordinates are the given measurements $f(Z_i)$, $1 \le i \le m$; that is,

$$X = (f(Z_1), \ldots, f(Z_m)) .$$

Any hope of identifying f from the given data vector X is based on the belief that X captures enough features of f; for example, that f can be distinguished from 0 with only X as data. Thus, one has to address the following question:

Question 1.1 If f is reasonably far away from 0, when is that fact exhibited by a typical realization of X?

An obvious way of exhibiting separation between f and 0 is through the Euclidean norm of X; specifically, by showing that, with high probability,

$$\frac{\|X\|_2^2}{m} = \frac{1}{m} \sum_{i=1}^m f^2(Z_i) \ge \kappa \|f\|_{L_2}^2 \tag{1.1}$$

for a suitable constant κ and for any $m \ge m_0$. Independence proves to be extremely useful in establishing (1.1). Indeed, under a weak *small-ball assumption*, that

$$\mathbb{P}(|f(Z)| \ge \kappa \|f\|_{L_2}) \ge \rho, \tag{1.2}$$

it is straightforward to verify that with probability at least $1 - 2\exp(-c\rho m)$,

$$|\{i : |f(Z_i)| \geq \kappa \|f\|_{L_2}\}| \geq \frac{\rho}{2}m. \tag{1.3}$$

Thus, with very high probability, a proportion of the coordinates of X are large—of the order of $\|f\|_{L_2}$.

While (1.3) clearly implies (1.1) and point separation, it says much more: under the small-ball assumption (1.2), *independent sampling leads to stable point separation*: not only is $\|X\|_2$ large, the reason that it is large is because many of its coordinates $|\langle X, e_i\rangle|$ are nontrivial—making point separation robust to noise. In particular, even if a (small) fraction of the measurements $f(Z_i)$ are corrupted maliciously, the fact that f is far away from 0 is still exhibited by the corrupted vector.

In a more geometric language, stable point separation is manifested by the fact that $(\langle X, e_i\rangle)_{i=1}^{N}$ is a well-spread vector, and obviously this significant additional information does not come for free: stable point separation is much harder to prove than point separation. At the same time, the importance of the notion is clear: intuitively, a sampling method can be useful in statistical recovery problems, where being robust to noise is of the utmost importance, only if it satisfies a *uniform version of stable point separation*. Indeed, at the heart of numerous statistical procedures is the fact that if F is a class of functions, then with high probability, for every $f, h \in F$ that are 'far enough'

$$|\{i : |(f - h)(X_i)| \geq \kappa \|f - h\|_{L_2}\}| \geq c(\rho)m, \tag{1.4}$$

which is a uniform version of stable point separation. It allows one to distinguish between any two functions in the given class that are sufficiently far apart using a typical sample, even when a proportional number of the given measurements are corrupted by noise (see Example 1.3 and Sect. 4.2).

Uniform stable point separation has played a central role in the recent progress on some key questions in learning theory and statistics. For example, it has led to the introduction of an optimal learning procedure in [14, 16]; to optimal vector mean estimation in [12, 13] and to optimal covariance estimation in [17, 18]—all of which in heavy-tailed situations.

Unfortunately, stable point separation and its uniform counterpart are well understood only for iid sampling, and the downside of iid sampling is that it leads to various computational difficulties. For example, consider a relatively simple recovery problem, where the goal is to identify an unknown $t_0 \in T \subset \mathbb{R}^n$ using linear measurements $(\langle Z_i, t_0\rangle)_{i=1}^{m}$ and $Z_1, \ldots, Z_m$ are independent copies of the standard Gaussian random vector in $\mathbb{R}^n$. Procedures that aim at recovering t_0 are based on vector multiplications with the matrix $\Gamma = \sum_{i=1}^{m} \langle Z_i, \cdot\rangle e_i$, but because Γ has independent Gaussian rows, vector multiplication is computationally expensive.

To address these and other computational difficulties of a similar nature, other sampling methods are often used in recovery problems. However, once the iid framework is abandoned, establishing the required point separation/stable point separation becomes a formidable task; in fact, it is often far from obvious that either one of the properties is true when the sample points are not independent.

Motivated by general sampling methods, the question we focus on is as follows: are point separation and stable point separation really the outcome of independence? Rather informally, the question we study is:

Question 1.2 Given a centred random vector $X \in \mathbb{R}^n$,

(a) What conditions on X are needed to ensure that for an arbitrary linear operator $T : \mathbb{R}^n \to \mathbb{R}^m$ and with high probability $\|TX\|_2$ is reasonably large?
(b) When is the fact that $\|TX\|_2$ is large exhibited by an arbitrary coordinate structure? In other words, given an arbitrary orthonormal basis $(u_i)_{i=1}^m$, are many of the values $|\langle TX, u_i \rangle|$ reasonably large?

Question 1.2 clearly extends the notions of point separation and stable point separation from the iid setup, where $m = n$ and $X = (f(Z_i))_{i=1}^m$: in the general framework of Question 1.2 the coordinates of X need not be independent or identically distributed, and X is further distorted by a linear operator T.

Let us illustrate how addressing the two parts of Question 1.2 can become unpleasant very quickly once independence is left behind. The example we focus on here and in what follows is the very popular *random sub-sampled convolutions* scheme, which is used in numerous applications, such as SAR radar imaging, optical imaging, channel estimation, etc. (see [6, 25] for more details on these and other applications).

Example 1.3 Let ξ be an isotropic random vector in $\mathbb{R}^n$ (that is, ξ is centred and for every $t \in \mathbb{R}^n$, $\mathbb{E}\langle \xi, t \rangle^2 = \|t\|_2^2$). Fix $a \in \mathbb{R}^n$ and let $W = a \circledast \xi$ be the discrete convolution of a and ξ; i.e., if $j \ominus i = j - i \mod n$ and τ_i is the shift operator defined by $(\tau_i x) = (x_{j \ominus i})_{j=1}^n$, then

$$a \circledast \xi = (\langle a, \tau_i \xi \rangle)_{i=1}^n.$$

The measurements of the vector a one receives come from a selection of a random subset of the coordinates of $a \circledast \xi$: let $\delta_1, \ldots, \delta_n$ be independent, $\{0, 1\}$-valued random variables with mean δ; set $I = \{i : \delta_i = 1\}$; and define $Z = (a \circledast \xi)_{i \in I}$.

Note that typically $|I| \sim \delta n$ and $\mathbb{E}\|Z\|_2^2 = \delta n \|a\|_2^2$. Therefore, this sampling method exhibits point separation of a and 0 if, with high probability,

$$\frac{1}{\delta n} \sum_{i \in I} Z_i^2 \geq c\|a\|_2^2 \tag{1.5}$$

for a suitable constant c (that should be independent of a and δ). And, it exhibits stable point separation of a and 0 with respect to the standard basis $(e_i)_{i=1}^n$ if with high probability,

$$\left|\{i \in I : |\langle Z, e_i \rangle| \geq c'\|a\|_2\}\right| \geq c''\delta n \qquad (1.6)$$

for suitable constants c' and c''. In particular, (1.6) means that the large Euclidean norm of Z is exhibited by the fact that many of the coordinates of Z (with respect to the standard basis) are large.

Stable point separation can be used to address a well-known sparse recovery problem that has been studied extensively in recent years (see, e.g. [6] and references therein and [19] for some recent progress). Assume that $t_0 \in \mathbb{R}^n$ is an unknown vector that is sufficiently sparse relative to the standard basis. One wishes to identify t_0 and to that end generates the random sub-sampled convolution of t_0 with ξ, i.e., the points $Z_i = \langle t_0 \circledast \xi, e_i \rangle$ for $i \in I$. However, the information one is actually given is a corrupted sample $(Y_i)_{i \in I}$, generated by a malicious adversary who has the freedom to change up to $(\frac{1}{2} - \eta)|I|$ of the points $(Z_i)_{i \in I}$ in any way they fit.

As we explain in what follows, under minimal assumptions of ξ and thanks to the stable point separation property we establish, one can still recover t_0 with the optimal number of measurements: if t_0 is supported on at most s coordinates with respect to the standard basis one requires $|I| \sim s \log(en/s)$—as if ξ were Gaussian and the sample were not contaminated maliciously. The one restriction is that s has to be sufficiently small, at most $\sim \sqrt{n/\log n}$ (see Theorem 1.11 and Sect. 4.2 for an exact formulation and proof).

Clearly, identifying when, or even if, (1.5) and (1.6) are true is considerably harder than establishing (1.1) and (1.3). And if they are, it has nothing to do with independence.

A wildly optimistic conjecture is that both parts of Question 1.2 are (almost) universally true under minimal assumptions on X. And deferring an accurate definition of what is meant by "reasonably large", the main result of this article is that this wildly optimistic conjecture is, in fact, true:

- Under the bare-minimum assumptions on X, for an arbitrary linear operator T, TX has a large Euclidean norm; moreover, that norm is exhibited by many large coordinates *with respect to an arbitrary orthonormal basis*.
- Both facts hold with high probability and are simply generic properties of X that have nothing to do with independence, nor with concentration of measure.
- In particular, almost any random vector TX exhibits both point separation and stable point separation with respect to an arbitrary orthonormal basis.

We show in what follows that the reason why both parts of Question 1.2 are universally true is a *small-ball assumption* which we now describe.

1.2 The Small-Ball Assumption

To have some intuition on the sort of quantitative answers to Question 1.2 one can hope for, assume for the time being that X is isotropic. Let $F \subset \mathbb{R}^n$ be a subspace of dimension k, and set P_F to be the orthogonal projection onto F. Thus, $\mathbb{E}\|P_F X\|_2^2 = k$ and at least intuitively, saying that $P_F X$ has a "reasonably large Euclidean norm can be taken to mean that $\|P_F X\|_2 \geq \varepsilon \sqrt{k}$ for some $0 < \varepsilon < 1$. Moreover, a "reasonably large coordinate" of such a k-dimensional vector should be at least of the order of $(\mathbb{E}\|P_F X\|_2^2)^{\frac{1}{2}}/\sqrt{k} = 1$.

Following the same path with a general linear operator $T : \mathbb{R}^n \to \mathbb{R}^m$ instead of P_F, the intuitive notion of being relatively large is that $\|TX\|_2$ is at least $\varepsilon \|T\|_{S_2} = \varepsilon (\mathbb{E}\|TX\|_2^2)^{\frac{1}{2}}$, where $\|T\|_{S_2}$ denotes the Hilbert-Schmidt norm of T; and given an orthonormal basis $(u_i)_{i=1}^m$, a "large coordinate" of TX satisfies that $|\langle TX, u_i \rangle| \gtrsim \|T\|_{S_2}/\sqrt{m}$.

Once the two notions are agreed upon, the answers to the two parts of Question 1.2 are given in the form of *small-ball estimates*, that is, upper bounds on

$$\mathbb{P}(\|TX\|_2 \leq y) \quad \text{for} \quad 0 < y \leq \|T\|_{S_2}$$

and *coordinate small-ball estimates* which are upper bounds on

$$\mathbb{P}(|\{i : |\langle TX, u_i \rangle| \leq y\}| \geq \ell) \quad \text{for} \quad 0 < y \leq \|T\|_{S_2}/\sqrt{m}.$$

Let us emphasize a fact, which at first glance, may be surprising:

> Small-ball estimates and coordinate small-ball estimates have nothing to do with concentration.

Indeed, although the notions of small-ball estimates and coordinate small-ball estimates may seem to be related to *concentration of measure*, they are actually based on a totally different phenomenon that has nothing to do with the way the random variable $\|TX\|$ concentrates around its mean $\mathbb{E}\|TX\|$—no matter what norm $\|\ \|$ is considered.

There are several reasons for that: firstly, two-sided concentration estimates of the form

$$\mathbb{P}(|\|TX\| - \mathbb{E}\|TX\|| \geq y)$$

are a combination of the upper estimate—that with high probability, $\|TX\| \leq \mathbb{E}\|TX\| + y$, and the lower one, that $\|TX\| \geq \mathbb{E}\|TX\| - y$. By now it is well understood (see, for example, the discussion in [15]) that the two estimates are totally different and are caused by unrelated features of the random vector X. Moreover, the upper tail is almost always the bottleneck in the two-sided estimate, while our interests lie in the lower one. Secondly, the scale one is interested in when studying the small-ball behaviour of TX corresponds to the lower tail with the choice of $y = (1 - s)\mathbb{E}\|TX\|$, for s close to 0. That is very different from the lower tail at the 'concentration scale' of $y = s\mathbb{E}\|TX\|$ for s close to 0.

Remark 1.4 As we explain in what follows, the behaviour of $\mathbb{P}(\|TX\|_2 \leq y)$ is more subtle than what this intuitive description may lead one to believe. In fact, $\mathbb{P}(\|TX\|_2 \leq \varepsilon\|T\|_{S_2})$ exhibits multiple phase transitions at different levels of ε in the small-ball regime.

The minimal assumption that is required for establishing small-ball and coordinate small-ball estimates is as follows:

Assumption 1.5 *The random vector X satisfies a* small ball assumption *(denoted from here on by SBA) with constant $\mathcal{L}$ if for every $1 \leq k \leq n-1$, every k dimensional subspace F, every $z \in \mathbb{R}^n$ and every $\varepsilon > 0$,*

$$\mathbb{P}\left(\|P_F X - z\|_2 \leq \varepsilon\sqrt{k}\right) \leq (\mathcal{L}\varepsilon)^k, \tag{1.7}$$

where P_F is the orthogonal projection onto the subspace F.

It is straightforward to verify (see, e.g., Proposition 2.2 in [26]) that X satisfies the *SBA* with constant $\mathcal{L}$ if and only if for every $1 \leq k \leq n$, the densities of all k-dimensional marginals of X are bounded by $\mathcal{L}^k$ (assuming, of course, that X has a density, and in which case f_X denotes that density).

Since, for most of our results, we will assume nothing about the mean (center) of X it is more natural to consider all translations with vectors z in (1.5).

There are numerous examples of generic random vectors that satisfy the *SBA* with an absolute constant; among them are vectors with iid coordinates that have a bounded density (see [26] and [11] for the optimal constant) as well as various log-concave random vectors.[1] For more details see Appendix A, where we list several examples of generic log-concave random vectors that satisfy Assumption 1.5.

Although Assumption 1.5 requires that X has a density, this is not essential and our main results remain true even under the following weaker assumption.

Assumption 1.6 *Let $\mathcal{L}$ and θ be such that $\mathcal{L}\theta < 1$. The random vector X satisfies the* weak small-ball assumption *(denoted from here on by wSBA) with constants*

---

[1] Recall that X is log-concave if it has a density f_X that satisfies that for every x, y in the support of f_X and every $0 \leq \lambda \leq 1$, $f_X((1 - \lambda)x + \lambda y) \geq f_X^{(1-\lambda)}(x)f_X^{\lambda}(y)$.

θ and $\mathcal{L}$ if for every $1 \leq k \leq n - 1$, every k dimensional subspace F, and every $z \in \mathbb{R}^n$,

$$\mathbb{P}\left(\|P_F X - z\|_2 \leq \theta\sqrt{k}\right) \leq (\mathcal{L}\theta)^k. \tag{1.8}$$

Clearly, if X satisfies the *SBA* with constant $\mathcal{L}$ then it satisfies the *wSBA* with constants θ and $\mathcal{L}$ for every $\theta > 0$. Moreover, it follows from [26] that if X has independent coordinates, and if each coordinate satisfies the *wSBA* with constants θ and $\mathcal{L}$, then X satisfies the *wSBA* with constants $C\theta$ and $\mathcal{L}$, where $C > 0$ is an absolute constant.

Remark 1.7 Most of the results presented in what follows hold under the *wSBA*. However, to simplify the presentation only one result is proved under that assumption—the coordinate small-ball estimate (Theorem 1.20); the other results are formulated using the *SBA* which leads to a proof that is less involved.

Before we formulate the main results, let us mention one of their outcomes: a stable point separation bound for the random sub-sampled convolutions scheme.

1.2.1   Example 1.3 Revisited

As it happens, the existing state of the art on point separation/stable point separation of the random sub-sampled convolutions scheme can be improved dramatically, as existing estimates are based on severe restrictions on the random vector ξ. The reason for those restrictions is a wasteful method of proof, as is explained in Sect. 5.1, and that leads to the following:

Theorem 1.8 ([19]) *For every constant $L \geq 1$ there exist constants c_0, c_1, c_2, c_3 and c_4 that depend only on L for which the following holds. Let x be a mean-zero, variance one, L-subgaussian random variable,[2] and set $\xi = (x_i)_{i=1}^n$, i.e., a vector whose coordinates are independent copies of x. Let $s \leq c_0 n / \log^4 n$ and consider $a \in S^{n-1}$ that is s-sparse[3] with respect to the standard basis $(e_i)_{i=1}^n$. Then with probability at least $1 - 2\exp(-c_1 \min\{n/s, \delta n\})$ with respect to both ξ and $(\delta_i)_{i=1}^n$,*

$$\sum_{i \in I} \langle a \circledast \xi, e_i \rangle^2 \geq c_2 \delta n \quad \text{and} \quad |\{i \in I : |\langle a \circledast \xi, e_i \rangle| \geq c_3\}| \geq c_4 \delta n. \tag{1.9}$$

We show that one can replace the wasteful parts of Theorem 1.8, leading to a sharp point separation and stable point separation that hold as long as ξ satisfies the *SBA*, and with a much better probability estimate.

---

[2] A centred random variable x is L-subgaussian if for every $p \geq 2$, $\|x\|_{L_p} \leq L\sqrt{p}\|x\|_{L_2}$.

[3] We say that a vector is s-sparse if it is supported on a set of cardinality no more than s.

In [19], a standard representation of Av where A is a complete circulant matrix via the Discrete Fourier Transform has been used. Following [19], if $\mathcal{F}$ is the discrete Fourier matrix,

$$a \circledast \xi = \mathcal{F}^{-1}\mathcal{F}(a \circledast \xi) = \mathcal{F}^{-1}\left((\mathcal{F}a)_i \cdot (\mathcal{F}\xi)_i\right)_{i=1}^n = \mathcal{F}^{-1}D_{\mathcal{F}a}\mathcal{F}\xi,$$

where $D_{\mathcal{F}a}$ is a diagonal matrix whose diagonal entries are $d_{ii} = (\mathcal{F}a)_i$. Setting

$$U = \mathcal{F}^{-1}/\sqrt{n}, \ W = \mathcal{F}/\sqrt{n}, \text{ and } O = \mathcal{F}/\sqrt{n}, \tag{1.10}$$

it follows that

$$a \circledast \xi = \sqrt{n}UD_{Wa}O\xi \equiv \Gamma_a\xi, \tag{1.11}$$

where U, W and O are unitary matrices with the property that

$$\max_{i \le n} \|W_i\|_\infty \le \frac{1}{\sqrt{n}}, \tag{1.12}$$

where $W_i, i \le n$ are the rows of W. However, note that Γ_a is a real valued matrix. Set $\hat{a} := \frac{1}{\sqrt{n}}\mathcal{F}a = Wa$. With the above notation we have the following

Theorem 1.9 *Let ξ satisfies the SBA with constant $\mathcal{L}$ and consider $a \in S^{d-1}$ that is s-sparse for $s \le c_0 n/\log n$. Then for any $0 < \varepsilon < 1$ and $q > 2$, with probability at least*

$$1 - (c_1\mathcal{L}\varepsilon)^{\frac{c_2}{\|\hat{a}\|_q^{2q/(q-2)}}} - \exp(-c_3\delta n),$$

$$\sum_{i \in I} \langle a \circledast \xi, e_i \rangle^2 \ge c_4\varepsilon^2\delta n \text{ and } |\{i \in I : |\langle a \circledast \xi, e_i \rangle| \ge \varepsilon\}| \ge c_5\delta n;$$

here, c_0, c_1 and c_2 are constants that depend on q and c_3, c_4, c_5 are absolute constants.

The differences between Theorems 1.8 and 1.9 are substantial. Firstly, the estimate in Theorem 1.9 holds for a random vector that satisfies the SBA rather than only for vectors that have iid subgaussian coordinates. Secondly, note that for any $a \in S^{n-1}$ and any $q > 2$, $1/\|\hat{a}\|_q^{2q/(q-2)} \ge 1/\|\hat{a}\|_\infty^2$; and for any $a \in S^{n-1}$ that is s-sparse, $1/\|\hat{a}\|_\infty^2 \ge n/s$. Thus, the probability estimate in Theorem 1.9 is always better than in Theorem 1.8, and often the gap between the two is significant.

Remark 1.10 It is possible to prove a version of Theorem 1.9 for X that satisfies the wSBA, but for the sake of a simpler presentation we shall not do that here.

Let us turn to the sparse recovery problem that was outlined previously. Consider an unknown vector $t_0 \in \mathbb{R}^n$ and let

$$Z_i = \langle t_0 \circledast \xi, e_i \rangle, \quad \text{for } i \in I$$

be the random sub-sampled convolution of t_0 with ξ.

Before the random sample $(Z_i)_{i \in I}$ is given, an adversary maliciously changes at most $(\frac{1}{2} - \eta)|I|$ of the sample points—knowing in advance the recovery strategy that has been chosen; the corrupted sample given as data is denoted by $(Y_i)_{i \in I}$. Despite the malicious changes, optimal recovery is possible as long as t_0 was sufficiently sparse:

Theorem 1.11 *Let $q > 1$ and assume that $s \leq c_0(q, \eta, \mathcal{L})\sqrt{n/\log n}$ and that $\delta n = c_1(q, \eta, \mathcal{L})s \log(en/s)$. There is a recovery procedure that, with probability at least*

$$1 - 2\exp(-c_2(\eta, q, \mathcal{L})s \log(en/s)) - c_3(\eta, \mathcal{L})/n^q,$$

for any unknown t_0 that is s-sparse, and upon receiving as data the corrupted sample $(Y_i)_{i \in I}$, is able to recover t_0.

Note that the number of measurements needed for recovery is the optimal one $\sim s \log(en/s)$, as if ξ where the standard Gaussian vector; however, the probability estimate is somewhat weaker than in the Gaussian case and is caused by the fact that linear forms $\langle \xi, t \rangle$ need not have any moments beyond the second one. The one major restriction is the fact that t_0 has to be very sparse, but even in that range, the best estimate (even in the noise-free scenario!) is from [19], where it is assumed that ξ is a subgaussian random vector with iid coordinates.

Theorem 1.11 is about the *existance* of a statistical recovery procedure even under strong adversarial corruption of the given data. This leaves open the interesting question of finding a computationally feasible procedure for such a problem. We are not dealing with the feasibility question in this paper.

1.3 Small-Ball Estimates

If one wants to highlight the crucial (and rather remarkable) feature of the small-ball estimate presented here, it is the following:

A Gaussian random vector is not the best case; actually, it is the worst one.

To explain what we mean by this, let X be a random vector taking values in $\mathbb{R}^n$ and for now assume that it satisfies the *SBA* with constant $\mathcal{L}$. Set $m \leq n$ and let $T : \mathbb{R}^n \to \mathbb{R}^m$ be a linear operator of full rank. Without loss of generality one may assume that T actually maps $\mathbb{R}^n$ into $\mathbb{R}^n$ and denote by $s_1, \cdots, s_m$ its nonzero singular values.

Recall that the p-Schatten norm of T is

$$\|T\|_{S_p} = \left(\sum_{i=1}^{m} s_i^p \right)^{1/p},$$

and following [21], for $2 < q \leq \infty$, let

$$\mathrm{srank}_q(T) = \left(\frac{\|T\|_{S_2}}{\|T\|_{S_q}} \right)^{\frac{2q}{q-2}}$$

be the *q-stable rank* of T. Clearly $\mathrm{srank}_q(T) \leq m = \mathrm{rank}(T)$ and the case $q = \infty$ corresponds to the standard notion of the stable rank, i.e.,

$$\mathrm{srank}(T) = \left(\frac{\|T\|_{S_2}}{\|T\|_{S_\infty}} \right)^2.$$

The current state of the art as far as small-ball estimates are concerned is due to Rudelson and Vershynin:

Theorem 1.12 ([26]) *There are absolute constants c_0 and c_1 for which the following holds. If X satisfies the SBA with constant $\mathcal{L}$ then for any $\varepsilon > 0$,*

$$\mathbb{P}\left(\|TX\|_2 \leq \varepsilon \|T\|_{S_2} \right) \leq (c_0 \mathcal{L} \varepsilon)^{c_1 \mathrm{srank}(T)}. \tag{1.13}$$

Remark 1.13 Although Theorem 1.12 is not stated explicitly in [26], it follows from the analysis presented there in a straightforward way. Previous estimates of the same flavour have been derived for a centred random vector X that has independent subgaussian entries in [10] and for an X that is isotropic, log-concave and subgaussian in [23].

One instance in which Theorem 1.12 can be applied is when T is an orthogonal projection of rank k (and in which case, $\mathrm{srank}(T) = k$). On the other hand, it is straightforward to verify that if (1.13) holds for any such orthogonal projection then X satisfies the *SBA* (though perhaps with a slightly different constant). Despite this equivalence, Theorem 1.12 is far from optimal—because of the loose probability estimate; it does not "see" phase transitions that occur as ε decreases.

In contrast to Theorem 1.12, our first main result is a comparison theorem which shows that the worst random vector in the context of small-ball estimates is actually the standard Gaussian. Then, in Corollary 1.17 and Theorem 1.18 one uses the Gaussian case to establish the right probability estimate at every scale.

Theorem 1.14 *Let X be an n-dimensional random vector that satisfies the SBA with constant $\mathcal{L}$, let $T : \mathbb{R}^n \to \mathbb{R}^m$ be a linear map and set G to be the standard Gaussian vector in $\mathbb{R}^n$. Then for every $1 \leq k < m = \mathrm{rank}(T)$,*

$$\mathbb{E}\left(\|TX\|_2^{-k}\right) \leq \mathbb{E}\left(\|TG/(\sqrt{2\pi}\mathcal{L})\|_2^{-k}\right). \qquad (1.14)$$

The connection between small-ball estimates and negative moments is an immediate corollary of Markov's inequality, which, combined with Theorem 1.14, implies that

$$\mathbb{P}\left(\|TX\|_2 \leq \varepsilon\right) = \mathbb{P}\left(\|TX\|_2^{-k} \geq \varepsilon^{-k}\right) \leq \varepsilon^k \mathbb{E}\left(\|TX\|_2^{-k}\right)$$

$$\leq \varepsilon^k \mathbb{E}\left(\|TG/(\sqrt{2\pi}\mathcal{L})\|_2^{-k}\right).$$

Remark 1.15 It is natural to ask whether Theorem 1.14 is sharp, as potentially there could be a significant gap between $(\mathbb{E}\|TX\|_2^{-k})^{-\frac{1}{k}}$ and $(\mathbb{E}\|TG\|_2^{-k})^{-\frac{1}{k}}$. However, the two happen to be equivalent for any centred log-concave vector (up to the SBA constant $\mathcal{L}$). Indeed, one can show that there is an absolute constant c_0 such that for any centred log-concave random vector X and $1 \leq k \leq \mathrm{rank}(T)$,

$$\left(\mathbb{E}\|TX\|_2^{-k}\right)^{-\frac{1}{k}} \leq c_0 \left(\mathbb{E}\|TG\|_2^{-k}\right)^{-\frac{1}{k}}. \qquad (1.15)$$

A sketch of the proof of this fact is presented in Appendix B.

Theorem 1.14 is a clear indication that the small-ball behaviour of a random vector has nothing to do with concentration or with tail estimates: to a certain extent, concentration exhibited by Gaussian vectors is the best one can hope for, but when it comes to small-ball estimates the situation is the complete opposite. Moreover, thanks to the lower bound from Theorem 1.14, the worst case scenario is actually very good and can be controlled. Indeed, to complement Theorem 1.14 one may estimate the negative moments of $\|TG\|_2$—which requires the following definition:

Definition 1.16 Let $T : \mathbb{R}^n \to \mathbb{R}^m$ and $\mathrm{rank}(T) = m$. For $1 \leq k \leq m - 1$ set,

$$a_k(T) = \left(\int_{\mathcal{G}_{m,k}} \det^{-\frac{1}{2}}[(P_F T)(P_F T)^*] dF\right)^{-\frac{1}{k}} \qquad (1.16)$$

where P_F is the orthogonal projection onto the subspace F and the integration takes place on the Grassmannian $\mathcal{G}_{m,k}$ with respect to the Haar measure. Also for $k = m$ put

$$a_m(T) = \det^{\frac{1}{2m}}(TT^*).$$

It is straightforward to verify that $a_k(T)$ has strong ties to the negative moments of $\|TG\|_2$. Indeed, as is shown in Sect. 2, for any linear operator T and $1 \le k < \mathrm{rank}(T) = m$,

$$\left(\mathbb{E}\|TG\|_2^{-k}\right)^{-\frac{1}{k}} = a_k(T)\left(\mathbb{E}\|G_m\|_2^{-k}\right)^{-\frac{1}{k}} \sim a_k(T)\sqrt{m}, \qquad (1.17)$$

where G_m is the standard Gaussian random vector in $\mathbb{R}^m$.

That, combined with Theorem 1.14, leads to the accurate small-ball behaviour of TX:

Corollary 1.17 *There is an absolute constant c such that the following holds. If X satisfies the SBA with constant $\mathcal{L}$ and $T : \mathbb{R}^n \to \mathbb{R}^m$ then for $1 \le k \le m = \mathrm{rank}(T)$ and every $\varepsilon > 0$,*

$$\mathbb{P}\left(\|TX\|_2 \le \varepsilon\sqrt{m}a_k(T)\right) \le (c\mathcal{L}\varepsilon)^k.$$

As it happens, one can control $a_k(T)$ in terms of $\|T\|_{S_2}$ as long as the operator T does not have a trivial q-stable rank:

Theorem 1.18 *For every $q > 2$ there are constants c_q and c_q' that depend only on q, and absolute constants c and c' such that the following holds. Let X be a random vector in $\mathbb{R}^n$ that satisfies the SBA with constant $\mathcal{L}$ and $T : \mathbb{R}^n \to \mathbb{R}^m$ with $m = \mathrm{rank}(T)$. For every $k \le c_q\mathrm{srank}_q(T)$,*

$$\left(\mathbb{E}\|TG\|_2^{-k}\right)^{-\frac{1}{k}} \ge c\|T\|_{S_2};$$

in particular, for every $\varepsilon > 0$

$$\mathbb{P}\left(\|TX\|_2 \le \frac{\varepsilon}{2e\mathcal{L}}\|T\|_{S_2}\right) \le (c'\varepsilon)^{c_q'\mathrm{srank}_q(T)}. \qquad (1.18)$$

The proofs of Theorem 1.14 and Theorem 1.18 are presented in Sect. 2.

1.4 Coordinate Small Ball

As we noted previously, the fact that the Euclidean norm $\|TX\|_2$ is likely to be large gives limited information on the geometry of the random vector TX. Most notably, it says nothing on the crucial feature that leads to stable point separation— the number of large coordinates TX has with respect to a fixed orthonormal basis $(u_i)_{i=1}^m$. The coordinate small-ball estimate we establish is based on the wSBA, and shows that indeed many of the coordinates $(\langle TX, u_i\rangle)_{i=1}^m$ are likely to be large. To see what sort of information on the coordinates $(\langle TX, u_i\rangle)_{i=1}^m$ one can hope for, let us return to the Gaussian case (which, based on Theorem 1.14, is a likely candidate to be the 'worst' random vector that satisfies the *wSBA*).

Example 1.19 Let $n = m$, set $X = G = (g_i)_{i=1}^m$ and consider the identity operator $T = Id : \mathbb{R}^m \to \mathbb{R}^m$. Given any orthonormal basis $(u_i)_{i=1}^m$ it follows from rotation invariance and independence that

$$\mathbb{P}\left(|\{i : |\langle G, u_i\rangle| \geq \varepsilon\}| \leq c_1 m\right) = \mathbb{P}\left(|\{i : |g_i| \geq \varepsilon\}| \leq c_1 m\right) \leq (c_2\varepsilon)^{c_3 m}$$

for absolute constants c_1, c_2 and c_3.

Thus, the fact that $\|G\|_2$ is likely to be $\gtrsim \sqrt{m}$ is exhibited by a proportional number of the coordinates $(\langle G, u_i\rangle)_{i=1}^m$ whose absolute values are larger than $\varepsilon\|Id\|_{S_2}/\sqrt{m} = \varepsilon$. However, in general, obtaining a coordinate small-ball estimate is a nontrivial task even when X has iid coordinates and T is the identity operator. Indeed, let $n = m$ and set $X = (x_i)_{i=1}^m$ where the x_i's are independent copies of a mean-zero random variable x. When $(u_i)_{i=1}^m$ is the standard basis, one has that $|\langle TX, u_i\rangle| = |x_i|$, and estimating

$$\mathbb{P}\left(\sum_{i=1}^m \mathbb{1}_{\{|x_i| \geq \varepsilon\}} \leq \ell\right)$$

is easy to do thanks to the independence of the $x_1, \ldots, x_m$. But when $(u_i)_{i=1}^m$ is a different orthonormal basis then the coordinates of $(\langle X, u_i\rangle)_{i=1}^m$ are likely to have strong dependencies and the wanted estimate is far from obvious.

We present two coordinate small-ball estimates: Theorem 1.20, when the linear operator T satisfies that $\|T^* u_i\|_2 = 1$ for every $1 \leq i \leq m$, and Theorem 3.5 for more general operators T.

Theorem 1.20 *There exists an absolute constant c such that the following holds. Let X satisfy the wSBA with constants θ and $\mathcal{L}$, set $(u_i)_{i=1}^m$ to be an orthonormal basis of $\mathbb{R}^m$ and consider $T : \mathbb{R}^n \to \mathbb{R}^m$ such that $\|T^* u_i\|_2 = 1$ for every $1 \leq i \leq m$. Let $q > 2$ and set $k_q = \mathrm{srank}_q(T)$. Then for $s \in (0, 1)$*

$$\mathbb{P}\left(|\{i \leq m : |\langle TX, u_i\rangle| \geq \theta\}| \leq (1 - s)m\right) \leq 2 \left(\frac{2}{s}\right)^{\frac{q}{q-2}} \frac{m}{k_q} \left(\frac{c_q \mathcal{L}\theta}{s}\right)^{\frac{1}{2}(s/2)^{\frac{q}{q-2}} k_q},$$

(1.19)

where $c_q \leq c(q/(q - 2))^{1/2}$.

To put Theorem 1.20 is some perspective, let us return to Example 1.19. Consider the case where $n = m$ and $T = Id$. Thus, $k_\infty = m$ and $\|T\|_{S_2}/\sqrt{m} = 1$. Recall that by Rudelson and Vershynin [26], if x is a random variable that has a density that is bounded by $\mathcal{L}$ and $X = (x_i)_{i=1}^m$ has iid coordinates distributed according to x, then X satisfies the *SBA* with constant $c\mathcal{L}$. If $(u_i)_{i=1}^m$ is an arbitrary orthonormal basis and $s = 0.01$, then by Theorem 1.20 one has that with probability at least $1 - 2(c_1 \mathcal{L} \varepsilon)^{c_2 m}$,

$$|\{i : |\langle X, u_i \rangle| \geq \varepsilon\}| \geq 0.99m, \tag{1.20}$$

which means that X exhibits the same coordinate small-ball behaviour with respect to an *arbitrary basis* as it would with respect to the standard basis; moreover, that behaviour is at least as good as that of the standard Gaussian vector.

Remark 1.21 The one place in which Theorem 1.20 is potentially loose is the factor m/k_q. It has an impact only in situations where the operator T is, in some sense, trivial—when the q-stable rank of T is smaller than $c_q \log m$.

The main applications of Theorem 1.20 are Theorems 1.9 and 1.11, showing that random sub-sampled convolutions exhibit stable point separation and can used in the recovery of sparse signals. In addition to that, a further application of Theorem 1.20 is an ℓ_p small-ball estimate.

Theorem 1.22 *There exists absolute constants c_1 and c_2 such that the following holds. Let X be a random vector in $\mathbb{R}^n$ that satisfies the SBA with constant $\mathcal{L}$. Set $a \in \mathbb{R}^n$ and let $k = (c_1 \|a\|_p / \|a\|_\infty)^p$. Then for any $0 < \varepsilon < 1$,*

$$\mathbb{P}\left(\left\| \sum_{i=1}^m a_i x_i e_i \right\|_p \leq \varepsilon \|a\|_p \right) \leq (c_2 \varepsilon \mathcal{L})^k. \tag{1.21}$$

Remark 1.23 To see that (1.21) is truly a small-ball estimate with respect to the ℓ_p norm, observe that by the *SBA*

$$\mathbb{E}\left\| \sum_{i=1}^m a_i x_i e_i \right\|_p^p = \sum_{i=1}^m |a_i|^p \cdot \mathbb{E}|x_i|^p \geq (c\mathcal{L})^p \|a\|_p^p;$$

therefore, under a suitable moment assumption, $\|a\|_p \sim \mathbb{E}\left\| \sum_{i=1}^m a_i x_i e_i \right\|_p$.

There is no obvious way of obtaining an upper bound on (1.21). If X has iid coordinates and satisfies the *SBA* with constant $\mathcal{L} = 1$, one may invoke [24], where it is shown that for any semi-norm $\| \ \|$ and any $u > 0$,

$$\mathbb{P}(\|X\| \leq u) \leq \mathbb{P}(\|Y\| \leq u), \tag{1.22}$$

and Y is the uniform measure on $[-\frac{1}{2}, \frac{1}{2}]^m$. However, similar comparison results of this kind for a general random vector X—whose coordinates need not be independent—are not known.

The proof of Theorem 1.22 is presented in Sect. 4.3.

We end the introduction with some notation. Throughout, c, c_1, c', etc., denote absolute constants. Their value may change from line to line. c_q and $c(q)$ denote constants that depend on the parameter q; $a \lesssim b$ means that there is an absolute constant c such that $a \leq cb$; and $a \lesssim_q b$ implies that c depends on the parameter q. The corresponding two-sided estimates are denoted by $a \sim b$ and $a \sim_q b$ respectively.

For a subspace $F \subset \mathbb{R}^n$ let P_F be the orthogonal projection onto F; $(e_i)_{i=1}^n$ is the standard basis of $\mathbb{R}^n$ and P_k is the orthogonal projection onto span$(e_1, \ldots, e_k)$. The standard Gaussian random vector in $\mathbb{R}^n$ is denoted by G, while G_m is the standard Gaussian random vector in $\mathbb{R}^m$. Finally, if f_X is the density of a random vector X, the density of $P_F X$ is denoted by $f_{P_F X}$.

2 Proofs: Small Ball Estimates

The starting point of the proof of Theorem 1.14 is the following equality (see [23], Proposition 4.6):

Proposition 2.1 *For every random vector W in $\mathbb{R}^m$ with bounded density and $1 \leq k \leq m - 1$,*

$$\frac{\left(\mathbb{E}\|W\|_2^{-k}\right)^{-\frac{1}{k}}}{\left(\mathbb{E}\|G_m\|_2^{-k}\right)^{-\frac{1}{k}}} = \frac{1}{\sqrt{2\pi}} \left(\int_{\mathcal{G}_{m,k}} f_{P_F W}(0) dF\right)^{-\frac{1}{k}}, \tag{2.1}$$

with integration taking place with respect to the Haar measure on the Grassmann manifold $\mathcal{G}_{m,k}$.

Proposition 2.1 indicates the path the proof of Theorem 1.14 follows: one obtains suitable lower bounds on the L_∞ norms the densities of typical projections of TX. This requires two straightforward volumetric observations that also explain the role of the Gaussian parameters $a_k(T)$.

Lemma 2.2 *Let X be a random vector with a density. Consider $S : \mathbb{R}^n \to \mathbb{R}^k$ for $k \leq n$ and with rank$(S) = k$, and let UDP_kV be the singular value decomposition of S. Then, for any compact subset $K \subset \mathbb{R}^k$*

$$\mathbb{P}(SX \in K) = \mathbb{P}\left(P_E X \in V^* D^{-1} U^* K\right), \tag{2.2}$$

where $E = V^*(\mathbb{R}^k)$.

Moreover,

$$\mathbb{P}(SX \in K) \leq \det(D^{-1})\text{vol}(K)\|f_{P_E X}\|_{L_\infty} = \frac{\text{vol}(K)}{\sqrt{\det(SS^*)}}\|f_{P_E X}\|_{L_\infty}. \tag{2.3}$$

Proof Since $S = UDP_kV$ and $P_kV = VP_E$, it follows that

$$\mathbb{P}(SX \in K) = \mathbb{P}(UDP_kVX \in K) = \mathbb{P}\left(P_kVX \in D^{-1}U^*K\right)$$

$$= \mathbb{P}\left(VP_EX \in D^{-1}U^*K\right) = \mathbb{P}\left(P_EX \in V^*D^{-1}U^*K\right) = (*);$$

and by a volumetric estimate,

$$(*) = \int_{V^*D^{-1}U^*K} f_{P_EX}(x)dx \le \mathrm{vol}(V^*D^{-1}U^*K)\|f_{P_EX}\|_\infty$$

$$= \det(D^{-1})\mathrm{vol}(K)\|f_{P_EX}\|_\infty.$$

□

The second observation yields an estimate on the L_∞ norm of the density of a projection of the random vector TX.

Lemma 2.3 *Let X be a random vector, set $1 \le k \le m - 1 \le n - 1$ and assume that for every $E \in \mathcal{G}_{n,k}$,*

$$\|f_{P_EX}\|_{L_\infty} \le \mathcal{L}^k. \tag{2.4}$$

Then, for every $F \in \mathcal{G}_{m,k}$ and $T : \mathbb{R}^n \to \mathbb{R}^m$,

$$\|f_{P_FTX}\|_{L_\infty} \le \frac{\mathcal{L}^k}{(\det[(P_FT)(P_FT)^*])^{\frac{1}{2}}}. \tag{2.5}$$

Proof Fix $F \in \mathcal{G}_{m,k}$ and observe that for every compact set $K \subset F$,

$$\frac{1}{\mathrm{vol}(K)} \int_K f_{P_FTX}(x)dx = \frac{1}{\mathrm{vol}(K)}\mathbb{P}(P_FTX \in K).$$

By (2.3) and the uniform estimate on $\|f_{P_EX}\|_{L_\infty}$ it follows that

$$\mathbb{P}(P_FTX \in K) \le \mathrm{vol}(K) \cdot \max_{E \in \mathcal{G}_{n,k}} \frac{\|f_{P_EX}\|_{L_\infty}}{(\det[(P_FT)(P_FT)^*])^{\frac{1}{2}}}$$

$$\le \mathrm{vol}(K) \cdot \frac{\mathcal{L}^k}{(\det[(P_FT)(P_FT)^*])^{\frac{1}{2}}}.$$

Therefore,

$$\frac{1}{\mathrm{vol}(K)} \int_K f_{P_FTX}(x)dx \le \frac{\mathcal{L}^k}{(\det[(P_FT)(P_FT)^*])^{\frac{1}{2}}},$$

and since the R.H.S. is independent of K the claim follows.

□

Proof of Theorem 1.14 Recall that G is the standard Gaussian random vector in $\mathbb{R}^n$ and G_m is the standard Gaussian random vector in $\mathbb{R}^m$. The proof follows by invoking Proposition 2.1 twice: it is used to compare negative moments of TG and G_m, and then to compare negative moments of G_m and TX.

Let $F \in \mathcal{G}_{m,k}$. Since $P_F TG$ is also a centred Gaussian vector, it standard to verify that

$$f_{P_F TG}^{\frac{1}{k}}(0) = \frac{1}{\sqrt{2\pi}} \left(\frac{1}{\det[(P_F T)(P_F T)^*]} \right)^{\frac{1}{2k}}. \tag{2.6}$$

Hence, by (2.1) and the definition of $a_k(T)$,

$$\frac{\left(\mathbb{E}\|TG\|_2^{-k} \right)^{-\frac{1}{k}}}{\left(\mathbb{E}\|G_m\|_2^{-k} \right)^{-\frac{1}{k}}} = \frac{1}{\sqrt{2\pi}} \left(\int_{\mathcal{G}_{m,k}} f_{P_F TG}(0) dF \right)^{-\frac{1}{k}}$$

$$= \left(\int_{\mathcal{G}_{m,k}} \det^{-\frac{1}{2}}[(P_F T)(P_F T)^*] dF \right)^{-\frac{1}{k}} = a_k(T). \tag{2.7}$$

On the other hand,

$$\frac{\left(\mathbb{E}\|TX\|_2^{-k} \right)^{-\frac{1}{k}}}{\left(\mathbb{E}\|G_m\|_2^{-k} \right)^{-\frac{1}{k}}} = \frac{1}{\sqrt{2\pi}} \left(\int_{\mathcal{G}_{m,k}} f_{P_F TX}(0) dF \right)^{-\frac{1}{k}};$$

by Lemma 2.3, for every $F \in \mathcal{G}_{n,k}$,

$$f_{P_F TX}(0) \le \|f_{P_F TX}\|_{L_\infty} \le \frac{\mathcal{L}^k}{(\det[(P_F T)(P_F T)^*])^{\frac{1}{2}}},$$

implying that

$$\frac{1}{\sqrt{2\pi}} \left(\int_{\mathcal{G}_{m,k}} f_{P_F TX}(0) dF \right)^{-\frac{1}{k}} \ge \frac{1}{\sqrt{2\pi}\mathcal{L}} \left(\int_{\mathcal{G}_{m,k}} \det^{-\frac{1}{2}}[(P_F T)(P_F T)^*] dF \right)^{-\frac{1}{k}}$$

$$= \frac{a_k(T)}{\sqrt{2\pi}\mathcal{L}}.$$

Therefore,

$$\frac{\left(\mathbb{E}\|TX\|_2^{-k}\right)^{-\frac{1}{k}}}{\left(\mathbb{E}\|TG/(\sqrt{2\pi}\mathcal{L})\|_2^{-k}\right)^{-\frac{1}{k}}} \geq 1,$$

as claimed.

Note that (2.1) and (2.6) imply that $\left(\mathbb{E}\|TG\|_2^{-k}\right)^{-\frac{1}{k}} = a_k(T)\left(\mathbb{E}\|G_m\|_2^{-k}\right)^{-\frac{1}{k}}$ as claimed in (1.17). $\qquad\square$

2.1 Proof of Theorem 1.18

Thanks to Theorem 1.14, it suffices to obtain a suitable lower bound on $(\mathbb{E}\|TG\|_2^{-k})^{-\frac{1}{k}}$ for $k \lesssim \mathrm{srank}_q(T)$ and $q > 2$.

Lemma 2.4 *There exists an absolute constant c for which the following holds. Let $0 < \theta < 1$ and $q > 2$, and set*

$$m = (c\theta)^{2q/(q-2)}\mathrm{srank}_q(T).$$

If $(s_i)_{i=1}^r$ are the non-zero singular values of T arranged in a non-increasing order and

$$\tilde{s}_i = \min\{s_i, \|T\|_{S_2}/\sqrt{m}\}$$

then

$$\sum_{i=1}^r \tilde{s}_i^2 \geq (1-\theta^2)\|T\|_{S_2}^2.$$

Proof Let $0 < \theta < 1$ and observe that for every $1 \leq i \leq m$, $s_i \leq \|T\|_{S_q}/i^{1/q}$. Therefore,

$$\sum_{i \leq m} s_i^2 \leq \|T\|_{S_q}^2 \sum_{i \leq m} \frac{1}{i^{2/q}} \leq \frac{cq}{q-2}\|T\|_{S_q}^2 m^{1-2/q} \leq \theta^2 \|T\|_{S_2}^2$$

provided that

$$m \leq (c\theta)^{2q/(q-2)}\mathrm{srank}_q(T).$$

At the same time, $s_{m+1} \leq \|T\|_{S_2}/\sqrt{m+1}$, implying that

$$|\{i : s_i \geq \|T\|_{S_2}/\sqrt{m}\}| \leq m.$$

Thus,

$$\sum_{i=1}^{r} \min \left\{ \|T\|_{S_2}/\sqrt{m}, s_i \right\}^2 \geq \sum_{i=m+1}^{r} s_i^2 \geq (1 - \theta^2)\|T\|_{S_2},$$

and the claim follows.   □

The proof of Theorem 1.18 is based on the following outcome of the so-called "B-Theorem" (see [5] for the proof of the "B-Theorem") and requires some additional notation.

For $a \in \mathbb{R}^n$ let $G_a = \langle G, a \rangle$, and for $A \subset \mathbb{R}^n$ set

$$d_*(A) = \left(\frac{\mathbb{E} \sup_{a \in A} G_a}{\sup_{a \in A}(\mathbb{E}G_a^2)^{1/2}} \right)^2.$$

Theorem 2.5 ([8, 9].) *There are absolute constants c_1 and c_2 such that for any $A \subset \mathbb{R}^n$ and any $0 < s < 1$,*

$$\mathbb{P}\left(\sup_{a \in A} G_a \leq s\mathbb{E} \sup_{a \in A} G_a \right) \leq (c_1 s)^{c_2 d_*(A)}.$$

Proof of Theorem 1.18 Let $(s_i)_{i=1}^{r}$ be the non-zero singular values of T and set $(\tilde{s}_i)_{i=1}^{r}$ to be as in the proof of Lemma 2.4. Using the notation of the lemma, let $\theta^2 = 3/4$. Note that if D is a diagonal operator that satisfies $d_{ii} = s_i$ for $i \leq r$ and 0 otherwise, and $\tilde{D}$ is a diagonal operator whose non-zero diagonal entries are $d_{ii} = \tilde{s}_i$ for $i \leq r$, then

$$\tilde{D}B_2^n \subset DB_2^n, \quad \|\tilde{D}\|_{S_\infty} \leq \frac{\|T\|_{S_2}}{\sqrt{m}}, \quad \text{and} \quad \|\tilde{D}\|_{S_2} \geq \frac{\|T\|_{S_2}}{2}.$$

By rotation invariance, for every k, $\mathbb{E}\|TG\|_2^{-k} = \mathbb{E}\|DG\|_2^{-k}$, and for every $x \in \mathbb{R}^n$, $\|\tilde{D}x\|_2 \leq \|Dx\|_2$. Hence,

$$(E\|TG\|_2^{-k})^{-\frac{1}{k}} = (\mathbb{E}\|DG\|_2^{-k})^{-\frac{1}{k}} \geq (\mathbb{E}\|\tilde{D}G\|_2^{-k})^{-\frac{1}{k}}.$$

Let $A = \tilde{D}B_2^n$ and observe that for $t \in \mathbb{R}^n$,

$$\sup_{a \in A} \langle a, t \rangle = \sup_{x \in B_2^n} \left\langle x, \tilde{D}t \right\rangle = \|\tilde{D}t\|_2;$$

therefore,

$$\mathbb{E}\sup_{a\in A}\langle a, G\rangle = \mathbb{E}\|\tilde{D}G\|_2 \geq \|\tilde{D}\|_{S_2} \geq \frac{\|T\|_{S_2}}{2}$$

and

$$\sup_{a\in A}\mathbb{E}\langle a, G\rangle^2 \leq \max_i d_{ii} \leq \frac{\|T\|_{S_2}}{\sqrt{m}}.$$

Finally, by Theorem 2.5, for every $0 < u < 1$,

$$\mathbb{P}(\|\tilde{D}G\|_2 \leq c_1 u \|T\|_{S_2}) \leq (u/2)^{c_2 m},$$

where c_1 and c_2 are suitable absolute constants. A straightforward tail integration argument shows that for $k \leq c_3 m$

$$(\mathbb{E}\|\tilde{D}G\|_2^{-k})^{-\frac{1}{k}} \geq c_4 \|T\|_{S_2},$$

as required. □

3   Proofs: Coordinate Small-Ball Estimates

Let us turn to the proof of Theorem 1.20. Recall that X is an n-dimensional random vector that satisfies the *wSBA* with constants θ and $\mathcal{L}$, let $(u_i)_{i=1}^m$ be an orthonormal basis of $\mathbb{R}^m$ and set $T : \mathbb{R}^n \to \mathbb{R}^m$ to be a linear operator that satisfies that for $1 \leq i \leq m$

$$\|T^* u_i\|_2 = 1. \tag{3.1}$$

The key component of the proof of Theorem 1.20 is a decomposition lemma. To formulate it, let $\sigma \subset \{1, \ldots, m\}$ and denote by $P_\sigma : \mathbb{R}^m \to \mathbb{R}^\sigma$ the orthogonal projection onto $\mathrm{span}(u_i)_{i\in\sigma}$. Thus, $P_\sigma^* : \mathbb{R}^\sigma \to \mathbb{R}^m$ is the formal identity operator with respect to the basis $(u_i)_{i=1}^m$.

Lemma 3.1 *Let $q > 2$ and set $c_q \sim (q/(q-2))^{1/2}$. Assume that for every $1 \leq i \leq m$, $\|T^* u_i\|_2 = 1$ and set $k_q = \mathrm{srank}_q(T)$. Then for any $\lambda \in (0, 1)$ there are disjoint subsets $\sigma_1, \ldots, \sigma_\ell \subset \{1, \ldots, m\}$ such that*

- *For $1 \leq j \leq \ell$, $|\sigma_j| \geq \lambda^{\frac{q}{q-2}} k_q/2$ and $\sum_{j=1}^\ell |\sigma_j| \geq (1-\lambda)m$; and*
- *$\|(T^* P_{\sigma_j}^*)^{-1}\|_{S_\infty} \leq c_q$.*

The proof of Lemma 3.1 is based on the idea of restricted invertibility. The version used here is Theorem 8 from [21]:

Theorem 3.2 *For* $q > 2$ *set* $c_q \sim (q/(q-2))^{1/2}$. *If* $A : \mathbb{R}^m \to \mathbb{R}^n$ *is a linear operator then there exists* $\sigma \subset \{1, \dots, m\}$ *of cardinality at least* $|\sigma| \geq \mathrm{srank}_q(A)/2$ *such that the map* $(AP_\sigma^*)^{-1}$ *is well defined and*

$$\|(AP_\sigma^*)^{-1}\|_{S_\infty} \leq c_q \frac{\sqrt{m}}{\|A\|_{S_2}}. \tag{3.2}$$

Proof of Lemma 3.1 The construction of the subsets $(\sigma_j)_{j=1}^\ell$ is performed inductively. First, let $k_q = \mathrm{srank}_q(T)$ and apply Theorem 3.2 to $A = T^*$. Thus, noting that $\|T^*\|_{S_2} = \sqrt{m}$, there is $\sigma_1 \subset \{1, \dots, m\}$ such that

$$|\sigma_1| \geq \frac{k_q}{2} \quad \text{and} \quad \|(T^*P_{\sigma_1}^*)^{-1}\|_{S_\infty} \lesssim \left(\frac{q}{q-2}\right)^{1/2}. \tag{3.3}$$

If $|\sigma_1| \geq (1-\lambda)m$ the lemma is proved. Otherwise, let $m_1 = m - |\sigma_1|$ and set $T_1 = P_{\sigma_1^c} T : \mathbb{R}^n \to \mathbb{R}^{m_1}$. Since $\|T_1^* u_i\|_2 = 1$ for all $i \in \{1, \dots, m\} \setminus \sigma_1$, it is evident that $\|T_1\|_{S_2}^2 \geq \lambda m$, and because $P_{\sigma_1^c}$ is a contraction one has that $\|T_1\|_{S_q} = \|P_{\sigma_1^c} T\|_{S_q} \leq \|T\|_{S_q}$. Set $k_q^{(1)} = \mathrm{srank}_q(T_1)$; thus,

$$k_q^{(1)} = \left(\frac{\|T_1\|_{S_2}}{\|T_1\|_{S_q}}\right)^{\frac{2q}{q-2}} \geq \lambda^{\frac{q}{q-2}} \left(\frac{\|T\|_{S_2}}{\|T\|_{S_q}}\right)^{\frac{2q}{q-2}} = \lambda^{\frac{q}{q-2}} k_q.$$

Invoking Theorem 3.2 again, this time for $A = T_1^*$, there is $\sigma_2 \subset \{1, \dots, m\} \setminus \sigma_1$, such that

$$|\sigma_2| \geq \frac{k_q^{(1)}}{2} \geq \frac{1}{2} \lambda^{\frac{q}{q-2}} k_q \quad \text{and} \quad \|(T^*P_\sigma^*)^{-1}\|_{S_\infty} \lesssim \left(\frac{q}{q-2}\right)^{1/2}.$$

Again, if $|\sigma_1| + |\sigma_2| \geq (1-\lambda)m$ the lemma is proved, and if not one may continue in the same way, constructing operators T_j and sets σ_j inductively until $\sum_{j=1}^\ell |\sigma_j| \geq (1-\lambda)m$. □

The fact that $\|(T^*P_\sigma^*)^{-1}\|_{S_\infty} \leq \gamma$ implies that the ellipsoid $P_\sigma T(B_2^n)$ contains the Euclidean ball $\gamma^{-1} B_2^\sigma$, which leads to a small-ball estimate.

Lemma 3.3 *There is an absolute constant c for which the following holds. Let* $\sigma \subset \{1, \dots, m\}$ *such that* $\|(T^*P_\sigma^*)^{-1}\|_{S_\infty} \leq \gamma$, *and set* $\gamma_0 = \max\{1, \gamma\}$. *If* X *satisfies the wSBA with constants* θ *and* $\mathcal{L}$, *then for any* $\tau \subset \sigma$,

$$\mathbb{P}(\|P_\tau T\|_2 \leq \theta\sqrt{|\tau|}) \leq (c\gamma_0 \theta \mathcal{L})^{|\tau|}.$$

An observation one needs for the proof of Lemma 3.3 is a monotonicity property for the *wSBA*. Its proof can be found in Proposition 2.1 in [26] and is based on a simple covering argument.

Lemma 3.4 *Let X satisfy the wSBA with constants θ and $\mathcal{L}$. Then for every $M > 1$, X also satisfies the wSBA with constants $M\theta$ and $3\mathcal{L}$.*

Proof of Lemma 3.3 Let $r = |\tau|$ and note that $B = P_\tau P_\sigma T : \mathbb{R}^n \to \mathbb{R}^r$ is a linear operator of rank r. As noted previously, there are $|\sigma|$ non-zero singular values of $P_\sigma T$, all of which are at least γ^{-1}; and since $\tau \subset \sigma$, it follows that $s_i(B) \geq \gamma^{-1}$ for $1 \leq i \leq r$. By the singular value decomposition theorem there are $U \in \mathcal{O}_r$, $V \in \mathcal{O}_n$ and a diagonal matrix $D = \text{diag}(s_1(B), \cdots, s_r(B))$ such that $B = UDP_rV$, where, as always, P_r denotes the orthogonal projection onto $\{e_1, \dots, e_r\}$. Setting $F = V^*(\mathbb{R}^r)$ one has that $P_r V = V P_F$.

Recall that X satisfies the wSBA with constants θ and $\mathcal{L}$. Since all the entries in the diagonal of D are at least γ^{-1}, invoking Lemma 3.4 it is evident that

$$\mathbb{P}\left(\|P_\tau T X\|_2 \leq \theta\sqrt{|\tau|}\right) = \mathbb{P}\left(\|UDVP_FX\|_2 \leq \theta\sqrt{|\tau|}\right) \leq \mathbb{P}\left(\|P_FX\|_2 \leq \gamma\theta\sqrt{|\tau|}\right)$$

$$\leq (c\theta\gamma_0\mathcal{L})^{|\tau|}.$$

$\square$

Proof of Theorem 1.20 Let $(\sigma_j)_{j=1}^\ell$ be the collection of subsets as in Lemma 3.1 and set

$$\sigma = \bigcup_{j \leq \ell} \sigma_j.$$

In particular, for $1 \leq j \leq \ell$,

$$|\sigma_j| \geq \lambda^{\frac{q}{q-2}} k_q/2.$$

Recall that $(u_i)_{i=1}^m$ is an orthonormal basis of $\mathbb{R}^m$ and for $\tau \subset \{1, \dots, m\}$ set

$$Q_\tau = \{x \in \mathbb{R}^m : \max_{i \in \tau} |\langle x, u_i \rangle| \leq 1\}.$$

Consider the random variables

$$\eta_i = \mathbb{1}_{\{z:|\langle Tz,u_i\rangle|\geq\theta\}}(X) \quad \text{and} \quad \zeta_i = \mathbb{1}_{\{z:|\langle Tz,u_i\rangle|<\theta\}}(X) \tag{3.4}$$

for $1 \leq i \leq m$.

It follows that for $0 < \alpha < 1$ and $1 \leq j \leq \ell$,

$$\mathbb{P}\left(\sum_{i \in \sigma_j} \zeta_i \geq \alpha|\sigma_j|\right) \leq \sum_{\tau \subset \sigma_j, \, |\tau|=\alpha|\sigma_j|} \mathbb{P}\left(\bigcap_{i \in \tau}\{|\langle TX, u_i\rangle| < \theta\}\right)$$

$$\leq \left(\frac{e}{\alpha}\right)^{\alpha|\sigma_j|} \max_{\tau \subset \sigma_j, \, |\tau|=\alpha|\sigma_j|} \mathbb{P}(P_\tau TX \in \theta Q_\tau)$$

$$\leq \left(\frac{e}{\alpha}\right)^{\alpha|\sigma_j|} \max_{\tau \subset \sigma_j, \, |\tau|=\alpha|\sigma_j|} \mathbb{P}\left(\|P_\tau TX\|_2 < \theta\sqrt{|\tau|}\right),$$

where the last inequality holds because $Q_\tau \subset \sqrt{|\tau|}B_2^\tau = \sqrt{|\tau|}B_2^m \subset \mathbb{R}^\tau$. And, by Lemma 3.3 one has that

$$\max_{\tau \subset \sigma_j, \, |\tau| = \alpha|\sigma_j|} \mathbb{P}\left(\|P_\tau TX\|_2 < \theta\sqrt{|\tau|}\right) \leq (c_q \mathcal{L}\theta)^{\alpha|\sigma_j|}$$

where $c_q \sim (q/(q-2))^{1/2}$; therefore,

$$\mathbb{P}\left(\sum_{i \in \sigma_j} \eta_i \leq (1-\alpha)|\sigma_j|\right) \leq \left(\frac{ec_q \mathcal{L}\theta}{\alpha}\right)^{\alpha|\sigma_j|}. \tag{3.5}$$

Now set $0 < s < 1$, let $\lambda = s/2$ and recall that $|\sigma| \geq (1-\lambda)m$. Let $1 - \alpha = (1-s)/(1-\lambda)$ and observe that $\alpha \geq s/2$. With this choice of λ, the union bound and (3.5),

$$\mathbb{P}\left(\sum_{i=1}^m \eta_i \leq (1-s)m\right) = \mathbb{P}\left(\sum_{i \in \sigma} \eta_i \leq (1-s) \cdot \frac{|\sigma|}{1-\lambda}\right)$$

$$\leq \mathbb{P}\left(\sum_{j \leq \ell}\sum_{i \in \sigma_\ell} \eta_i \leq (1-\alpha)\sum_{j \leq \ell}|\sigma_j|\right) \leq \sum_{j \leq \ell}\mathbb{P}\left(\sum_{i \in \sigma_j} \eta_i \leq (1-\alpha)|\sigma_j|\right)$$

$$\leq \sum_{j \leq \ell}\mathbb{P}\left(\sum_{i \in \sigma_j} \eta_i \leq \left(1 - \frac{s}{2}\right)|\sigma_j|\right) \leq \sum_{j \leq \ell}\left(\frac{ec_q \mathcal{L}\theta}{s/2}\right)^{|\sigma_j|/(s/2)} = (*).$$

Finally, since

$$|\sigma_j| \geq c_1 \lambda^{\frac{q}{q-2}} k_q \sim s^{q/(q-2)} k_q,$$

it is evident that $\ell \lesssim s^{-q/(q/2)} m / k_q$ and the claim follows. $\qquad\square$

3.1   Coordinate Small-Ball for General Operators

The assumption that $\|T^*u_i\|_2 = 1$ for every $1 \leq i \leq m$ is not essential and can be replaced by a considerably weaker condition. If instead one assumes that there are constants $\delta_1 > 0$ and $\delta_2 \geq 1$ such that

$$\left(\frac{1}{m}\sum_{i=1}^m \|T^*u_i\|_2^{2+\delta_1}\right)^{\frac{1}{2+\delta_1}} \leq \delta_2 \frac{\|T\|_{S_2}}{\sqrt{m}}, \tag{3.6}$$

then the following version of Theorem 1.20 can be established:

Theorem 3.5 *Let X satisfy the SBA with constant $\mathcal{L}$. Consider $T : \mathbb{R}^n \to \mathbb{R}^m$, an orthonormal basis $(u_i)_{i=1}^m$ of $\mathbb{R}^m$ such that (3.6) is satisfied, and for $q > 2$ set $k_q = \mathrm{srank}_q(T)$. Then, for any $\varepsilon \in (0, 1)$ one has that*

$$\mathbb{P}\left(\left|\left\{i \le m : |\langle TX, u_i\rangle| \ge \varepsilon \frac{\|T\|_{S_2}}{\sqrt{m}}\right\}\right| \le c_0 m\right) \le c_1 \frac{m}{k_q}(c_2\mathcal{L}\varepsilon)^{c_3 k_q} \qquad (3.7)$$

where $c_0 = c_0(\delta_1, \delta_2)$,

$$c_1 = 5^{\frac{q}{q-2}}, \quad c_2 \sim \left(\frac{q}{q-2}\right)^{1/2} \quad \text{and} \quad c_3 = \left(\frac{1}{2\delta_2}\right)^{\frac{2+\delta_1}{\delta_1}} \in (0, 1).$$

Because the proof of Theorem 3.5 follows a similar path to that of Theorem 1.20 we will only outline the necessary modifications.

Sketch of Proof Note that (3.6) and the Paley-Zygmund inequality imply that

$$\left|\left\{i \le m : \|T^*u_i\|_2 \ge \frac{\|T\|_{S_2}}{2\sqrt{m}}\right\}\right| \ge c_0(\delta_1, \delta_2)m, \qquad (3.8)$$

and without loss of generality one may assume that $c_0 m$ is an integer. In particular, let $\sigma_0 \subset \{1, \ldots, m\}$ to be of cardinality $c_0 m$ and for every $i \in \sigma_0$,

$$\|T^*u_i\|_2 \ge \frac{\|T\|_{S_2}}{2\sqrt{m}}.$$

We may assume that $\sigma_0 = \{1, \ldots, c_0 m\}$ and let $T_0 = P_{\sigma_0}T$ where P_{σ_0} is the orthogonal projection onto $\mathrm{span}(u_i)_{i\in\sigma_0}$. Therefore,

$$\|T\|_{S_2} \ge \|T_0\|_{S_2} \ge \frac{\sqrt{c_0}}{2}\|T\|_{S_2},$$

and for any $\sigma \subset \sigma_0$,

$$\|P_\sigma T_0\|_{S_2} \ge \frac{1}{2}\sqrt{\frac{|\sigma|}{m}}\|T\|_{S_2} \ge \frac{\sqrt{c_0}}{2}\sqrt{\frac{|\sigma|}{|\sigma_0|}}\|T_0\|_{S_2}.$$

Let $\delta = \sqrt{c_0}/2$ and set

$$k_{q,0} = \mathrm{srank}_q(T_0) \ge \left(\frac{c_0}{4}\right)^{\frac{q}{q-2}} k_q.$$

Following the argument used in the proof of Lemma 3.1, it is evident that there are disjoint subsets $\sigma_1, \ldots, \sigma_\ell \subset \{1, \ldots, c_0 m\}$ such that

- For $1 \leq j \leq \ell$, $|\sigma_j| \geq (\delta^2 \lambda)^{\frac{q}{q-2}} k_{q,0}/2$ and $\sum_{j=1}^{\ell} |\sigma_j| \geq (1 - \lambda)|\sigma_0| = (1 - \lambda)c_0 m$; and
- $\|(T^* P_{\sigma_j}^*)^{-1}\|_{S_\infty} \leq \frac{c_0}{\delta^2}$.

From here on the proof is identical to that of Theorem 1.20 with the choice of $\lambda = 1/2$; the details are omitted.   □

Remark A version of Theorem 3.5 holds true under the *wSBA* as well. We leave the details of the proof to the reader.

Theorems 1.20 and 3.5 imply that under mild assumptions on T, the coordinate small-ball estimate exhibits the standard small-ball one. Indeed, as an example, set $s = 1/2$, let k_4 denote the q stable rank for $q = 4$ and observe that if $\|T^* e_i\|_2 = 1$ for every $1 \leq i \leq m$ then $\|T\|_{S_2} = \sqrt{m}$. Hence,

$$\left| \left\{ i : |\langle TX, u_i \rangle| \geq (\theta/\sqrt{2})\frac{\|T\|_{S_2}}{\sqrt{m}} \right\} \right| \geq \frac{m}{2}$$

with probability at least

$$1 - (m/k_4) \cdot (c\theta\mathcal{L})^{c'k_4}$$

where c and c' are absolute constants. Therefore, if $k_4 \gtrsim \log m$ and $\theta \lesssim 1/\mathcal{L}$,

$$\mathbb{P}\left(\|TX\|_2 \leq \frac{\theta}{8}\|T\|_{S_2} \right) \leq (c''\theta\mathcal{L})^{c'k_4/2},$$

which recovers the small-ball estimate (and obviously similar bounds hold for any $q > 2$ at the price of modified constants).

At the same time, the difference between the two estimates cannot be overstated: Theorem 1.20 implies that for *any* choice of a coordinate basis $(u_i)_{i=1}^m$, a typical realization of the vector $(\langle TX, u_i \rangle)_{i=1}^m$ will have $\sim m$ large coordinates, which is a significantly stronger statement than the standard small-ball estimate. Indeed, there are many examples in which the coordinate structure dictated by the orthonormal basis is a feature of the problem and a small-ball estimate is simply not good enough. The case of random sub-sampled convolutions, which we now turn to, is one such example.

4 Proofs: Applications

Here we present the proofs of the applications that follow from Theorem 1.20, starting with point separation and stable point separation for random sub-sampled convolutions.

4.1   Random Sub-Sampled Convolutions and Stable Point Separation

Proof of Theorem 1.9 Recall that ξ is an isotropic random vector in $\mathbb{R}^n$ and that $(\delta_i)_{i=1}^n$ are independent $\{0, 1\}$-valued random variables with mean δ. If $I = \{i : \delta_i = 1\}$, the question of point separation is whether with high probability

$$\frac{1}{\delta n} \sum_{i \in I} (a \circledast \xi)_i^2 \geq c_0 \|a\|_2^2 \tag{4.1}$$

for a suitable constant c_0 that is independent of a and of δ; and, as far as stable point separation is concerned, whether

$$|\{i \in I : (a \circledast \xi)_i \geq c_1 \|a\|_2\}| \geq c_2 \delta n. \tag{4.2}$$

Recall that

$$a \circledast \xi = \sqrt{n} U D_{W_a} O \xi \equiv \Gamma_a \xi, \tag{4.3}$$

where U, W and O are unitery matrices defined in (1.10). Note that ξ satisfies the *SBA* with constant $\mathcal{L}$; that $\hat{a} = Wa$. Hence, for every $q > 2$,

$$\|\Gamma_a\|_{S_q} = \sqrt{n} \|U D_{W_a} O\|_{S_q} = \sqrt{n} \|D_{W_a}\|_{S_q} = \sqrt{n} \|\hat{a}\|_q.$$

Therefore, if $\|a\|_2 = 1$,

$$\text{srank}_q(\Gamma_a) = \left(\frac{\|\hat{a}\|_2}{\|\hat{a}\|_q} \right)^{\frac{2q}{q-2}} = \left(\frac{1}{\|\hat{a}\|_q} \right)^{\frac{2q}{q-2}}.$$

By Theorem 1.20 for $(u_i)_{i=1}^n = (e_i)_{i=1}^n$, there is an event $\mathcal{A}$ of probability at least

$$1 - 2 (c_1)^{\frac{q}{q-2}} \frac{n}{k_q} (c_2(q)\mathcal{L}\varepsilon)^{c_3^{\frac{q}{q-2}} k_q} = (*)$$

with respect to ξ, on which

$$|\{i : |\langle \Gamma_a \xi, e_i \rangle| \geq \varepsilon\}| \geq 0.99n.$$

If a is s-sparse with respect to the standard basis then, by (1.12)

$$\|\hat{a}\|_q^q = \sum_{i=1}^n |\langle W_i, a \rangle|^q \leq n \cdot \left(\frac{s}{n} \right)^{q/2} = \frac{s^{q/2}}{n^{(q-2)/2}},$$

implying that

$$(*) \geq 1 - (c_4(q)\varepsilon\mathcal{L})^{c_5(q)/\|\hat{a}\|_q^{2q/(q-2)}}$$

as long as $s \leq c_6(q)n/\log n$ and $\varepsilon \leq c_7(q)$.

Finally, for every realization of ξ in the event $\mathcal{A}$, with probability at least $1 - 2\exp(-c_8\delta n)$ with respect to $(\delta_i)_{i=1}^n$ one has that

$$|\{i \in I : |\langle \Gamma_a\xi, e_i \rangle| \geq \varepsilon\}| \geq 0.98\delta n, \tag{4.4}$$

and in particular,

$$\sum_{i \in I} |\langle \Gamma_a\xi, e_i \rangle|^2 \geq 0.98\varepsilon^2\delta n. \tag{4.5}$$

A Fubini argument completes the proof. □

Remark 4.1 Observe that by modifying the constants from absolute ones to constants that depend on $0 < \eta < 1$, (4.4) and (4.5) hold with $1 - \eta$ replacing 0.98.

4.2 Sparse Recovery Under Malicious Noise

Proof of Theorem 1.11 Recall that for the recovery problem of the unknown (s-sparse) vector $t_0 \in \mathbb{R}^n$, the corrupted data $(Y_i)_{i\in I}$ is generated by the malicious adversary by changing at most $(1/2 - \eta)|I|$ of the original random sub-sampled convolution $(\langle t_0 \circledast \xi, e_i \rangle)_{i\in I}$. The procedure we use selects t that is s-sparse which satisfies that the median of the numbers $(|Y_i - \langle t \circledast \xi, e_i \rangle|)_{i\in I}$ is 0. The key to the success of the procedure is showing that with high probability, more than half of the values of $(|Y_i - \langle t_0 \circledast \xi, e_i \rangle|)_{i\in I}$ are 0, while if $t \neq t_0$, then the more than half of the values $(|Y_i - \langle t \circledast \xi, e_i \rangle|)_{i\in I}$ are strictly positive.

The first step in the proof of Theorem 1.11 is (4.4), combined with the fact that $1/\|\hat{a}\|_q^{2q/(q-2)} \geq 1/\|\hat{a}\|_\infty^2$; in particular, for $0 < \eta < 1/10$, with probability at least

$$1 - (c_1(\eta)\mathcal{L}\varepsilon)^{c_2(\eta)/\|\hat{a}\|_\infty^2},$$

one has that

$$|\{i : |\langle \Gamma_a\xi, e_i \rangle| \geq \varepsilon\}| \geq (1-\eta)n. \tag{4.6}$$

Next, for any $a \in S^{n-1}$ that is s-sparse one has that $1/\|\hat{a}\|_2^2 \geq n/s$; therefore, (4.6) holds with probability at least $1 - (c_1(\eta)\mathcal{L}\varepsilon)^{c_2(\eta)n/s}$. Let J_a be the set of indices in

(4.6) and observe that with probability at least $1 - 2\exp(-c_3(\eta)\delta n)$ with respect to the selectors $(\delta_i)_{i=1}^n$, $|J_a \cap I| \geq (1 - 4\eta)|I|$. In particular, with probability at least

$$1 - (c_1(\eta)\mathcal{L}\varepsilon)^{c_2(\eta)n/s} - 2\exp(-c_3(\eta)\delta n)$$

$$|\{i \in I : |\langle \Gamma_a \xi, e_i \rangle| \geq \varepsilon\}| \geq (1 - 4\eta)|I|. \tag{4.7}$$

Once this individual estimate has been established, let us turn to the uniform estimate; to that end we shall make use of three simple observations:

(1) Let U_s be the set of s-sparse vectors in the Euclidean unit ball and consider $\gamma \geq 1$. For $\rho > 0$ the number of translates of ρB_2^n needed to cover U_{2s} satisfies

$$\mathcal{N}(U_{2s}, \rho B_2^n) \leq \binom{n}{2s}\left(\frac{5}{\rho}\right)^{c_4 s} \leq \left(\frac{n}{\rho s}\right)^{c_5 s},$$

where c_4 and c_5 are absolute constants; Therefore, there is a ρ-cover of U_{2s} of cardinality $\exp(c_5(\gamma + 1)s \log(n/s))$ and for ρ at most $(s/n)^\gamma$.

Denote by V_{2s} that ρ-cover, and for $u \in U_{2s}$ let $\pi u \in V_{2s}$ the best approximation to u in V_{2s} with respect to the ℓ_2 norm. Hence,

$$\sup_{u \in U_{2s}} \max_{1 \leq i \leq n} |\langle (u - \pi u) \circledast \xi, e_i \rangle| \leq \sup_{u \in U_{2s}} \|u - \pi u\|_2 \|\xi\|_2 \leq \|\xi\|_2 \left(\frac{s}{n}\right)^\gamma.$$

(2) Since ξ is isotropic it follows that for $t \geq 1$,

$$\mathbb{P}(\|\xi\|_2 \geq t\sqrt{n}) \leq 1/t^2.$$

Hence, on that event,

$$\sup_{u \in U_{2s}} \max_{1 \leq i \leq n} |\langle (u - \pi u) \circledast \xi, e_i \rangle| \leq t\sqrt{n}\left(\frac{s}{n}\right)^\gamma \leq \frac{t}{n^{(\gamma-1)/2}},$$

provided that $s \leq \sqrt{n}$.

(3) Set ε to satisfy $c_1(\eta)\varepsilon\mathcal{L} = 1/2$. If $\delta n \lesssim_{\gamma,\eta,\mathcal{L}} s \log(en/s)$ and $s \gtrsim_{\gamma,\eta} \sqrt{n/\log n}$, then (4.7) holds for all the points in the set V_{2s} uniformly, and the outcome holds with probability $1 - 2\exp(-c_6(\gamma, \eta, \mathcal{L})s \log(en/s))$.

Now fix $u \in U_{2s}$, let πu be its best approximation in V_{2s} and set $J_u \subset I$ for which (4.7) holds. Recall that $\varepsilon \sim_\eta 1/\mathcal{L}$ and observe that for $j \in J_u$,

$$|\langle u \circledast \xi, e_j \rangle| \geq |\langle \pi u \circledast \xi, e_j \rangle| - |\langle (u - \pi u) \circledast \xi, e_j \rangle| \geq \varepsilon - \frac{t}{n^{(\gamma-1)/2}} \geq \frac{\varepsilon}{2},$$

provided that $t \leq \varepsilon n^{(\gamma-1)/2}/2$. As a result, with probability at least

$$1 - 2\exp(-c_6 s \log(en/s)) - \frac{2}{\varepsilon n^{(\gamma-1)/2}}, \qquad (4.8)$$

for every $u \in U_{2s}$ there is $J_u \subset I$ of cardinality at least $(1 - 4\eta)|I|$ such that for any $j \in J_u$

$$|\langle u \circledast \xi, e_j \rangle| \geq \frac{\varepsilon}{2}. \qquad (4.9)$$

Finally, consider an s-sparse vector $t \in \mathbb{R}^n$, $t \neq t_0$, and set $u = (t - t_0)/\|t - t_0\|_2 \in U_{2s}$. By (4.9),

$$|\langle t - t_0 \circledast \xi, e_j \rangle| \geq \frac{\varepsilon}{2}\|t - t_0\|_2 \qquad (4.10)$$

for more than $(1 - \eta/4)|I|$ indices. Therefore, even if one changes no more than $(1/2 - \eta)|I|$ values, there are still at least $(1/2 + 3\eta/4)|I| > |I|/2$ indices for which (4.10) holds. As a result, for any $t \neq t_0$ the median of $(|Y_j - \langle t \circledast \xi, e_j \rangle|)_{i \in I}$ is strictly positive. Moreover, it is straightforward to verify that the median of $(|Y_i - \langle t_0 \circledast \xi, e_i \rangle|)_{i \in I}$ is 0, implying that with probability as in (4.8), t_0 is the only point in $\mathbb{R}^n$ for which the median of $(|Y_j - \langle t \circledast \xi, e_i \rangle|)_{i \in I}$ is 0, as claimed. As a result, the recovery procedure will be able to recover t_0 for any t_0 that is s-sparse. $\square$

4.3 Small-Ball Estimates for the ℓ_p-Norm

Proof of Theorem 1.22 Let $X = (x_i)_{i=1}^n$ satisfy the *SBA* with constant $\mathcal{L}$ and fix $a \in \mathbb{R}^n$. The goal here is to use Theorem 1.20 and the information it provides on the distribution of the coordinates of a random vector X to control the probability

$$\mathbb{P}\left(\left\|\sum_{i=1}^n a_i x_i e_i\right\|_p \geq \varepsilon \|a\|_p\right);$$

here, as always, $(e_i)_{i=1}^n$ denotes the standard basis in $\mathbb{R}^n$ and $\| \|_p$ is the ℓ_p norm. Without loss of generality assume that $a_1 \geq a_2 \ldots \geq 0$ and set

$$I_j = \left\{i : \frac{1}{2^{j+1}} < \frac{a_i}{a_1} \leq \frac{1}{2^j}\right\}. \qquad (4.11)$$

For every integer ℓ let

$$\Lambda_\ell = \{j : |I_j| = \ell\} \qquad (4.12)$$

and note that it is possible that some of the sets Λ_ℓ's are empty. For every ℓ define

$$j(\ell) = \min \Lambda_\ell \tag{4.13}$$

and if Λ_ℓ is empty let $j(\ell) = 0$.

The idea behind this decomposition of $\{1, \ldots, n\}$ to the union of the sets I_j is that if $\Lambda_\ell \neq \emptyset$ then the contribution to $\|a\|_p$ that comes from $\bigcup_{j \in \Lambda_\ell} I_j$ is equivalent to the contribution of $I_{j(\ell)}$. Indeed, for any I_j,

$$\frac{1}{2^p}|I_j|\frac{a_1^p}{2^{jp}} < \sum_{i \in I_j} a_i^p \leq |I_j|\frac{a_1^p}{2^{jp}} \tag{4.14}$$

and by comparing the sum to an appropriate geometric progression, there are absolute constants c_1 and c_2 such that

$$c_1^p \sum_{i \in I_{j(\ell)}} a_i^p \leq \sum_{j \in \Lambda_\ell} \sum_{i \in I_j} a_i^p \leq c_2^p \sum_{i \in I_{j(\ell)}} a_i^p.$$

As a result, there are disjoint coordinate blocks, each one of different cardinality, such that

$$\|a\|_p^p \sim \sum_{\{\ell \geq 1 : \Lambda_\ell \neq \emptyset\}} \sum_{i \in I_{j(\ell)}} a_i^p.$$

Fix an index ℓ such that $\Lambda_\ell \neq \emptyset$ and consider $j = j(\ell)$. One has that

$$\left\{i \in I_j : |x_i a_i| \leq \varepsilon \frac{a_1}{2^{j+1}}\right\} \subset \left\{i \in I_j : |x_i| \leq \varepsilon\right\},$$

and by Theorem 1.20 for the orthogonal projection onto $\text{span}(e_i : i \in I_j)$, denoted in what follows by P_{I_j}, there are absolute constants c_3 and c_4 such that

$$\mathbb{P}\left(\left|\{i \in I_{j(\ell)} : |x_i| \leq \varepsilon\}\right| \leq \frac{\ell}{2}\right) \leq (c_3 \mathcal{L}\varepsilon)^{c_4 \ell}.$$

Hence, with probability at least $1 - (c_3 \mathcal{L}\varepsilon)^{c_4 \ell}$, there are at least $\ell/2$ indices $i \in I_{j(\ell)}$ such that

$$|x_i a_i| \geq \varepsilon \frac{a_1}{2^{j+1}},$$

and in particular,

$$\sum_{i \in I_{j(\ell)}} |x_i a_i|^p \geq c_5^p \varepsilon^p \ell \frac{a_1^p}{2^{jp}} \geq c_6^p \varepsilon^p \sum_{i \in I_{j(\ell)}} a_i^p.$$

Set

$$\phi^p(k) = \sum_{\{\ell \geq k, \Lambda_\ell \neq \emptyset\}} \sum_{i \in I_{j(\ell)}} a_i^p,$$

note that

$$(c_7 \|a\|_p)^p \leq \phi^p(1) \leq \|a\|_p^p$$

and that by the union bound, for every integer k, with probability at least $1 - (c_8 \mathcal{L}\varepsilon)^{c_9 k}$,

$$\left\| \sum_{i=1}^n a_i x_i e_i \right\|_p \geq \left(\sum_{\{\ell \geq k, \Lambda_\ell \neq \emptyset\}} \sum_{i \in I_{j(\ell)}} |a_i x_i|^p \right)^{1/p} \geq c_6 \varepsilon \phi(k).$$

All that is left to show is that for a well chosen absolute constant c and for $k = (c\|a\|_p/\|a\|_\infty)^p$ one has that $\phi(k) \gtrsim \|a\|_p^p$. To that end, and because $\phi^p(1) \geq c_7^p \|a\|_p^p$, the claim follows if

$$\sum_{\ell \leq k} \sum_{i \in I_{j(\ell)}} a_i^p \leq \frac{c_7^p}{2} \|a\|_p^p.$$

By the exponential decay of $\|P_{I_j} a\|_\infty$,

$$\sum_{\{\ell < k, \Lambda_\ell \neq \emptyset\}} \sum_{i \in I_{j(\ell)}} a_i^p \leq \sum_{\{\ell < k, \Lambda_\ell \neq \emptyset\}} |I_{j(\ell)}| \|P_{I_{j(\ell)}} a\|_\infty^p$$

$$= \sum_{\{\ell < k, \Lambda_\ell \neq \emptyset\}} |\ell| \|P_{I_{j(\ell)}} a\|_\infty^p \leq c_{10}^p k \|a\|_\infty^p$$

from which the wanted estimate follows immediately for our choice of k. $\qquad\square$

5 Concluding Remarks

Finally, let us describe how coordinate small-ball estimates *should not be established*. Unfortunately, up to this point, the only known way of obtaining such estimates was this (suboptimal) way.

5.1 The Wrong Way

The standard way in which coordinate small-ball estimates have been established was based on the following simple observation. Consider a vector $x \in \mathbb{R}^n$ that satisfies $\|x\|_2 \geq \alpha\sqrt{n}$ for some $\alpha > 0$ and for the sake of simplicity assume that $x_1 \geq x_2 \geq \ldots \geq x_n \geq 0$. Clearly, having any estimate on $\|x\|_2$ says nothing about the number of large coordinates that x has; however, if the contribution to $\|x\|_2$ made by the $k = \beta n$ largest coordinates of x is smaller than $\alpha\sqrt{n}/2$ then x is "well spread". Indeed, on the one hand

$$\sum_{i=k+1}^{n} x_i^2 \geq \frac{3}{4}\alpha^2 n,$$

and on the other

$$x_k^2 \leq \frac{\alpha^2 n}{4k}.$$

Therefore, by a Paley-Zygmund type argument, a proportional number ($\sim_{\alpha,\beta} n$) of the x_i's are at least $c(\alpha, \beta)$.

Obtaining a coordinate small-ball estimate in this way is particularly appealing in light of Theorem 1.14: because we know that $\|TX\|_2$ is likely to be large, it seems like half the job is already done. However, there are two crucial reasons why, despite the appeal, this is the wrong approach. Firstly, it gives no flexibility: one has no control on the proportion of nontrivial coordinates that the vector has, nor on the lower bound on the absolute values of these coordinates; in particular, there is no hope of proving Theorem 1.20 using this type of argument. Secondly, while lower bounds are, in some sense, universal, upper bounds—which play an integral part in the argument and are based on tail estimates—are clearly not. In this case, given an orthonormal basis $(u_i)_{i=1}^{m}$ the necessary upper bound is on $\|(\langle TX, u_i \rangle)_{i=1}^{m}\|_{[k]}$, where we set

$$\|x\|_{[k]} = \max_{|I|=k} \left(\sum_{i \in I} x_i^2 \right)^{\frac{1}{2}}.$$

Upper estimates of this kind hold with reasonable probability only in very special cases. In fact, even when X is a Gaussian vector, the resulting probability estimate is weaker than, say, the one in Theorem 1.20; and for more heavy-tailed random vectors estimates on $\|(\langle TX, u_i \rangle)\|_{[k]}$ are completely useless.

As a general principle,

> It is wrong to try to establish coordinate small-ball estimates (which are lower bounds) using an argument that is based on "large deviations". Such a method may lead to nontrivial bounds only for very nice random vectors, and the bounds will be suboptimal even in those cases.

A   Examples of Vectors That Satisfy The *SBA*

Here we give examples of several generic random vectors that satisfy the *SBA*. This is far from being an exhaustive list and should be viewed only as an indication to the fact that the *SBA* is a property shared by many natural random vectors.

(1) Let $X = (\xi_1, \cdots, \xi_n)$ where the ξ_i's are independent random variables with densities bounded by $\mathcal{L}$. It was shown in [26] that X satisfies the *SBA* with constant $c\mathcal{L}$, where $c > 0$ is an absolute constant.

This fact was further extended in [11, 24]; most notably, it was shown in [24] that if the coordinates of $X = (\xi_i)_{i=1}^n$ are independent random variables with densities bounded by 1 and the coordinates of $Y = (\eta_i)_{i=1}^n$ are uniformly distributed in $[-\frac{1}{2}, \frac{1}{2}]$, then for every semi-norm $\| \cdot \|$ and $t > 0$,

$$\mathbb{P}(\|X\| \leq t) \leq \mathbb{P}(\|Y\| \leq t). \tag{A.1}$$

In particular, among all such vectors the 'worse' small-ball behaviour—with respect to *any* semi-norm—is exhibited by the uniform measure on the cube $[-\frac{1}{2}, \frac{1}{2}]^n$.

Observe that for the Euclidean norm, the small-ball behaviour of Y and of the standard Gaussian vector G is the same up to absolute constants.

(2) *Perturbations:* It is standard to verify that if X satisfies the *SBA* with a constant $\mathcal{L}$ and W is an arbitrary random vector that is independent of X, then $W + \delta X$ satisfies *SBA* with a constant depending on δ and $\mathcal{L}$.

(3) The question of whether there is a constant $\mathcal{L}$ such that *any* isotropic log-concave random vector satisfies the *SBA* with constant $\mathcal{L}$ is equivalent to Bourgain's celebrated Hyperplane Conjecture (see [2] and the discussion in [7] and [4])).

Thanks to the extensive study of log-concave measures and the connection the SBA has with the Hyperplane conjecture for such measures, there are some important examples of isotropic, log-concave random vectors that are known to satisfy the *SBA* with an absolute constant:

- If X is also 1-unconditional (see [20], section 8.2);
- If X is also subgaussian ([2, 3]);
- If X is also supergaussian (this follows from results of [22]).

B Proof of Remark 1.15

The proof requires some additional notation. Let X be a random vector in $\mathbb{R}^n$ and let $p \geq 1$. The Z_p body of X is defined as the (centrally-symmetric) convex body whose support function is

$$h_{Z_p(X)}(\theta) = \left(\mathbb{E}|\langle X, \theta\rangle|^p\right)^{\frac{1}{p}}, \quad \theta \in S^{n-1}. \tag{B.1}$$

It is straightforward to verify that if $T : \mathbb{R}^n \to \mathbb{R}^m$ is a linear operator then

$$Z_p(TX) = TZ_p(X). \tag{B.2}$$

Lemma B.1 *There are absolute constants c_1 and c_2 for which the following holds. Let X be a centred log-concave random vector in $\mathbb{R}^n$ that satisfies the SBA with constant $\mathcal{L}$. For any $T \in GL_n$ and $F \in \mathcal{G}_{n,k}$ one has*

$$\frac{c_1}{|\det[(P_F T)(P_F T)^*]|^{\frac{1}{2k}}} \leq f_{P_F TX}^{\frac{1}{k}}(0) \leq \frac{c_2 \mathcal{L}}{(\det[(P_F T)(P_F T)^*])^{\frac{1}{2k}}}, \tag{B.3}$$

where the left-hand side holds true under the additional assumption that X is isotropic.

The proof of Lemma B.1 is based on two facts. The first is a standard observation from linear algebra: let $T : \mathbb{R}^n \to \mathbb{R}^k$, set $E = \ker(T)^\perp = \operatorname{im}(T^*)$ and denote by $T|_E$ the restriction of T to E. Then for any compact set $K \subset \mathbb{R}^n$,

$$\operatorname{vol}(TK) = \det(TT^*) \cdot \operatorname{vol}(P_E K). \tag{B.4}$$

The second observation is Proposition 3.7 from [23]: If X is a centred, log-concave random vector then

$$f_X^{\frac{1}{n}}(0) \sim \operatorname{vol}^{-\frac{1}{n}}(Z_n(X)). \tag{B.5}$$

Proof of Lemma B.1 By the Prekopá-Leindler inequality, for every linear operator S, the random vector SX is also log-concave and centred. Hence, using (B.2), (B.5) and (B.4), it is evident that

$$f_{P_F TX}^{\frac{1}{k}}(0) \sim (\text{vol}(Z_k(P_F TX)))^{-\frac{1}{k}} \sim (\text{vol}(P_F T Z_k(X)))^{-\frac{1}{k}}$$

$$\sim \frac{1}{(\det[(P_F T)(P_F T)^*])^{\frac{1}{2k}}} (\text{vol}(P_E Z_k(X)))^{-\frac{1}{k}}$$

$$\sim \frac{1}{(\det[(P_F T)(P_F T)^*])^{\frac{1}{2k}}} (\text{vol}(Z_k(P_E X)))^{-\frac{1}{k}}$$

$$\sim \frac{f_{P_E X}^{\frac{1}{k}}(0)}{(\det[(P_F T)(P_F T)^*])^{\frac{1}{2k}}}.$$

Clearly, $f_{P_E X}^{\frac{1}{k}}(0) \leq \mathcal{L}$, which proved the right-hand side inequality in (B.3). Moreover if X is an isotropic log-concave random vector in $\mathbb{R}^n$ then $f_X^{\frac{1}{n}}(0) \geq c$, where c is an absolute constant (see, e.g. [1]). And since $P_F X$ is also isotropic when X is, the left-hand side inequality in (B.3) follows.   □

Combining (B.3) and (2.1) it is evident that:

Proposition B.2 *There are absolute constants c_1 and c_2 for which the following holds. Let X be an isotropic log-concave random vector in $\mathbb{R}^n$ that satisfies the SBA with constant $\mathcal{L}$ and let $T : \mathbb{R}^n \to \mathbb{R}^m$ be a linear operator. Then*

$$\left(\mathbb{E}\|TG/(c_2\mathcal{L})\|_2^{-k}\right)^{-\frac{1}{k}} \leq \left(\mathbb{E}\|TX\|_2^{-k}\right)^{-\frac{1}{k}} \leq \left(\mathbb{E}\|TG/c_1\|_2^{-k}\right)^{-\frac{1}{k}}. \tag{B.6}$$

References

1. K. Ball, Logarithmically concave functions and sections of convex sets in $\mathbf{R}^n$. Stud. Math. **88**(1), 69–84 (1988)
2. J. Bourgain, On the distribution of polynomials on high-dimensional convex sets, in *Geometric Aspects of Functional Analysis (1989–90)*. Lecture Notes in Mathematics, vol. 1469 (Springer, Berlin, 1991), pp. 127–137
3. J. Bourgain, On the isotropy-constant problem for "PSI-2"-bodies, in *Geometric Aspects of Functional Analysis*. Lecture Notes in Mathematics, vol. 1807 (Springer, Berlin, 2003), pp. 114–121
4. S. Brazitikos, A. Giannopoulos, P. Valettas, B.-H. Vritsiou, *Geometry of Isotropic Convex Bodies*. Mathematical Surveys and Monographs, vol. 196 (American Mathematical Society, Providence, 2014)

5. D. Cordero-Erausquin, M. Fradelizi, B. Maurey, The (B) conjecture for the Gaussian measure of dilates of symmetric convex sets and related problems. J. Funct. Anal. **214**(2), 410–427 (2004)
6. S. Foucart, H. Rauhut, *A Mathematical Introduction to Compressive Sensing*. Applied and Numerical Harmonic Analysis (Birkhäuser/Springer, New York, 2013)
7. B. Klartag, On convex perturbations with a bounded isotropic constant. Geom. Funct. Anal. **16**(6), 1274–1290 (2006)
8. B. Klartag, R. Vershynin, Small ball probability and Dvoretzky's theorem. Isr. J. Math. **157**, 193–207 (2007)
9. R. Latal, K. Oleszkiewicz, Small ball probability estimates in terms of widths. Stud. Math. **169**(3), 305–314 (2005)
10. R. Latala, P. Mankiewicz, K. Oleszkiewicz, N. Tomczak-Jaegermann, Banach-Mazur distances and projections on random subgaussian polytopes. Discrete Comput. Geom. **38**(1), 29–50 (2007)
11. G. Livshyts, G. Paouris, P. Pivovarov, On sharp bounds for marginal densities of product measures. Isr. J. Math. **216**(2), 877–889 (2016)
12. G. Lugosi, S. Mendelson, Near-optimal mean estimators with respect to general norms. Probab. Theory Relat. Fields **175**(3–4), 957–973 (2019)
13. G. Lugosi, S. Mendelson, Sub-Gaussian estimators of the mean of a random vector. Ann. Stat. **47**(2), 783–794 (2019)
14. G. Lugosi, S. Mendelson, Risk minimization by median-of-means tournaments. J. Eur. Math. Soc. **22**(3), 925–965 (2020)
15. S. Mendelson, Learning without concentration. J. ACM **62**(3), Article 21, 25 (2015)
16. S. Mendelson, An unrestricted learning procedure. J. ACM **66**(6), Article 42, 42 (2019)
17. S. Mendelson, Approximating the covariance ellipsoid. Commun. Contemp. Math. **22**(8), 1950089, 24 (2020)
18. S. Mendelson, N. Zhivotovskiy, Robust covariance estimation under L_4-L_2 norm equivalence. Ann. Stat. **48**(3), 1648–1664 (2020)
19. S. Mendelson, H. Rauhut, R. Ward, Improved bounds for sparse recovery from subsampled random convolutions. Ann. Appl. Probab. **28**(6), 3491–3527 (2018)
20. S. Mendelson, E. Milman, G. Paouris, Generalized dual Sudakov minoration via dimension-reduction—a program. Stud. Math. **244**(2), 159–202 (2019)
21. A. Naor, P. Youssef, Restricted invertibility revisited, in *A Journey Through Discrete Mathematics* (Springer, Cham, 2017), pp. 657–691
22. G. Paouris, On the isotropic constant of marginals. Stud. Math. **212**(3), 219–236 (2012)
23. G. Paouris, Small ball probability estimates for log-concave measures. Trans. Am. Math. Soc. **364**(1), 287–308 (2012)
24. G. Paouris, P. Pivovarov, Randomized isoperimetric inequalities, in *Convexity and Concentration*. IMA Vol. Mathematics Applied, vol. 161 (Springer, New York, 2017), pp. 391–425
25. J. Romberg, Compressive sensing by random convolution. SIAM J. Imaging Sci. **2**(4), 1098–1128 (2009)
26. M. Rudelson, R. Vershynin, Small ball probabilities for linear images of high-dimensional distributions. Int. Math. Res. Not. IMRN (19), 9594–9617 (2015)

On the Lipschitz Properties
of Transportation Along Heat Flows

Dan Mikulincer and Yair Shenfeld

Abstract We prove new Lipschitz properties for transport maps along heat flows, constructed by Kim and Milman. For (semi)-log-concave measures and Gaussian mixtures, our bounds have several applications: eigenvalues comparisons, dimensional functional inequalities, and domination of distribution functions.

1 Introduction and Main Results

In recent years, the study of Lipschitz transport maps has emerged as an important line of research, with applications in probability and functional analysis. Let us fix a measure μ on $\mathbb{R}^d$. It is often desirable to write μ as a push-forward $\mu = \varphi_* \eta$, for a well-behaved measure η and a Lipschitz map $\varphi : \mathbb{R}^d \to \mathbb{R}^d$. The main advantage of this approach lies in the fact that one can use the regularity of φ to transfer known analytic properties from η to μ, compensating for the potential complexity of μ.

Perhaps the most well-known result in this direction is due to Caffarelli [7], which states that if γ_d is the standard Gaussian in $\mathbb{R}^d$, and μ is more log-concave than γ_d, then there exists a 1-Lipschitz map φ^{opt} such that $\varphi^{\mathrm{opt}}_* \gamma_d = \mu$. The map φ^{opt} is known as the *optimal transport map* [6]. Crucially, the Lipschitz constant of φ^{opt} does not depend on the dimension d and, consequently, φ^{opt} transfers functional inequalities from γ_d to μ, in a dimension-free fashion. For example, the optimal bounds on the Poincaré and log-Sobolev constants are recovered for the class of strongly log-concave measures [10]. The main goal of this work is to establish quantitative generalizations of this fact, for measures that are not

D. Mikulincer (✉) · Y. Shenfeld
Department of Mathematics, Massachusetts Institute of Technology, Cambridge, MA, USA
e-mail: danmiku@mit.edu; shenfeld@mit.edu

© The Author(s), under exclusive license to Springer Nature Switzerland AG 2023
R. Eldan et al. (eds.), *Geometric Aspects of Functional Analysis*, Lecture Notes in Mathematics 2327, https://doi.org/10.1007/978-3-031-26300-2_9

necessarily strongly log-concave. To this end, we shall use a different transport map, φ^{flow}, along the heat flow, of Kim and Milman [15], which was previously used, in the context of functional inequalities, by Otto and Villani [24].[1]

In general, there is no reason to expect that an arbitrary measure could be represented as a push-forward of γ_d by a Lipschitz map. Indeed, in line with the above discussion, such measures must satisfy certain functional inequalities with constants that are determined by the regularity of the mapping. Thus, we restrict our attention to classes of measures that contain, among others, log-concave measures with bounded support and Gaussian mixtures.

We now turn to discuss, in greater detail, the types of measures for which our results shall hold. First, we consider log-concave measures with support contained in a ball of radius D. It is a classical fact that these measures satisfy Poincaré [25] and log-Sobolev [11] inequalities with constants of order D. For this reason, Kolesnikov raised the question of whether, in this setting, the optimal transport map φ^{opt} is $O(D)$-Lipschitz [17, Problem 4.3]. Up to now, the best known estimate, in [17, Theorem 4.2], gave a Lipschitz constant that is of order $\sqrt{d}D$. One of our main contributions is to close this gap, for the map φ^{flow}. In fact, we prove a stronger result that captures a trade-off between the convexity of μ and the size of its support.

In the sequel, for $\kappa \in \mathbb{R}$ (possibly negative), we say that μ is κ-log-concave if its support is convex and, for μ-almost every x, its density satisfies,

$$-\nabla^2 \log \left(\frac{d\mu}{dx}(x) \right) \succeq \kappa I_d.$$

Theorem 1 *Let μ be a κ-log-concave probability measure on $\mathbb{R}^d$, and set $D := \text{diam}(\text{supp}(\mu))$. Then, for the map $\varphi^{\text{flow}} : \mathbb{R}^d \to \mathbb{R}^d$, which satisfies $\varphi^{\text{flow}}_* \gamma_d = \mu$, the following holds:*

1. If $\kappa > 0$ then,

$$\|\nabla \varphi^{\text{flow}}(x)\|_{\text{op}} \leq \frac{1}{\sqrt{\kappa}},$$

for μ-almost every x.
2. If $\kappa D^2 < 1$ then,

$$\|\nabla \varphi^{\text{flow}}(x)\|_{\text{op}} \leq e^{\frac{1-\kappa D^2}{2}} D,$$

for μ-almost every x.

---
[1] In general, the maps φ^{flow} and φ^{opt} are not the same, see [28].

Item 1 of Theorem 1 follows from the result of Kim and Milman [15], and is analogous to Caffarelli's result [7]. Item 2 improves and generalizes the bound in Item 1 in two ways:

- When $\kappa > 0$ and $\kappa D^2 < 1$, since $e^{\frac{1-\kappa D^2}{2}} D < \frac{1}{\sqrt{\kappa}}$, Item 2 offers a strict improvement of the Lipschitz constant in Caffarelli's result.
- When $\kappa \leq 0$, Theorem 1 provides a Lipschitz transport map for measures that are not strongly log-concave. In particular, the case $\kappa = 0$ is precisely the setting of Kolesnikov's question [17, Problem 4.3].

Theorem 1 may also be compared with [9, Theorem 1.1], which studies Lipschitz properties of the optimal transport map when the target measure is a semi-log-concave perturbation of γ_d. We point out that the two results apply in different regimes: while our result applies to semi-log-concave measures with bounded support, the result of [9] requires that the support of the measure is the entire $\mathbb{R}^d$.

The other type of measures we consider are Gaussian mixtures of the form $\mu = \gamma_d \star \nu$, where ν has bounded support. It was recently shown that these measures satisfy several dimension-free functional inequalities [3, 8, 30]. As we shall show, this phenomenon can be better understood and further strengthened by establishing the existence of a Lipschitz transport map.

Theorem 2 *Let ν be a probability measure on $\mathbb{R}^d$ with $\mathrm{diam}(\mathrm{supp}(\nu)) \leq R$ and consider $\mu = \gamma_d \star \nu$. Then, for the map $\varphi^{\mathrm{flow}} : \mathbb{R}^d \to \mathbb{R}^d$, which satisfies $\varphi_*^{\mathrm{flow}} \gamma_d = \mu$,*

$$\|\nabla \varphi^{\mathrm{flow}}(x)\|_{\mathrm{op}} \leq e^{\frac{R^2}{2}},$$

for almost every $x \in \mathbb{R}^d$.

As mentioned above, the proofs of Theorems 1 and 2 follow from the analysis of Kim and Milman [15]. The main result of [15] is a generalization of Caffarelli's result that establishes Lipschitz properties of φ^{flow}, under an appropriate symmetry assumption. We shall extend the analysis to the classes of measures considered in Theorems 1 and 2. A similar, but in some sense orthogonal to this work, extension was recently performed by Klartag and Putterman [16, Section 3] where the authors considered transportation from μ to $\mu \star \gamma_d$. We also mention the concurrent work of Neeman in [22], which, using a similar method to one presented here, studied Lipschitz properties of bounded perturbations of the Gaussians, generalizing [9]. In the broader context, a similar map was recently used in [1].

Both of the results presented above deal with Lipschitz transport maps that push-forward the standard Gaussian. As discussed, and as we shall demonstrate, the existence of such maps is important for applications. However, one could also ask the reverse question: for which measures μ do we have $\gamma_d = \varphi_* \mu$, with φ Lipschitz?

To answer this question we introduce the class of β-semi-log-convex measures, as measures μ on $\mathbb{R}^d$, which satisfy,

$$-\nabla^2 \log\left(\frac{d\mu}{dx}(x)\right) \preceq \beta I_d,$$

for some $\beta > 0$. It follows from the definition that $\operatorname{supp}(\mu) = \mathbb{R}^d$ (which is why $\beta > 0$). In some sense, this is a complementary notion to being κ-log-concave. Our next result makes this intuition precise.

Theorem 3 *Let $\beta > 0$ and let μ be a β-semi-log-convex probability measure on $\mathbb{R}^d$. Then, for the inverse map $(\varphi^{\mathrm{flow}})^{-1} : \mathbb{R}^d \to \mathbb{R}^d$, which satisfies $(\varphi^{\mathrm{flow}})^{-1}_* \mu = \gamma_d$,*

$$\|\nabla(\varphi^{\mathrm{flow}})^{-1}(x)\|_{\mathrm{op}} \leq \sqrt{\beta},$$

for almost every $x \in \mathbb{R}^d$.

Let us remark that the same question was previously addressed in [17, Theorem 2.2], which expanded upon Caffarelli's original proof, and obtained the same Lipschitz bounds, for $(\varphi^{\mathrm{opt}})^{-1}$. Thus, Theorem 3 gives a more complete picture by proving the analogous result for the map $(\varphi^{\mathrm{flow}})^{-1}$.

Transport Along Heat Flows and the Brownian Transport Map It is tempting to compare Theorems 1 and 2 to the recent construction in [20] of the Brownian transport map. The results apply in similar settings, and the asymptotic dependencies on all parameters are essentially the same. However, as we shall explain, the results are not strictly comparable.

The constructions are qualitatively different: the domain of the Brownian transport map is the infinite-dimensional Wiener space, in contrast to the finite-dimensional domain afforded by the above theorems. Since the Gaussian measure, also in infinite dimensional Wiener space, satisfies numerous functional inequalities with dimension-free constants, realizing a measure on $\mathbb{R}^d$ as a push-forward of the Wiener measure turns out to be satisfactory for many applications. However, there are some applications that require a map between equal dimensions, which explains the need for the present work. We expand on such applications below.

On the other hand, as demonstrated by Mikulincer and Shenfeld [20, Theorem 1.5], in several interesting cases, the Brownian transport map is provably 'Lipschitz on average'. Bounding the averaged derivative of a transport map is an important property (related to the Kannan-Lovász-Simonovits conjecture [14] and to quantitative central limit theorems [20, Theorem 1.7]) that seems to be out of reach for current finite-dimensional constructions.

Having said the above, we do note that for log-concave measures, the Lipschitz constants of the Brownian transport map [20, Theorem 1.1] are usually better than the ones provided by Theorem 1. For Gaussian mixtures, the roles seem to reverse, at least when R is large, as Theorem 2 can be better than [20, Theorem 1.4].

1.1 Applications

As mentioned in the previous section, for some applications it is essential that the domain and image of the transport map coincide. Here we review such applications and state several new implications of Theorems 1 and 2. To keep the statements concise, we will not cover applications that could be obtained by previous results, as in [10, 20, 21].

1.1.1 Eigenvalues Comparisons

A measure μ is said to satisfy a Poincaré inequality if, for some constant $C_p(\mu) \geq 0$ and every test function g,

$$\text{Var}_\mu(g) \leq C_p(\mu) \int_{\mathbb{R}^d} \|\nabla g\|^2 d\mu.$$

We implicitly assume that, when it exists, $C_p(\mu)$ denotes the optimal constant. According to the Gaussian Poincaré inequality [2], $C_p(\gamma_d) = 1$. If $\mu = \varphi_* \gamma_d$ and φ is L-Lipschitz, this immediately implies $C_p(\mu) \leq L^2$. Indeed,

$$\text{Var}_\mu(g) = \text{Var}_{\gamma_d}(g \circ \varphi) \leq \int_{\mathbb{R}^d} \|\nabla(g \circ \varphi)\|^2 d\gamma_d$$

$$\leq \int_{\mathbb{R}^d} \|\nabla \varphi\|_{\text{op}}^2 (\|\nabla g\| \circ \varphi)^2 d\gamma_d \leq L^2 \int_{\mathbb{R}^d} \|\nabla g\|^2 d\mu. \tag{1}$$

Note that the same argument works even if φ is a map between spaces of different dimensions. However, for certain generalizations of the Poincaré inequality, as we now explain, it turns out that it is beneficial for the domain of φ to be the same as the domain of μ. If $\frac{d\mu}{dx} = e^{-V}$ and we define the weighted Laplacian $\mathcal{L}_\mu = \Delta - \langle \nabla, \nabla V \rangle$, then $C_p(\mu)$ corresponds to the inverse of the first non-zero eigenvalue of $\mathcal{L}_\mu$. In [21, Theorem 1.7], E. Milman showed that a similar argument to (1) works for higher order eigenvalues of $\mathcal{L}_\mu$ and $\mathcal{L}_{\gamma_d}$.

Since for $\mathcal{L}_{\gamma_d}$ the multiplicities of the eigenvalues grow with the dimension d, the full power of Milman's argument requires that φ is a map from $\mathbb{R}^d$ to $\mathbb{R}^d$. Thus, by considering the map φ^{flow} from Theorems 1 and 2 and applying Milman's contraction principle, we immediately obtain:

Corollary 4 *Let μ be a probability measure on $\mathbb{R}^d$ and let $\lambda_i(\mathcal{L}_\mu)$ (resp. $\lambda_i(\mathcal{L}_{\gamma_d})$) stand for the ith eigenvalue of $\mathcal{L}_\mu$ (resp. $\mathcal{L}_{\gamma_d}$). Then,*

1. If μ is κ-log-concave, $D := \mathrm{diam}(\mathrm{supp}(\mu))$, and $\kappa D^2 < 1$,

$$\frac{1}{e^{1-\kappa D^2} D^2} \lambda_i(\mathcal{L}_{\gamma_d}) \le \lambda_i(\mathcal{L}_\mu).$$

2. If $\mu = \gamma_d \star \nu$ and $\mathrm{diam}(\mathrm{supp}(\nu)) \le R$, then

$$\frac{1}{e^{R^2}} \lambda_i(\mathcal{L}_{\gamma_d}) \le \lambda_i(\mathcal{L}_\mu).$$

1.1.2 Dimensional Functional Inequalities

Another direction for improving and generalizing the Poincaré inequality goes through dimensional functional inequalities, as in [4].

Let us give a first example, in the form of the dimensional Gaussian log-Sobolev inequality [2], which is a strict improvement over the logarithmic Sobolev inequality. For $g : \mathbb{R}^d \to \mathbb{R}_+$ we define its entropy relative to μ as

$$\mathrm{Ent}_\mu(g) := \int_{\mathbb{R}^d} \log(g) g \, d\mu - \log\left(\int_{\mathbb{R}^d} g \, d\mu\right) \int_{\mathbb{R}^d} g \, d\mu.$$

For γ_d, the following holds:

$$\mathrm{Ent}_{\gamma_d}(g) \le \frac{d}{2} \log\left(1 + \frac{1}{d} \int_{\mathbb{R}^d} \frac{\|\nabla g\|^2}{g} d\gamma_d\right).$$

With the same argument as in (1), and since the logarithm is monotone, we have the corollary:

Corollary 5 *Let μ be a probability measure on $\mathbb{R}^d$ and let $g : \mathbb{R}^d \to \mathbb{R}_+$ be a test function. Then,*

1. If μ is κ-log-concave, $D := \mathrm{diam}(\mathrm{supp}(\mu))$, and $\kappa D^2 < 1$,

$$\mathrm{Ent}_\mu(g) \le \frac{d}{2} \log\left(1 + \frac{e^{1-\kappa D^2} D^2}{d} \int_{\mathbb{R}^d} \frac{\|\nabla g\|^2}{g} d\mu\right).$$

2. If $\mu = \gamma_d \star v$ and $\mathrm{diam}(\mathrm{supp}(v)) \leq R$, then

$$\mathrm{Ent}_\mu(g) \leq \frac{d}{2} \log \left(1 + \frac{e^{R^2}}{d} \int_{\mathbb{R}^d} \frac{\|\nabla g\|^2}{g} d\mu \right).$$

Another example is the dimensional weighted Poincaré inequality which appears in [5, Corrolary 5.6], according to which,

$$\mathrm{Var}_{\gamma_d}(g) \leq \frac{d(d+3)}{d-1} \int_{\mathbb{R}^d} \frac{\|\nabla g(x)\|^2}{1 + \|x\|^2} d\gamma_d(x). \tag{2}$$

For certain test functions, this is a strict improvement of the Gaussian Poincaré inequality. When the target measure is symmetric, we can adapt the argument in (1), and obtain:

Corollary 6 *Let μ be a symmetric probability measure on $\mathbb{R}^d$. Then, for any test function $g : \mathbb{R}^d \to \mathbb{R}$,*

1. *If μ is κ-log-concave, $D := \mathrm{diam}(\mathrm{supp}(\mu))$, and $\kappa D^2 < 1$,*

$$\mathrm{Var}_\mu(g) \leq \frac{d(d+3)}{d-1} e^{1-\kappa D^2} D^2 \int_{\mathbb{R}^d} \frac{\|\nabla g(x)\|^2}{1 + \frac{e^{\kappa D^2 - 1}}{D^2} \|x\|^2} d\mu(x).$$

2. *If $\mu = \gamma_d \star v$ and $\mathrm{diam}(\mathrm{supp}(v)) \leq R$,*

$$\mathrm{Var}_\mu(g) \leq \frac{d(d+3)}{d-1} e^{R^2} \int_{\mathbb{R}^d} \frac{\|\nabla g(x)\|^2}{1 + e^{-R^2}\|x\|^2} d\mu(x).$$

Proof Suppose that $\mu = \varphi_* \gamma_d$ where $\varphi : \mathbb{R}^d \to \mathbb{R}^d$ is L-Lipschitz and satisfies $\varphi(0) = 0$. Then, by (2),

$$\mathrm{Var}_\mu(g) = \mathrm{Var}_{\gamma_d}(g \circ \varphi) \leq \frac{d(d+3)}{d-1} \int_{\mathbb{R}^d} \frac{\|\nabla(g \circ \varphi(x))\|^2}{1 + \|x\|^2} d\gamma_d$$

$$\leq \frac{d(d+3)L^2}{d-1} \int_{\mathbb{R}^d} \frac{(\|\nabla g\| \circ \varphi(x))^2}{1 + \|x\|^2} d\gamma_d.$$

To handle the integral on the right hand side, we invoke the disintegration theorem [13, Theorems 1 and 2] to decompose γ_d along the fibers of φ in the following way: there exists a family of probability measures $\{\gamma_x\}_{x\in\mathbb{R}^d}$, such that $\mathrm{supp}(\gamma_x) \subset \varphi^{-1}(\{x\})$, and satisfies

$$\int_{\mathbb{R}^d} h(x)d\gamma_d(x) = \int_{\mathbb{R}^d} \int_{\varphi^{-1}(\{x\})} h(y)d\gamma_x(y)d\mu(x),$$

for every test function h. Hence, taking $h(x) = \frac{(\|\nabla g\|\circ\varphi(x))^2}{1+\|x\|^2}$,

$$\int_{\mathbb{R}^d} \frac{(\|\nabla g\| \circ \varphi(x))^2}{1 + \|x\|^2}d\gamma_d(x) = \int_{\mathbb{R}^d} \int_{\varphi^{-1}(\{x\})} \frac{(\|\nabla g\| \circ \varphi(y))^2}{1 + \|y\|^2}d\gamma_x(y)d\mu(x)$$

$$= \int_{\mathbb{R}^d} \int_{\varphi^{-1}(\{x\})} \frac{\|\nabla g(x)\|^2}{1 + \|y\|^2}d\gamma_x(y)d\mu(x) \leq \int_{\mathbb{R}^d} \int_{\varphi^{-1}(\{x\})} \frac{\|\nabla g(x)\|^2}{1 + L^{-2}\|x\|^2}d\gamma_x(y)d\mu(x)$$

$$= \int_{\mathbb{R}^d} \frac{\|\nabla g(x)\|^2}{1 + L^{-2}\|x\|^2} \left(\int_{\varphi^{-1}(\{x\})} d\gamma_x(y) \right) d\mu(x) = \int_{\mathbb{R}^d} \frac{\|\nabla g(x)\|^2}{1 + L^{-2}\|x\|^2}d\mu(x)$$

where in the inequality we have used the estimate $\|y\| \geq \frac{1}{L}\|x\|$ for any y such that $\varphi(y) = x$. Indeed, by assumption, $\varphi(0) = 0$ and φ is L-Lipschitz, which immediately yields $\|\varphi(y)\| \leq L\|y\|$.

Finally, when μ is symmetric, our transport map, $\varphi := \varphi^{\mathrm{flow}}$, will turn out to be odd and, hence, satisfies $\varphi^{\mathrm{flow}}(0) = 0$ (see Remark 8). The result follows by combining the above calculations with Theorems 1 and 2. $\square$

1.1.3 Majorization

For an absolutely continuous measure μ, define its distribution function by

$$F_\mu(\lambda) = \mathrm{Vol}\left(\left\{x : \frac{d\mu}{dx}(x) \geq \lambda\right\}\right).$$

We say that μ majorizes η, denoted as $\eta \prec \mu$, if for every $t \in \mathbb{R}$,

$$\int_t^\infty F_\eta(\lambda)d\lambda \leq \int_t^\infty F_\mu(\lambda)d\lambda.$$

In [19, Lemma 1.4], the following assertion is proven: If $\mu = \varphi_*\eta$ for some $\varphi : \mathbb{R}^d \to \mathbb{R}^d$, and $|\det(\nabla\varphi(x))| \leq 1$ for every $x \in \mathbb{R}^d$, then $\eta \prec \mu$.

We use the singular value decomposition to deduce the identity $|\det(\nabla\varphi(x))| = \prod_{i=1}^{d} \sigma_i(\nabla\varphi(x))$, where $\sigma_i(\nabla\varphi(x))$ stands for the ith singular value of $\nabla\varphi(x)$. So, we have the implication,

$$\|\nabla\varphi(x)\|_{op} \leq 1 \implies |\det(\nabla\varphi(x))| \leq 1.$$

By using Theorems 1 and 2 we can find regimes of parameters where φ^{flow} is 1-Lipschitz as required by the computation above. For log-concave measures it is enough to have a sufficiently bounded support, while for Gaussian mixtures one needs to both re-scale the variance and bound the support of the mixing measure. With this in mind, we get the following corollary:

Corollary 7 *Let μ be a probability measure on $\mathbb{R}^d$.*

1. If μ is κ-log-concave, $D := \mathrm{diam}(\mathrm{supp}(\mu))$, $\kappa D^2 < 1$, and $e^{\frac{1-\kappa D^2}{2}} D \leq 1$, then,

$$\gamma_d \prec \mu.$$

2. If $\mu = \gamma_d^a \star \nu$, where γ_d^a stands for the Gaussian measure with covariance aI_d, and $\sqrt{a}e^{\frac{\mathrm{diam}(\mathrm{supp}(\nu))^2}{2a}} \leq 1$ then,

$$\gamma_d \prec \mu.$$

Proof For the first part, the condition $e^{\frac{1-\kappa D^2}{2}} D \leq 1$, along with Theorem 1, ensures that the transport map φ^{flow} is 1-Lipschitz. The claim follows from [19, Lemma 1.4].

For the second part, let $a > 0$ and $X \sim \gamma_d^a \star \nu$, where $\mathrm{diam}(\mathrm{supp}(\nu)) = R$. Then, $\frac{1}{\sqrt{a}}X \sim \gamma_d \star \tilde{\nu}$, and $\mathrm{diam}(\mathrm{supp}(\tilde{\nu})) \leq \frac{R}{\sqrt{a}}$. Let φ^{flow} be the $e^{\frac{R^2}{2a}}$-Lipschitz map, from Theorem 2, that transports γ_d to $\gamma_d \star \tilde{\nu}$. The above argument shows that $\sqrt{a}\varphi^{flow}$ transports γ_d to $\gamma_d^a \star \nu$ and the map is $\sqrt{a}e^{\frac{R^2}{2a}}$-Lipschitz. Thus, if $\sqrt{a}e^{\frac{R^2}{2a}} \leq 1$, there exists a 1-Lipschitz transport map, which implies the result. $\square$

The fact that a measure majorizes the standard Gaussian has some interesting consequences. We state here one example, which appears in the proof of [19, Corollary 2.14]. If $\gamma_d \prec \mu$, then

$$h_q(\gamma_d) \leq h_q(\mu),$$

where, for $q > 0$,

$$h_q(\mu) := \frac{\log\left(\int_{\mathbb{R}^d} \left(\frac{d\mu}{dx}(x)\right)^q dx\right)}{1-q},$$

is the q-Rényi entropy. So, Corollary 1 allows us to bound the q-Rényi entropy from below for some measures.

2   Proofs

2.1   Preliminaries

Before proving the main results, we briefly recall the construction of the transport map from [15, 24]. We take an informal approach and provide a rigorous statement at the end of the section.

Let $(Q_t)_{t \geq 0}$ stand for the Orenstein-Uhlenbeck semi-group, acting on functions $g : \mathbb{R}^d \to \mathbb{R}$ by,

$$Q_t g(x) = \int_{\mathbb{R}^d} g(e^{-t}x + \sqrt{1 - e^{-2t}}\,y)d\gamma_d(y).$$

For sufficiently integrable g, we have, for almost every $x \in \mathbb{R}^d$,

$$Q_0 g(x) = g(x) \text{ and } \lim_{t \to \infty} Q_t g(x) = \mathbb{E}_{\gamma_d}[g].$$

Now, fix μ, a measure on $\mathbb{R}^d$, with $f(x) := \frac{d\mu}{d\gamma_d}(x)$, and consider the measure-valued path $\mu_t := (Q_t f)\gamma_d$. We have $\mu_0 = \mu$ and, for well-behaved measures, we also have $\mu_t \xrightarrow{t \to \infty} \gamma_d$. Thus, there exists a time-dependent vector field V_t, for which the continuity equation holds (see [29, Chapter 8] and [26, Section 4.1.2]):

$$\frac{d}{dt}\mu_t + \nabla \cdot (V_t \mu_t) = 0.$$

In other words, by differentiating under the integral sign, for any test function g,

$$\int_{\mathbb{R}^d} g\left(\frac{d}{dt}Q_t f\right)d\gamma_d = \int_{\mathbb{R}^d} \langle \nabla g, V_t \rangle (Q_t f)d\gamma_d.$$

We now turn to computing V_t. Observe that, by the definition of Q_t,

$$\frac{d}{dt}Q_t f(x) = \Delta Q_t f(x) - \langle x, \nabla Q_t f(x) \rangle.$$

Hence, integrating by parts with respect to the standard Gaussian shows,

$$\int_{\mathbb{R}^d} g\left(\frac{d}{dt}Q_t f\right) d\gamma_d = -\int_{\mathbb{R}^d} \langle \nabla g, \nabla Q_t f\rangle d\gamma_d,$$

whence it follows that $V_t = -\frac{\nabla Q_t f}{Q_t f} = -\nabla \log Q_t f$. Now consider the maps $\{S_t\}_{t\geq 0}$, obtained as the solution to the differential equation

$$\frac{d}{dt}S_t(x) = V_t(S_t(x)), \quad S_0(x) = x. \tag{3}$$

The map S_t turns out to be a diffeomorphism which transports μ_0 to μ_t and we denote $T_t := S_t^{-1}$, which transports μ_t to μ_0. We define the transport maps T and S as the limits

$$T := \lim_{t\to\infty} T_t, \quad S := \lim_{t\to\infty} S_t,$$

in which case, we have $T_*\gamma_d = \mu$ and $S_*\mu = \gamma_d$. These are our transport maps

$$\varphi^{\text{flow}} := T \quad \text{and} \quad (\varphi^{\text{flow}})^{-1} := S.$$

Remark 8 It is clear that if $f(x) = f(-x)$, then V_t and, consequently, S_t (see the discussion following [15, Lemma 3.1]) are odd functions. Hence, if the target measure is symmetric, $T(0) = 0$.

The above arguments are heuristic and require a rigorous justification (as in [15, Section 3]). For the sake of completeness, below, in Lemma 10, we prove sufficient conditions for the existence of the diffeomorphisms $\{S_t\}_{t\geq 0}$, $\{T_t\}_{t\geq 0}$ and for the existence of the transport maps S and T.

We shall require the following approximation lemma, adapted from [22, Lemma 2.1] (a generalization of [15, Lemma 3.2]), which we shall repeatedly use.

Lemma 9 *Let η and η' be two probability measures on $\mathbb{R}^d$, and let $\{\eta_k\}_{k\geq 0}$, $\{\eta'_k\}_{k\geq 0}$ be two sequences of probability measures which converge to η and η' in distribution. Suppose that for every k there exists an L_k-Lipschitz map φ_k with $(\varphi_k)_*\eta_k = \eta'_k$. Then, if $L := \limsup_{k\to\infty} L_k < \infty$, there exists an L-Lipschitz map φ with $\varphi_*\eta = \eta'$. Moreover, by passing to a sub-sequence, we have that for η-almost every x,*

$$\lim_{k\to\infty} \varphi_k(x) = \varphi(x).$$

Proof Under the assumptions of the lemma, the existence of the limiting map φ is assured by the proof of [22, Lemma 2.1]. We are left with showing that φ is L-Lipschitz. Let $r > 0$, and observe that, since $\limsup L_k < \infty$, there exists a sub-sequence, still denoted φ_k, such that, for every $k \geq 0$, φ_k is $(L + r)$-Lipschitz. It follows from [22, Lemma 2.1] that φ is $(L + r)$-Lipschitz. Since r is arbitrary the proof is complete.   □

We are now ready to state our main technical lemma.

Lemma 10 *Assume that μ has a smooth density.*

- *Suppose that, for every $t \geq 0$, there exists $a_t < \infty$ such that,*

$$\sup_{s \in [0,t]} \|\nabla V_s\|_{\mathrm{op}} \leq a_t. \tag{4}$$

Then, there exists a solution, $\{S_t\}_{t \geq 0}$, to (3), which is a diffeomorphism, for every $t \geq 0$.

- *As $t \to \infty$, μ_t converges weakly to γ_d.*
- *Suppose (4) holds, and that, for every $t \geq 0$, T_t (resp. S_t) is L_t-Lipschitz. Then, if $L := \limsup_{t \to \infty} L_t < \infty$, the map T (resp. S) is well-defined and T (resp. S) is L-Lipschitz.*

Proof Combining the assumption on the smoothness of $\frac{d\mu}{dx}$ with (4) gives that, for every $T < \infty$, V is a smooth, spatially Lipschitz, function on $[0, T] \times \mathbb{R}^d$. Thus, by the Picard–Lindelöf theorem, [23, Theorem 3.1], there exists a unique global smooth (see [12, Chapter 1, Theorem 3.3] and the subsequent discussion) solution S_t to (3). By inverting the flow, one may see that the maps S_t are invertible. Indeed, for fixed $t > 0$, consider, for $0 \leq s \leq t$,

$$\frac{d}{ds} T_{t,s}(x) = -V_{t-s}(T_{t,s}(x)), \quad T_{t,0}(x) = x.$$

Then, $S_t^{-1} := T_t := T_{t,t}$, which establishes the first item.

For the second item, note that the Orenstein-Uhlenbeck process is ergodic (see, for example, [15, Lemma 3.2]) and, hence,

$$\lim_{t \to \infty} \|Q_t f - \mathbb{E}_{\gamma_d}[f]\|_{L_1(\gamma_d)} = \lim_{t \to \infty} \|Q_t f - 1\|_{L_1(\gamma_d)} = 0.$$

Thus, μ_t converges to γ_d in total variation, implying weak convergence.

To see the third item, note that the first item establishes the existence of maps S_t which satisfy, $(S_t)_*\mu = \mu_t$, [26, Section 4.1.2]. The second item shows that, as $t \to \infty$, we may approximate γ_d by μ_t. These conditions allow us to invoke Lemma 9, which shows that there exists a sequence $t_k \xrightarrow{k \to \infty} 1$, such that, for μ-almost every x,

$$S(x) := \lim_{k \to \infty} S_{t_k}(x),$$

is well-defined and such that $S_*\mu = \gamma_d$. Since S_t is invertible, for every $t \geq 0$, the same argument, applied to T_t, shows the existence of T.

Finally, let us address the Lipschitz constants of S and T. We shall prove the claim for S; the proof for T is identical. The previous argument shows that there exists a null set $E \subset \operatorname{supp}(\mu)$, such that, for every $z \in \operatorname{supp}(\mu) \setminus E$, $\lim_{k \to \infty} S_{t_k}(z)$ exists. So, for any $x, y \in \operatorname{supp}(\mu) \setminus E$,

$$\|S(x) - S(y)\| = \lim_{k \to \infty} \|S_{t_k}(x) - S_{t_k}(y)\| \le \limsup_{k \to \infty} L_{t_k} \|x - y\| \le L \|x - y\|.$$

This shows $\|S(x) - S(y)\| \le L \|x - y\|$, μ-almost everywhere, which finishes the proof. $\qquad\square$

We shall also require the following lemma, which explains how to deduce global Lipschitz bounds from estimates on the derivatives of the vector fields V_t.

Lemma 11 *Let the above notation prevail and assume that μ has a smooth density. For every $t \ge 0$, let $\theta_t^{\max}, \theta_t^{\min}$ be such that*

$$\theta_t^{\max} \ge \lambda_{\max}\left(-\nabla V_t(x)\right) \ge \lambda_{\min}\left(-\nabla V_t(x)\right) \ge \theta_t^{\min},$$

for almost every $x \in \mathbb{R}^d$. Then,

1. The Lipschitz constant of S is at most $\exp\left(-\int_0^\infty \theta_t^{\min} dt\right)$.

2. The Lipschitz constant of T is at most $\exp\left(\int_0^\infty \theta_t^{\max} dt\right)$.

Proof We begin with the first item. For every $t \ge 0$, we will show that

$$\|S_t(x) - S_t(y)\| \le \exp\left(-\int_0^t \theta_s^{\min} ds\right) \|x - y\| \text{ for every } x, y \in \mathbb{R}^d. \tag{5}$$

The desired result will be obtained by taking $t \to \infty$ and invoking Item 3 of Lemma 10.

Towards (5), it will suffice to show that, for every unit vector $w \in \mathbb{R}^d$,

$$\|\nabla S_t(x) w\| \le \exp\left(-\int_0^t \theta_s^{\min} ds\right).$$

Fix $x, w \in \mathbb{R}^d$ with $\|w\| = 1$, and define the function $\alpha_w(t) := \nabla S_t(x)w$. To understand the evolution of $\|\alpha_w(t)\|$, recall that S_t satisfies the differential equation in (3). Thus,

$$
\begin{aligned}
\frac{d}{dt}\|\alpha_w(t)\| &= \frac{1}{\|\alpha_w(t)\|}\alpha_w(t)^\mathsf{T} \cdot \frac{d}{dt}\alpha_w(t) \\
&= \frac{1}{\|\alpha_w(t)\|}w^\mathsf{T}\nabla S_t(x)^\mathsf{T}\nabla V_t(S_t(x))\nabla S_t(x)w \\
&\leq -\theta_t^{\min}\frac{1}{\|\alpha_w(t)\|}w^\mathsf{T}\nabla S_t(x)^\mathsf{T}\nabla S_t(x)w = -\theta_t^{\min}\|\nabla S_t(x)w\| \\
&= -\theta_t^{\min}\|\alpha_w(t)\|.
\end{aligned}
$$

Since $\|\alpha_w(0)\| = 1$, from Gronwall's inequality we deduce,

$$
\|\nabla S_t(x)w\| = \|\alpha_w(t)\| \leq \exp\left(-\int_0^t \theta_s^{\min}ds\right).
$$

Thus, (5) is established, as required.

The proof of the second part is similar, but this time we will need to show, for every unit vector $w \in \mathbb{R}^d$,

$$
\|\nabla S_t(x)w\| \geq \exp\left(-\int_0^t \theta_s^{\max}ds\right).
$$

Indeed, this would imply $\nabla S_t(x)\nabla S_t(x)^\mathsf{T} \succeq \exp\left(-2\int_0^t \theta_s^{\max}ds\right)\mathrm{I}_d$. Since S_t is a diffeomorphism, and $T_t = S_t^{-1}$, by the inverse function theorem, the local expansiveness of S_t implies

$$
\nabla T_t(x)\nabla T_t(x)^\mathsf{T} \preceq \exp\left(2\int_0^t \theta_s^{\max}ds\right)\mathrm{I}_d.
$$

So, for almost every $x \in \mathbb{R}^d$, $\|\nabla T_t(x)\|_{\mathrm{op}} \leq \exp\left(\int_0^t \theta_s^{\max}ds\right)$, and the claim is proven by, again, invoking Item 3 in Lemma 10.

Let $\alpha_w(t)$ be as above. Then,

$$
\begin{aligned}
\frac{d}{dt}\|\alpha_w(t)\| &= \frac{1}{\|\alpha_w(t)\|}\alpha_w(t)^{\mathsf{T}} \cdot \frac{d}{dt}\alpha_w(t) \\
&= \frac{1}{\|\alpha_w(t)\|} w^{\mathsf{T}}\nabla S_t(x)^{\mathsf{T}}\nabla V_t(S_t(x))\nabla S_t(x)w \\
&\overset{'}{\geq} -\theta_t^{\max}\frac{1}{\|\alpha_w(t)\|} w^{\mathsf{T}}\nabla S_t(x)^{\mathsf{T}}\nabla S_t(x)w = -\theta_t^{\max}\|\nabla S_t(x)w\| \\
&= -\theta_t^{\max}\|\alpha_w(t)\|.
\end{aligned}
$$

As before, Gronwall's inequality implies

$$
\|\nabla S_t(x)w\| = \|\alpha_w(t)\| \geq \exp\left(-\int_0^t \theta_s^{\max}ds\right),
$$

which concludes the proof. $\qquad\qquad\square$

2.2 Lipschitz Properties of Transportation Along Heat Flows

2.2.1 Transportation from the Gaussian

Our proofs of Theorems 1 and 2 go through bounding the derivative, $\nabla V_t = -\nabla^2 \log Q_t f$, of the vector field constructed above, and then applying Lemma 11. Our main technical tools are uniform estimates on $\nabla^2 \log Q_t f$, when the measures satisfy some combination of convexity and boundedness assumptions.

Lemma 12 Let $\mu = f\gamma_d$ and let $D := \mathrm{diam}(\mathrm{supp}(\mu))$. Then, for μ-almost every x,

$$
-\nabla V_t(x) \succeq -\frac{e^{-2t}}{1 - e^{-2t}}\mathrm{I}_d.
$$

Furthermore,

1. For every $t \geq 0$,

$$
-\nabla V_t(x) \preceq e^{-2t}\left(\frac{D^2}{(1 - e^{-2t})^2} - \frac{1}{1 - e^{-2t}}\right)\mathrm{I}_d.
$$

2. *Let $\kappa \in \mathbb{R}$ and suppose that μ is κ-log-concave. Then,*

$$- \nabla V_t(x) \preceq e^{-2t} \frac{1-\kappa}{\kappa(1-e^{-2t})+e^{-2t}},$$

where the inequality holds for any $t \geq 0$ when $\kappa \geq 0$, and for $t \in \left[0, \frac{1}{2}\log\left(\frac{\kappa-1}{\kappa}\right)\right]$ if $\kappa < 0$.

3. *If $\mu := \gamma_d \star \nu$, with $\mathrm{diam}(\mathrm{supp}(\nu)) \leq R$, then, for $t \geq 0$,*

$$- \nabla V_t(x) \preceq e^{-2t} R^2 \mathrm{I}_d.$$

Proof Let $(P_t)_{t\in[0,1]}$ stand for the heat semi-group, related to Q_t by $Q_t f(x) = P_{1-e^{-2t}} f(e^{-t}x)$. In particular,

$$- \nabla V_t(x) = \nabla^2 \log Q_t f(x) = e^{-2t} \nabla^2 \log P_{1-e^{-2t}} f(e^{-t}x).$$

The desired result is now an immediate consequence of [20, Lemma 3.3 and Equation (3.3)], where the paper uses the notation $v(t, x) := \nabla \log P_{1-t} f(x)$. □

By integrating Lemma 12 and plugging the result into Lemma 11 we can now prove Theorems 1 and 2. We begin with the proof of Theorem 2, which is easier.

Proof of Theorem 2 Recall that φ^{flow} is the transport map T, constructed in Sect. 2.1. Remark that the conditions of Lemma 10 are satisfied for the measures we consider: Lemma 12 ensures that (4) holds and μ has a smooth density.

If $\mu := \gamma_d \star \nu$, and ν is supported on a ball of radius R, then, by Lemma 12, we may take $\theta_t^{\mathrm{max}} = e^{-2t} R^2$ in Lemma 11. Compute

$$\int_0^\infty \theta_t^{\mathrm{max}} dt = \frac{R^2}{2}.$$

Thus, φ^{flow} is Lipschitz with constant $e^{\frac{R^2}{2}}$.   □

The proof of Theorem 1 is similar, but the calculations involved are more tedious, even if elementary.

Proof of Theorem 1 We begin by assuming that μ has a smooth density, and handle the general case later with an approximation argument. Thus, as in the proof of Theorem 2, the conditions of Lemma 10 are satisfied, and we recall that φ^{flow} is the transport map T. The first item of the Theorem is covered by Kim and Milman [15, Theorem 1.1] (the authors actually prove it for $\kappa = 1$; the general case follows by a re-scaling argument), so we may assume $\kappa D^2 < 1$. Set $t_0 = \frac{1}{2} \log \left(\frac{D^2(\kappa-1)-1}{\kappa D^2-1} \right)$. By optimizing over the first and second estimates in Lemma 12 we define,

$$\theta_t^{\max} = \begin{cases} \dfrac{e^{-2t}(1-\kappa)}{\kappa(1-e^{-2t})+e^{-2t}} & \text{if } t \in [0, t_0] \\[3mm] e^{-2t}\left(\dfrac{D^2}{(1-e^{-2t})^2} - \dfrac{1}{1-e^{-2t}}\right) & \text{if } t > t_0 \end{cases}$$

Remark that when $\kappa < 0$, $t_0 < \frac{1}{2}\log\left(\frac{\kappa-1}{\kappa}\right)$, so the second bound of Lemma 12 remains valid in this case.

We compute,

$$\int_0^\infty \theta_t^{\max} dt = \int_0^{t_0} \theta_t^{\max} dt + \int_{t_0}^\infty \theta_t^{\max} dt$$

$$= \int_0^{t_0} \frac{e^{-2t}(1-\kappa)}{\kappa(1-e^{-2t})+e^{-2t}} dt + \int_{t_0}^\infty e^{-2t}\left(\frac{D^2}{(1-e^{-2t})^2} - \frac{1}{1-e^{-2t}}\right) dt$$

$$= -\frac{1}{2}\log(\kappa(1-e^{-2t})+e^{-2t})\Big|_0^{t_0} + \frac{1}{2}\left(-\frac{D^2}{1-e^{-2t}} - \log(1-e^{-2t})\right)\Big|_{t_0}^\infty$$

$$= \frac{1}{2}\log\left(1 - D^2(\kappa-1)\right) + \frac{1-\kappa D^2}{2} + \frac{1}{2}\log(D^2)$$

$$- \frac{1}{2}\log(1 - D^2(\kappa-1))$$

$$= \frac{1-\kappa D^2}{2} + \frac{1}{2}\log(D^2).$$

By Lemma 11, the Lipschitz constant of φ^{flow} is at most

$$\exp\left(\int_0^\infty \theta_t^{\max} dt\right) = e^{\frac{1-\kappa D^2}{2}} D.$$

If μ does not have a smooth density, by Lemma 9, it will be enough to show that μ can be approximated in distribution by $\{\mu_k\}_{k\geq 0}$, where each μ_k is log-concave with bounded support and

$$\lim_{k\to\infty} \text{diam}(\text{supp}(\mu_k)) = D.$$

For $\varepsilon > 0$, let $h_\varepsilon(x) = e^{-\frac{1}{1-\|\frac{x}{\varepsilon}\|^2}} \mathbf{1}_{\{\|x\|\leq\varepsilon\}}$ and define the measure ξ_ε with density proportional to h_ε. Then, ξ_ε is a log-concave measure with smooth density and $\text{diam}(\text{supp}(\xi_\varepsilon)) = \varepsilon$. It is straightforward to verify that, for every $\varepsilon > 0$, $\mu_\varepsilon := \xi_\varepsilon \star \mu$

is κ-log-concave with smooth density. Further, as $\varepsilon \to 0$, μ_ε converges to μ, in distribution, and $\lim_{\varepsilon \to 0} \mathrm{diam}(\mathrm{supp}(\mu_\varepsilon)) = D$. The claim is proven. $\square$

2.2.2 Transportation to the Gaussian

To prove Theorem 3 we will need an analogue of Lemma 12 with bounds in the other direction. This is done in the following lemma which shows that the evolution of log-convex functions along the heat flow is dominated by the evolution of Gaussian functions. The proof of the lemma is similar to the proof that strongly log-concave measures are preserved under convolution, [27, Theorem 3.7(b)]. The only difference between the proofs is that the use of the Prékopa-Leindler inequality is replaced by the fact that a mixture of log-convex functions is log-convex.

Lemma 13 (Semi-Log-Convexity Under the Heat Flow) *Let $d\mu = f d\gamma$ be a β-semi-log-convex probability measure on $\mathbb{R}^d$. Then, for almost every x,*

$$-\nabla V_t(x) \succeq \frac{e^{-2t}(1-\beta)}{(1-e^{-2t})(\beta-1)+1} \mathrm{I}_d.$$

Proof We let $(P_t)_{t\geq 0}$ stand for the heat semi-group, defined by

$$P_t f(x) = \int_{\mathbb{R}^d} f(x+\sqrt{t}y) d\gamma_d(y).$$

Since $-\nabla V_t(x) = \nabla^2 \log Q_t f(x) = e^{-2t}\nabla^2 \log P_{1-e^{-2t}} f(e^{-t}x)$, it will be enough to prove,

$$\nabla^2 \log P_t f(x) \succeq \frac{(1-\beta)}{t(\beta-1)+1} \mathrm{I}_d. \tag{6}$$

We first establish the claim in the special case when $f(x) := \psi_\beta(x) \propto e^{-\frac{1}{2}(\beta-1)\|x\|^2}$, where the symbol $\propto$ signifies equality up to a constant which does not depend on x, which corresponds to $\mu = \mathcal{N}(0, \frac{1}{\beta}\mathrm{I}_d)$. This case is facilitated by the fact that P_t acts on f by convolving it with a Gaussian kernel. The result follows since a convolution of Gaussians is a Gaussian and since $\nabla^2 \log$ applied to a Gaussian yields the covariance matrix. To elucidate what comes next, we provide below the full calculation.

For convenience denote $\beta_t = (t(\beta - 1) + 1)$, and compute,

$$P_t \psi_\beta(x) \propto \int_{\mathbb{R}^d} e^{-\frac{1}{2}(\beta - 1)\|x + \sqrt{t}y\|^2} e^{-\frac{\|y\|^2}{2}} dy$$

$$= \int_{\mathbb{R}^d} \exp\left(-\frac{1}{2}\left((\beta - 1)\|x\|^2 + 2(\beta - 1)\sqrt{t}\langle x, y\rangle\right.\right.$$

$$\left.\left. + (t(\beta - 1) + 1)\|y\|^2\right)\right) dy$$

$$= \int_{\mathbb{R}^d} \exp\left(-\frac{\beta_t}{2}\left(\frac{\beta - 1}{\beta_t}\|x\|^2 + 2\sqrt{t}\frac{\beta - 1}{\beta_t}\langle x, y\rangle + \|y\|^2\right)\right) dy$$

$$= \exp\left(-\frac{\beta_t}{2}\left(\frac{\beta - 1}{\beta_t}\left(1 - t\frac{\beta - 1}{\beta_t}\right)\right)\|x\|^2\right)$$

$$\times \int \exp\left(-\frac{\beta_t}{2}\left\|\sqrt{t}\frac{\beta - 1}{\beta_t}x + y\right\|^2\right) dy.$$

The integrand in the last line is proportional to the density of a Gaussian. Hence, the value of the integral does not depend on x, and

$$P_t \psi_\beta(x) \propto \exp\left(-\frac{\beta_t}{2}\left(\frac{\beta - 1}{\beta_t}\left(1 - t\frac{\beta - 1}{\beta_t}\right)\right)\|x\|^2\right)$$

$$= \exp\left(-\frac{1}{2}\left((\beta - 1)\left(1 - t\frac{\beta - 1}{\beta_t}\right)\right)\|x\|^2\right)$$

$$= \exp\left(-\frac{1}{2}\left(\frac{\beta - 1}{t(\beta - 1) + 1}\|x\|^2\right)\right).$$

So,

$$\nabla^2 \log P_t \psi_\beta(x) = \frac{(1 - \beta)}{t(\beta - 1) + 1} I_d, \tag{7}$$

which gives equality in (6).

For the general case, the log-convexity assumption means that we can write $\frac{d\mu}{dx} = e^{V(x)-\beta\frac{\|x\|^2}{2}}$, for a convex function V. Hence, $f(x) \propto e^{V(x)-\frac{1}{2}(\beta-1)\|x\|^2}$. With analogous calculations to the ones made above, we get,

$$P_t f(x) \propto \int_{\mathbb{R}^d} e^{V(x+\sqrt{t}y)-\frac{1}{2}(\beta-1)\|x+\sqrt{t}y\|^2} e^{-\frac{y^2}{2}} dy$$

$$= \exp\left(-\frac{1}{2}\left(\frac{\beta-1}{t(\beta-1)+1}\|x\|^2\right)\right)$$

$$\times \int_{\mathbb{R}^d} \exp\left(V(x+\sqrt{t}y) - \frac{\beta_t}{2}\left\|\sqrt{t}\frac{\beta-1}{\beta_t}x+y\right\|^2\right) dy$$

$$\propto P_t\psi_\beta(x) \int_{\mathbb{R}^d} \exp\left(V(x+\sqrt{t}y) - \frac{\beta_t}{2}\left\|\sqrt{t}\frac{\beta-1}{\beta_t}x+y\right\|^2\right) dy.$$

Write $H_t(x) := \int_{\mathbb{R}^d} \exp\left(V(x+\sqrt{t}y) - \frac{\beta_t}{2}\left\|\sqrt{t}\frac{\beta-1}{\beta_t}x+y\right\|^2\right) dy$ and observe by (7),

$$\nabla^2 \log P_t f(x) = \nabla^2 \log P_t\psi_\beta(x) + \nabla^2 \log(H_t(x)) = \frac{(1-\beta)}{t(\beta-1)+1}\mathrm{Id}$$

$$+ \nabla^2 \log(H_t(x)). \tag{8}$$

To finish the proof we will show that $\nabla^2 \log(H_t(x)) \succeq 0$, or , equivalently, that H_t is log-convex. By applying a linear change of variables, we can re-write H_t as,

$$H_t(x) = \int_{\mathbb{R}^d} \exp\left(V\left(\left(1-t\frac{\beta-1}{\beta_t}\right)x+\sqrt{t}y\right)\right) e^{-\frac{\beta_t\|y\|^2}{2}} dy.$$

As V is convex, for every $t \geq 0$ and $y \in \mathbb{R}^d$, the function $x \mapsto V\left(\left(1-t\frac{\beta-1}{\beta_t}\right)x+\sqrt{t}y\right)$ is convex. So, $H_t(x)$ is a mixture of log-convex functions. Since a mixture of log-convex functions is also log-convex (see [18, Chapter 16.B]), the proof is complete. □

We now prove Theorem 3.

Proof of Theorem 3 Recall that $(\varphi^{\mathrm{flow}})^{-1}$ is the transport map S, constructed in Sect. 2.1. Again, we begin by assuming that μ has a smooth density, and one may verify that the conditions of Lemma 10 are satisfied, which makes S well-defined.

Let $\theta_t^{\min} = e^{-2t}\frac{(1-\beta)}{(1-e^{-2t})(\beta-1)+1}$. Combining Lemma 13 with Lemma 11 shows that S is $\exp\left(\int_0^\infty -\theta_t^{\min}dt\right)$-Lipschitz. Compute,

$$\int_0^\infty -\theta_t^{\min}dt = \int_0^\infty -e^{-2t}\frac{(1-\beta)}{(1-e^{-2t})(\beta-1)+1}dt$$
$$= \frac{1}{2}\log\left((1-e^{-2t})(\beta-1)+1\right)\Big|_0^\infty = \frac{\log(\beta)}{2}.$$

Hence, S is $\exp\left(\frac{\log(\beta)}{2}\right) = \sqrt{\beta}$-Lipschitz.

To finish the proof, we shall construct a family $\{\mu_\varepsilon\}_{\varepsilon>0}$ of β_ε-log-convex measures which converge to μ in distribution as $\varepsilon \to 0$, and such that

$$\lim_{\varepsilon\to 0}\beta_\varepsilon = \beta.$$

The claim then follows by invoking Lemma 9.

Let $\gamma_{d,\varepsilon}$ stand for the d-dimensional Gaussian measure with covariance εI_d, and set $\mu_\varepsilon = \mu \star \gamma_{d,\varepsilon}$. It is clear that, as $\varepsilon \to 0$, μ_ε converges to μ in distribution. Moreover, if we replace f by $\frac{d\mu}{dx}$, in (8), we see that μ_ε is β_ε-log-convex, with,

$$\beta_\varepsilon := \frac{\beta}{\varepsilon\beta+1} \xrightarrow{\varepsilon\to 0} \beta.$$

$\square$

Acknowledgments We wish to thank Max Fathi, Larry Guth, Emanuel Milman, and Ramon van Handel for several enlightening comments and suggestions. We also thank the anonymous referee for carefully reading the paper and providing many helpful comments that improved this manuscript. This material is based upon work supported by the National Science Foundation under Award Number 2002022.

References

1. L. Ambrosio, M. Goldman, D. Trevisan, On the quadratic random matching problem in two-dimensional domains. Electron. J. Probab. **27**, (2022)
2. D. Bakry, I. Gentil, M. Ledoux, *Analysis and Geometry of Markov Diffusion Operators*, vol. 348 (Springer Science & Business Media, New York, 2013)
3. J.-B. Bardet, N. Gozlan, F. Malrieu, P.-A. Zitt, Functional inequalities for Gaussian convolutions of compactly supported measures: explicit bounds and dimension dependence. Bernoulli **24**(1), 333–353 (2018)
4. F. Bolley, I. Gentil, A. Guillin, Dimensional improvements of the logarithmic Sobolev, Talagrand and Brascamp-Lieb inequalities. Ann. Probab. **46**(1), 261–301 (2018)
5. M. Bonnefont, A. Joulin, Y. Ma, Spectral gap for spherically symmetric log-concave probability measures, and beyond. J. Funct. Anal. **270**(7), 2456–2482 (2016)

6. Y. Brenier, Polar factorization and monotone rearrangement of vector-valued functions. Commun. Pure Appl. Math. **44**(4), 375–417 (1991)
7. L. A. Caffarelli, Monotonicity properties of optimal transportation and the FKG and related inequalities. Commun. Math. Phys. **214**(3), 547–563 (2000)
8. H.-B. Chen, S. Chewi, J. Niles-Weed, Dimension-free log-Sobolev inequalities for mixture distributions. *J. Funct. Anal.*, 281(11):Paper No. 109236, 17, 2021.
9. M. Colombo, A. Figalli, Y. Jhaveri. Lipschitz changes of variables between perturbations of log-concave measures. Ann. Sc. Norm. Super. Pisa Cl. Sci. (5) **17**(4), 1491–1519 (2017)
10. D. Cordero-Erausquin, Some applications of mass transport to Gaussian-type inequalities. Arch. Ration. Mech. Anal. **161**(3), 257–269 (2002)
11. A. Frieze, R. Kannan, Log-Sobolev inequalities and sampling from log-concave distributions. Ann. Appl. Probab. **9**(1), 14–26 (1999)
12. J.K. Hale, *Ordinary Differential Equations*, 2nd edn. (Robert E. Krieger Publishing, Huntington, 1980)
13. J. Hoffmann-Jørgensen, Existence of conditional probabilities. Math. Scand. **28**(2), 257–264 (1971)
14. R. Kannan, L. Lovász, M. Simonovits, Isoperimetric problems for convex bodies and a localization lemma. Discrete Comput. Geom. **13**(3–4), 541–559 (1995)
15. Y.-H. Kim, E. Milman, A generalization of Caffarelli's contraction theorem via (reverse) heat flow. Math. Ann. **354**(3), 827–862 (2012)
16. B. Klartag, E. Putterman, Spectral monotonicity under Gaussian convolution, to appear in Ann. Fac. Sci. Toulouse Math. (2021)
17. A.V. Kolesnikov, Mass transportation and contractions. Preprint (2011). arXiv:1103.1479
18. A.W. Marshall, I. Olkin, B.C. Arnold, *Inequalities: Theory of Majorization and Its Applications*, 2nd edn. Springer Series in Statistics (Springer, New York, 2011)
19. J. Melbourne, C. Roberto, Transport-majorization to analytic and geometric inequalities. J. Func. Anal. **284**(1): Paper No, 109717, 36, 2023.
    J. Melbourne, C. Roberto, Transport-majorization to analytic and geometric inequalities. Preprint (2021). arXiv:2110.03641
20. D. Mikulincer, Y. Shenfeld, The Brownian transport map. Preprint (2021). arXiv:2111.11521
21. E. Milman, Spectral estimates, contractions and hypercontractivity. J. Spectr. Theory **8**(2), 669–714 (2018)
22. J. Neeman, Lipschitz changes of variables via heat flow. Preprint (2022). arXiv:2201.03403
23. D. O'Regan, *Existence Theory for Nonlinear Ordinary Differential Equations*. Mathematics and Its Applications, vol. 398 (Kluwer Academic Publishers Group, Dordrecht, 1997)
24. F. Otto, C. Villani, Generalization of an inequality by Talagrand and links with the logarithmic Sobolev inequality. J. Funct. Anal. **173**(2), 361–400 (2000)
25. L.E. Payne, H.F. Weinberger, An optimal Poincaré inequality for convex domains. Arch. Ration. Mech. Anal. **5**, 286–292 (1960)
26. F. Santambrogio, *Optimal Transport for Applied Mathematicians*. Progress in Nonlinear Differential Equations and Their Applications, vol. 87 (Birkhäuser/Springer, Cham, 2015). Calculus of variations, PDEs, and modeling
27. A. Saumard, J. A. Wellner, Log-concavity and strong log-concavity: a review. Stat. Surv. **8**, 45–114 (2014)
28. A. Tanana, Comparison of transport map generated by heat flow interpolation and the optimal transport Brenier map. Commun. Contemp. Math. **23**(6), Paper No. 2050025, 7 (2021)
29. C. Villani, *Topics in Optimal Transportation*. Graduate Studies in Mathematics, vol. 58 (American Mathematical Society, Providence, 2003)
30. F.-Y. Wang, J. Wang, Functional inequalities for convolution probability measures. Ann. Inst. Henri Poincaré Probab. Stat. **52**(2), 898–914 (2016)

A Short Direct Proof of the Ivanisvili-Volberg Inequality

Piotr Nayar and Jacek Rutkowski

Abstract According to the inequality of Ivanisvili and Volberg, for any $n \geq 1$ and any function $f : \{-1, 1\}^n \to \mathbb{R}$ we have

$$\operatorname{Re} \mathbb{E}(f + i|\nabla f|)^{3/2} \leq \operatorname{Re}(\mathbb{E}f)^{3/2},$$

where $z^{3/2}$ for a complex number z is taken with principal branch and Re denotes the real part. Here expectation in taken with respect to the uniform measure on $\{-1, 1\}^n$. We provide a short and direct proof of this inequality.

Keywords Discrete cube · Poincaré inequality · Beckner inequality

1 Introduction

In this note we give a short and direct proof of the Poincaré type inequality on the Hamming cube proved by Ivanisvili and Volberg in [2]. For a function $f : \{-1, 1\}^n \to \mathbb{R}$ and $1 \leq i \leq n$ let us define its directional derivatives $\nabla_i f$ via

$$\nabla_i f(x) = \frac{1}{2}(f(x) - f(\sigma_i(x))),$$

where $\sigma_i(x_1, \ldots, x_n) = (x_1, \ldots, x_{i-1}, -x_i, x_{i+1}, \ldots, x_n)$.

P.N. was supported by National Science Centre Poland grant 2015/18/A/ST1/00553.

P. Nayar (✉) · J. Rutkowski
Institute of Mathematics, University of Warsaw, Warsaw, Poland
e-mail: nayar@mimuw.edu.pl; jr371580@students.mimuw.edu.pl

We further set $\nabla f = (\nabla_1 f, \ldots, \nabla_n f)$ and $|\nabla f|^2 = \sum_{i=1}^n |\nabla_i f|^2$. The inequality states that for any $n \geq 1$ and any $f : \{-1, 1\}^n \to \mathbb{R}$ we have

$$\mathrm{Re}\,\mathbb{E}(f + i|\nabla f|)^{3/2} \leq \mathrm{Re}(\mathbb{E}f)^{3/2}, \tag{1}$$

where $z^{3/2}$ for a complex number z is taken with principal branch, Re denotes the real part, and $\mathbb{E}$ is the expectation with respect to the uniform measure on the Hamming cube. In other words (1) reads as

$$\mathbb{E}M(f, |\nabla f|) \leq M(\mathbb{E}f, 0), \tag{2}$$

where M is defined by

$$M(x, y) = \mathrm{Re}(x + iy)^{3/2} = \frac{1}{\sqrt{2}}(2x - \sqrt{x^2 + y^2})\sqrt{\sqrt{x^2 + y^2} + x}.$$

The main motivation for considering inequality (2) is that it provides a strengthening of the well known Beckner inequality valid in the Gaussian space, see [1] for this and further applications.

The proof of (2) given in [2] relies on the following four point inequality.

Lemma 1 (Main Inequality) *For any real numbers x, y, a, b, we have*

$$2M(x, y) \geq M(x + a, \sqrt{a^2 + (y + b)^2}) + M(x - a, \sqrt{a^2 + (y - b)^2}). \tag{3}$$

This inequality is then proved using a very technical reasoning involving multiple term concrete polynomials with large integer coefficients. In [3] the authors provide a short proof of (3) using certain minimax principle (this reasoning is also reproduced in [2]). This beautiful argument is quite indirect as it involves making certain non-trivial guesses. Here we give a short and direct proof of Lemma 1.

The fact that (3) implies (2) is an abstract principle observed by the authors in [2]. Let us state it as a lemma for completeness.

Lemma 2 *Suppose $M : \mathbb{R}^2 \to \mathbb{R}$ satisfies (3) and the function $t \to M(x, t)$ is non-increasing for $t \geq 0$, for any given x in $\mathbb{R}$. Then (2) holds true for any $f : \{-1, 1\}^n \to \mathbb{R}$.*

In our case M verifies the second assumption due to (4) (see the next section).

The rest of this article is organized as follows. In Sect. 2 we give a proof of the Main inequality (Lemma 1). The proof of Lemma 2 is given in Sect. 3.

2 Proof of the Main Inequality

Without loss of generality we assume that $y \neq 0$ (the case $y = 0$ will follow by continuity). We shall fix x and y and consider the right hand side of (3) as a function Ψ of a, b and prove that the only critical point is $a = b = 0$. We have

$$M_x = \frac{3}{2\sqrt{2}} \sqrt{\sqrt{x^2 + y^2} + x}, \qquad M_y = -\frac{3}{2\sqrt{2}} \frac{y}{\sqrt{\sqrt{x^2 + y^2} + x}}. \qquad (4)$$

Note that equivalently $M_y = -\frac{3}{2\sqrt{2}} \operatorname{sgn}(y) \sqrt{\sqrt{x^2 + y^2} - x}$, which is clearly decreasing in y and thus $y \mapsto M(x, y)$ is concave. Let $P_\pm = \sqrt{x \pm a + \sqrt{(x \pm a)^2 + a^2 + (y \pm b)^2}}$. If $P_+ = 0$ or $P_- = 0$ then necessarily $a = 0$ and thus our inequality reduces to

$$2M(x, y) \geq M(x, |y + b|) + M(x, |y - b|)$$

which follows by concavity and monotonicity of $M(x, y)$ for $y \geq 0$, namely

$$\frac{1}{2} M(x, |y + b|) + \frac{1}{2} M(x, |y - b|) \leq M\left(x, \frac{|y + b| + |y - b|}{2}\right)$$

$$\leq M(x, |y|) = M(x, y).$$

Assume that P_+ and P_- are non-zero. The equations for the critical point of Ψ read

$$\frac{2\sqrt{2}}{3} \frac{\partial \Psi}{\partial b} = -\frac{y + b}{P_+} + \frac{y - b}{P_-} = 0, \qquad (5)$$

$$\frac{2\sqrt{2}}{3} \frac{\partial \Psi}{\partial a} = P_+ - \frac{a}{P_+} - P_- - \frac{a}{P_-} = 0. \qquad (6)$$

The first equation yields $P_+(y - b) = P_-(y + b)$, which is equivalent to $y(P_+ - P_-) = b(P_+ + P_-)$. Using this we rewrite the second equation (multiplied by y)

$$0 = y\left(P_+ - \frac{a}{P_+} - P_- - \frac{a}{P_-}\right) = y(P_+ - P_-) - \frac{ay}{P_+ P_-}(P_+ + P_-)$$

$$= (P_+ + P_-)\left(b - \frac{ay}{P_+ P_-}\right).$$

This yields $bP_+ P_- = ay$. As a consequence

$$ay(y + b) = b(y + b)P_+ P_- = b(y - b)P_+^2$$

and

$$ay(y - b) = b(y - b)P_+ P_- = b(y + b)P_-^2.$$

Plugging in the expressions for P_+ and P_-, moving $x \pm a$ to the left hand side and then squaring, we get

$$(ay(y + b) - b(y - b)(x + a))^2 = b^2(y - b)^2((x + a)^2 + a^2 + (y + b)^2)$$

$$(ay(y - b) - b(y + b)(x - a))^2 = b^2(y + b)^2((x - a)^2 + a^2 + (y - b)^2).$$

This is equivalent to

$$a^2 y^2(y + b)^2 - 2aby(y^2 - b^2)(x + a) = b^2(y - b)^2(a^2 + (y + b)^2)$$

$$a^2 y^2(y - b)^2 - 2aby(y^2 - b^2)(x - a) = b^2(y + b)^2(a^2 + (y - b)^2).$$

Subtracting these equations gives

$$4a^2 y^3 b - 4a^2 by(y^2 - b^2) = -4a^2 b^3 y.$$

This gives $a^2 b^3 y = 0$, which is the same as $ab = 0$ since we assume $y \neq 0$. Now if $a = 0$ then from (6) we get $P_+ = P_-$ and then (5) yields $b = 0$. In this case our inequality reduces to equality. If $b = 0$ then $P_+ = P_-$ (since $y \neq 0$) and thus (6) gives $a = 0$.

We are left with checking the validity of our inequality in the limit $a^2 + b^2 \to \infty$. Let $r^2 = a^2 + b^2$. We shall show $\lim_{r \to \infty} \Psi(a, b) = -\infty$. By using Taylor expansion it is straightforward to observe that

$$\sqrt{2}\Psi(a, b) = (2a - \sqrt{2a^2 + b^2})\sqrt{a + \sqrt{2a^2 + b^2}}$$

$$+ (-2a - \sqrt{2a^2 + b^2})\sqrt{-a + \sqrt{2a^2 + b^2}} + O(\sqrt{r}).$$

Now it suffices to show that

$$(2a - \sqrt{2a^2 + b^2})\sqrt{a + \sqrt{2a^2 + b^2}} + (-2a - \sqrt{2a^2 + b^2})\sqrt{-a + \sqrt{2a^2 + b^2}}$$

$$\leq -C(2a^2 + b^2)^{3/4}, \tag{7}$$

for some absolute constant $C > 0$, since this implies $\Psi(a, b) \leq -Cr^{3/2} + O(\sqrt{r})$, which tends to $-\infty$ when $r \to \infty$. To establish (7) we observe that by homogeneity one can assume $2a^2 + b^2 = 1$ and $|a| \leq 1/\sqrt{2}$. In this case we shall verify the inequality

$$(2a - 1)\sqrt{a + 1} + (-2a - 1)\sqrt{-a + 1} \leq -C, \qquad |a| \leq 1/\sqrt{2}, \tag{8}$$

which is easy (e.g., the only roots of the left hand side are given by $|a| = \sqrt{3}/2 > 1/\sqrt{2}$).

3 Proof of Lemma 2

Step 1. We first claim that for any $a, x \in \mathbb{R}$ and $B, Y \in \mathbb{R}^n$ we have

$$2M(x, |Y|) \geq M(x+a, \sqrt{a^2 + |Y + B|^2}) + M(x-a, \sqrt{a^2 + |Y - B|^2}). \quad (9)$$

Indeed, if we apply (3) with $b = \frac{1}{2}|Y + B| - \frac{1}{2}|Y - B|$ and $y = \frac{1}{2}|Y + B| + \frac{1}{2}|Y - B|$, we get

$$M(x + a, \sqrt{a^2 + |Y + B|^2}) + M(x - a, \sqrt{a^2 + |Y - B|^2})$$
$$\leq 2M(x, y) \leq 2M(x, |Y|),$$

where the last inequality follows from the fact that $y \geq |Y|$ (triangle inequality) and the assumption that $t \to M(x, t)$ is non-increasing for $t \geq 0$.

Step 2. We prove Lemma 2 by induction on n. Suppose we want to prove it for $n + 1$ assuming it is true for n. Take $f : \{-1, 1\}^{n+1} \to \mathbb{R}$. For $x \in \{-1, 1\}^n$ we define $h(x) = \frac{1}{2}f(x, 1) - \frac{1}{2}f(x, -1)$ and $g(x) = \frac{1}{2}f(x, 1) + \frac{1}{2}f(x, -1)$. Then $f(x, x_{n+1}) = g(x) + x_{n+1}h(x)$. Moreover,

$$\nabla_i f(x, x_{n+1}) = \nabla_i g(x) + x_{n+1}\nabla_i h(x), \quad 1 \leq i \leq n \quad \text{and}$$
$$\nabla_{n+1} f(x, x_{n+1}) = x_{n+1}h(x).$$

Thus

$$|\nabla f(x, 1)|^2 = \sum_{i=1}^{n} |\nabla_i g(x) + \nabla_i h(x)|^2 + |h(x)|^2 = |\nabla g(x) + \nabla h(x)|^2 + |h(x)|^2$$

and

$$|\nabla f(x, -1)|^2 = \sum_{i=1}^{n} |\nabla_i g(x) - \nabla_i h(x)|^2 + |h(x)|^2 = |\nabla g(x) - \nabla h(x)|^2 + |h(x)|^2.$$

Denoting by $\mathbb{E}_n$ the expectation with respect to the uniform measure on $\{-1, 1\}^n$ and using (9) with $a = h$, $x = g$, $B = \nabla g$ and $Y = \nabla h$, we get

$$\mathbb{E}_{n+1} M(f, |\nabla f|) = \frac{1}{2} \mathbb{E}_n M(g - h, \sqrt{|\nabla g - \nabla h|^2 + h^2})$$
$$+ \frac{1}{2} \mathbb{E}_n M(g + h, \sqrt{|\nabla g + \nabla h|^2 + h^2})$$
$$\leq \mathbb{E}_n M(g, |\nabla g|) \leq M(\mathbb{E}_n g, 0) = M(\mathbb{E}_{n+1} f, 0),$$

where the second inequality follows from the induction hypothesis and the last step from the fact that $\mathbb{E}_{n+1} f = \frac{1}{2} \mathbb{E}_n (g + h) + \frac{1}{2} \mathbb{E}_n (g - h) = \mathbb{E}_n g$. The case $n = 1$ is included in the above estimate with g, h being constants. Is this case the last inequality becomes equality.

References

1. W. Beckner, A generalized Poincaré inequality for Gaussian measures. Proc. Am. Math. Soc. **105**(2), 397–400 (1989)
2. P. Ivanisvili, A. Volberg, Poincaré inequality 3/2 on the Hamming cube. Rev. Mat. Iberoam. **36**(1), 79–97 (2020)
3. P. Ivanisvili, F. Nazarov, A. Volberg, Square function and the Hamming cube: duality. Discrete Anal. **1**, 18 pp. (2018). https://doi.org/10.19086/da.3113

The Anisotropic Total Variation and Surface Area Measures

Liran Rotem

Abstract We prove a formula for the first variation of the integral of a log-concave function, which allows us to define the surface area measure of such a function. The formula holds in complete generality with no regularity assumptions, and is intimately related to the notion of anisotropic total variation and to anisotropic coarea formulas. This improves previous partial results by Colesanti and Fragalà, by Cordero-Erausquin and Klartag and by the author.

1 Introduction

One of the basic constructions in convex geometry is the surface area measure of a convex body. To recall the definition, let $K \subseteq \mathbb{R}^n$ be a convex body, i.e. a compact convex set with non-empty interior. Then the Gauss map $n_K : \partial K \to \mathbb{S}^{n-1}$ exists $\mathcal{H}^{n-1}$-almost everywhere, where $\mathcal{H}^{n-1}$ denotes the $(n-1)$-dimensional Hausdorff measure and $\mathbb{S}^{n-1} = \{x \in \mathbb{R}^n : |x| = 1\}$ denotes the unit sphere. We then define the surface area measure S_K as the push-forward $S_K = (n_K)_\sharp \left(\mathcal{H}^{n-1} \big|_{\partial K} \right)$. More explicitly, for every measurable function $\varphi : \mathbb{S}^{n-1} \to \mathbb{R}$ we have

$$\int_{\mathbb{S}^{n-1}} \varphi \mathrm{d}S_K = \int_{\partial K} (\varphi \circ n_K) \, \mathrm{d}\mathcal{H}^{n-1}.$$

The surface area measure can be equivalently defined using the first variation of volume. For convex bodies K and L, let $K + L$ denotes the usual Minkowski sum,

The author is partially supported by ISF grant 1468/19 and BSF grant 2016050.

L. Rotem (✉)
Faculty of Mathematics, Technion - Israel Institute of Technology, Haifa, Israel
e-mail: lrotem@technion.ac.il

© The Author(s), under exclusive license to Springer Nature Switzerland AG 2023
R. Eldan et al. (eds.), *Geometric Aspects of Functional Analysis*, Lecture Notes in Mathematics 2327, https://doi.org/10.1007/978-3-031-26300-2_11

and let $|K|$ denote the (Lebesgue) volume of K. Then we have

$$\lim_{t \to 0^+} \frac{|K + tL| - |K|}{t} = \int_{\mathbb{S}^{n-1}} h_L \mathrm{d}S_K, \tag{1.1}$$

where $h_L : \mathbb{S}^{n-1} \to \mathbb{R}$ is the support function of L which is defined by $h_L(\theta) = \max_{x \in L} \langle x, \theta \rangle$. For a proof of this fact the reader may consult a standard reference book in convex geometry such as [19] or [11].

In this paper we will be interested in functional extensions of the formula (1.1). Recall that a function $f : \mathbb{R}^n \to [0, \infty)$ is called log-concave if

$$f\left((1 - \lambda)x + \lambda y\right) \geq f(x)^{1-\lambda} f(y)^\lambda$$

for all $x, y \in \mathbb{R}^n$ and $0 \leq \lambda \leq 1$. We denote by LC_n the class of all upper semi-continuous log-concave functions. Note that the class of convex bodies in $\mathbb{R}^n$ embeds naturally into LC_n using the map

$$K \hookrightarrow \mathbf{1}_K(x) = \begin{cases} 1 & x \in K \\ 0 & \text{otherwise.} \end{cases}$$

We would like to consider log-concave functions as "generalized convex bodies", and extend the notion of the surface area measure to this setting. To achieve this goal we first need to recall the standard operations on log-concave functions: The sum of two log-concave functions is given by the sup-convolution, i.e.

$$(f \star g)(x) = \sup_{y \in \mathbb{R}^n} \left(f(y)g(x - y)\right).$$

The associated dilation operation is given by $(\lambda \cdot f)(x) = f\left(\frac{x}{\lambda}\right)^\lambda$ – note that we have for example $f \star f = 2 \cdot f$. Finally, the "volume" of f will be given by the Lebesgue integral $\int f$. Using these constructions we may define:

Definition 1.1 Fix $f, g \in \mathrm{LC}_n$. The first variation of the integral of f in the direction of g is given by

$$\delta(f, g) = \lim_{t \to 0^+} \frac{\int f \star (t \cdot g) - \int f}{t}. \tag{1.2}$$

The study of log-concave functions as geometric objects has become a major idea in convex geometry, with useful applications even if eventually one is only interested in convex bodies. Due to the very large number of papers in this direction we will not survey all of them, but only mention the ones that directly deal with the first variation $\delta(f, g)$. In the case when $f = e^{-\frac{|x|^2}{2}}$ is a Gaussian, $\delta(f, g)$ was studied under the name "the mean width of g" by Klartag and Milman [13] in one of the papers that began the geometric study of log-concave functions. This mean width was further studied in [16]. In particular it was proved there that in this case we have

$$\delta(f, g) = \int_{\mathbb{R}^n} h_g(x) e^{-|x|^2/2} dx.$$

Here $h_g : \mathbb{R}^n \to \mathbb{R}$ is the support function of g, defined by $h_g = (-\log g)^*$ where

$$\varphi^*(x) = \sup_{y \in \mathbb{R}^n} (\langle x, y \rangle - \varphi(y))$$

is the Legendre transform.

The case of a general function f was studied by Colesanti and Fragalà in [4]. In particular they showed that the limit in (1.2) always exists when $0 < \int f < \infty$, though it may be equal to $+\infty$. To further explain their results we will need some important definitions:

Definition 1.2 Fix a log-concave function $f : \mathbb{R}^n \to \mathbb{R}$ with $0 < \int f < \infty$, and write $f = e^{-\varphi}$ for a convex function $\varphi : \mathbb{R}^n \to (-\infty, \infty]$. Then:

(1) The measure μ_f is a measure on $\mathbb{R}^n$ defined as the push-forward $\mu_f = (\nabla \varphi)_{\sharp} (f dx)$.

(2) The measure ν_f is a measure on the sphere $\mathbb{S}^{n-1}$, defined as the push-forward $\nu_f = (n_{K_f})_{\sharp} (f d\mathcal{H}^{n-1}|_{\partial K_f})$. Here K_f is a shorthand notation for the support of f, i.e. $K_f = \{x \in \mathbb{R}^n : f(x) > 0\}$, and n_{K_f} denotes the Gauss map $n_{K_f} : \partial K_f \to \mathbb{S}^{n-1}$.

For example, for $f(x) = e^{-|x|^2/2}$ we have $\mu_f = e^{-|x|^2/2} dx$ and $\nu_f \equiv 0$, as $\partial K_f = \partial \mathbb{R}^n = \emptyset$. For a convex body K we have $\mu_{1_K} = |K| \delta_0$ and $\nu_{1_K} = S_K$, the usual surface area measure.

It will be important for us to observe that no regularity is required for the definitions of μ_f and ν_f. Indeed, as $\varphi = -\log f$ is a convex function it is differentiable Lebesgue-almost-everywhere on the set $\{x : \varphi(x) < \infty\} = K_f$. Therefore the push-forward $(\nabla \varphi)_{\sharp} (f dx)$ is well-defined. Similarly since K_f is a closed convex set its boundary ∂K_f is a Lipschitz manifold, so in particular the Gauss map n_{K_f} is defined $\mathcal{H}^{n-1}$-almost-everywhere and the push-forward is again well-defined.

While regularity is not needed for the definitions of μ_f and ν_f, it was definitely needed for the representation theorem of [4]:

Theorem 1.3 (Colesanti–Fragalà) *Fix $f, g \in LC_n$, and assume that:*

(1) *The supports K_f, K_g are C^2 smooth convex bodies with everywhere positive Gauss curvature.*

(2) *The functions $\psi = -\log f$ and $\varphi = -\log g$ are continuous in K_f and K_g respectively, C^2 smooth in the interior of these sets, and have strictly positive-definite Hessians.*
(3) *We have $\lim_{x \to \partial K_f} |\nabla \psi(x)| = \lim_{x \to \partial K_g} |\nabla \varphi(x)| = \infty$.*
(4) *The difference $h_f - c \cdot h_g$ is convex for small enough $c > 0$.*

Then

$$\delta(f, g) = \int_{\mathbb{R}^n} h_g \mathrm{d}\mu_f + \int_{\mathbb{S}^{n-1}} h_{K_g} \mathrm{d}\nu_f.$$

Based on Theorem 1.3 we can make the following definition:

Definition 1.4 Given $f \in \mathrm{LC}_n$ with $0 < \int f < \infty$, we call the pair (μ_f, ν_f) the surface area measures of the function f.

We emphasize that unlike a convex body, a log-concave function has two surface area measures: one defined on $\mathbb{R}^n$, and one defined on $\mathbb{S}^{n-1}$.

While the regularity assumptions of Theorem 1.3 are sufficient, it has always been clear that they are not necessary. For example, we already saw that for the function $f(x) = e^{-|x|^2/2}$ no regularity assumptions on g are needed. In fact, it was proved in [17] that if $0 < \int f < \infty$ and $\nu_f = 0$ then we have

$$\delta(f, g) = \int_{\mathbb{R}^n} h_g \mathrm{d}\mu_f$$

with no regularity assumptions. Note that since f is log-concave and upper semi-continuous it is only discontinuous at points $x \in \partial K_f$ such that $f(x) \neq 0$. Therefore the condition $\nu_f = 0$ is equivalent to the statement that f is continuous $\mathcal{H}^{n-1}$-almost everywhere. This property was dubbed essential continuity by Cordero-Erausquin and Klartag [5]. In their paper, the authors [5] studied the moment measure of a convex function φ, which in our terminology is simply the surface area measure $\mu_{e^{-\varphi}}$. One of the main results of their paper is a functional analogue of Minkowski's existence theorem: Given a measure μ on $\mathbb{R}^n$, they provide a necessary and sufficient condition for the existence of a function $f \in \mathrm{LC}_n$ with $\mu_f = \mu$ and $\nu_f = 0$. They also prove the uniqueness of such an f. We remark that when the functions involved are not necessarily essentially continuous, but are sufficiently regular in the sense of Theorem 1.3, a similar uniqueness result was previously proved by Colesanti–Fragalà in [4]. We will further discuss this issue of uniqueness in Sect. 3 after proving our main theorem, and explain the results of both papers. We also remark that another proof of the same existence theorem was given by Santambrogio in [18].

The main goal of this paper is to prove the most general form of Theorem 1.3, which requires no regularity assumptions:

Theorem 1.5 *Fix* $f, g \in LC_n$ *such that* $0 < \int f < \infty$. *Then*

$$\delta(f, g) = \int_{\mathbb{R}^n} h_g \mathrm{d}\mu_f + \int_{\mathbb{S}^{n-1}} h_{K_g} \mathrm{d}\nu_f. \tag{1.3}$$

While the improvement over previous results is simply the elimination of the various technical assumptions, we do believe Theorem 1.5 is of real value. For example, we will see as a corollary that the pair of measures (μ_f, ν_f) determines f uniquely, and for this result it is very useful not to have any technical conditions for the validity of formula (1.3). Moreover, we believe our proof sheds some light on the reason this formula holds. In particular, we will see an interesting connection between Theorem 1.5 and the notions of anisotropic total variation and anisotropic perimeter. The main point will be that when $g = 1_L$ and $f \in LC_n$ is arbitrary, Theorem 1.5 can be viewed as an anisotropic version of the coarea formula.

The rest of this note is dedicated to the proof of the theorem. In Sect. 2 we will introduce the anisotropic coarea formula, and explain why it is in fact equivalent to Theorem 1.5 in the case when $g = 1_L$ is the indicator of a convex body. Then in Sect. 3 we will discuss the case of a general function g, and conclude the proof. Some of the ingredients that were used in previous results (mostly in [17]) can be used in the proof of Theorem 1.5 with few changes, and in these cases we will either give an exact reference or briefly sketch the argument.

Before we start with the main proof let us prove a finiteness property of the measures μ_f and ν_f:

Proposition 1.6 *Assume* $f \in LC_n$ *and* $0 < \int f < \infty$. *Then the measure* μ_f *is finite with a finite first moment. The measure* ν_f *is also finite.*

Proof μ_f is finite by definition, as $\int_{\mathbb{R}^n} \mathrm{d}\mu_f = \int f \mathrm{d}x$. Moreover, we have

$$\int |x| \mathrm{d}\mu_f = \int |\nabla\varphi| \, f \mathrm{d}x = \int |\nabla f| \, \mathrm{d}x < \infty,$$

where the last inequality is part of Lemma 4 of [5].

Next we show that ν_f is finite. Note that this is not entirely trivial since $\int \mathrm{d}\nu_f = \int_{\partial K_f} f \mathrm{d}\mathcal{H}^{n-1}$, and while f is clearly bounded we can have $\mathcal{H}^{n-1}(\partial K_f) = \infty$. We therefore adapt a simple argument of Ball [2]. Since f is log-concave and integrable, there exists constants $A, c > 0$ such that $f(x) \leq Ae^{-c|x|}$ (see e.g. Lemma 2.1 of [12]). Note that for all $x \in \mathbb{R}^n$ we have

$$e^{-c|x|} = c \int_0^\infty e^{-ct} 1_{tB}(x) \mathrm{d}t,$$

where $B = \{x : |x| \le 1\}$ is the unit ball. We may therefore compute

$$
\begin{aligned}
\int \mathrm{d}\nu_f &= \int_{\partial K_f} f \mathrm{d}\mathcal{H}^{n-1} \le A \int_{\partial K_f} e^{-c|x|} \mathrm{d}\mathcal{H}^{n-1} \\
&= Ac \int_0^\infty \int_{\partial K_f} e^{-ct} \mathbf{1}_{tB}(x) \mathrm{d}\mathcal{H}^{n-1}(x) \mathrm{d}t \\
&= Ac \cdot \int_0^\infty e^{-ct} \mathcal{H}^{n-1} \left(\partial K_f \cap tB \right) \mathrm{d}t \\
&\le Ac \cdot \int_0^\infty e^{-ct} \mathcal{H}^{n-1} \left(\partial \left(K_f \cap tB \right) \right) \mathrm{d}t \\
&\le Ac \cdot \int_0^\infty e^{-ct} \mathcal{H}^{n-1} \left(\partial \left(tB \right) \right) \mathrm{d}t \\
&= Ac \cdot \mathcal{H}^{n-1} \left(\mathbb{S}^{n-1} \right) \cdot \int_0^\infty t^{n-1} e^{-ct} \mathrm{d}t < \infty,
\end{aligned}
$$

which is what we wanted to prove. Note that the last inequality holds since $K_f \cap tB \subseteq tB$ and surface area is monotone for convex bodies. $\qquad\square$

In particular, it follows that the equality in (1.3) is an equality of finite quantities whenever g is compactly supported, as in this case $h_g \le A|x| + B$ and h_{K_g} is bounded on $\mathbb{S}^{n-1}$. When g is not compactly supported it is possible to have $\delta(f, g) = \infty$, as already mentioned.

2 Anisotropic Total Variations

In order to start our proof, we need the notion of the anisotropic total variation. First recall the classical (isotropic) total variation: An integrable function $f : \mathbb{R}^n \to \mathbb{R}$ is said to have bounded variation if

$$
\sup \left\{ \int_{\mathbb{R}^n} f \operatorname{div} \Phi \mathrm{d}x : \begin{array}{l} \Phi : \mathbb{R}^n \to \mathbb{R}^n \text{ is } C^1\text{-smooth, compactly} \\ \text{supported and } |\Phi(x)| \le 1 \text{ for all } x \in \mathbb{R}^n \end{array} \right\} < \infty.
$$

This supremum is then known as the total variation of f, which we will denote by $\mathrm{TV}(f)$. Moreover, if f is of bounded variation then there exists a vector-valued measure Df on $\mathbb{R}^n$ such that

$$
\int_{\mathbb{R}^n} f \operatorname{div} \Phi \mathrm{d}x = - \int_{\mathbb{R}^n} \langle \Phi, \mathrm{d}(Df) \rangle ,
$$

and $\mathrm{TV}(f) = |Df|(\mathbb{R}^n)$. Here $|Df|$ denotes the total variation (in the sense of measures) of Df.

A set $A \subseteq \mathbb{R}^n$ is said to have finite perimeter if 1_A has finite variation, and we define $\mathrm{Per}(A) = \mathrm{TV}(1_A)$. Finally, the coarea formula states that if f has bounded variation then $F_s = \{x : f(x) \geq s\}$ has finite perimeter for almost every s, and $\mathrm{TV}(f) = \int_{-\infty}^{\infty} \mathrm{Per}(F_s) \mathrm{d}s$. All of these facts are standard—see e.g. Chapter 5 of [7] for proofs.

It is less well known that the role of Euclidean norm in the definition of $\mathrm{TV}(f)$ is not essential. Fix a convex body $L \subseteq \mathbb{R}^n$ and assume that 0 belongs to the interior of L. Then the (non-symmetric) norm

$$\|x\|_L = \inf\left\{\lambda > 0 : \frac{x}{\lambda} \in L\right\}$$

is equivalent to the Euclidean norm. We then define:

Definition 2.1

(1) Let $f : \mathbb{R}^n \to \mathbb{R}^n$ be an integrable function. Then the L-total variation of f is given by

$$\mathrm{TV}_L(f)$$
$$= \sup\left\{\int_{\mathbb{R}^n} f \,\mathrm{div}\,\Phi \mathrm{d}x : \begin{array}{l} \Phi : \mathbb{R}^n \to \mathbb{R}^n \text{ is } C^1\text{-smooth, compactly} \\ \text{supported and } \|\Phi(x)\|_L \leq 1 \text{ for all } x \in \mathbb{R}^n \end{array}\right\}.$$

(2) Let $A \subseteq \mathbb{R}^n$ be a measurable set. The L-perimeter of A is defined by $\mathrm{Per}_L(A) = \mathrm{TV}_L(1_A)$.

Since $\|\cdot\|_L$ and $|\cdot|$ are equivalent the notion of "bounded variation" does not depend on L, and $\mathrm{TV}_L(f) < \infty$ if and only if $\mathrm{TV}(f) < \infty$. Of course, the variation itself does depend on L.

The theory of anisotropic total variations is analogous to the standard theory. We now cite two results that we will require. We were only able to find as reference the technical report [10], where these results are proven by Grasmair, but the results can be proved in the same way as the classical proofs that appear e.g. in [7]:

Proposition 2.2 *Let $f : \mathbb{R}^n \to \mathbb{R}$ be of bounded variation. Write the vector valued measure Df as $Df = \sigma\mu$ where μ is a positive measure and $\sigma : \mathbb{R}^n \to \mathbb{R}^n$ satisfies $h_L(\sigma(x)) = 1$ for all $x \in \mathbb{R}^n$. Then $\mathrm{TV}_L(f) = \mu(\mathbb{R}^n)$.*

Note that when L is the Euclidean ball the measure μ is exactly $|Df|$, the usual total variation of Df. In the general case μ can be considered as "total variation of Df with respect to L". Also note that $\mathrm{TV}_L(f)$ was defined using the norm $\|\cdot\|_L$, but in Proposition 2.2 the norm that appears is the dual norm h_L.

We will also need the anisotropic coarea formula:

Theorem 2.3 *Fix an integrable* $f : \mathbb{R}^n \to \mathbb{R}$, *and denote its level sets by* $F_s = \{x : f(x) \geq s\}$. *Then*

$$\mathrm{TV}_L(f) = \int_{-\infty}^{\infty} \mathrm{Per}_L(F_s)\mathrm{d}s.$$

Our goal for this section is to prove Theorem 1.5 when $g = 1_L$, the indicator of a convex body. We will do so by proving that in this case Theorem 1.5 is essentially equivalent to Theorem 2.3. We begin with finding an alternative formula for $\delta(f, g)$ in this case. Recall that if $K, L \subseteq \mathbb{R}^n$ are convex bodies then the volume $|K + tL|$ for $t \geq 0$ can be written as a polynomial,

$$|K + tL| = \sum_{k=0}^{n} \binom{n}{k} W_k(K, L)t^k. \tag{2.1}$$

The non-negative coefficients $W_k(K, L)$ are known in our normalization as the relative quermassintegrals of K with respect to L. Formula (2.1) is a special case of the celebrated Minkowski theorem, and for the proof and basic properties of the relative quermassintegrals we refer the reader again to [19] or [11]. For now we just note that $W_0(K, L) = |K|$.

We now prove:

Proposition 2.4 *Fix* $f \in \mathrm{LC}_n$ *with* $0 < \int f < \infty$ *and fix a compact convex body* $L \subseteq \mathbb{R}^n$. *For every* $s > 0$ *we write* $F_s = \{x \in \mathbb{R}^n : f(x) \geq s\}$. *Then*

$$\delta(f, 1_L) = n \int_0^{\infty} W_1(F_s, L)\mathrm{d}s.$$

This result essentially appears in [3], at least in the case when L is the unit ball. Nonetheless we present its short proof:

Proof For brevity we define $f_t = f \star (t \cdot 1_L) = f \star 1_{tL}$ and $F_s^{(t)} = \{x \in \mathbb{R}^n : f_t(x) \geq s\}$. It is immediate from the definition of f_t that $F_s^{(t)} = F_s + tL$. By layer cake decomposition we have

$$\delta(f, g) = \lim_{t \to 0^+} \frac{\int f_t - \int f}{t} = \lim_{t \to 0^+} \frac{\int_0^{\infty} \left|F_s^{(t)}\right| \mathrm{d}s - \int_0^{\infty} |F_s| \, \mathrm{d}s}{t}$$

$$= \lim_{t \to 0^+} \int_0^{\infty} \frac{|F_s + tL| - |F_s|}{t} \mathrm{d}s.$$

Using (2.1) we see that for $0 < t < 1$ we have

$$0 \le \frac{|F_s + tL| - |F_s|}{t} = \sum_{k=1}^{n} \binom{n}{k} W_k(F_s, L) t^{k-1}$$

$$\le \sum_{k=1}^{n} \binom{n}{k} W_k(F_s, L) = |F_s + L| - |F_s|.$$

Since $\int_0^\infty (|F_s + L| - |F_s|)\, ds = \int f_1 - \int f < \infty$, we can use the dominated convergence theorem to conclude that

$$\delta(f, g) = \int_0^\infty \left(\lim_{t \to 0^+} \frac{|F_s + tL| - |F_s|}{t} \right) ds = n \int_0^\infty W_1(F_s, L)\, ds.$$

$\square$

We will also need one more identity, a version of the divergence (or Gauss–Green) theorem, which was proved in [17] as part of Theorem 3.2:

Proposition 2.5 *Let* $\Phi : \mathbb{R}^n \to \mathbb{R}^n$ *denote a* C^1*-smooth compactly supported vector field. Fix* $f \in LC_n$ *with* $0 < \int f < \infty$. *Then*

$$\int_{\mathbb{R}^n} f \operatorname{div} \Phi\, dx = -\int_{\mathbb{R}^n} \langle \nabla f, \Phi \rangle\, dx + \int_{\partial K_f} f \langle \Phi, n_{K_f} \rangle\, d\mathcal{H}^{n-1}.$$

The divergence theorem was extended beyond the smooth setting by many authors, most notably Federer [8, 9]. His result does not formally imply Proposition 2.5 as log-concave functions are not necessarily Lipschitz, but this is handled using an approximation argument.

Armed with these tools, we can finally prove:

Proposition 2.6 *Fix* $f \in LC_n$ *with* $0 < \int f < \infty$, *and set* $g = e^c 1_L$ *for a compact convex set* $L \subseteq \mathbb{R}^n$ *and* $c \in \mathbb{R}$. *Then*

$$\delta(f, g) = \mathrm{TV}_L(f) + c \int f = \int_{\mathbb{R}^n} h_g\, d\mu_f + \int_{\mathbb{S}^{n-1}} h_{K_g}\, d\nu_f.$$

Proof First note that if the result holds for a function g, it also holds for $\tilde{g}(x) = e^c \cdot g(x)$ for all $c \in \mathbb{R}$. Indeed, it is easy to check by the chain rule that $\delta(f, \tilde{g}) = \delta(f, g) + c \int f$ (see Proposition 3.4 in [17]). Since $h_{\tilde{g}} = h_g + c$ and $h_{K_g} = h_{K_{\tilde{g}}}$, we have

$$\delta(f, \widetilde{g}) = \delta(f, g) + c \int f = \int_{\mathbb{R}^n} h_g d\mu_f + \int_{\mathbb{S}^{n-1}} h_{K_g} dv_f + c \int f$$

$$= \int_{\mathbb{R}^n} (h_g + c) d\mu_f + \int_{\mathbb{S}^{n-1}} h_{K_g} dv_f = \int_{\mathbb{R}^n} h_{\widetilde{g}} d\mu_f + \int_{\mathbb{S}^{n-1}} h_{K_{\widetilde{g}}} dv_f$$

as claimed. Therefore from now on we can (and will) assume that $g = 1_L$.

Next, assume that 0 belongs to the interior of L so the theory of anisotropic total variations applies. Proposition 2.5 immediately implies that

$$d(Df) = -\nabla f dx + f n_{\partial K_f} d\mathcal{H}^{n-1}\Big|_{\partial K_f}.$$

Therefore the measure μ from Proposition 2.2 is $d\mu = h_L(-\nabla f) dx + f h_L(n_{\partial K_f}) d\mathcal{H}^{n-1}\big|_{\partial K_f}$. By the same proposition we then have

$$TV_L(f) = \int d\mu = \int_{\mathbb{R}^n} h_L(-\nabla f) dx + \int_{\partial K_f} f h_L(n_{\partial K_f}) d\mathcal{H}^{n-1} \tag{2.2}$$

$$= \int_{\mathbb{R}^n} h_L(\nabla \varphi) f dx + \int_{\partial K_f} h_L(n_{\partial K_f}) f d\mathcal{H}^{n-1}$$

$$= \int h_L d\mu_f + \int h_L dv_f = \int h_g d\mu_f + \int h_{K_g} dv_f,$$

where of course $\varphi = -\log f$.

As (2.2) holds for all $f \in LC_n$ with $0 < \int f < \infty$ we can in particular apply this formula to the indicator 1_F of a convex body F. We then obtain

$$Per_L(F) = TV_L(1_F) = \int h_L d(|F|\delta_0) + \int h_L dS_F = \int h_L dS_F = nW_1(F, L), \tag{2.3}$$

where the last equality is a standard (and follows immediately from (1.1) and (2.1)).

Therefore, using in order Proposition 2.4, Eq. (2.3), Theorem 2.3 and Eq. (2.2) we have

$$\delta(f, g) = n \int_0^\infty W_1(F_s, L) ds = \int_0^\infty Per_L(F_s) ds$$

$$= TV_L(f) = \int h_g d\mu_f + \int h_{K_g} dv_f,$$

where $F_s = \{x \in \mathbb{R}^n : f(x) \geq s\}$ as before. This concludes the proof in the case $0 \in int(L)$.

For the general case, fix a large Euclidean ball B centered at the origin such that $B + L$ contains the origin in its interior. From Proposition 2.4 and standard properties of quermassintegrals it follows that $\delta(f, \mathbf{1}_L)$ is linear in L with respect to the Minkowski addition. Therefore

$$\delta(f, \mathbf{1}_L) = \delta(f, \mathbf{1}_{L+B}) - \delta(f, \mathbf{1}_B)$$

$$= \left(\int h_{L+B} \mathrm{d}\mu_f + \int h_{L+B} \mathrm{d}v_f \right) - \left(\int h_B \mathrm{d}\mu_f + \int h_B \mathrm{d}v_f \right)$$

$$= \int h_L \mathrm{d}\mu_f + \int h_L \mathrm{d}v_f$$

and the proof is complete. □

Note that as a corollary we obtain the following result:

Corollary 2.7 *For $f \in LC_n$ with $0 < \int f < \infty$ the sum $\mu_f + v_f$ is centered, i.e. for all $v \in \mathbb{R}^n$ we have*

$$\int_{\mathbb{R}^n} \langle x, v \rangle \, \mathrm{d}\mu_f + \int_{\mathbb{S}^{n-1}} \langle x, v \rangle \, \mathrm{d}v_f = 0.$$

Proof Simply take $g = \mathbf{1}_{\{v\}}$ in Proposition 2.6, and note that $\delta(f, g) = 0$. □

The fact that μ_f is centered when $v_f = 0$ was observed already in [5].

3 Completing the Proof

In this section we finish the proof of Theorem 1.5. We start with the case of compactly supported g. The following lemma from [17] will be crucial:

Lemma 3.1 *Fix $f, g \in LC_n$ such that $0 < \int f < \infty$ and g is compactly supported. Then for (Lebesgue) almost every $x \in \mathbb{R}^n$ we have*

$$\lim_{t \to 0^+} \frac{(f \star (t \cdot g))(x) - f(x)}{t} = h_g(\nabla\varphi(x)) f(x).$$

Here $\varphi = -\log f$, and the right hand side is interpreted at 0 whenever $f(x) = 0$.

This lemma is proved in [17] as part of the proof of Lemma 3.7 (The condition $h_g(y) \le m|y| + c$ in the statement of that lemma is exactly equivalent to g being compactly supported). If all functions involved are sufficiently regular Lemma 3.1 follows from the standard formula for the first variation of the Legendre transform (see more information in [17]). To prove this result with no regularity assumptions does take a bit of work which we will not reproduce here.

We will now prove:

Proposition 3.2 *Fix $f, g \in LC_n$ such that $0 < \int f < \infty$ and g is compactly supported. Then*

$$\delta(f, g) = \int_{\mathbb{R}^n} h_g d\mu_f + \int_{\mathbb{S}^{n-1}} h_{K_g} dv_f.$$

Proof To simplify our notation let us define $f_t = f \star (t \cdot g)$. We also choose $A > 0$ such that $0 \le g(x) \le A$ for all $x \in \mathbb{R}^n$, and we define $\widetilde{g} = A \cdot \mathbf{1}_{K_g}$ and $\widetilde{f}_t = f \star (t \cdot \widetilde{g})$. Note that $g \le \widetilde{g}$, so $f_t \le \widetilde{f}_t$ for all $t > 0$. Also note that

$$\lim_{t \to 0^+} \frac{\widetilde{f}_t - f_t}{t} = \lim_{t \to 0^+} \left(\frac{\widetilde{f}_t - f}{t} - \frac{f_t - f}{t} \right) = h_{\widetilde{g}}(\nabla\varphi) f - h_g(\nabla\varphi) f$$

almost everywhere, where we used Lemma 3.1 twice. We may therefore apply Fatou's lemma and deduce that

$$\liminf_{t \to 0^+} \left(\frac{\int \widetilde{f}_t - \int f}{t} - \frac{\int f_t - \int f}{t} \right)$$

$$= \liminf_{t \to 0^+} \int \frac{\widetilde{f}_t - f_t}{t} dx \ge \int \left(h_{\widetilde{g}}(\nabla\varphi) f - h_g(\nabla\varphi) f \right) dx$$

$$\doteq \int \left(h_{\widetilde{g}} - h_g \right) d\mu_f.$$

However, by Proposition 2.6 we know that

$$\lim_{t \to 0^+} \frac{\int \widetilde{f}_t - \int f}{t} = \delta(f, \widetilde{g}) = \int h_{\widetilde{g}} d\mu_f + \int h_{K_g} dv_f,$$

where we used the fact that $K_{\widetilde{g}} = K_g$. Combining the last two formulas we see that

$$\int h_{\widetilde{g}} d\mu_f + \int h_{K_g} dv_f - \limsup_{t \to 0^+} \left(\frac{\int f_t - \int f}{t} \right) \ge \int \left(h_{\widetilde{g}} - h_g \right) d\mu_f,$$

so

$$\limsup_{t \to 0^+} \frac{\int f_t - \int f}{t} \le \int h_g d\mu_f + \int h_{K_g} dv_f. \qquad (3.1)$$

Note that we were allowed to cancel $\int h_{\widetilde{g}} d\mu_f$ from both sides since this expression is finite by Proposition 1.6.

The proof of the opposite inequality is similar. Fix $m \in \mathbb{N}$ and consider $K_m = \left\{ x \in K_g : g(x) \geq \frac{1}{m} \right\}$. This time we define $\widetilde{g} = \frac{1}{m} \mathbf{1}_{K_m}$ and $\widetilde{f}_t = f \star (t \cdot \widetilde{g})$, and we have the opposite inequality $f_t \geq \widetilde{f}_t$. Applying Fatou's lemma in the same way we have

$$\liminf_{t \to 0^+} \left(\frac{\int f_t - \int f}{t} - \frac{\int \widetilde{f}_t - \int f}{t} \right)$$

$$= \liminf_{t \to 0^+} \int \frac{f_t - \widetilde{f}_t}{t} \geq \int \left(h_g \left(\nabla \varphi \right) f - h_{\widetilde{g}} \left(\nabla \varphi \right) f \right)$$

$$= \int \left(h_g - h_{\widetilde{g}} \right) d\mu_f,$$

and this time we have

$$\lim_{t \to 0^+} \frac{\int \widetilde{f}_t - \int f}{t} = \delta(f, \widetilde{g}) = \int h_{\widetilde{g}} d\mu_f + \int h_{K_m} dv_f,$$

so we obtain

$$\liminf_{t \to 0^+} \frac{\int f_t - \int f}{t} \geq \int h_g d\mu_f + \int h_{K_m} dv_f.$$

Since $K_m \subseteq K_{m+1}$ for all m and $\overline{\bigcup_{m=1}^{\infty} K_m} = K_g$, we have $h_{K_g} = \sup_m h_{K_m} = \lim_{m \to \infty} h_{K_m}$. We may therefore let $m \to \infty$ in the last formula and deduce that

$$\liminf_{t \to 0^+} \frac{\int f_t - \int f}{t} \geq \int h_g d\mu_f + \int h_{K_g} dv_f,$$

which together with (3.1) completes the proof.   □

Now we can finally prove Theorem 1.5. The final step is an approximation argument, which is essentially the same as the one in [17]. Therefore we repeat the argument briefly without repeating some of the computations:

Proof of Theorem 1.5 Define a sequence $\{g_m\}_{m=1}^{\infty} \subseteq LC_n$ by

$$g_m(x) = \begin{cases} g(x) & |x| \leq m \\ 0 & \text{otherwise.} \end{cases}$$

A computation shows that $h_{g_m} \nearrow h_g$ as $m \to \infty$. Moreover, since $K_{g_m} \subseteq K_{g_{m+1}}$ for all m and $\overline{\bigcup_{m=1}^{\infty} K_{g_m}} = K_g$ we also have $h_{K_{g_m}} \nearrow h_{K_g}$. If we also define $f_t = f \star (t \cdot g)$ and $f_{t,m} = f \star (t \cdot g_m)$ then another computation shows that $f_{t,m}(x) \nearrow f_t(x)$ for all $t > 0$ and $x \in \mathbb{R}^n$. This implies that $\int f_{t,m} \nearrow \int f_t$ (see e.g. Lemma 3.2 of [1]).

Using the chain rule for derivatives we may write

$$\delta(f, g) = \left(\int f \right) \cdot \lim_{t \to 0^+} \frac{\log \int f_t - \log \int f}{t}.$$

This formula has the advantage that by the Prékopa-Leindler inequality [14, 15] the function $t \mapsto \log \int f_t$ is concave, so we may replace the limit by a supremum. It follows that

$$\lim_{m \to \infty} \delta(f, g_m) = \sup_m \delta(f, g_m) = \left(\int f \right) \cdot \sup_m \sup_{t > 0} \frac{\log \int f_{t,m} - \log \int f}{t}$$

$$= \left(\int f \right) \cdot \sup_{t > 0} \sup_m \frac{\log \int f_{t,m} - \log \int f}{t}$$

$$= \left(\int f \right) \cdot \sup_{t > 0} \frac{\log \int f_t - \log \int f}{t} = \delta(f, g).$$

Therefore, applying Proposition 3.2 and the monotone convergence theorem we conclude that

$$\delta(f, g) = \lim_{m \to \infty} \delta(f, g_m) = \lim_{m \to \infty} \left(\int h_{g_m} d\mu_f + \int h_{K_{g_m}} d\nu_f \right)$$

$$= \int h_g d\mu_f + \int h_{K_g} d\nu_f,$$

and the proof is complete. $\qquad \square$

As a corollary of the theorem we now prove that the measures μ_f and ν_f characterize the function f uniquely up to translations:

Corollary 3.3 *Fix $f, g \in LC_n$ with $0 < \int f, \int g < \infty$ and assume that $\mu_f = \mu_g$ and $\nu_f = \nu_g$. Then there exists $x_0 \in \mathbb{R}^n$ such that $f(x) = g(x - x_0)$.*

Proof Corollary 5.3 of [4] states that if $f, g \in LC_n$ satisfy $0 < \int f = \int g < \infty$, $\delta(f, g) = \delta(g, g)$ and $\delta(g, f) = \delta(f, f)$, then there exists $x_0 \in \mathbb{R}^n$ such that $f(x) = g(x - x_0)$. This is proved by showing that we have equality in the Prékopa-Leindler inequality, and using a characterization of the equality case by Dubuc [6]. A similar strategy was used in [5], and indeed in the classical proof that the surface area measure S_K determines the body K uniquely.

In our case we have $\int f = \mu_f(\mathbb{R}^n) = \mu_g(\mathbb{R}^n) = \int g$, and since

$$\delta(f, g) = \int_{\mathbb{R}^n} h_g d\mu_f + \int_{S^{n-1}} h_{K_g} d\nu_f$$

we clearly have $\delta(f, g) = \delta(g, g)$ and similarly $\delta(g, f) = \delta(f, f)$. The result follows immediately. $\qquad \square$

In [4] the same argument was used but with Theorem 1.3 replacing Theorem 1.5, so uniqueness was only proved under the regularity assumptions of that theorem. In [5] there was no explicit representation formula for $\delta(f, g)$, but a weaker statement that in the essentially continuous case was also sufficient in order to reduce the uniqueness result to the equality case of Prékopa-Leindler inequality (see also [17] for an explanation of why the result of [5] is a weak representation theorem for $\delta(f, g)$). We see that in order to get a clean uniqueness result in the general case one indeed needs the full strength of Theorem 1.5.

Of course, Corollary 3.3 raises the question of existence: Given measures μ and ν, when is there a function $f \in LC_n$ with $\mu_f = \mu$ and $\nu_f = \nu$? We believe this question can be answered by essentially the same argument as the argument of [5], which handled the case $\nu_f \equiv 0$, but the full details are beyond the scope of this paper.

Acknowledgments I would like to thank the anonymous referee for their useful suggestions and corrections.

References

1. S. Artstein-Avidan, B. Klartag, V. Milman, The Santaló point of a function, and a functional form of the Santaló inequality. Mathematika **51**(1–2), 33–48 (2010)
2. K. Ball, The reverse isoperimetric problem for Gaussian measure. Discrete Comput. Geom. **10**(4), 411–420 (1993)
3. S. Bobkov, A. Colesanti, I. Fragala, Quermassintegrals of quasi-concave functions and generalized Prékopa-Leindler inequalities. Manuscripta Math. **143**(1–2), 131–169 (2014)
4. A. Colesanti, I. Fragalà, The first variation of the total mass of log-concave functions and related inequalities. Adv. Math. **244**, 708–749 (2013)
5. D. Cordero-Erausquin, B. Klartag, Moment measures. J. Funct. Anal. **268**(12), 3834–3866 (2015)
6. S. Dubuc, Critères de convexité et inégalités intégrales. Ann. Inst. Fourier **27**(1), 135–165 (1977)
7. L.C. Evans, R.F. Gariepy, *Measure Theory and Fine Properties of Functions* (CRC Press, New York, 1992)
8. H. Federer, The Gauss-Green theorem. Trans. Am. Math. Soc. **58**(1), 44 (1945)
9. H. Federer, A note on the Gauss-Green theorem. Proc. Am. Math. Soc. **9**(3), 447 (1958)
10. M. Grasmair, A Coarea Formula for Anisotropic Total Variation Regularisation. Technical report, FWF National Research Network S92, No. 103 (2010)
11. D. Hug, W. Weil, *Lectures on Convex Geometry*, vol. 286. Graduate Texts in Mathematics (Springer International Publishing, Cham, 2020)
12. B. Klartag, Uniform almost sub-gaussian estimates for linear functionals on convex sets. Algebra i Analiz **19**(1), 109–148 (2007)
13. B. Klartag, V. Milman, Geometry of log-concave functions and measures. Geom. Dedicata **112**(1), 169–182 (2005)
14. L. Leindler, On a certain converse of Hölder's inequality II. Acta Sci. Math. **33**(3–4), 217–223 (1972)
15. A. Prékopa, Logarithmic concave measures with application to stochastic programming. Acta Sci. Math. **32**(3–4), 301–316 (1971)

16. L. Rotem, On the mean width of log-concave functions, in *Geometric Aspects of Functional Analysis, Israel Seminar 2006–2010*, vol. 2050, ed. by B. Klartag, S. Mendelson, V. Milman. Lecture Notes in Mathematics (Springer, Berlin, 2012), pp. 355–372
17. L. Rotem, Surface area measures of log-concave functions. J. Anal. Math. **147**, 373–400 (2022)
18. F. Santambrogio, Dealing with moment measures via entropy and optimal transport. J. Funct. Anal. **271**(2), 418–436 (2016)
19. R. Schneider, *Convex Bodies: The Brunn-Minkowski Theory*, 2nd edn. Encyclopedia of Mathematics and its Applications (Cambridge University Press, Cambridge, 2014)

Chasing Convex Bodies Optimally

Mark Sellke

Abstract In the *chasing convex bodies* problem, an online player receives a request sequence of N convex sets $K_1, \ldots, K_N$ contained in a normed space X of dimension d. The player starts at $x_0 = 0 \in X$, and at time n observes the set K_n and then moves to a new point $x_n \in K_n$, paying a cost $||x_n - x_{n-1}||$. The player aims to ensure the total cost exceeds the minimum possible total cost by at most a bounded factor α_d independent of N, despite x_n being chosen without knowledge of the future sets $K_{n+1}, \ldots, K_N$. The best possible α_d is called the competitive ratio. Finiteness of the competitive ratio for convex body chasing was proved for $d = 2$ in Friedman and Linial (Discrete Comput. Geom. 9(3):293–321, 1993.) and conjectured for all d. Bubeck et al. (Proceedings of the 51st Annual ACM SIGACT Symposium on Theory of Computing, pp. 861–868, 2019) recently resolved this conjecture, proving an exponential $2^{O(d)}$ upper bound on the competitive ratio.

We give an improved algorithm achieving competitive ratio d in any normed space, which is *exactly* tight for ℓ^∞. In Euclidean space, our algorithm also achieves competitive ratio $O(\sqrt{d \log N})$, nearly matching a $\sqrt{d}$ lower bound when N is subexponential in d. Our approach extends that of Bubeck et al. (Proceedings of the Fourteenth Annual ACM-SIAM Symposium on Discrete Algorithms, pp. 1496–1508. SIAM, 2020.) for *nested* convex bodies, which is based on the classical Steiner point of a convex body. We define the *functional* Steiner point of a convex function and apply it to the associated work function.

1 Introduction

Let X be a d-dimensional normed space and $K_1, K_2, \ldots, K_N \subseteq X$ a finite sequence of convex bodies. In the *chasing convex bodies* problem, a player starting at $x_0 = 0 \in X$ learns the sets K_n one at a time, and after observing K_n moves to a point

M. Sellke (✉)
Stanford University, Stanford, CA, USA
e-mail: msellke@stanford.edu; msellke@ias.edu

© The Author(s), under exclusive license to Springer Nature Switzerland AG 2023
R. Eldan et al. (eds.), *Geometric Aspects of Functional Analysis*, Lecture Notes in Mathematics 2327, https://doi.org/10.1007/978-3-031-26300-2_12

$x_n \in K_n$. The player's cost is the total path length

$$\text{cost}(x_1, \ldots, x_N) = \sum_{n=1}^{N} ||x_n - x_{n-1}||.$$

Denote the smallest cost (in hind-sight) among all such sequences by

$$\text{cost}(K_1, \ldots, K_N) = \min_{(y_n \in K_n)_{n \leq N}} \sum_{n=1}^{N} ||y_n - y_{n-1}||.$$

The player's goal is to ensure that

$$\text{cost}(x_1, \ldots, x_N) \leq \alpha_d \cdot \text{cost}(K_1, \ldots, K_N) \qquad (1)$$

holds for any sequence $K_1, \ldots, K_N$, where α_d is as small as possible and is independent of N. The challenge is that the points $x_n = x_n(K_1, \ldots, K_n)$ must depend only on the sets revealed so far. To encapsulate this requirement we say the player's path must be *online*, as opposed to the optimal *offline* path which can depend on future information. An online algorithm achieving (1) for some finite α_d is said to be α_d-*competitive*, and the smallest possible α_d among all online algorithms is the *competitive ratio* of chasing convex bodies.

In the most general sense, the problem of asking a player to choose an online path $x_1, \ldots, x_N$ through a sequence of subsets $S_1, \ldots, S_N$ in a metric space $\mathcal{X}$ is known as *metrical service systems*. These sets are typically called "requests". When arbitrary subsets $S_i \subseteq \mathcal{X}$ can be requested, the competitive ratio possible is $|\mathcal{X}| - 1$ in any metric space [28]. One also considers the slightly more general *metrical task systems* problem in which requests are non-negative cost functions $f_n : \mathcal{X} \to \mathbb{R}^+$ rather than sets and the cost takes the form

$$\text{cost}(x_1, \ldots, x_N) = \sum_{n=1}^{N} d_{\mathcal{X}}(x_n, x_{n-1}) + f_n(x_n)$$

where $\sum_{n=1}^{N} d_{\mathcal{X}}(x_n, x_{n-1})$ is called the *movement cost* while $\sum_{n=1}^{N} f_n(x_n)$ is the *service cost*. As in (1), one aims to ensure

$$\text{cost}(x_1, \ldots, x_N) \leq \alpha \cdot \text{cost}(f_1, \ldots, f_n) = \alpha \cdot \inf_{(y_n \in \mathcal{X})_{n \leq N}} \text{cost}(y_1, \ldots, y_N). \qquad (2)$$

The competitive ratio of metrical task systems is always $2|\mathcal{X}| - 1$ [12]. Actually both competitive ratios just stated are for deterministic algorithms; one may also allow external randomness, so that one chooses $x_n = x_n(S_1, \ldots, S_n, \omega)$ for some random variable ω independent of the sets S_i. One then aims for the same guarantee as in (1), (2) with the expected cost of the player on the left-hand side, for any fixed sequence $(S_1, \ldots, S_N)$. With randomization the competitive ratio of metrical task or service systems sharply drops and is known to be in the range $\left[\frac{c_1 \log |\mathcal{X}|}{\log \log |\mathcal{X}|}, c_2 (\log |\mathcal{X}|)^2\right]$, and to be $\Theta(\log |\mathcal{X}|)$ in some specific cases [9, 10, 13, 18]. However this is not the end of the story as a wide range of problems, including chasing convex bodies, result from restricting which subsets are allowed as requests. The literature on such problems is vast and includes scheduling [21], self-organizing lists [33], efficient covering [1], safely using machine-learned advice [11, 24, 27, 35], and the famous k-server problem [6, 22, 23, 28].

Chasing convex bodies was proposed in [19] to study the interaction between convexity and metrical task systems. Of course the general upper bounds above are of no use as $|X| = \infty$, while the lower bounds also do not apply due to the convexity constraint. Friedman and Linial [19] gave an algorithm with finite competitive ratio for the already non-trivial $d = 2$ case and conjectured that the competitive ratio is finite for any $d \in \mathbb{N}$. The best known asymptotic lower bounds come from requesting the faces of a hypercube by taking $K_n = (\varepsilon_1, \varepsilon_2, \ldots, \varepsilon_n) \times [-1, 1]^{d-n}$ for $\varepsilon_i \in \{-1, 1\}$ uniformly random and $n \leq d$. This construction implies that the competitive ratio is at least $\sqrt{d}$ in Euclidean space and at least d for $X = \ell^\infty$—see [15, Lemma 5.4] for more on lower bounds. Unlike in many competitive analysis problems, randomization is useless for chasing convex bodies and we may freely restrict attention to deterministic algorithms. This is because $\text{cost}(x_1, \ldots, x_N)$ is convex on X^N, and so randomized paths are no better than their (deterministic) pointwise expectations.

Following a lack of progress on the full conjecture, restricted cases such as chasing subspaces were studied, e.g. [2]. A notable restriction is chasing *nested* convex bodies, where the convex sets $K_1 \supseteq K_2 \supseteq \ldots$ are required to decrease. Nested chasing was introduced in [8] and solved rather comprehensively in [3] and then [15]. The latter work gave an algorithm with optimal competitive ratio up to $O(\log d)$ factors for all ℓ^p spaces based on Gaussian-weighted centroids. Moreover it gave a d-competitive memoryless algorithm based on the Steiner point which we discuss later.

Some time after chasing convex bodies was posed, an equivalent problem called *chasing convex functions* emerged. This is a metrical task systems problem in which requests are convex functions $f_n : X \to \mathbb{R}^+$ instead of convex sets. As described above the total cost

$$\text{cost}(x_1, \ldots, x_N) = \sum_{n=1}^{N} ||x_n - x_{n-1}|| + f_n(x_n)$$

decomposes as a movement cost plus a service cost. Chasing convex functions subsumes chasing convex bodies by replacing the body K_n with the function $f_n = 2 \cdot d(x, K_n)$. This is because an arbitrary algorithm for the requests f_n is improved by projecting x_n onto K_n—actually the same argument shows more generally that metrical task systems subsumes metrical service systems. Conversely as observed in [14], convex function chasing in X can be reduced to convex body chasing in $X \oplus \mathbb{R}$ up to a constant factor by alternating requests of the epigraphs $\{(x, y) \in X \times \mathbb{R} : y \geq f_n(x)\}$ with the hyperplane $X \times \{0\}$. As with chasing convex bodies, randomized algorithms are no better than deterministic algorithms since $\mathrm{cost}(x_1, \ldots, x_N)$ remains convex on X^N.

Convex function requests allow one to model many practical problems. Indeed chasing convex functions was originally considered as a model for efficient power management in cooling data centers [26]. In light of this, restricted or modified versions of chasing convex function have also been studied. For example, [7] determines the exact competitive ratio in 1 dimension, while works such as [16, 20] show dimension-independent competitive ratios for similar problems with further restrictions on the cost functions.

Main Result In prior joint work with S. Bubeck, Y.T. Lee, and Y. Li [14] we gave the first algorithm achieving a finite competitive ratio for convex body chasing. Unfortunately this algorithm used an induction on dimension that led to a exponential competitive ratio $2^{O(d)}$. We give an upper bound of d for the competitive ratio of chasing convex bodies in a general normed space, which is tight for ℓ^∞. In Euclidean space, our algorithm has competitive ratio $O(\sqrt{d \log N})$, nearly matching the lower bound $\sqrt{d}$ when the number of requests N is sub-exponential in d. The statement following combines Theorems 4.1 and 5.3.

Theorem 1.1 *In any d-dimensional normed space there is a $d + 1$ competitive algorithm for chasing convex functions and a d competitive algorithm for chasing convex bodies. Moreover in Euclidean space this algorithm is $O(\sqrt{d \log N})$-competitive.*

The proof is inspired by our joint work with S. Bubeck, B. Klartag, Y.T. Lee, and Y. Li [15] on chasing nested convex bodies. It is shown there that moving to the new body's *Steiner point*, a stable center point of any convex body defined in [34], gives total movement at most d starting from the unit ball in d dimensions. (The argument in [15] is restricted to Euclidean space but the proof works in general as we will explain.) We extend their argument by defining the *functional Steiner point* of a convex function. Our algorithm follows the functional Steiner point of the so-called *work function* which encodes at any time the effective total cost of all requests so far.

We remark that given the form of (1), chasing convex bodies may be viewed as an online version of a Lipschitz selection problem. In the broadest generality, for some family $\mathcal{S} \subseteq 2^X$ of subsets of a set X, a selector takes sets $S \in \mathcal{S}$ to elements $s \in S$. Of course the relevant comparison for us is when $\mathcal{S}$ consists of all convex bodies in X. Continuity and Lipschitz properties of general selectors have received significant attention [17, 25, 31, 32]. Taking the Hausdorff metric on convex sets, the Steiner point is d-Lipschitz in any normed space. Moreover as explained in [29, Section 4], it achieves the exact optimal Lipschitz constant (of order $\Theta(\sqrt{d})$) when X is Euclidean due to a beautiful symmetrization argument. We find it appealing that this in some sense optimal Hausdorff-Lipschitz selector also solves an online version of Lipschitz selection.

Concurrently with this work, C.J. Argue, A. Gupta, G. Guruganesh, and Z. Tang obtained similar results for chasing convex bodies in Euclidean space [4]. Their algorithm is based on Steiner points of level sets of the work function; these turn out to be almost the same as the functional Steiner point as we show in Sect. 6.

2 Problem Setup

2.1 Notations and Conventions

The variables T, t, s denote real times while N, n denote integer times. $\fint_{x \in S} f(x) dx$ denotes the average value $\frac{\int_{x \in S} f(x) dx}{\int_{x \in S} 1 dx}$ of $f(x)$ on the set S. Denote by $B_1 \subseteq X$ the unit ball and $B_1^* \subseteq X^*$ the dual unit ball. The symbol ∂ denotes boundary, and $\langle \cdot, \cdot \rangle$ denotes the natural pairing between X, X^*.

2.2 Continuous Time Formulation

Our proof is more natural in continuous time, so we first solve the problem in this setting and then specialize to discrete time. In continuous time chasing convex functions, we receive a locally bounded family of non-negative convex functions $(f_t : X \to \mathbb{R}^+)_{t \in [0,T]}$. We assume that $f_t(x)$ is piece-wise continuous in t with a locally finite set of continuities. The player constructs a bounded variation path (x_t) online, so that x_s depends only on $(f_t)_{t \le s}$. We will assume f_t and x_t are cadlag (right-continuous with left-limits) in the time variable t. The cost is again the sum of movement and service costs given by

$$\text{cost}((x_t)_{t \in [0,T]}) = \int_0^T f_t(x_t) + ||x_t'|| dt.$$

Here and throughout, the integral of $||x_t'||$ is understood to mean the total variation of the path x_t. As before the goal is to achieve a small competitive ratio against the optimal offline path. Given a sequence $f_1, f_2, \ldots, f_N$ of convex requests, one readily obtains a corresponding continuous-time problem instance by choosing, for each $t \in [0, N]$, the function $f_t = f_n$ for $t \in (n-1, n]$. The next proposition shows that solving this continuous problem suffices to solve the discrete problem.

Proposition 2.1 *Any discrete-time instance of chasing convex function has the same offline optimal cost as its continuous-time counterpart. Meanwhile for any continuous-time online algorithm there exists a discrete-time online algorithm achieving both smaller movement and smaller service cost on every sequence of functions $f_1, \ldots, f_N$.*

Proof It is easy to see that the continuous and discrete time problems have the same offline optimum value. Given a solution for continous-time convex function chasing, suppose the player sees a discrete time request f_n. The player then computes the continuous time path $(x_t)_{t\in(n-1,n]}$ and moves to some x_{t_n} with $t_n \in (n-1, n]$ and

$$f_n(x_{t_n}) \le \int_{n-1}^{n} f_n(x_t)\mathrm{d}t.$$

The discretized sequence $(x_{t_1}, \ldots, x_{t_N})$ has a smaller movement cost than the continuous path $(x_t)_{t\in[0,T]}$ because the triangle inequality implies

$$\sum_{n=1}^{N} ||x_{t_n} - x_{t_{n-1}}|| \le \sum_{n=1}^{N} \int_{t_{n-1}}^{t_n} ||x'_s||\mathrm{d}s$$

$$= \int_0^{t_N} ||x'_s||\mathrm{d}s$$

$$\le \int_0^N ||x'_s||\mathrm{d}s.$$

The discretized path also has smaller service cost by construction, hence the result.
$\square$

3 Functional Steiner Point and Work Function

We begin by recalling the definition of the Steiner point in a d-dimensional normed space X. For a convex body $K \subseteq X$ and $v \in X^*$, define

$$f_K(v) = \arg\max_{x\in K}\langle v, x\rangle,$$

$$h_K(v) = \max_{x\in K}\langle v, x\rangle = \langle f_K(v), x\rangle.$$

h_K is commonly known as the *support function* of K. Let μ denote the cone measure on ∂B_1^*, which can be sampled from by choosing a uniformly random $z \in B_1^*$ and normalizing to $\theta = \frac{z}{||z||}$. For $\theta \in \partial B_1^*$ define $n(\theta) \in X$ to be the outward unit normal defined (for μ-almost all θ) by $||n(\theta)|| = 1$ and $\langle n(\theta), \theta\rangle = 1$.

Definition 3.1 ([29, Chapter 6]) The Steiner point $s(K) \in X$ is

$$s(K) = \fint_{v\in B_1^*} f_K(v)\mathrm{d}v. \tag{3}$$

$$= d\fint_{\theta\in\partial B_1^*} h_K(\theta)n(\theta)\mathrm{d}\mu(\theta). \tag{4}$$

The equivalence of the two definitions follows from the divergence theorem and the identity $\nabla h_K = f_K$. The factor d comes from the discrepancy in total measure of the ball and the sphere. See [29, Chapter 6] for a careful derivation.

Using Definition 3.1, the upper bound d for nested chasing in [15] immediately extends to any normed space. We recall the main result here. It is not phrased as a competitive ratio because some apriori reductions are possible in nested chasing— roughly speaking we stay inside the unit ball B_1 and treat the offline optimum cost as being 1. Note that both (3) and (4) are essential in the argument below.

Theorem 3.2 ([15, Theorem 2.1]) *Let* $B_1 \supseteq K_1 \supseteq K_2 \supseteq \cdots \supseteq K_N$ *be convex bodies in* X, *with* $x_n = s(K_n)$ *for each* n. *Then* $x_n \in K_n$ *for each* n *and*

$$\sum_{n=1}^{N} \|x_n - x_{n-1}\| \leq d.$$

Proof It follows from (3) that $s(K) \in K$, so it remains to estimate the total movement. For convenience take $K_0 = B_1$ the unit ball so that $x_0 = (0, 0, \ldots, 0) = s(K_0)$. From $K_n \subseteq K_{n-1}$ it follows that $h_{K_n}(\theta) \leq h_{K_{n-1}}(\theta)$ for each $n \leq N$ and $\theta \in \partial B_1^*$. Combining with (4) yields:

$$\sum_{n=1}^{N} \|s(K_n) - s(K_{n-1})\| \leq d \fint_{\theta \in \partial B_1^*} \sum_{n=1}^{N} |h_{K_n}(\theta) - h_{K_{n-1}}(\theta)| d\mu(\theta)$$

$$= d \fint_{\theta \in \partial B_1^*} \sum_{n=1}^{N} h_{K_{n-1}}(\theta) - h_{K_n}(\theta) d\mu(\theta)$$

$$= d \fint_{\theta \in \partial B_1^*} 1 - h_{K_N}(\theta) d\mu(\theta)$$

$$\leq d.$$

Here the last inequality follows from $h_{K_N}(\theta) + h_{K_N}(-\theta) \geq 0$. $\qquad\square$

We now extend the definition of Steiner point to convex functions. The idea is to replace the support function by the concave conjugate (also known as the Fenchel-Legendre transform). Recall that for a convex function $W : X \to \mathbb{R}^+$, the concave conjugate $W^* : X^* \to \mathbb{R} \cup \{-\infty\}$ is defined by

$$W^*(v) = \inf_{w \in X} (W(w) - \langle v, w \rangle) \tag{5}$$

Let us assume W is not only convex but also 1-Lipschitz, and that $W(w) - \|w\|$ is uniformly bounded. We will refer to such a W as an (abstract) work function. Note $W^*(v)$ is finite whenever $\|v\| < 1$ by the last assumption, and moreover the infimum in (5) is attained. We denote this point attaining this infimum by

$$v^* = \arg\min_{w \in X} (W(w) - \langle v, w \rangle),$$

the conjugate point to v with respect to W. It satisfies $\nabla W(v^*) = v$ and is well-defined for almost every $v \in B_1^*$ by Alexandrov's theorem. Moreover we have $\nabla W^*(v) = -v^*$. Combining this latter relation with the divergence theorem yields another identity, from which the functional Steiner point is defined.

Definition 3.3 Let X be an arbitrary d-dimensional normed space, and $W : X \to \mathbb{R}^+$ a work function as defined above. The functional Steiner point $s(W) \in X$ is:

$$s(W) = \fint_{v \in B_1^*} v^* dv. \tag{6}$$

$$= -d \fint_{\theta \in \partial B_1^*} W^*(\theta) n(\theta) d\mu(\theta). \tag{7}$$

We remark that if a convex body K is identified with the function $f(x) = d(x, K)$, then the definitions above agree. We call (3), (6) the *primal* definitions and (4), (7) the *dual* definitions.

3.1 The Work Function

The work function is a central object in online algorithms; in general it records the smallest cost required to satisfy an initial sequence of requests while ending in a given state. Work function based algorithms are essentially optimal among deterministic algorithms for general metrical task systems [12] as well as the k-server problem [23].

Definition 3.4 Given requests $(f_s)_{s \leq t}$, the work function $W_t(x)$ is the offline-optimal cost among paths satisfying $x_t = x$:

$$W_t(x) = \inf_{\substack{x_s:[0,t] \to X \\ x_t = x}} \|x_0\| + \int_0^t f_s(x_s) + \|x_s'\| ds \tag{8}$$

$$= \inf_{\substack{x_s:[0,t] \to X \\ x_t = x}} \text{cost}_t(x_s). \tag{9}$$

Here we allow $x_s : [0, t] \to X$ to be any path of bounded variation, and as before interpret $\int_0^t \|x_s'\| ds$ to mean the total variation of the path. Likewise for a discrete-time request sequence $(f_1, \ldots, f_n)$, the work function $W_n(x)$ is defined as above with $f_t = f_n$ for $t \in (n-1, n]$ or more simply by

$$W_n(x) = \min_{x_1,\ldots,x_n \in X} ||x - x_n|| + \sum_{n=1}^{N} ||x_n - x_{n-1}|| + f_n(x_n).$$

For a sequence $(K_1, \ldots, K_n)$ of convex set requests the work function W_n is defined analogously.

In the case that $f_s(x)$ is piecewise constant in s (which is all we need for the original discrete-time problem), the best offline continuous time strategy clearly coincides with the best offline discrete time strategy. The infimum is attained in (9) in general because the paths $(x_s)_{s \le t}$ of variation at most C satisfying $x_t = T$ are compact in the usual topology on cadlag functions for any C, and cost_t is lower semicontinuous.

Denote by $W_t^*(\cdot)$ the concave conjugate of W_t, and v_t^* the point with $\nabla W_t(v_t^*) = v$. We record the following proposition summarizing the properties of the work function and its dual.

Proposition 3.5 *In either discrete or continuous time, W_t and W_t^* satisfy:*

1. $W_0(x) = ||x||$.
2. $W_0^*(v) = 0$ whenever $||v|| \le 1$.
3. $W_t(x)$ is increasing in t and is convex for all fixed t.
4. $W_t^*(x)$ is increasing in t and concave in x.
5. $W_t(x)$ is an abstract work function.
6. $W_t^*(v)$ is non-negative and finite whenever $||v|| \le 1$.
7. $\text{cost}((f_s)_{s \in [0,t]}) = \min_{x \in X} W_t(x)$.

Proof It is clear that $W_0(x) = ||x||$, and that $W_t(x)$ is increasing in t. The computation of W_0^* is clear. Convexity of $W_t(\cdot)$ holds by convexity of $\text{cost}_t(\cdot)$— given paths $x_s^0 : [0, t] \to X$ and $x_s^1 : [0, t] \to X$ the path $x_s^q : [0, t] \to X$ given by

$$x_s^q = q x_s^1 + (1-q) x_s^0$$

satisfies for any $q \in [0, 1]$,

$$\text{cost}_t(x_s^q) \le q \cdot \text{cost}_t(x_s^1) + (1-q) \cdot \text{cost}_t(x_s^0).$$

Convexity of W_t implies that W_t^* is concave by general properties of the Fenchel-Legendre transform. Because W_t is increasing in t, the definition (5) implies that W_t^* is increasing in t as well. It is easy to see that W_t is 1-Lipschitz; to show

$$W_t(x) \le W_t(y) + ||x - y||$$

it suffices to take the lowest cost path to y and then move from y to x. Similarly $W_t(x) - ||x||$ is bounded, making W_t an abstract work function. It follows from this that $W_t^*(v)$ is finite when $||v|| \leq 1$. $\qquad\qquad\qquad\qquad\qquad\qquad\square$

Lemma 3.6 *For all t,*

$$\max_{||\theta|| \leq 1} W_t^*(\theta) \leq 2 \cdot \min_x W_t(x),$$

$$\fint_{\theta \in \partial B_1^*} W_t^*(\theta) d\mu(\theta) \leq \min_x W_t(x),$$

$$\fint_{v \in B_1} W_t^*(v) dv \leq \min_x W_t(x).$$

Proof Set

$$OPT_t = \arg\min_x W_t(x).$$

The definition (5) of W_t^* implies

$$W_t^*(\theta) \leq W_T(OPT_t) - \theta \cdot OPT_t.$$

Finally

$$|W_t(OPT_t)| = \inf_{\substack{x_s:[0,t] \to X \\ x_t = OPT_t}} ||x_0|| + \int_0^t f_s(x_s) + ||x_s'|| ds$$

$$\geq \inf_{\substack{x_s:[0,t] \to X \\ x_t = OPT_t}} ||x_0|| + \int_0^t ||x_s'|| ds$$

$$\geq |OPT_t|$$

holds where the triangle inequality was used in the last line. All assertions now follow. $\qquad\qquad\qquad\qquad\qquad\qquad\qquad\qquad\qquad\qquad\qquad\qquad\qquad\square$

We next compute the time derivative of $W_t^*(v)$ for fixed v with $|v| < 1$. The proof, a simple exercise, is left to the appendix.

Lemma 3.7 *For any $\varepsilon > 0$ suppose $f_s(x)$ is jointly continuous in (s, x) and convex in x for $(s, x) \in [t, t + \varepsilon] \times X$. Then for almost all v with $||v|| < 1$,*

$$\frac{d}{dt} W_t^*(v) = f_t(v_t^*).$$

4 Linear Competitive Ratio

Our algorithm for continuous-time convex function chasing is defined by setting $x_t = s(W_t)$. In its analysis, the primal definition (6) controls the service cost while the dual definition (7) controls the movement cost.

Theorem 4.1 $x_t = s(W_t)$ *is* $d + 1$ *competitive for continuous-time convex function chasing in any* d-*dimensional normed space* X. *In particular:*

1. The movement cost of x_t *is* d-*competitive:*

$$\int_0^T \|x_t'\|dt \le d \cdot \min_x W_t(x).$$

2. The service cost of x_t *is* 1-*competitive:*

$$\int_0^T f_t(x_t)dt \le \min_x W_t(x).$$

Proposition 2.1 yields an induced algorithm for chasing bodies/functions in discrete time which we call the discrete-time functional Steiner point.

Corollary 4.2 *The discrete-time functional Steiner point is* $d + 1$ *competitive for chasing convex functions and* d *competitive for chasing convex bodies.*

Proof of Corollary 4.2 This follows from Proposition 2.1 and the fact that chasing convex bodies has 0 service cost. ☐

Proof of Theorem 4.1 We begin with part 1. From the dual definition (7) of $s(W_t)$ and the fact that W_t^* increases with t from $W_0^* = 0$,

$$\int_0^T \|x_t'\|dt = d \cdot \int_0^T \left\| \frac{d}{dt}\fint_{\theta \in \partial B_1^*} W_t^*(\theta)\theta d\mu(\theta) \right\|$$

$$\le d \cdot \int_0^T \fint_{\theta \in \partial B_1^*} \left| \frac{d}{dt} W_t^*(\theta) \right| d\mu(\theta)$$

$$= d \cdot \fint_{\theta \in \partial B_1^*} W_T^*(\theta)d\mu(\theta).$$

Lemma 3.6 implies

$$d \cdot \fint_{\theta \in \partial B_1^*} W_T^*(\theta)d\mu(\theta) \le d \min_x W_T(x).$$

This completes the proof of part 1 and we turn to part 2. From the primal definition (6) and convexity of f_t it follows that

$$f_t(s(W_t)) \le \fint_{v \in B_1^*} \cdot f_t(v_t^*) dv.$$

Integrating in time and using Lemmas 3.7 and 3.6 yields:

$$\int_0^T f_t(s(W_t)) dt \le \fint_{v \in B_1} \int_0^T f_t(v_t^*) dt d\mu(\theta)$$

$$= \fint_{v \in B_1^*} \int_0^T \frac{d}{dt} W_t^*(v) dt dv$$

$$= \fint_{v \in B_1^*} W_T^*(v) - W_0^*(v) dv$$

$$= \fint_{v \in B_1^*} W_T^*(v) dv$$

$$\le \min_x W_T(x).$$

$\square$

Remark In the continuous time setting, only $f_t(x_t)$ and $\nabla f_t(x_t)$ are actually necessary to solve convex function chasing. This is because the player can always lower bound f_t by

$$f_t(x) \ge \tilde{f}_t(x) \equiv \max\left(f_t(x_t) + \langle \nabla f_t(x_t), x - x_t \rangle, 0\right).$$

As $\tilde{f}_t(x_t) = f_t(x_t)$, by simply pretending the requests are $\tilde{f}_t$, any competitive algorithm can be transformed into one which only uses the values $f_t(x_t)$ and $\nabla f_t(x_t)$ and which obeys the same guarantees.

In the discrete time setting, if we are given $f_n(x_{n-1})$ and $\nabla f_n(x_{n-1})$ before choosing x_n, there is another source of error because $f_n(x_n)$ is totally unknown. However this error is easily controlled when the f_n are uniformly Lipschitz. Let $(x_n)_{n \le N}$ be the discrete-time functional Steiner point sequence for the functions recursively defined by

$$\tilde{f}_n(x) = \max\left(f_n(x_{n-1}) + \langle \nabla f_n(x_{n-1}), x - x_{n-1} \rangle, 0\right)$$

and let W_N be the discrete-time work function. We obtain:

$$\sum_{n=1}^{N} f_n(x_n) + ||x_n - x_{n-1}|| \leq \sum_{n=1}^{N} \tilde{f}_n(x_n) + ||x_n - x_{n-1}|| + \left(\sum_{n=1}^{N} f_n(x_n) - \tilde{f}_n(x_n) \right)$$

$$\leq (d+1) \min_x W_N(x) + \left(\sum_{n=1}^{N} f_n(x_n) - \tilde{f}_n(x_n) \right).$$

Suppose now that each f_n is L-lipschitz. Then the equality $f_n(x_{n-1}) = \tilde{f}_n(x_{n-1})$ implies $|f_n(x_n) - \tilde{f}_n(x_n)| \leq 2L||x_n - x_{n-1}||$. Because Theorem 4.1 and Proposition 2.1 imply

$$\sum_{n=1}^{N} ||x_n - x_{n-1}|| \leq d \min_x W_N(x),$$

it follows that the resulting competitive ratio is at most $(2L+1)d + 1$. Similar remarks apply to the result of Theorem 5.3.

5 Competitive Ratio $O(\sqrt{d \log N})$ in Euclidean Space

In this section we prove the discrete-time functional Steiner point has competitive ratio $O(\sqrt{d \log N})$ in Euclidean space (whose norm is denoted by $|| \cdot ||_2$). The same technique applies in any normed space given a suitable concentration result, however we restrict to the Euclidean case for convenience. The idea is as follows. Suppose that the average dual work function increase

$$\fint_{\theta \in \partial B_1^*} W_n^*(\theta) - W_{n-1}^*(\theta) d\mu(\theta)$$

at time-step n is significant. Then by (7) the movement from $s(W_{n-1}) \rightarrow s(W_n)$ is an integral of pushes by different vectors θ. By concentration of measure, these pushes decorrelate unless the total amount of pushing is exponentially small.

Lemma 5.1 ([5, Lemma 2.2]) *For any* $0 \leq \varepsilon < 1$ *and* $|w| \leq 1$ *in Euclidean space, the set*

$$\{\theta \in \partial B_1 : \langle w, \theta \rangle \geq \varepsilon\}$$

occupies at most $e^{-d\varepsilon^2/2}$ *fraction of* ∂B_1.

Lemma 5.2 *Suppose that* $|W_n^*(\theta) - W_{n-1}^*(\theta)| \leq C$ *for all* $\theta \in \partial B_1$, *and set*

$$\lambda = \oint_{v \in B_1} W_n^*(v) - W_{n-1}^*(v) dv.$$

Then the functional Steiner point movement is at most

$$\|s(W_n) - s(W_{n-1})\|_2 = O\left(\lambda \sqrt{d\left(1 + \log\left(\frac{C}{\lambda}\right)\right)}\right).$$

Proof Observe that

$$\|s(W_n) - s(W_{n-1})\|_2 = \max_{\|w\|_2=1} \langle w, s(W_n) - s(W_{n-1})\rangle.$$

Fixing a unit vector w, we estimate the inner product on the right-hand side. Set

$$g_n(\theta) = W_n^*(\theta) - W_{n-1}^*(\theta) \geq 0,$$

$$I_z = \oint_{\theta \in \partial B_1^*} g_n(\theta) \cdot 1_{\langle w, \theta\rangle \geq z} d\mu(\theta).$$

Then $g_n(\theta) \in [0, C]$ for all θ and $\oint_{\theta \in \partial B_1^*} g_n d\mu(\theta) = \lambda$. Consequently by Lemma 5.1,

$$I_z \leq \min\left(\lambda, C e^{-dz^2/2}\right). \tag{10}$$

We thus find

$$\langle w, s(W_n) - s(W_{n-1})\rangle = d \oint_{\theta \in \partial B_1^*} g_n(\theta)\langle w, \theta\rangle d\mu(\theta)$$

$$\leq d \oint_{\substack{\theta \in \partial B_1^* \\ \langle w,\theta\rangle \geq 0}} g_n(\theta)\langle w, \theta\rangle d\mu(\theta)$$

$$= d \int_0^1 I_z dz$$

$$\leq d \int_0^1 \min\left(\lambda, C e^{-dz^2/2}\right) dz. \tag{11}$$

Here the second equality is the tail-sum integral formula. To estimate the resulting integral, set

$$A = \sqrt{\frac{2\log(C/\lambda)}{d}}.$$

so that $Ce^{-dA^2/2} = \lambda$. We will assume $A \leq 1$; if $A > 1$ then the expression (11) is at most $d\lambda \leq dA\lambda$ and it suffices to mimic the below without the second term. We estimate

$$\int_0^1 \min\left(\lambda, Ce^{-dz^2/2}\right) dz = A\lambda + C \int_A^1 e^{-dz^2/2} dz.$$

and use the simple bounds

$$\int_A^1 e^{-dz^2/2} dz \leq \int_0^1 e^{-dz^2/2} dz \leq O(d^{-1/2}),$$

$$\int_A^1 e^{-dz^2/2} dz \leq e^{-dA^2/2} \int_A^\infty e^{-dA(z-A)} dz = \frac{e^{-dA^2/2}}{dA}.$$

Combining,

$$\langle w, s(W_n) - s(W_{n-1}) \rangle \leq d \int_0^1 \min\left(\lambda, Ce^{-dz^2/2}\right) dz$$

$$\leq dA\lambda + \min\left(C\sqrt{d}, \frac{Ce^{-dA^2/2}}{A}\right)$$

$$= O\left(\lambda\sqrt{d\log\left(\frac{C}{\lambda}\right)}\right) + \min\left(C\sqrt{d}, \lambda\sqrt{\frac{d}{2\log(C/\lambda)}}\right).$$

With $u = \lambda/C \in [0, 1]$, the last term is

$$C\sqrt{d} \cdot \min\left(1, \frac{u}{\sqrt{2\log(1/u)}}\right)$$

For $u \leq [0, 1/2]$, we have $\frac{u}{\sqrt{2\log(1/u)}} \leq O(u)$ giving the bound $O(\lambda\sqrt{d})$. For $u \geq 1/2$ we have $C\sqrt{d} \leq 2\lambda\sqrt{d}$. Hence in both cases,

$$\langle w, s(W_n) - s(W_{n-1}) \rangle \leq O\left(\lambda\sqrt{d\left(1 + \log\left(\frac{C}{\lambda}\right)\right)}\right)$$

as desired. □

Theorem 5.3 *The discrete time functional Steiner point algorithm is $O(\sqrt{d \log N})$ competitive for chasing convex functions in Euclidean space.*

Proof Call $(x_t)_{t\in[0,N]}$ the continuous path and $(x_{t_n})_{n\leq N}$ the discrete path for $t_n \in (n-1, n]$ as in Proposition 2.1. Since the service cost for the discrete path is at most that of the continuous path, we only need to establish the $O(\sqrt{d \log N})$ competitive ratio on the movement of the discrete path. By Lemma 3.6,

$$\max_{|\theta| \le 1} W_N^*(\theta) \le 2 \cdot \min_x W_N(x).$$

Set

$$\lambda_n = \int_{\theta \in \partial B_1^*} W_{t_n}^*(\theta) - W_{t_{n-1}}^*(\theta) \mathrm{d}\mu(\theta).$$

Applying Lemma 5.2 with $C = 2 \cdot \min_x W_N(x)$ to the movement $||x_{t_n} - x_{t_{n-1}}||_2$ at each step yields:

$$\sum_{n=1}^{N} ||x_{t_n} - x_{t_{n-1}}||_2 \le O(Cd^{1/2}) \cdot \sum_{n \le N} \frac{\lambda_n}{C} \sqrt{1 + \log\left(\frac{C}{\lambda_n}\right)}. \tag{12}$$

Here the values λ_n are all non-negative and sum to $\int_{\theta \in \partial B_1^*} W_N^*(\theta) \mathrm{d}\mu(\theta) \le C$. Letting $h(u) = u\sqrt{1 + \log(1/u)}$, one readily computes that for $u \in (0, 1)$,

$$h'(u) = \frac{2\log(1/u) + 1}{2(1 + \log(1/u))^{1/2}} \ge 0, \qquad h''(u) = \frac{-2\log(1/u) - 3}{4u(1 + \log(1/u))^{3/2}} \le 0.$$

Jensen's inequality therefore implies that setting $\lambda_n = \frac{C}{N}$ for all $n \le N$ in (12) gives an upper bound. It follows that the movement cost is at most $O(C\sqrt{d \log(N+1)})$.

$\square$

6 Steiner Points of Level Sets

6.1 A Simplification for Chasing Convex Bodies

Here we show that for chasing convex bodies in discrete time, it suffices to simply set $x_n = s(W_n)$ instead of reducing from a continuous-time problem via Proposition 2.1. This simplification does not seem possible for chasing convex functions. The movement cost estimates continue to hold with no changes in the proof, however establishing $s(W_n) \in K_n$ requires a short additional argument. Define the support set $\mathrm{Supp}(W) \subseteq \mathbb{R}_d$ of an abstract work function W to be the set of points x possessing a subgradient $v \in \nabla W(x)$ with $|v| < 1$. For a work function W and convex body K, set

$$W^K(x) = \min_{y \in K} W(y) + ||y - x||.$$

If W is the work function for some sequence of requests, then making an additional request of K results in the new work function W^K.

Proposition 6.1 $Supp(W^K) \subseteq K$ *holds for any work function W and convex body K.*

Proof Suppose $x \notin K$ and set

$$y \in \arg\min_{y_0 \in K}(W(y_0) + ||y_0 - x||).$$

For any z on the segment $\overline{yx}$, it follows that $W(x) - W(z) = ||x - z||$. This implies that no v with $|v| < 1$ can be a subgradient in $\nabla W_n(x)$. $\square$

Corollary 6.2 *The algorithm $x_n = s(W_n)$ is d competitive for chasing convex bodies, and $O(\sqrt{d \log N})$ competitive in Euclidean space.*

Proof Proposition 6.1 and the primal definition (6) together imply $s(W_n) \in K_n$, i.e. the algorithm is valid. The d-competitiveness follows from Theorem 4.1 and the argument of Proposition 2.1 while the $O(\sqrt{d \log N})$ competitive ratio in Euclidean space follows from the argument of Theorem 5.3. $\square$

6.2 Steiner Points of Level Sets

This final subsection has two main objectives. Theorem 6.3 states that the functional Steiner point of any work function can be expressed as the Steiner point of large level sets. Corollary 6.6 states that the Steiner point of any level set of the work function W_n is inside K_n for convex body chasing. As we discuss at the end, Corollary 6.6 is related to the algorithm for chasing convex bodies given by Argue et al. [4]. Denote level sets by

$$\Omega_{W,R} = \{x : W(x) \le R\}.$$

It is easy to see that for any work function W and $R \ge \min_x W(x)$,

$$W^{\Omega_{W,R}}(x) = \begin{cases} W(x), & \text{for } x \in \Omega_{W,R} \\ d(x, \Omega_{W,R}) + R, & \text{for } x \notin \Omega_{W,R}. \end{cases}$$

Theorem 6.3 *For any work function W and $R \ge \min_x W(x)$, it holds that $s(\Omega_{W,R}) = s(W^{\Omega_{W,R}})$ and $\lim_{R\to\infty} s(\Omega_{W,R}) = s(W)$. Moreover if $Supp(W) \subseteq \Omega_{W,R}$ then $s(\Omega_{W,R}) = s(W)$.*

Proof The dual definitions (4), (7) imply

$$s(\Omega_{W,R}) - s(W) = d\oint_{\theta \in \partial B_1^*} \left(W^*(\theta) + h_{\Omega_{W,R}}(\theta)\right) n(\theta) d\mu(\theta). \tag{13}$$

Also for any $\theta \in \partial B_1^*$,

$$\left(W^{\Omega_{W,R}}\right)^*(\theta) = \inf_{w \in X} \left(W^{\Omega_{W,R}}(w) - \langle w, \theta \rangle\right)$$

$$= \inf_{w \in \partial \Omega_{W,R}} \left(W^{\Omega_{W,R}}(w) - \langle w, \theta \rangle\right)$$

$$= R - h_{\Omega_{W,R}}(\theta).$$

It follows from the symmetry $\theta \leftrightarrow -\theta$ that

$$\fint_{\theta \in \partial B_1^*} n(\theta) d\mu(\theta) = 0.$$

Combining the above yields

$$s(\Omega_{W,R}) = s\left(W^{\Omega_{W,R}}\right).$$

We proceed similarly for the second claim. For any $\theta \in \partial B_1^*$,

$$W^*(\theta) = \inf_{w \in X}(W(w) - \langle \theta, w \rangle)$$

$$= \lim_{R \to \infty} \inf_{w \in \Omega_{W,R}} (W(w) - \langle \theta, w \rangle)$$

$$= \lim_{R \to \infty} \inf_{w \in \partial \Omega_{W,R}} (W(w) - \langle \theta, w \rangle)$$

$$= \lim_{R \to \infty} \left(R - h_{\Omega_{W,R}}(\theta)\right).$$

Because $W(x) - \|x\|$ is uniformly bounded it follows that the expression

$$W^*(\theta) + h_{\Omega_{W,R}}(\theta) - R$$

is uniformly bounded for $(\theta, R) \in (\partial B_1^* \times \mathbb{R}^+)$. As just shown it tends to 0 as $R \to \infty$. The bounded convergence theorem therefore implies

$$\lim_{R \to \infty} \fint_{\theta \in \partial B_1^*} \left|W^*(\theta) + h_{\Omega_{W,R}}(\theta) - R\right| d\mu(\theta) = 0.$$

Combining with Eq. (13) shows that $\lim_{R \to \infty} \|s(\Omega_{W,R}) - s(W)\| = 0$, proving the second assertion. The last assertion is proved similarly after observing that

$\mathrm{Supp}(W) \subseteq \Omega_{W,R}$ implies

$$
\begin{aligned}
W^*(\theta) &= \inf_{w \in X}(W(w) - \langle \theta, w \rangle) \\
&= \lim_{\lambda \uparrow 1} \inf_{w \in X}(W(w) - \langle \lambda\theta, w \rangle) \\
&= \lim_{\lambda \uparrow 1} \inf_{w \in \Omega_{W,R}}(W(w) - \langle \lambda\theta, w \rangle) \\
&= R - h_{\Omega_{W,R}}(\theta).
\end{aligned}
$$

$\square$

Proposition 6.4 $Supp(W^{\Omega_{W,R}}) \subseteq Supp(W)$ *holds for any* $R \geq \min_x W(x)$.

Proof Because $\Omega_{W,R}$ is a level set,

$$
W^{\Omega_{W,R}}(x) = \begin{cases} W(x), & \text{for } x \in \Omega_{W,R} \\ d(x, \Omega_{W,R}) + R, & \text{for } x \notin \Omega_{W,R} \end{cases}
$$

Proposition 6.1 combined with the fact that W and $W^{\Omega_{W,R}}$ agree inside $\Omega_{W,R}$ imply that the only possible new support points are on the boundary $\partial\Omega_{W,R}$. Fix a boundary point $y \in \partial\Omega_{W,R} \setminus \mathrm{Supp}(W)$. Because $y \notin \mathrm{Supp}(W)$, there exists a sequence $(y_i)_{i \in \mathbb{N}} \to y$ satisfying

$$
W(y) - W(y_i) \geq (1 - o(1))\|y - y_i\|.
$$

Such a sequence of points y_i must eventually satisfy $W(y_i) \leq W(y)$ and therefore $y_i \in \Omega_{W,R}$, implying $W(y_i) = W^{\Omega_{W,R}}(y_i)$. Hence

$$
W^{\Omega_{W,R}}(y) - W^{\Omega_{W,R}}(y_i) \geq (1 - o(1))\|y - y_i\|.
$$

This implies $y \notin \mathrm{Supp}(W^{\Omega_{W,R}})$, completing the proof. $\square$

Corollary 6.5 *Let* $W = \widehat{W}^K$ *for a work function* $\widehat{W}$ *and convex body* K. *For any* $R \geq \min_x W(x)$,

$$
s(\Omega_{W,R}) = s\left(W^{\Omega_{W,R}}\right) \in K.
$$

Proof Propositions 6.1 and 6.4 show that

$$
\mathrm{Supp}\left(W^{\Omega_{W,R}}\right) \subseteq \mathrm{Supp}(W) \subseteq K.
$$

The primal definition (6) of the functional Steiner point now implies $s(W^{\Omega_{W,R}}) \in K$.

$\square$

Corollary 6.6 *Let W_n be the work function for convex body requests $(K_1, \ldots, K_n)$. Then*

$$s(W_n^{\Omega_{W_n,R}}) \in K_n$$

for any $R \geq \min_x W_n(x)$.

Proof Immediate from Corollary 6.5 with $\widehat{W} = W_{n-1}$ and $K = K_n$. $\qquad\square$

Remark [4] solved chasing convex bodies in Euclidean space by taking $x_n = s\left(W_n^{\Omega_{W_n,R_n}}\right)$ with $R_n = 2^{\lceil \log_2 (\min_x W_n(x)) \rceil}$. This defines a selector by Corollary 6.6. Estimating the movement cost is not difficult because the sets $W_n^{\Omega_{W_n,R}}$ decrease for fixed R. Note that $diam(\Omega_{W_n,R}) \leq 2R$ because of the inequality $W_t(x) \geq ||x||$ (recall Proposition 3.5). Using Theorem 3.2, the movement from each fixed R value is at most $O(\min(dR, R\sqrt{d \log T}))$. Summing over the geometric sequence of R values yields the same upper bound as in Theorems 4.1 and 5.3 up to a constant factor.

Argue et al. [4] prove that $s\left(W_n^{\Omega_{W_n,R_n}}\right) \in K_n$ using reflectional symmetries that may not exist in arbitrary normed spaces. Corollary 6.6 thus implies that their algorithm works for general norms.

Appendix: Proof of Lemma 3.7

Proof We prove the result for all $v \in B_1^*$ where $\nabla W_t^*(v)$ exists. This includes almost all v by Alexandrov's theorem. Moreover it ensures the conjugate point $v_t^* = \arg\min_{w \in X} W(w) - \langle v, w \rangle$ is well-defined and that W_t is strictly convex at v_t^* [30, Corollary 25.1.2]. We write:

$$W_{t+\delta}(v) = \min_{x_s:[0,t+\delta] \to X} \left(\int_0^{t+\delta} (f_s(x_s) + ||x_s'||) \mathrm{d}s - \langle v, x_{t+\delta} \rangle \right)$$

$$= \min_{x_s:[t,t+\delta] \to X} \left(W_t(x_t) + \int_t^{t+\delta} f_s(x_s) + ||x_s'|| \mathrm{d}s - \langle v, x_{t+\delta} \rangle \right)$$

For small $\delta \in (0, \varepsilon)$, we show $W_{t+\delta}^*(v) = W_t^*(v) + \delta f_t(v_t^*) + o(\delta)$. For the upper bound,

$$W_{t+\delta}(v_t^*) \leq W_t(v_t^*) + \int_t^{t+\delta} f_s(v_t^*) \mathrm{d}s$$

$$= W_t(v_t^*) + \delta f_t(v_t^*) + o(\delta)$$

holds by taking $x_s = v_t^*$ constant for $s \in [t, t + \delta)$ and recalling the assumption that $f_s(x)$ is continuous on $s \in [t, t + \delta)$. Since $v_t^* = \arg \min_x \left(W_t(x) - \langle x, v \rangle \right)$, the upper bound follows from

$$
\begin{aligned}
W_{t+\delta}^*(v) &\leq W_{t+\delta}(v_t^*) - \langle v, v_t^* \rangle \\
&\leq W_t(v_t^*) + \delta f_t(v_t^*) + o(\delta) - \langle v, v_t^* \rangle \\
&= W_t^*(v) + \delta f_t(v_t^*) + o(\delta).
\end{aligned}
$$

For the lower bound, the strict convexity of W_t at v_t^* implies

$$
W_t(x) = W_t(v_t^*) + \langle v, x - v_t^* \rangle + \gamma(||x - v_t^*||)
$$

where $\gamma : \mathbb{R}^+ \to \mathbb{R}^+$ is continuous and increasing with unique minimum $F(0) = 0$. Therefore any path $x_s : [0, t + \delta] \to X$ satisfies:

$$
W_t(x_t) + \int_t^{t+\delta} f_s(x_s) + ||x_s'|| ds - \langle v, x_{t+\delta} \rangle \geq W_t(v_t^*) + \langle v, x_t - v_t^* \rangle
$$

$$
+ \gamma(||x_t - v_t^*||) + \int_t^{t+\delta} f_s(x_s) + ||x_s'|| ds - \langle v, x_{t+\delta} \rangle.
$$

The observation $\int_t^{t+\delta} ||x_s'|| ds \geq ||x_{t+\delta} - x_t|| \geq \langle v, x_{t+\delta} - x_t \rangle$ implies

$$
\begin{aligned}
W_t(x_t) + \int_t^{t+\delta} f_s(x_s) + ||x_s'|| ds - \langle v, x_{t+\delta} \rangle &\geq W_t(x_t) - \langle v, v_t^* \rangle + f(||x_t - v_t^*||) \\
&\quad + \int_t^{t+\delta} f_s(x_s) ds \\
&\geq W_t(v_t^*) - \langle v, v_t^* \rangle + \gamma(||x_t - v_t^*||) \\
&\quad + \int_t^{t+\delta} f_s(x_s) ds \\
&\geq W_t^*(v) + \gamma(||x_t - v_t^*||) \\
&\quad + \int_t^{t+\delta} f_s(x_s) ds.
\end{aligned}
$$

Because $W_{t+\delta}(v) = W_t(v) + O(\delta)$, we see that for $\delta \to 0$ small we must have $||x_t - v_t^*|| = o_{\delta \to 0}(1)$ for any optimal trajectory x_s witnessing the correct value $W_{t+\delta}$. Additionally,

$$
\int_t^{t+\delta} ||x_s'|| ds + \langle v, x_t - x_{t+\delta} \rangle \geq (1 - |v|) \int_t^{t+\delta} ||x_s'|| ds \geq (1 - |v|) \sup_{s \in [t, t+\delta]} |x_t - x_s|.
$$

which similarly implies $\sup_{s \in [t,t+\delta]} ||x_t - x_s|| = o(1)$ for any optimal trajectory since $||v|| < 1$. It follows that all optimal trajectories satisfy $\int_t^{t+\delta} f_s(x_s)\mathrm{d}s = \delta f_t(v_t^*) + o(\delta)$. This concludes the proof. $\qquad\Box$

Acknowledgments The author thanks Sébastien Bubeck, Bo'az Klartag, Yin Tat Lee, and Yuanzhi Li for the introduction to convex body chasing and the Steiner point, and many stimulating discussions. He thanks Ethan Jaffe, Felipe Hernandez, and Christian Coester for discussions about properties of the work function, and the anonymous referee for several suggestions. He additionally thanks Sébastien for feedback on previous drafts and gratefully acknowledges the support of an NSF graduate fellowship and a Stanford graduate fellowship.

References

1. N. Alon, B. Awerbuch, Y. Azar, The online set cover problem, in *Proceedings of the Thirty-Fifth Annual ACM Symposium on Theory of Computing* (2003), pp. 100–105
2. A. Antoniadis, N. Barcelo, M. Nugent, K. Pruhs, K. Schewior, M. Scquizzato, Chasing convex bodies and functions, in *LATIN 2016: Theoretical Informatics* (Springer, Berlin, 2016), pp. 68–81
3. C.J. Argue, S. Bubeck, M.B. Cohen, A. Gupta, Y.T. Lee, A nearly-linear bound for chasing nested convex bodies, in *Proceedings of the Thirtieth Annual ACM-SIAM Symposium on Discrete Algorithms* (SIAM, 2019), pp. 117–122
4. C.J. Argue, A. Gupta, G. Guruganesh, Z. Tang, Chasing convex bodies with linear competitive ratio. J. ACM **68**(5), 1–10 (2021)
5. K. Ball, An elementary introduction to modern convex geometry. Flavors Geom. **31**, 1–58 (1997)
6. N. Bansal, N. Buchbinder, A. Madry, J. Naor, A polylogarithmic-competitive algorithm for the k-server problem. J. ACM **62**(5), 1–49 (2015)
7. N. Bansal, A. Gupta, R. Krishnaswamy, K. Pruhs, K. Schewior, C. Stein, A 2-competitive algorithm for online convex optimization with switching costs, in *Approximation, Randomization, and Combinatorial Optimization. Algorithms and Techniques (APPROX/RANDOM 2015)*. Schloss Dagstuhl-Leibniz-Zentrum fuer Informatik (2015)
8. N. Bansal, M. Böhm, M. Eliáš, G. Koumoutsos, S.W. Umboh, Nested convex bodies are chaseable, in *Proceedings of the Twenty-Ninth Annual ACM-SIAM Symposium on Discrete Algorithms* (SIAM, 2018), pp. 1253–1260
9. Y. Bartal, N. Linial, M. Mendel, A. Naor, On metric Ramsey-type phenomena. Ann. Math. **162**, 643–710 (2005)
10. Y. Bartal, B. Bollobás, M. Mendel, Ramsey-type theorems for metric spaces with applications to online problems. J. Comput. Syst. Sci. **72**(5), 890–921 (2006)
11. A. Blum, C. Burch, On-line learning and the metrical task system problem. Mach. Learn. **39**(1), 35–58 (2000)
12. A. Borodin, N. Linial, M.E. Saks, An optimal on-line algorithm for metrical task system. J. ACM **39**(4), 745–763 (1992)
13. S. Bubeck, M.B. Cohen, J.R. Lee, Y.T. Lee, Metrical task systems on trees via mirror descent and unfair gluing, in *Proceedings of the Thirtieth Annual ACM-SIAM Symposium on Discrete Algorithms* (SIAM, 2019), pp. 89–97
14. S. Bubeck, Y.T. Lee, Y. Li, M. Sellke, Competitively chasing convex bodies, in *Proceedings of the 51st Annual ACM SIGACT Symposium on Theory of Computing* (2019), pp. 861–868

15. S. Bubeck, B. Klartag, Y.T. Lee, Y. Li, M. Sellke, Chasing nested convex bodies nearly optimally, in *Proceedings of the Fourteenth Annual ACM-SIAM Symposium on Discrete Algorithms* (SIAM, 2020), pp. 1496–1508
16. N. Chen, G. Goel, A. Wierman, Smoothed online convex optimization in high dimensions via online balanced descent, in *Conference On Learning Theory* (PMLR, 2018), pp. 1574–1594
17. C. Fefferman, P. Shvartsman, Sharp finiteness principles for Lipschitz selections. Geom. Funct. Anal. **28**(6), 1641–1705 (2018)
18. A. Fiat, M. Mendel, Better algorithms for unfair metrical task systems and applications. SIAM J. Comput. **32**(6), 1403–1422 (2003)
19. J. Friedman, N. Linial, On convex body chasing. Discrete Comput. Geom. **9**(3), 293–321 (1993)
20. G. Goel, Y. Lin, H. Sun, A. Wierman, Beyond online balanced descent: an optimal algorithm for smoothed online optimization. Adv. Neural Inf. Process. Syst. **32**, 1875–1885 (2019)
21. R.L Graham, Bounds for certain multiprocessing anomalies. Bell Syst. Tech. J. **45**(9), 1563–1581 (1966)
22. E.F. Grove, The harmonic online k-server algorithm is competitive, in *Proceedings of the twenty-third annual ACM symposium on Theory of computing* (1991), pp. 260–266
23. E. Koutsoupias, C.H. Papadimitriou, On the k-server conjecture. J. ACM **42**(5), 971–983 (1995)
24. R. Kumar, M. Purohit, Z. Svitkina, Improving Online Algorithms via ML Predictions, in *Proceedings of the 32nd International Conference on Neural Information Processing Systems* (2018), pp. 9684–9693
25. I. Kupka, Continuous selections for Lipschitz multifunctions. Acta Math. Univ. Comenianae **74**(1), 133–141 (2005)
26. M. Lin, A. Wierman, L.L.H. Andrew, E. Thereska, Dynamic right-sizing for power-proportional data centers. IEEE/ACM Trans. Netw. **21**(5), 1378–1391 (2013)
27. T. Lykouris, S. Vassilvtiskii, Competitive caching with machine learned advice, in *International Conference on Machine Learning* (PMLR, 2018), pp. 3296–3305
28. M.S. Manasse, L.A. McGeoch, D.D. Sleator, Competitive algorithms for server problems. J. Algorithms **11**(2), 208–230 (1990)
29. K. Przesławski, D. Yost, Continuity properties of selectors. Mich. Math. J. **36**(1), 13 (1989)
30. R.T. Rockafellar, *Convex Analysis*, vol. 36 (Princeton University Press, Princeton, 1970)
31. P. Shvartsman, Lipshitz selections of multivalued mappings and traces of the Zygmund class of functions to an arbitrary compact. Dokl. Acad. Nauk SSSR **276**, 559–562 (1984). English translation in Soviet Math. Dokl, volume 29, pages 565–568, 1984
32. P. Shvartsman. Lipschitz selections of set-valued mappings and Helly's theorem. J. Geom. Anal. **12**(2), 289–324 (2002)
33. D.D. Sleator, R.E. Tarjan, Amortized efficiency of list update and paging rules. Commun. ACM **28**(2), 202–208 (1985)
34. J. Steiner. From the center of curvature of plane curves. J. Pure Appl. Math. **21**, 33–63 (1840)
35. A. Wei, F. Zhang, Optimal Robustness-Consistency Trade-offs for Learning-Augmented Online Algorithms, in *Advances in Neural Information Processing Systems*, vol. 33 (2020), pp. 8042–8053

Shephard's Inequalities, Hodge-Riemann Relations, and a Conjecture of Fedotov

Ramon van Handel

Abstract A well-known family of determinantal inequalities for mixed volumes of convex bodies were derived by Shephard from the Alexandrov-Fenchel inequality. The classic monograph *Geometric Inequalities* by Burago and Zalgaller states a conjecture on the validity of higher-order analogues of Shephard's inequalities, which is attributed to Fedotov. In this note we disprove Fedotov's conjecture by showing that it contradicts the Hodge-Riemann relations for simple convex polytopes. Along the way, we make some expository remarks on the linear algebraic and geometric aspects of these inequalities.

Keywords Mixed volumes · Alexandrov-Fenchel inequality · Shephard's inequalities · Hodge-Riemann relations for convex polytopes

1 Introduction

1.1. Let $K_1, \ldots, K_m$ be convex bodies in $\mathbb{R}^n$ and $\lambda_1, \ldots, \lambda_m > 0$. One of the most basic facts of convex geometry, due to H. Minkowski, is that the volume of convex bodies is a homogeneous polynomial in the sense that

$$\mathrm{Vol}(\lambda_1 K_1 + \cdots + \lambda_m K_m) = \sum_{i_1, \ldots, i_n = 1}^{m} \mathsf{V}(K_{i_1}, \ldots, K_{i_n}) \lambda_{i_1} \cdots \lambda_{i_n}.$$

The coefficients $\mathsf{V}(K_1, \ldots, K_n)$, called mixed volumes, define a large family of natural geometric parameters of convex bodies, and play a central role in convex geometry [5, 14]. Mixed volumes are always nonnegative, are symmetric in their arguments, and are additive and homogeneous in each argument.

R. van Handel (✉)
Princeton University, Princeton, NJ, USA
e-mail: rvan@math.princeton.edu

© The Author(s), under exclusive license to Springer Nature Switzerland AG 2023
R. Eldan et al. (eds.), *Geometric Aspects of Functional Analysis*, Lecture Notes in Mathematics 2327, https://doi.org/10.1007/978-3-031-26300-2_13

The fundamental inequality in the theory of mixed volumes is the following.

Theorem 1.1 (Alexandrov-Fenchel) *For convex bodies* $K, L, C_1, \ldots, C_{n-2}$ *in* $\mathbb{R}^n$

$$V(K, L, C_1, \ldots, C_{n-2})^2 \geq V(K, K, C_1, \ldots, C_{n-2}) \, V(L, L, C_1, \ldots, C_{n-2}).$$

Numerous inequalities in convex geometry may be derived from the Alexandrov-Fenchel inequality, cf. [5, §20] and [14, §7.4]. The starting point for this note is a well-known family of determinantal inequalities, due to Shephard [18], that extend the Alexandrov-Fenchel inequality to more than n bodies.

Theorem 1.2 (Shephard) *Given convex bodies* $K_1, \ldots, K_m, C_1, \ldots, C_{n-2}$ *in* $\mathbb{R}^n$, *define the* $m \times m$ *symmetric matrix* M *by setting*

$$M_{ij} := V(K_i, K_j, C_1, \ldots, C_{n-2}).$$

Then

$$(-1)^m \det M \leq 0.$$

The special case $m = 2$ of Theorem 1.2 is just a reformulation of the Alexandrov-Fenchel inequality, and Shephard's inequalities may thus be viewed as a considerable generalization of the Alexandrov-Fenchel inequality. However, as is shown by Shephard (and as we will explain later in this note), the general inequalities may in fact be deduced from the $m = 2$ case by a simple linear algebraic argument. In the case $m = 3$, this result dates back already to Minkowski [12, p. 478].

1.2. The classic monograph *Geometric Inequalities* by Burago and Zalgaller states a conjecture on the validity of a higher-order generalization of Theorem 1.2, which is attributed to Fedotov [5, §20.6]. Let us recall the statement of this conjecture. In the sequel, we will frequently employ the notation

$$V(K_1[m_1], K_2[m_2], \ldots, K_r[m_r]) := V(\underbrace{K_1, \ldots, K_1}_{m_1}, \underbrace{K_2, \ldots, K_2}_{m_2}, \ldots, \underbrace{K_r, \ldots, K_r}_{m_r})$$

when convex bodies are repeated multiple times in the arguments of a mixed volume.

Conjecture 1.3 (Fedotov) *Let* $k \leq n/2$, *and let* $K_1, \ldots, K_m, C_1, \ldots, C_{n-2k}$ *be convex bodies in* $\mathbb{R}^n$. *Define the* $m \times m$ *symmetric matrix* M *by setting*

$$M_{ij} := V(K_i[k], K_j[k], C_1, \ldots, C_{n-2k}).$$

Then

$$(-1)^m \det M \le 0.$$

If true, this conjecture would entail a considerable generalization of Shephard's inequalities. The conjecture is rather appealing, as it is easily verified to be true in two extreme cases that have a different flavor.

Lemma 1.4 *Conjecture 1.3 is valid in the following two cases:*

a. *When $k = 1$ and m is arbitrary.*
b. *When $m = 2$ and k is arbitrary.*

Proof Case *a* is nothing other than Theorem 1.2. To prove *b*, it suffices to note that iterating the Alexandrov-Fenchel inequality yields [14, (7.63)]

$$V(K_1[k], K_2[l], C_1, \dots, C_{n-k-l})^{k+l} \ge$$

$$V(K_1[k+l], C_1, \dots, C_{n-k-l})^k V(K_2[k+l], C_1, \dots, C_{n-k-l})^l$$

for any $k, l \ge 1, k+l \le n$. The case $k = l$ is readily seen to be equivalent to *b*. □

The main purpose of this note is to explain that Conjecture 1.3 fails when one goes beyond the special cases of Lemma 1.4. More precisely, we will prove:

Theorem 1.5 *For every $k > 1$, Conjecture 1.3 is false for some $m > 2$.*

1.3. In order to explain how we will disprove Conjecture 1.3, it is useful to first briefly recall some of its history.

Despite the fundamental nature of the Alexandrov-Fenchel inequality, no really elementary proof of it is known. Alexandrov gave two different (but closely related) proofs in the 1930s: a combinatorial proof using strongly isomorphic polytopes [2], and an analytic proof using elliptic operators [3]. Further remarks on its history and on more modern proofs may be found in [14, 15].

In the 1970s, unexpected connections were discovered between the theory of mixed volumes and algebraic geometry. In particular, a remarkable identity due to Bernstein and Kushnirenko [5, Theorem 27.1.2] shows that the number of solutions $z \in (\mathbb{C} \setminus \{0\})^n$ of a generic system of polynomial equations $p_1(z) = 0, \dots, p_n(z) = 0$ with given monomials coincides with the mixed volume of an associated family of lattice polytopes in $\mathbb{R}^n$ (i.e., polytopes with vertices in $\mathbb{Z}^n$).

Motivated by these developments, Fedotov [7] proposed a simple proof of the Alexandrov-Fenchel inequality using only basic properties of polynomials. Fedotov further notes that his method even yields the more general Conjecture 1.3, which is stated in [7] as a theorem. These results were included in the Russian edition of the monograph of Burago and Zalgaller. Unfortunately, Fedotov's elementary approach turns out to contain a serious flaw, which renders his method of proof invalid. A correct algebraic proof of the Alexandrov-Fenchel inequality was given by Teissier and Khovanskii using nontrivial machinery, namely a reduction to the Hodge index theorem of algebraic geometry. The latter proof is included in the English translation of Burago-Zalgaller [5, §27], but does not settle the validity of Fedotov's higher-order analogue of Shephard's inequalities [5, §20.6].

On the other hand, the algebraic connection yields other higher-order inequalities. The Alexandrov-Fenchel inequality is analogous to a Hodge-Riemann relation of degree 1 in the cohomology ring of a smooth projective variety [6, 8]. Hodge-Riemann relations of higher degree give rise to new inequalities in convex geometry. Such inequalities were first stated by McMullen [11] for strongly isomorphic simple polytopes as a byproduct of his work on the g-conjecture. Their geometric significance was greatly clarified by Timorin [20], whose formulation is readily interpreted in terms of explicit inequalities for mixed volumes. Very recently, some special cases were extended also to smooth convex bodies in [1, 9, 10].

The proof of Theorem 1.5 may now be explained as follows. Using the properties of hyperbolic quadratic forms, we will first reformulate Conjecture 1.3 as a higher-order Alexandrov-Fenchel inequality. In this equivalent formulation, it will be evident that this inequality contradicts the Hodge-Riemann relation of degree 2. Thus the results of McMullen and Timorin imply that Conjecture 1.3 is false. Beside disproving the conjecture, a more expository aim of this note is to draw attention to some basic linear algebraic and geometric aspects of the above inequalities (none of which are really new here) in the context of classical convexity.

Remark It should be noted that Fedotov's conjecture as stated in [5, §20.6] is somewhat more general than Conjecture 1.3: the matrix M considered there is

$$\mathsf{M}_{ij} := \mathsf{V}(K_i[k], K_j[l], C_1, \ldots, C_{n-k-l})$$

for any $k, l \geq 1$ such that $k + l \leq n$. Lemma 1.4 extends to this setting: the case $k = l = 1$ and general m reduces to Shephard's inequalities, while the case $m = 2$ and general k, l is obtained by multiplying the inequality [14, (7.63)] used in the proof of Lemma 1.4 by the same inequality with the roles of k, l reversed. When $k \neq l$, however, the matrix M is not symmetric, and the spectral interpretation of the conjecture becomes unclear. Given that we show the conjecture fails for general m already in the symmetric case $k = l$, it seems implausible that the nonsymmetric case $k \neq l$ has any merit, and we do not consider it further in this note.

1.4. The remainder of this note is organized as follows. In Sect. 2, we recall some basic properties of hyperbolic quadratic forms that will be used in the sequel. We also briefly discuss Shephard's inequalities and clarify their equality cases. In Sect. 3 we formulate the Hodge-Riemann relations for strongly isomorphic simple polytopes, due to McMullen and Timorin, entirely in the language of classical convexity. Finally, Sect. 4 completes the proof of Theorem 1.5.

While the proof of Theorem 1.5 explains clearly *why* Fedotov's conjecture must fail, the construction is rather indirect. Once the proof has been understood, however, it is not difficult to engineer an explicit counterexample, which will be done in Sect. 5. Beside further illustrating the basic construction, this example will show that we may in fact choose $m = 3$ in Theorem 1.5.

We conclude this note by highlighting a puzzling aspect of the Hodge-Riemann relations: even though their statement makes sense in principle for arbitrary convex bodies, the Hodge-Riemann relations have only been proved for special classes of bodies (e.g., strongly isomorphic simple polytopes). In Sect. 6, we will illustrate by means of a simple example that the Hodge-Riemann relations may fail for general convex bodies. This highlights the rather unusual nature of the Hodge-Riemann relations as compared to other inequalities in convex geometry.

2   Linear Algebra

The aim of this section is to explain that the connection between the Alexandrov-Fenchel and Shephard inequalities has nothing to do with convexity, but is rather a simple linear-algebraic fact. The results of this section are known in various forms, see, e.g., [4, Theorem 4.4.6], [15, Lemma 2.9], or [16, Lemma 3.1], but we provide simple self-contained proofs for the variants needed here.

2.1   Hyperbolic Matrices

We begin by giving a spectral interpretation of the Alexandrov-Fenchel inequality. In the sequel, a matrix M will be called positive if $M_{ij} > 0$ for all i, j. For $y \in \mathbb{R}^m$, we write $y \geq 0$ ($y > 0$) if $y_i \geq 0$ ($y_i > 0$) for all i. The linear span of all eigenvectors of a symmetric matrix M with positive eigenvalues will be called the positive eigenspace of M.

Lemma 2.1 *Let* M *be a symmetric positive matrix. The following are equivalent:*

1. *The positive eigenspace of* M *is one-dimensional.*
2. $\langle x, My \rangle^2 \geq \langle x, Mx \rangle \langle y, My \rangle$ *for all $x \geq 0$ and $y \geq 0$.*
3. $\langle x, My \rangle = 0$ *implies* $\langle x, Mx \rangle \leq 0$ *for all x and $y \geq 0$, $y \neq 0$.*

Proof As M is a positive matrix, the Perron-Frobenius theorem implies that it has at least one eigenvector $v > 0$ with positive eigenvalue.

$3 \Rightarrow 1$: Let $x \perp v$ be any other eigenvector of M. Then $\langle x, Mv \rangle = 0$, so *3* implies $\langle x, Mx \rangle \leq 0$. Thus the eigenvalue associated to x must be nonpositive.

$1 \Rightarrow 2$: It follows from *1* that M is negative semidefinite on $v^\perp$. Fix $x, y \geq 0$; we may assume $y \neq 0$ (else the inequality is trivial), so that $\langle y, v \rangle > 0$ and $\langle y, My \rangle > 0$. If we define $z = x - ay$ with $a = \langle x, v \rangle / \langle y, v \rangle$, then $z \in v^\perp$, so

$$0 \geq \langle z, Mz \rangle = \langle x, Mx \rangle - 2a\langle x, My \rangle + a^2 \langle y, My \rangle \geq \langle x, Mx \rangle - \frac{\langle x, My \rangle^2}{\langle y, My \rangle}.$$

$2 \Rightarrow 3$: We first show that 2 remains valid for any x (not just $x \geq 0$). Suppose first that $y > 0$. Then $x + by \geq 0$ when b is chosen sufficiently large, so 2 implies

$$\langle x + by, My \rangle^2 \geq \langle x + by, M(x + by) \rangle \langle y, My \rangle.$$

Expanding both sides of this inequality shows that all terms involving b cancel, so $\langle x, My \rangle^2 \geq \langle x, Mx \rangle \langle y, My \rangle$ for any x and $y > 0$. This conclusion remains valid for any $y \geq 0$ by applying the above argument with $y \leftarrow y + \varepsilon v$ and letting $\varepsilon \to 0$. Now 3 follows immediately once we note that $y \geq 0$, $y \neq 0$ implies $\langle y, My \rangle > 0$.

$\square$

In the sequel, a symmetric (but not necessarily positive) matrix that has a one-dimensional positive eigenspace will be called *hyperbolic*.

2.2 Shephard's Inequalities

An $m \times m$ hyperbolic matrix M has 1 positive and $m - 1$ nonpositive eigenvalues. It is therefore immediately obvious that such a matrix satisfies $(-1)^m \det M \leq 0$ (as the determinant is the product of the eigenvalues). Shephard's inequalities follow directly from this observation.

Proof of Theorem 1.2 We may assume without loss of generality that all the convex bodies have nonempty interior, so that M is a positive matrix (otherwise we may replace $K_i \leftarrow K_i + \varepsilon B$, $C_i \leftarrow C_i + \varepsilon B$ for any body B with nonempty interior, and take $\varepsilon \to 0$ in the final inequality.) Condition 2 of Lemma 2.1 is immediate from the Alexandrov-Fenchel inequality (Theorem 1.1 with $K = \sum_i x_i K_i$ and $L = \sum_i y_i K_i$). Thus M is hyperbolic by Lemma 2.1, which implies $(-1)^m \det M \leq 0$.

$\square$

While this is only tangentially related to the rest of this note, let us take the opportunity to clarify the cases of equality in Shephard's inequalities.

Proposition 2.2 *In the setting and notations of Theorem 1.2, we have* $\det M = 0$ *if and only if there are linearly independent vectors* $x, y > 0$ *such that* $K = \sum_i x_i K_i$, $L = \sum_i y_i K_i$ *yield equality in the Alexandrov-Fenchel inequality of Theorem 1.1.*

Proof We must show $\det M = 0$ if and only if $\langle x, My \rangle^2 = \langle x, Mx \rangle \langle y, My \rangle$ for some linearly independent $x, y > 0$. We may assume $M \neq 0$ (else the result is trivial).

Suppose first that $\det M = 0$. Then there exists $z \in \ker M$, $z \neq 0$. Choose any $y > 0$ that is linearly independent of z. Evidently $\langle z, My \rangle^2 = \langle z, Mz \rangle \langle y, My \rangle$. But as this identity is invariant under the replacement $z \leftarrow z + by$ (as in the proof of $2 \Rightarrow 3$ of Lemma 2.1), we may choose $x = z + by > 0$ for b sufficiently large.

Now suppose $\langle x, My \rangle^2 = \langle x, Mx \rangle \langle y, My \rangle$ for linearly independent $x, y > 0$. Then

$$q(v) := \langle x + v, My \rangle^2 - \langle x + v, M(x + v) \rangle \langle y, My \rangle$$

satisfies $q(0) = 0$, and $q(v) \geq 0$ for all v in a neighborhood of 0 by the Alexandrov-Fenchel inequality. Thus $\nabla q(0) = 0$, which yields $z = \langle y, My \rangle x - \langle x, My \rangle y \in \ker M$. Moreover, $z \neq 0$ as x, y are linearly independent. Thus $\det M = 0$.   □

Proposition 2.2 reduces the equality cases of Shephard's inequalities to those of the Alexandrov-Fenchel inequality. The characterization of the latter is a long-standing open problem [14, §7.6], which was recently settled in several important cases in [16, 17]. This problem remains open in full generality.

2.3 A Sylvester Criterion

While any hyperbolic $m \times m$ matrix M trivially satisfies $(-1)^m \det M \leq 0$, the converse implication clearly does not hold: the sign of the determinant does not determine the number of positive eigenvalues. However, the implication can be reversed if the determinant condition holds for all principal submatrices of M. This hyperbolic analogue of the classical Sylvester criterion may be proved in essentially the same manner.[1]

In the following, we denote for any $m \times m$ symmetric matrix M and subset $I \subseteq [m]$ by $M_I := (M_{ij})_{i,j \in I}$ the associated principal submatrix.

Lemma 2.3 *For a symmetric positive $m \times m$ matrix, the following are equivalent:*

1. *The positive eigenspace of* M *is one-dimensional.*
2. $(-1)^{|I|} \det M_I \leq 0$ *for all* $I \subseteq [m]$.

Proof To prove $1 \Rightarrow 2$, note first that condition 2 of Lemma 2.1 is inherited by all its principal submatrices M_I (as one may restrict to x, y supported on I). The conclusion therefore follows immediately from Lemma 2.1.

To prove $2 \Rightarrow 1$, we argue by induction on m. For $m = 2$, it suffices to note that as M has at least one positive eigenvalue by the Perron-Frobenius theorem, $\det M \leq 0$ implies that its other eigenvalue must be nonpositive.

Now let $m > 2$ and assume the result has been proved in dimensions up to $m - 1$. Then 2 implies that M_I is hyperbolic for all $I \subsetneq [m]$. By the Perron-Frobenius theorem, M has an eigenvector v with positive eigenvalue. Now suppose 1 fails, that is, M is not hyperbolic. Then there must be another eigenvector $w \perp v$ with positive eigenvalue. As $(-1)^m \det M \leq 0$, there must then be a third eigenvector $u \perp \{v, w\}$ with nonnegative eigenvalue. Choose any $i \in [m]$ such that $u_i \neq 0$ and

---

[1] The author learned the elementary approach used here from lecture notes of M. Hladík.

let $I = [m]\backslash\{i\}$. Choose $a, b \in \mathbb{R}$ so that $x := v - au$ and $y := w - bu$ satisfy $x_i = y_i = 0$. By construction, x, y are linearly independent and $\langle z, Mz \rangle > 0$ for all $z \in \text{span}\{x, y\}, z \neq 0$. As x, y are supported on I, this implies M_I has a positive eigenspace of dimension at least two, contradicting the induction hypothesis. □

It follows immediately from Lemma 2.3 that Conjecture 1.3 is equivalent to the statement that the matrix M is hyperbolic. This observation will form the basis for the proof of Theorem 1.5 in Sect. 4: we will show that hyperbolicity of M contradicts the Hodge-Riemann relations for simple convex polytopes.

3 Hodge-Riemann Relations

The Hodge-Riemann relations in algebraic geometry give rise to higher order analogues of the Alexandrov-Fenchel inequality [11, 20]. While these inequalities are not usually stated in this form in the literature, they may be equivalently formulated as explicit inequalities between mixed volumes. The aim of this section is to draw attention to this elementary formulation of the Hodge-Riemann relations in terms of familiar objects from classical convex geometry.

Recall that a convex polytope in $\mathbb{R}^n$ is called *simple* if it has nonempty interior and each vertex is contained in exactly n facets. In the following, let us fix an arbitrary simple polytope Λ in $\mathbb{R}^n$, and denote by $\mathcal{P}(\Lambda)$ the collection of polytopes that are *strongly isomorphic* to Λ: that is, $P \in \mathcal{P}(\Lambda)$ if and only if

$$\dim F(P, u) = \dim F(\Lambda, u) \quad \text{for all } u \in S^{n-1},$$

where $F(P, u)$ denotes the face of P with normal direction u. For the basic properties of simple and strongly isomorphic polytopes, the reader is referred to [14, §2.4]. For the purposes of this note, the only significance of these definitions is that they are needed for the validity of the following theorem (see Sect. 6).

Theorem 3.1 (McMullen-Timorin) *Fix $n \geq 2$ and a simple polytope $\Lambda \in \mathbb{R}^n$, and let $m \geq 1$, $k \leq n/2$, $K_1, \ldots, K_m, L, C_1, \ldots, C_{n-2k} \in \mathcal{P}(\Lambda)$, and $x \in \mathbb{R}^m$. If*

$$\sum_i x_i \, \mathsf{V}(K_i[k], M[k-1], L, C_1, \ldots, C_{n-2k}) = 0 \tag{3.1}$$

holds for every $M \in \mathcal{P}(\Lambda)$, then

$$(-1)^k \sum_{i,j} x_i x_j \, \mathsf{V}(K_i[k], K_j[k], C_1, \ldots, C_{n-2k}) \geq 0. \tag{3.2}$$

Moreover, the statement is nontrivial in the sense that for any $n \geq 2$ and $k \leq n/2$, there is a simple polytope $\Lambda = L = C_1 = \cdots = C_{n-2k}$ in $\mathbb{R}^n$, $m \geq 1$,

$K_1, \ldots, K_m \in \mathcal{P}(\Lambda)$, and $x \in \mathbb{R}^m$ so that (3.1) holds and the inequality in (3.2) is strict.

The case $k = 1$ of Theorem 3.1 is nothing other than the Alexandrov-Fenchel inequality. To see why this is so, assume without loss of generality that $L = \sum_i y_i K_i$ for some $y \geq 0$, $y \neq 0$ (otherwise let $m \leftarrow m + 1$ and $K_{m+1} \leftarrow L$), and define

$$M_{ij} = V(K_i, K_j, C_1, \ldots, C_{n-2}).$$

Then the statement of Theorem 3.1 for $k = 1$ may be formulated as

$$\langle x, My \rangle = 0 \quad \text{implies} \quad \langle x, Mx \rangle \leq 0$$

for any x and $y \geq 0$, $y \neq 0$. Thus by Lemma 2.1, the inequality of Theorem 3.1 in the case $k = 1$ is equivalent to the Alexandrov-Fenchel inequality for convex bodies in $\mathcal{P}(\Lambda)$. As any collection of convex bodies can be approximated by simple strongly isomorphic polytopes [14, Theorem 2.4.15], the general case of the Alexandrov-Fenchel inequality is further equivalent to this special case.

For $k > 1$, the statement of Theorem 3.1 may be viewed as an analogue of the Alexandrov-Fenchel inequality for $M_{ij} = V(K_i[k], K_j[k], C_1, \ldots, C_{n-2k})$. Thus the Hodge-Riemann relations are reminiscent of Conjecture 1.3, but their formulation is considerably more subtle. In Sect. 4, we will show that the Hodge-Riemann relations in fact contradict Conjecture 1.3, disproving the latter.

The aim of the rest of this section is to convince the reader that the statement of Theorem 3.1 given here in terms of mixed volumes is equivalent to the statement of the Hodge-Riemann relations as given in [20]. The reader who is primarily interested in Theorem 1.5 may safely jump ahead to Sect. 4.

To explain the formulation of [20], we must first introduce some additional notation. Let $u_1, \ldots, u_N \in S^{n-1}$ be the normal directions of the facets of Λ. For any $P \in \mathcal{P}(\Lambda)$, we denote by $h_P \in \mathbb{R}^N$ its support vector

$$(h_P)_i := \sup_{y \in P} \langle y, u_i \rangle.$$

Then there is a homogenous polynomial $V : \mathbb{R}^N \to \mathbb{R}$ of degree n, called the *volume polynomial*, so that $\text{Vol}(P) = V(h_P)$ for every $P \in \mathcal{P}(\Lambda)$ [14, §5.2]. Moreover, as $\mathcal{P}(\Lambda)$ is closed under addition [14, §2.4], it follows immediately from the definition of mixed volumes that we have for any $P_1, \ldots, P_n \in \mathcal{P}(\Lambda)$

$$V(P_1, \ldots, P_n) = \frac{1}{n!} D_{h_{P_1}} \cdots D_{h_{P_n}} V,$$

where D_h denotes the directional derivative in direction h. In this notation, the Hodge-Riemann relations are formulated in [20, p. 385] as follows:

Theorem 3.2 *Let* $k \le n/2$, $L, C_1, \ldots, C_{n-2k} \in \mathcal{P}(\Lambda)$, *and let* $\alpha = \sum_{|I|=k} \alpha_I D^I$ *be a homogeneous differential operator of order k with constant coefficients. If*

$$\alpha D_{h_L} D_{h_{C_1}} \cdots D_{h_{C_{n-2k}}} V = 0, \tag{3.3}$$

then

$$(-1)^k \alpha^2 D_{h_{C_1}} \cdots D_{h_{C_{n-2k}}} V \ge 0. \tag{3.4}$$

Moreover, equality is attained if and only if $\alpha V = 0$.

To write Theorem 3.2 in terms of mixed volumes, we need the following.

Lemma 3.3 *For any homogeneous differential operator* $\alpha = \sum_{|I|=k} \alpha_I D^I$, *there exist* $m \ge 1$, $K_1, \ldots, K_m \in \mathcal{P}(\Lambda)$, *and* $x \in \mathbb{R}^m$ *so that* $\alpha = \sum_i x_i (D_{h_{K_i}})^k$.

Proof We first recall that for any $z \in \mathbb{R}^N$, $h_\Lambda + \varepsilon z$ is the support vector of some polytope $K \in \mathcal{P}(\Lambda)$ for sufficiently small ε (as Λ is simple, cf. [14, Lemma 2.4.13]). We may therefore write $z = h_L - h_{L'}$ where $L = \varepsilon^{-1} K$ and $L' = \varepsilon^{-1}\Lambda$.

Now denote by $e_1, \ldots, e_N$ the standard coordinate basis of $\mathbb{R}^N$. By the above observation, we may write $e_i = h_{L_i} - h_{L'_i}$ for $L_i, L'_i \in \mathcal{P}(\Lambda)$. We can therefore write

$$\alpha = \sum_{i_1 \le \cdots \le i_k} \alpha_{i_1, \ldots, i_k} (D_{h_{L_{i_1}}} - D_{h_{L'_{i_1}}}) \cdots (D_{h_{L_{i_k}}} - D_{h_{L'_{i_k}}}).$$

By expanding the product, we may evidently express α as a linear combination of differential operators of the form $D_{h_{R_{i_1}}} \cdots D_{h_{R_{i_k}}}$ with $R_i \in \mathcal{P}(\Lambda)$. But as

$$D_{h_{R_{i_1}}} \cdots D_{h_{R_{i_k}}} = \frac{1}{k!} \sum_{\delta \in \{0,1\}^k} (-1)^{k+\delta_1+\cdots+\delta_k} (D_{h_{\delta_1 R_{i_1} + \cdots + \delta_k R_{i_k}}})^k$$

by the polarization formula [5, p. 137], the proof is readily concluded. □

We are now ready to show that the Hodge-Riemann relations expressed by Theorems 3.1 and 3.2 are equivalent. First, note that $\alpha D_{h_L} D_{h_{C_1}} \cdots D_{h_{C_{n-2k}}} V$ in (3.3) is a homogeneous polynomial of degree $k - 1$. Thus (3.3) is equivalent to the statement that $\beta \alpha D_{h_L} D_{h_{C_1}} \cdots D_{h_{C_{n-2k}}} V = 0$ for every homogeneous differential operator β of order $k - 1$. By Lemma 3.3, the statement of Theorem 3.2 (without the equality case) may be equivalently formulated as follows: if

$$\alpha (D_{h_M})^{k-1} D_{h_L} D_{h_{C_1}} \cdots D_{h_{C_{n-2k}}} V = 0$$

for all $M \in \mathcal{P}(\Lambda)$, then

$$(-1)^k \alpha^2 D_{hC_1} \cdots D_{hC_{n-2k}} V \geq 0.$$

That (3.3)–(3.4) imply (3.1)–(3.2) follows immediately by choosing the differential operator $\alpha = \sum_i x_i (D_{hK_i})^k$. Conversely, that (3.1)–(3.2) imply (3.3)–(3.4) follows as any α can be expressed as $\alpha = \sum_i x_i (D_{hK_i})^k$ by Lemma 3.3.

It remains to check that the Hodge-Riemann relations are nontrivial. This is certainly not obvious at first sight: the condition (3.1) is a very strong one (as it must hold for *any* $M \in \mathcal{P}(\Lambda)$), and it is not clear *a priori* that it can be satisfied in any nontrivial situation. To show this is the case, consider the special case where $L = C_1 = \cdots = C_{n-2k} = \Lambda$, and define the spaces

$$P_k := \{\alpha : \alpha(D_{h\Lambda})^{n-2k+1} V = 0\}, \qquad I := \{\alpha : \alpha V = 0\}.$$

The remarkable combinatorial theory underlying the Hodge-Riemann relations enables us to compute [20, Corollary 5.3.4]

$$\dim(P_k/I) = h_k - h_{k-1},$$

where $(h_1, \ldots, h_n)$ is the so-called h-vector of Λ. To show the Hodge-Riemann relations are nontrivial, it suffices to construct a simple polytope Λ in $\mathbb{R}^n$ whose h-vector satisfies $h_k > h_{k-1}$ for $k \leq n/2$, as by Theorem 3.2 this ensures the existence of α so that (3.3) holds and the inequality in (3.4) is strict (by Lemma 3.3, this implies the corresponding statement of Theorem 3.1 for some $m, K_1, \ldots, K_m, x$). But such an example is easily identified: e.g., we may choose Λ to be the unit cube in $\mathbb{R}^n$, whose h-vector is given by $h_k = \binom{n}{k}$ by the computations in [20, p. 387] (note that $\Lambda = [0,1] \times \cdots \times [0,1]$ and use the product formula for H-polynomials).

4 Proof of Theorem 1.5

We first consider the special case that $k = 2$.

Proof of Theorem 1.5 for $k = 2$ Fix any $n \geq 4$ and let $k = 2$. By the second part of Theorem 3.1, we may choose a simple polytope $\Lambda = L = C_1 = \cdots = C_{n-4}$ in $\mathbb{R}^n$, $m \geq 1$, polytopes $K_1, \ldots, K_m \in \mathcal{P}(\Lambda)$, and $x \in \mathbb{R}^m$ so that (3.1) holds and the inequality in (3.2) is strict. In the following, we will denote $K_{m+1} := \Lambda$.

Now define the $(m+1) \times (m+1)$ matrix

$$M_{ij} := V(K_i[2], K_j[2], \Lambda[n-4]),$$

and let $y = e_{m+1}$. Then (3.1) with $M = \Lambda$ implies

$$\langle x, My \rangle = 0,$$

while the strict inequality in (3.2) may be written as

$$\langle x, Mx \rangle > 0.$$

Note that M is a positive matrix, as all bodies in $\mathcal{P}(\Lambda)$ are full-dimensional. Thus M is not hyperbolic by Lemma 2.1. In particular, by Lemma 2.3, there exists $I \subseteq [m + 1]$ so that $(-1)^{|I|} \det M_I > 0$. The latter contradicts Conjecture 1.3.   □

Informally, the above proof works as follows. By Lemma 2.3, Fedotov's Conjecture 1.3 is equivalent to the statement that the matrix M is hyperbolic. However, when $k = 2$, the Hodge-Riemann relation (3.2) yields an inequality in the opposite direction from the one that holds for hyperbolic matrices by Lemma 2.1. Thus the Hodge-Riemann relation contradicts Fedotov's conjecture.

Precisely the same argument works whenever $k \geq 2$ is even. Curiously, however, the argument fails when k is odd, as then (3.2) and hyperbolicity yield inequalities in the same direction. To prove Theorem 1.5 for arbitrary k, we will use a different argument: rather than applying the Hodge-Riemann relation of degree k, we will instead reduce the problem for any $k > 2$ back to the case $k = 2$.

Proof of Theorem 1.5 for General k Fix any $n \geq 6$ and $2 < k \leq n/2$. Choose Λ, m, $K_1, \ldots, K_{m+1}$, x, y, and M as in the proof of the $k = 2$ case. Note first that

$$
\begin{aligned}
M_{ij} &:= V(K_i[2], K_j[2], \Lambda[n - 4]) \\
&= V(K_i[2], \Lambda[k - 2], K_j[2], \Lambda[k - 2], \Lambda[n - 2k]) \\
&= \frac{1}{(k!)^2} \sum_{\delta, \varepsilon \in \{0,1\}^k} (-1)^{k+\delta_1+\cdots+\delta_k}(-1)^{k+\varepsilon_1+\cdots+\varepsilon_k} V(K_{i\delta}[k], K_{j\varepsilon}[k], \Lambda[n - 2k])
\end{aligned}
$$

by the polarization formula [5, p. 137], where

$$K_{i\delta} := (\delta_1 + \delta_2)K_i + (\delta_3 + \cdots + \delta_k)\Lambda.$$

Define the $(m + 1)(2^k - 1) \times (m + 1)(2^k - 1)$ positive matrix

$$\tilde{M}_{i\delta, j\varepsilon} := V(K_{i\delta}[k], K_{j\varepsilon}[k], \Lambda[n - 2k])$$

for $i, j \in [m + 1]$, $\delta, \varepsilon \in \{0, 1\}^k \backslash (0, \ldots, 0)$, and define $\tilde{x}, \tilde{y} \in \mathbb{R}^{(m+1)(2^k-1)}$ as

$$\tilde{x}_{i\delta} = \frac{(-1)^{k+\delta_1+\cdots+\delta_k} x_i}{k!}, \qquad \tilde{y}_{i\delta} = 1_{i=m+1} 1_{\delta=(1,0,\ldots,0)}.$$

Then

$$\langle \tilde{x}, \tilde{M}\tilde{y} \rangle = \langle x, My \rangle = 0, \qquad \langle \tilde{x}, \tilde{M}\tilde{x} \rangle = \langle x, Mx \rangle > 0,$$

so $\tilde{M}$ cannot be hyperbolic. The latter contradicts Conjecture 1.3 for the given value of k as in the proof of the case $k = 2$. □

5 An Explicit Example

The proof of Theorem 1.5 shows that counterexamples to Fedotov's conjecture are prevalent: any simple polytope Λ whose Hodge-Riemann relation of degree 2 is nontrivial (that is, whose h-vector satisfies $h_2 > h_1$, cf. Sect. 3) gives rise to a counterexample to Conjecture 1.3 with $C_1 = \cdots = C_{n-2k} = \Lambda$ and some $K_1, \ldots, K_m$ strongly isomorphic to Λ. However, the construction itself is rather indirect. The aim of this section is to illustrate the construction by means of a simple explicit example in the case that Λ is the unit cube.

Let $\Lambda = [0, e_1] + \cdots + [0, e_n]$ be the unit cube in $\mathbb{R}^n$. Then any $M \in \mathcal{P}(\Lambda)$ is a parallelepiped of the form $M = M_a + v$ for some $a_1, \ldots, a_n > 0$ and $v \in \mathbb{R}^n$, where

$$M_a := a_1[0, e_1] + \cdots + a_n[0, e_n].$$

By translation-invariance of mixed volumes, it suffices to consider $v = 0$. We can compute mixed volumes of parallelepipeds using that

$$n! \, V([0, e_{i_1}], \ldots, [0, e_{i_n}]) = 1_{i_1 \neq \cdots \neq i_n}$$

by [14, (5.77)], so that by additivity of mixed volumes

$$n! \, V(M_{a^{(1)}}, \ldots, M_{a^{(n)}}) = \sum_{i_1 \neq \cdots \neq i_n} a_{i_1}^{(1)} \cdots a_{i_n}^{(n)}.$$

Using this simple expression, it is not difficult to generate explicit examples. For example, for the case $n = 4, k = 2$, let us define

$$K_1 := [0, e_1] + [0, e_2],$$
$$K_2 := [0, e_3] + [0, e_4],$$
$$K_3 := [0, e_1] + \cdots + [0, e_4] = \Lambda.$$

Then it is readily verified by means of the above formula that

$$3\, V(K_1[2], M, \Lambda) + 3\, V(K_2[2], M, \Lambda) - V(K_3[2], M, \Lambda) = 0$$

for all $M \in \mathcal{P}(\Lambda)$, that is, (3.1) holds with $x_1 = x_2 = 3$ and $x_3 = -1$. (This is most easily seen by using $\Lambda = K_1 + K_2$ and $V(K_1[3], M) = V(K_2[3], M) = 0$ for all M.) On the other hand, we compute

$$\sum_{i,j} x_i x_j \, V(K_i[2], K_j[2]) = 18 \, V(K_1[2], K_2[2]) - 6 \, V(K_1[2], K_3[2])$$

$$- 6 \, V(K_2[2], K_3[2]) + V(K_3[2], K_3[2]) = 2,$$

so that (3.2) holds with strict inequality. It therefore follows from the argument in the proof of Theorem 1.5 that Conjecture 1.3 must fail for $n = 4, k = 2, m = 3$ when K_1, K_2, K_3 are chosen as above. The author is indebted to the anonymous referee of this note for suggesting this example.

Remark 5.1 Technically speaking the above example does not verify the assumptions of Theorem 3.1, as K_1, K_2 have empty interior and are therefore not strongly isomorphic to Λ. However, the example remains valid if we replace K_1, K_2, x_3 by $K_1' = K_1 + \varepsilon K_2$, $K_2' = K_2 + \varepsilon K_1$, and $x_3' = -1 - 4\varepsilon - \varepsilon^2$ for any $\varepsilon > 0$.

Of course, given any explicit example, one can readily verify directly that Conjecture 1.3 fails without any reference to the Hodge-Riemann relations. However, this obscures the fundamental reason for the failure of Fedotov's conjecture which was essential for the discovery of such counterexamples. On the other hand, the above explicit example provides additional information beyond our main result as stated in Theorem 1.5: it shows that Fedotov's conjecture fails already when $k = 2$ and $m = 3$, that is, in the smallest case that is not covered by Lemma 1.4. The example is readily modified to extend this conclusion to any k.

Lemma 5.2 *For every $k \geq 2$ and $n \geq 2k$, Conjecture 1.3 fails for $m = 3$.*

Proof Define the following bodies:

$$K_1 = [0, e_1] + \cdots + [0, e_k],$$

$$K_2 = [0, e_{k+1}] + \cdots + [0, e_{2k}],$$

$$K_3 = [0, e_1] + \cdots + [0, e_{2k}],$$

$$C_1, \ldots, C_{n-2k} = [0, e_{2k+1}] + \cdots + [0, e_n].$$

Then we can compute $\mathsf{M}_{ij} := V(K_i[k], K_j[k], C_1, \ldots, C_{n-2k})$ explicitly as

$$\mathsf{M} = \begin{bmatrix} 0 & a & a \\ a & 0 & a \\ a & a & b \end{bmatrix}, \qquad a = \frac{(k!)^2 (n-2k)!}{n!}, \qquad b = \frac{(2k)!(n-2k)!}{n!}.$$

Therefore

$$\det M = a^2(2a - b) = \frac{(k!)^4((n - 2k)!)^3}{(n!)^3}(2(k!)^2 - (2k)!) < 0$$

whenever $k \geq 2$, contradicting Conjecture 1.3. $\qquad\qquad\qquad\qquad\square$

Remark 5.3 The explicit expression for $n! V(M_{a^{(1)}}, \ldots, M_{a^{(n)}})$ given above is nothing other than the permanent of the matrix whose columns are $a^{(1)}, \ldots, a^{(n)}$. It is well known [5, §25.4] that the permanent of a matrix is not only a special case of mixed volumes, but also of mixed discriminants (the linear-algebraic analogue of mixed volumes). The above example therefore shows that the analogue of Fedotov's conjecture for mixed discriminants is also invalid. This should not come as a surprise, as mixed discriminants also satisfy Hodge-Riemann relations [19] and thus the arguments behind Theorem 1.5 extend to this situation.

6 Hodge-Riemann Relations Fail for General Convex Bodies

Beside the disproof of Fedotov's conjecture, an expository aim of this note has been to highlight that the Hodge-Riemann relations of McMullen and Timorin may be interpreted entirely in terms of familiar objects from classical convex geometry: they provide inequalities between mixed volumes that generalize the Alexandrov-Fenchel inequality. From the viewpoint of classical convexity, however, the formulation of Theorem 3.1 exhibits a puzzling aspect. In principle, the statements of the relations (3.1) and (3.2) make sense when K_i, C_i, M, L are arbitrary convex bodies, but the statement of Theorem 3.1 requires these bodies to be strongly isomorphic simple polytopes. It is not immediately clear why the latter is important: most classical inequalities in convex geometry are either valid for arbitrary convex bodies, or involve geometric quantities that do not make sense in the absence of regularity conditions (such as uniform bounds on the principal curvatures).

We have shown in Sect. 3 that the Hodge-Riemann relation of degree $k = 1$ is equivalent to the Alexandrov-Fenchel inequality for strongly isomorphic polytopes. The inequality then extends readily to arbitrary convex bodies by approximation. This is possible because for $k = 1$ the relations (3.1) and (3.2) can be combined into a single inequality by Lemma 2.1, and this *inequality* is preserved by taking limits. However, a natural analogue of Lemma 2.1 does not hold for $k \geq 2$. It is therefore unclear how to apply an approximation argument, as the *equality* (3.1) need not be stable under approximation (that is, if (3.1) holds for a given collection of convex bodies, they might not be approximated by simple strongly isomorphic polytopes in such a way that (3.1) remains valid for the approximations).

We will presently show by means of a simple example that the Hodge-Riemann relation of degree $k = 2$ can in fact fail for general convex bodies.

Example 6.1 Let B be the Euclidean unit ball in $\mathbb{R}^4$, and let $L = \text{conv}\{B, x\}$ for some $x \notin B$, that is, L is a cap body of B. It is a classical fact, which dates back essentially to Minkowski, that [14, Theorem 7.6.17]

$$V(L, L, B, L) = V(B, L, B, L) = V(B, B, B, L) > V(B, B, B, B).$$

In particular, this gives rise to a nontrivial equality case of the Alexandrov-Fenchel inequality of Theorem 1.1 with $n = 4$, $K = C_1 = B$, $C_2 = L$. The latter implies

$$V(M, B, B, L) = V(M, L, B, L)$$

for all convex bodies M, cf. [14, Theorem 7.4.3] or [17, Lemma 3.12].

We will now use these observations to construct a counterexample to the Hodge-Riemann relation of degree $k = 2$ for general convex bodies. Define

$$K_1 := B, \qquad K_2 := L, \qquad K_3 := B + L.$$

Then

$$3\,V(K_1[2], M, L) + V(K_2[2], M, L) - V(K_3[2], M, L) =$$
$$2\,V(M, B, B, L) - 2\,V(M, L, B, L) = 0$$

for all convex bodies M; that is, (3.1) is satisfied with $x_1 = 3$, $x_2 = 1$, and $x_3 = -1$. On the other hand, we can compute

$$\sum_{i,j} x_i x_j \, V(K_i[2], K_j[2]) = 4\,V(B, B, B, B) - 4\,V(L, B, B, B) < 0,$$

contradicting the validity of (3.2).

Remark 6.2 There is nothing special about the particular choice of the Euclidean ball in this example: the conclusion remains valid when B is replaced by an arbitrary convex body K and L is a cap body of K as defined in [14, p. 87]. For example, we may take L to be the unit cube in $\mathbb{R}^4$ and K to be the same cube with one of its corners sliced off. The latter variant of the example shows that the Hodge-Riemann relations can fail for polytopes that are not strongly isomorphic.

The above example suggests that the validity of Hodge-Riemann relations of degree $k \geq 2$ is related to the study of the equality cases of the Alexandrov-Fenchel inequality: indeed, the assumption (3.1) is reminiscent of the equality condition of the Alexandrov-Fenchel inequality (cf. [14, Theorem 7.4.2]), which is precisely what was used to construct the above counterexample. Even though the Alexandrov-Fenchel inequality is stable under approximation, this cannot be used to study its nontrivial equality cases as the latter are destroyed by approximation [13, 16, 17].

The above example shows that for Hodge-Riemann relations of degree $k \geq 2$, this instability is manifested even by the inequality itself.

On the other hand, it is expected that the validity of Hodge-Riemann relations should extend to "ample" families of convex bodies other than simple strongly isomorphic polytopes. In particular, one may conjecture that the statement of Theorem 3.1 remains valid if the class $\mathcal{P}(\Lambda)$ is replaced by the class C_+^∞ of convex bodies whose boundaries are smooth and have strictly positive curvature. Some initial progress in this direction may be found in the recent papers [1, 9, 10].

Acknowledgments This work was supported in part by NSF grants DMS-1811735 and DMS-2054565, and by the Simons Collaboration on Algorithms & Geometry. The author is grateful to Jan Kotrbatý for bringing Fedotov's conjecture to his attention, and to the anonymous referee for very helpful comments on the first version of this note and for suggesting the explicit example of Sect. 5.

References

1. S. Alesker, Kotrbaty's theorem on valuations and geometric inequalities for convex bodies. **247**, 361–378 (2022). https://doi.org/10.1007/s11856-021-2269-z
2. A.D. Alexandrov, Zur Theorie der gemischten Volumina von konvexen Körpern II. Mat. Sbornik N.S. **2**, 1205–1238 (1937)
3. A.D. Alexandrov, Zur Theorie der gemischten Volumina von konvexen Körpern IV. Mat. Sbornik N.S. **3**, 227–251 (1938)
4. R.B. Bapat, T.E.S. Raghavan, *Nonnegative Matrices and Applications* (Cambridge University Press, Cambridge, 1997)
5. Y.D. Burago, V.A. Zalgaller, *Geometric Inequalities* (Springer, Berlin, 1988)
6. T.-C. Dinh, V.-A. Nguyên, The mixed Hodge-Riemann bilinear relations for compact Kähler manifolds. Geom. Funct. Anal. **16**(4), 838–849 (2006)
7. V.P. Fedotov, A new method for the proof of the inequalities between mixed volumes and the generalization of Aleksandrov-Fenchel-Shephard inequalities. Dokl. Akad. Nauk SSSR **245**(1), 31–34 (1979)
8. J. Huh, Combinatorial applications of the Hodge-Riemann relations, in *Proceedings of the International Congress of Mathematicians—Rio de Janeiro 2018. Vol. IV. Invited Lectures* (World Scientific Publishing, Hackensack, 2018), pp. 3093–3111
9. J. Kotrbatý, On Hodge-Riemann relations for translation-invariant valuations. Adv. Math. **390**, 107914 (2021). https://doi.org/10.1016/j.aim.2021.107914
10. J. Kotrbatý, T. Wannerer, From harmonic analysis of translation-invariant valuations to geometric inequalities for convex bodies, **33**, 541–592 (2023). https://doi.org/10.1007/s00039-023-00630-1
11. P. McMullen, On simple polytopes. Invent. Math. **113**(2), 419–444 (1993)
12. H. Minkowski, Volumen und Oberfläche. Math. Ann. **57**(4), 447–495 (1903)
13. R. Schneider, On the Aleksandrov-Fenchel inequality, in *Discrete Geometry and Convexity*, vol. 440. Annals of the New York Academy of Sciences (New York Academy of Sciences, New York, 1985), pp. 132–141
14. R. Schneider, *Convex Bodies: The Brunn-Minkowski Theory*, expanded edition (Cambridge University Press, Cambridge, 2014)
15. Y. Shenfeld, R. van Handel, Mixed volumes and the Bochner method. Proc. Am. Math. Soc. **147**(12), 5385–5402 (2019)

16. Y. Shenfeld, R. Van Handel, The extremals of Minkowski's quadratic inequality. Duke Math.
    J. **171**, 957–1027 (2021)
17. Y. Shenfeld, R. Van Handel, The extremals of the Alexandrov-Fenchel inequality for convex
    polytopes. Acta Math. (2022, to appear)
18. G.C. Shephard, Inequalities between mixed volumes of convex sets. Mathematika **7**, 125–138
    (1960)
19. V.A. Timorin, The mixed Hodge-Riemann bilinear relations in the linear situation. Funct.
    Anal. Appl. **32**(4), 268–272 (1998)
20. V.A. Timorin, An analogue of the Hodge-Riemann relations for simple convex polytopes.
    Russ. Math. Surv. **54**(2), 381–426 (1999)

The Local Logarithmic Brunn-Minkowski Inequality for Zonoids

Ramon van Handel

Abstract The aim of this note is to show that the local form of the logarithmic Brunn-Minkowski conjecture holds for zonoids. The proof uses a variant of the Bochner method due to Shenfeld and the author.

Keywords Logarithmic Brunn-Minkowski inequality · Zonoids · Mixed volumes · Bochner method

1 Introduction

1.1 The classical Brunn-Minkowski inequality states that

$$\mathrm{Vol}((1-t)K + tL)^{1/n} \geqslant (1-t)\,\mathrm{Vol}(K)^{1/n} + t\,\mathrm{Vol}(L)^{1/n} \tag{1.1}$$

for all $t \in [0, 1]$ and convex bodies K, L in $\mathbb{R}^n$, where

$$aK + bL := \{ax + by : x \in K, y \in L\}$$

denotes Minkowski addition. Its importance, both to convexity and to other areas of mathematics, can hardly be overstated; cf. [10]. As is well known, (1.1) is equivalent to the apparently weaker inequality

$$\mathrm{Vol}((1-t)K + tL) \geqslant \mathrm{Vol}(K)^{1-t}\,\mathrm{Vol}(L)^t. \tag{1.2}$$

where the arithmetic mean on the right-hand side has been replaced by the geometric mean. Clearly (1.1) implies (1.2), as the geometric mean is smaller than the arithmetic mean; the converse implication follows by rescaling K, L [10, §4].

R. van Handel (✉)
Princeton University, Princeton, NJ, USA
e-mail: rvan@math.princeton.edu

© The Author(s), under exclusive license to Springer Nature Switzerland AG 2023
R. Eldan et al. (eds.), *Geometric Aspects of Functional Analysis*, Lecture Notes in Mathematics 2327, https://doi.org/10.1007/978-3-031-26300-2_14

As part of their study of the Minkowski problem for cone volume measures, Böröczky et al. [4] asked whether one could replace also the "arithmetic mean" $(1-t)K + tL$ on the left-hand side of the Brunn-Minkowski inequality by a certain kind of "geometric mean": that is, whether

$$\text{Vol}(K^{1-t}L^t) \overset{?}{\geqslant} \text{Vol}(K)^{1-t}\text{Vol}(L)^t, \qquad (1.3)$$

where the meaning of $K^{1-t}L^t$ must be carefully defined (see (1.7) below). As the geometric mean is smaller than the arithmetic mean, this would yield an improvement of the classical Brunn-Minkowski inequality. While such an improved inequality turns out to be false for general convex bodies, it was conjectured in [4] that such an improved inequality holds whenever K, L are symmetric convex bodies (that is, $K = -K$ and $L = -L$), which they proved to be true in dimension 2. In higher dimensions, this *logarithmic Brunn-Minkowski conjecture* remains open.

1.2 It is readily seen that the Brunn-Minkowski inequality (1.1) and the logarithmic Brunn-Minkowski conjecture (1.3) are equivalent to concavity of the functions

$$\varphi : t \mapsto \text{Vol}((1-t)K + tL)^{1/n} \qquad \text{and} \qquad \psi : t \mapsto \log\text{Vol}(K^{1-t}L^t)$$

for all convex bodies K, L and symmetric convex bodies K, L in $\mathbb{R}^n$, respectively. We can therefore obtain equivalent formulations of (1.1) and (1.3) by considering the first- and second-order conditions for concavity of φ and ψ.

In order to formulate the resulting inequalities, we must first recall some additional notions (we refer to [21] for a detailed treatment). It was shown by Minkowski that the volume of convex bodies is a polynomial in the sense that for any convex bodies $K_1, \ldots, K_m$ in $\mathbb{R}^n$ and $\lambda_1, \ldots, \lambda_m > 0$, we have

$$\text{Vol}(\lambda_1 K_1 + \cdots + \lambda_m K_m) = \sum_{i_1,\ldots,i_n=1}^{m} \text{V}(K_{i_1}, \ldots, K_{i_n})\lambda_{i_1}\cdots\lambda_{i_n}.$$

The coefficients $\text{V}(K_1, \ldots, K_n)$, called *mixed volumes*, are nonnegative, symmetric in their arguments, and homogeneous and additive in each argument under Minkowski addition. Moreover, mixed volumes admit the integral representation

$$\text{V}(K_1, \ldots, K_n) = \frac{1}{n}\int h_{K_1} dS_{K_2,\ldots,K_n}, \qquad (1.4)$$

where the *mixed area measure* $S_{K_2,\ldots,K_n}$ is a finite measure on S^{n-1} and $h_K(x) :=$ $\sup_{z \in K}\langle z, x \rangle$ denotes the support function of a convex body K.

In view of the above definitions, it is now straightforward to obtain equivalent formulations of the Brunn-Minkowski inequality in terms of mixed volumes; see, e.g., [21, pp. 381–382 and 406]. In the sequel, we denote by $\mathcal{K}^n$ ($\mathcal{K}^n_s$) the family of all (symmetric) convex bodies in $\mathbb{R}^n$ with nonempty interior.

Lemma 1.1 (Minkowski) *The following are equivalent:*

1. For all $K, L \in \mathcal{K}^n$ and $t \in [0, 1]$, the Brunn-Minkowski inequality (1.1) holds.
2. For all $K \in \mathcal{K}^n$, we have

$$V(L, K, \ldots, K) \geqslant \mathrm{Vol}(L)^{1/n} \mathrm{Vol}(K)^{1-1/n} \quad \forall L \in \mathcal{K}^n. \tag{1.5}$$

3. For all $K \in \mathcal{K}^n$, we have

$$V(L, K, \ldots, K)^2 \geqslant V(L, L, K, \ldots, K) \mathrm{Vol}(K) \quad \forall L \in \mathcal{K}^n. \tag{1.6}$$

Proof If we apply (1.5)–(1.6) with $K \leftarrow (1 - s)K + sL$ and $L \leftarrow (1 - r)K + rL$, then a simple computation shows that Minkowski's first inequality (1.5) is nothing other than the first-order concavity condition $\varphi(r) \leqslant \varphi(s) + \varphi'(s)(r - s)$, while Minkowski's second inequality (1.6) is the second-order condition $\varphi''(s) \leqslant 0$. $\square$

Before we state an analogous reformulation of (1.3), we must first give a precise definition of $K^{1-t}L^t$. To motivate this definition, recall that the arithmetic mean of convex bodies is characterized by its support function $h_{(1-t)K+tL} = (1 - t)h_K + th_L$. We may therefore attempt to define $K^{1-t}L^t$ as the convex body whose support function is the geometric mean $h_K^{1-t}h_L^t$. However, the latter need not be the support function of any convex body. We therefore define $K^{1-t}L^t$ in general as the largest convex body whose support function is dominated by $h_K^{1-t}h_L^t$, that is,

$$K^{1-t}L^t := \{z \in \mathbb{R}^n : \langle z, x \rangle \leqslant h_K(x)^{1-t}h_L(x)^t \text{ for all } x \in \mathbb{R}^n\}. \tag{1.7}$$

We can now formulate the following analogue of Lemma 1.1.

Theorem 1.2 ([4, 7, 8, 14, 15, 17]) *The following are equivalent:*

1. For all $K, L \in \mathcal{K}_s^n$ and $t \in [0, 1]$, the log-Brunn-Minkowski inequality (1.3) holds.
2. For all $K \in \mathcal{K}_s^n$, we have

$$\int h_K \log\left(\frac{h_L}{h_K}\right) dS_{K,\ldots,K} \geqslant \mathrm{Vol}(K) \log\left(\frac{\mathrm{Vol}(L)}{\mathrm{Vol}(K)}\right) \quad \forall L \in \mathcal{K}_s^n. \tag{1.8}$$

3. For all $K \in \mathcal{K}_s^n$, we have

$$\frac{V(L, K, \ldots, K)^2}{\mathrm{Vol}(K)} \geqslant \frac{n-1}{n} V(L, L, K, \ldots, K)$$

$$+ \frac{1}{n^2} \int \frac{h_L^2}{h_K} dS_{K,\ldots,K} \quad \forall L \in \mathcal{K}_s^n. \tag{1.9}$$

The difficulty in the proof of Theorem 1.2 is that the map $t \mapsto K^{1-t}L^t$ can be nonsmooth: if it were the case that $h_{K^{1-t}L^t} = h_K^{1-t} h_L^t$ for all $t \in [0, 1]$, the result would follow easily from the first- and second-order conditions for concavity of ψ. That the conclusion remains valid using the correct definition (1.7) is a nontrivial fact that has been established through the combined efforts of several groups.

Remark 1.3 The notation (1.7) is nonstandard: $K^{1-t}L^t$ is often denoted in the literature as $(1 - t)K +_0 tL$, as it coincides with the $q \to 0$ limit of L^q-Minkowski addition. As the latter notation is somewhat confusing (the geometric mean is not defined by the rescaled bodies $(1 - t)K$ and tL), and as only geometric means are used in this paper, we have chosen a nonstandard but more suggestive notation.

1.3 It was shown in [4] that the logarithmic Brunn-Minkowski conjecture holds in dimension $n = 2$. In dimensions $n \geqslant 3$, however, the conjecture has been proved to date only under special symmetry assumptions: when K, L are complex [19] or unconditional [20] bodies (see also [3] for a generalization). In both cases the conjecture is established by replacing the geometric mean (1.7) by a smaller set whose construction requires the special symmetries, which yields strictly stronger inequalities than are conjectured for general bodies.

Even for a fixed reference body K, the validity of the inequalities (1.8) and (1.9) for *all* $L \in \mathcal{K}_s^n$ (i.e., in the absence of additional symmetries) appears to be unknown except in one very special family of examples: it follows from [14, 15] that (1.8) and (1.9) hold when K is the ℓ_p^n-ball with $2 \leqslant p < \infty$ and sufficiently large n, as well as for affine images and sufficiently small perturbations of these bodies. Note, however, that the analysis of these examples shows that they satisfy even stronger inequalities that cannot hold for general bodies (local L^q-Brunn-Minkowski inequalities with $q = -\frac{1}{4}$ [14, Theorem 10.4]), so that they do not approach the extreme cases of the logarithmic Brunn-Minkowski conjecture.[1]

The aim of this note is to contribute some further evidence toward the validity of the logarithmic Brunn-Minkowski conjecture. Recall that a convex body $K \in \mathcal{K}_s^n$ is called a *zonoid* if it is the limit of Minkowski sums of segments. The first main result of this note is the following theorem.

Theorem 1.4 *Let $K \in \mathcal{K}_s^n$ be a zonoid. Then the local logarithmic Brunn-Minkowski inequality (1.9) holds for all $L \in \mathcal{K}_s^n$.*

Our second main result settles the equality cases of Theorem 1.4.

Definition A vector $u \in S^{n-1}$ is called an *r-extreme normal vector* of a convex body K if there do not exist linearly independent normal vectors $u_1, \ldots, u_{r+2}$ at a boundary point of K such that $u = u_1 + \cdots + u_{r+2}$.

---

[1] For one extreme case, the ℓ_∞^n-ball, the validity of (1.9) may be verified by an explicit computation, see, e.g., [14, Theorem 10.2]. This does not follow as a limiting case of the general result [14, Theorem 10.4] on ℓ_p^n-balls, however, as the latter only holds for $n \geqslant n_0(p) \to \infty$ as $p \to \infty$.

Theorem 1.5 *Let $K \in \mathcal{K}_s^n$ be a zonoid. Then equality holds in (1.9) if and only if*

1. *$K = C_1 + \cdots + C_m$ for some $1 \leqslant m \leqslant n$ and zonoids $C_1, \ldots, C_m$ such that $\dim(C_1) + \cdots + \dim(C_m) = n$; and*
2. *there exist $a_1, \ldots, a_m \geqslant 0$ such that L and $a_1 C_1 + \cdots + a_m C_m$ have the same supporting hyperplanes in all 1-extreme normal directions of K.*

Theorem 1.4 does not suffice to conclude that the logarithmic Brunn-Minkowski inequality (1.3) holds when K, L are zonoids, as $K^{1-t}L^t$ is generally not a zonoid. Nonetheless, by combining Theorems 1.4–1.5 with [15, Theorem 2.1] we can deduce validity of the logarithmic Minkowski inequality (1.8), albeit without its equality cases. Some further implications will be given in Sect. 5.

Corollary 1.6 *Let $K \in \mathcal{K}_s^n$ be a zonoid. Then the logarithmic Minkowski inequality (1.8) holds for all $L \in \mathcal{K}_s^n$.*

It appears somewhat unlikely that our results make major progress in themselves toward the full resolution of the logarithmic Brunn-Minkowski conjecture; as is the case for other well-known conjectures in convex geometry (see, e.g., [11]), zonoids form a very special class of convex bodies that provide only modest insight into the behavior of general convex bodies. Nonetheless, let us highlight several interesting features of the main results of this note:

- Theorem 1.4 possesses many nontrivial equality cases; therefore, in contrast to the setting of previous results in dimensions $n \geqslant 3$ for general $L \in \mathcal{K}_s^n$, the class of zonoids includes many extreme cases of the logarithmic Brunn-Minkowski conjecture. (Theorem 1.5 supports the conjectured equality cases in [3].)
- Unlike in dimensions $n \geqslant 3$, every planar symmetric convex body is a zonoid. The $n = 2$ case of the logarithmic Brunn-Minkowski conjecture that was settled in [4] may therefore be viewed in a new light as a special case of our results (modulo the nontrivial Theorem 1.2). In fact, the proof of Theorem 1.4 will work in a completely analogous manner for $n = 2$ and $n \geqslant 3$.
- The ℓ_p^n-ball is a zonoid for every n and $2 \leqslant p \leqslant \infty$ [2, Theorem 6.6]. Our results therefore capture as special cases all explicit examples of convex bodies K for which (1.8) and (1.9) were previously known to hold.[2]

Before we proceed, let us briefly sketch some key ideas behind the proofs.

1.4 It was a fundamental insight of Hilbert [12, Chapter XIX] that mixed volumes of sufficiently smooth convex bodies admit a spectral interpretation. To this end, given any sufficiently smooth convex body $K \in \mathcal{K}^n$, Hilbert constructs an elliptic

---

[2] However, the methods of [14, 15] provide complementary information that does not follow from our results. For example, the estimates of [14] imply that for any $2 < p < \infty$ and $n \geqslant n_0(p)$, all $K \in \mathcal{K}_s^n$ that are sufficiently close to the ℓ_p^n-ball in a quantitative sense satisfy the L^q-Minkowski inequality with $q = -\frac{1}{4}$. More generally, it is shown in [15] that for any $K \in \mathcal{K}_s^n$, there exists $K' \in \mathcal{K}_s^n$ with $K \subseteq K' \subseteq 8K$ so that K' satisfies the L^q-Minkowski inequality with $q = -\frac{1}{4}$.

differential operator $\mathscr{A}_K$ (see Sect. 2.2 for a precise definition) and a measure $d\mu_K := \frac{1}{nh_K}dS_{K,\ldots,K}$ on S^{n-1} with the following properties:

- $\mathscr{A}_K$ defines a self-adjoint operator on $L^2(\mu_K)$ with discrete spectrum.
- $\mathscr{A}_K h_K = h_K$, that is, h_K is an eigenfunction with eigenvalue 1.
- $V(L, M, K, \ldots, K) = \langle h_L, \mathscr{A}_K h_M \rangle$ for all $L, M \in \mathcal{K}^n$.

Using these properties, it is readily verified that (1.6) is equivalent to

$$\langle f, \mathscr{A}_K f \rangle \leqslant 0 \quad \text{for} \quad f = h_L - \frac{\langle h_L, h_K \rangle}{\|h_K\|^2} h_K, \tag{1.10}$$

where we denote by $\langle \cdot, \cdot \rangle$ and $\| \cdot \|$ the inner product and norm of $L^2(\mu_K)$. As f in (1.10) is the projection of h_L on $\{h_K\}^\perp$, we obtain:

Lemma 1.7 (Hilbert) *Items 1–3 of Lemma 1.1 are equivalent to:*

4. *For every sufficiently smooth convex body* $K \in \mathcal{K}^n$, *any eigenfunction* $\mathscr{A}_K f = \lambda f$ *with* $\langle f, h_K \rangle = 0$ *has eigenvalue* $\lambda \leqslant 0$.

The condition of Lemma 1.7 is optimal, as $\mathscr{A}_K$ always has eigenfunctions with eigenvalue 0: any linear function $\ell(x) = h_{\{v\}}(x) = \langle v, x \rangle$ satisfies $\mathscr{A}_K \ell = 0$. If we restrict attention to *symmetric* convex bodies $K, L \in \mathcal{K}_s^n$, however, only *even* functions $f(x) = f(-x)$ arise in (1.10), and it is certainly possible that all even eigenfunctions of $\mathscr{A}_K$ have strictly negative eigenvalues. It was observed by Kolesnikov and Milman [14] that the logarithmic Brunn-Minkowski conjecture may be viewed as a quantitative form of this phenomenon: as (1.9) is equivalent to

$$\langle f, \mathscr{A}_K f \rangle \leqslant -\frac{1}{n-1}\|f\|^2 \quad \text{for} \quad f = h_L - \frac{\langle h_L, h_K \rangle}{\|h_K\|^2} h_K,$$

the following conclusion follows readily.

Lemma 1.8 (Kolesnikov-Milman) *Items 1–3 of Theorem 1.2 are equivalent to:*

4. *For every sufficiently smooth symmetric convex body* $K \in \mathcal{K}_s^n$, *any even eigenfunction* $\mathscr{A}_K f = \lambda f$ *with* $\langle f, h_K \rangle = 0$ *has eigenvalue* $\lambda \leqslant -\frac{1}{n-1}$.

It is verified in [14, Theorem 10.4] that when K is the ℓ_p^n-ball for $2 \leqslant p < \infty$, any even eigenfunction of $\mathscr{A}_K$ orthogonal to h_K has eigenvalue λ with $n\lambda \to -\infty$ as $n \to \infty$. This shows that such K satisfy the condition of Lemma 1.8 for large n, but does not explain the significance of the threshold $-\frac{1}{n-1}$.

A new approach to the study of the spectral properties of $\mathscr{A}_K$ was discovered by Shenfeld and the author in [22]. This approach, called the *Bochner method* in view of its analogy to the classical Bochner method in differential geometry, has found several surprising applications both inside and outside convex geometry. The Bochner method was already used in [22] to provide new proofs of the Alexandrov-Fenchel inequality, a much deeper result of which (1.6) is a special case, and of the Alexandrov mixed discriminant inequality. Subsequent applications outside convexity include the proof of certain properties of Lorentzian polynomials in [5, 9] and the striking results of [6], where the method is used to prove numerous combinatorial inequalities. The paper [15] contains another application to the study of isomorphic variants of the L^q-Minkowski problem.

The proofs of Theorems 1.4–1.5 provide yet another illustration of the utility of the Bochner method. By using a variation on the method of [22], we obtain a "Bochner identity" which relates the spectral condition of Lemma 1.8 in dimension n to the inequality (1.9) in dimension $n - 1$. The conclusion then follows by induction on the dimension. One interesting feature of this proof is that it provides an explanation for the appearance of the mysterious value $-\frac{1}{n-1}$ in Lemma 1.8.[3] While the specific formulas derived in this note rely on the zonoid assumption, our approach may provide some hope that other variations on the Bochner method could lead to further progress toward the logarithmic Brunn-Minkowski conjecture.

1.5 The rest of this note is organized as follows. In Sect. 2, we briefly recall some background from convex geometry that will be needed in the proofs, and we recall the basic idea behind the Bochner method as developed in [22]. Theorem 1.4 is proved in Sect. 3, and Theorem 1.5 is proved in Sect. 4. Finally, Sect. 5 spells out some implications of Theorem 1.4, including the proof of Corollary 1.6.

2 Preliminaries

Throughout this note, we will use without comment the standard properties of mixed volumes and mixed area measures: that they are nonnegative, symmetric and multilinear in their arguments, and continuous under Hausdorff convergence. We refer to the monograph [21] for a detailed treatment, or to [22, §2], [23, §4] for a brief review of such basic properties. The aim of this section is to recall some further notions that will play a central role in the sequel: the behavior of mixed volumes under projections, the construction of the Hilbert operator $\mathscr{A}_K$ for sufficiently smooth convex bodies, and the Bochner method of [22].

The following notation will often be used: if $f = h_K - h_L$ is a difference of support functions of convex bodies, then we define [21, §5.2]

$$V(f, C_1, \ldots, C_{n-1}) := V(K, C_1, \ldots, C_{n-1}) - V(L, C_1, \ldots, C_{n-1}),$$

$$S_{f, C_1, \ldots, C_{n-2}} := S_{K, C_1, \ldots, C_{n-2}} - S_{L, C_1, \ldots, C_{n-2}}.$$

---

[3] As was pointed out in [14], the eigenvalue $-\frac{1}{n-1}$ is attained when K is the cube, so that Lemma 1.8 may be interpreted as stating that the second eigenvalue of $\mathscr{A}_K$ is maximized by the cube. This interpretation does not explain, however, *why* this should be the case. In any case, there are many maximizers other than cubes, as is already illustrated by Theorem 1.5.

We similarly define $V(f, g, C_1, \ldots, C_{n-2})$ by linearity when f, g are differences of support functions, etc. Mixed volumes and area measures of differences of support functions are still symmetric and multilinear, but need not be nonnegative.

2.1 Projections and Zonoids

Let $E \subseteq \mathbb{R}^n$ be a linear subspace of dimension k, and let $C_1, \ldots, C_k$ be convex bodies in E. Then we denote by $V(C_1, \ldots, C_k)$ and $S_{C_1, \ldots, C_{k-1}}$ the mixed volume and mixed area measure computed in $E \simeq \mathbb{R}^k$. We will often view $S_{C_1, \ldots, C_{k-1}}$ as a measure on $\mathbb{R}^n$ that is supported in E. The projection of a convex body C in $\mathbb{R}^n$ onto E will be denoted as $\mathbf{P}_E C$.

The following basic formulas relate mixed volumes and mixed area measures of convex bodies to those of their projections.

Lemma 2.1 *For any $u \in S^{n-1}$ and $C_1, \ldots, C_{n-1} \in \mathcal{K}^n$, we have*

$$\frac{n}{2} V([-u, u], C_1, \ldots, C_{n-1}) = V(\mathbf{P}_{u^\perp} C_1, \ldots, \mathbf{P}_{u^\perp} C_{n-1}),$$

$$\frac{n-1}{2} S_{[-u,u], C_1, \ldots, C_{n-2}} = S_{\mathbf{P}_{u^\perp} C_1, \ldots, \mathbf{P}_{u^\perp} C_{n-2}}.$$

Proof The first identity is [21, (5.77)]. To prove the second identity, note that the first identity may be rewritten using (1.4) as

$$\frac{1}{2} \int f \, dS_{[-u,u], C_1, \ldots, C_{n-2}} = \frac{1}{n-1} \int f \, dS_{\mathbf{P}_{u^\perp} C_1, \ldots, \mathbf{P}_{u^\perp} C_{n-2}}$$

for $f = h_{C_{n-1}}$, where we used that $h_{\mathbf{P}_E C}(u) = h_C(u)$ for $u \in E$. The identity extends by linearity to any difference of support functions $f = h_K - h_L$. But as any $f \in C^2(S^{n-1})$ is of this form (cf. Lemma 2.4 below), the conclusion follows. □

A body $K \in \mathcal{K}_s^n$ is called a *zonoid* if it is the limit of Minkowski sums of segments $[-u, u]$. The following equivalent definition [21, Theorem 3.5.3] is well known.

Definition 2.2 $K \in \mathcal{K}_s^n$ is called a *zonoid* if

$$h_K(x) = \int h_{[-u,u]}(x) \, \eta(du)$$

for some finite even measure η on S^{n-1}, called the *generating measure* of K.

The significance of zonoids for our purposes is that mixed volumes of zonoids can be expressed in terms of mixed volumes of projections by Lemma 2.1 and linearity. One simple illustration of this is the following fact.

Lemma 2.3 *In dimension* 2, *any symmetric convex body* $K \in \mathcal{K}_s^2$ *is a zonoid with*

$$h_K(x) = \frac{1}{4} \int h_{[-u^\dagger, u^\dagger]}(x) \, S_K(du),$$

where for $u \in S^1$ *we denote by* $u^\dagger \in S^1$ *the clockwise rotation of* u *by the angle* $\frac{\pi}{2}$.

Proof Note first that $h_{[-u^\dagger, u^\dagger]}(x) = |\langle u^\dagger, x \rangle| = |\langle u, x^\dagger \rangle| = h_{[-x^\dagger, x^\dagger]}(u)$. Thus by Lemma 2.1, we can write for any $K \in \mathcal{K}^2$

$$\frac{1}{2} \int h_{[-u^\dagger, u^\dagger]}(x) \, S_K(du) = \mathsf{V}([-x^\dagger, x^\dagger], K) = \mathrm{Vol}(\mathbf{P}_{\mathrm{span}\{x\}} K) = h_K(x) + h_K(-x).$$

As $K \in \mathcal{K}_s^2$ is symmetric, we have $h_K(x) + h_K(-x) = 2h_K(x)$. □

2.2 Smooth Bodies and the Hilbert Operator

A support function h_K may be viewed either as a function on S^{n-1}, or as a 1-homogeneous function on $\mathbb{R}^n$. In particular, if h_K is a C^2 function on S^{n-1}, then its gradient ∇h_K in $\mathbb{R}^n$ is 0-homogeneous, and thus its Hessian $\nabla^2 h_K(x)$ in $\mathbb{R}^n$ is a linear map from $x^\perp$ to itself. We denote the restriction of $\nabla^2 h_K(x)$ to $x^\perp$ as $D^2 h_K(x)$. For a general function $f \in C^2(S^{n-1})$, the restricted Hessian $D^2 f$ is defined analogously by applying the above construction to the 1-homogeneous extension of f.

We now recall the following basic facts [22, §2.1]. Here we write $A \geq 0 \, (A > 0)$ to indicate that a symmetric matrix A is positive semidefinite (positive definite).

Lemma 2.4 *Let* $f \in C^2(S^{n-1})$. *Then the following hold:*

a. $f = h_K$ *for some convex body* K *if and only if* $D^2 f \geq 0$.
b. *For any convex body* L *such that* $h_L \in C^2(S^{n-1})$ *and* $D^2 h_L > 0$, *there is a convex body* K *and* $a > 0$ *so that* $f = a(h_K - h_L)$.

A particularly useful class of bodies is the following.

Definition 2.5 $K \in \mathcal{K}^n$ is of class C_+^k ($k \geq 2$) if $h_K \in C^k(S^{n-1})$ and $D^2 h_K > 0$.

For our purposes, the importance of C_+^k bodies is that they admit certain explicit representations of mixed volumes and mixed area measures. To define these, let us first recall that the mixed discriminant $\mathsf{D}(A_1, \ldots, A_{n-1})$ of $(n-1)$-dimensional matrices $A_1, \ldots, A_{n-1}$ is defined by the formula

$$\det(\lambda_1 A_1 + \cdots + \lambda_m A_m) = \sum_{i_1, \ldots, i_{n-1}=1}^{m} \mathsf{D}(A_{i_1}, \ldots, A_{i_{n-1}}) \lambda_{i_1} \cdots \lambda_{i_{n-1}}$$

in analogy with the definition of mixed volumes. Mixed discriminants are symmetric and multilinear in their arguments, and $D(A_1, \ldots, A_{n-1}) > 0$ for $A_1, \ldots, A_{n-1} > 0$. Moreover, we have the Alexandrov mixed discriminant inequality

$$D(A, B, M_1, \ldots, M_{n-3})^2 \geqslant D(A, A, M_1, \ldots, M_{n-3}) D(B, B, M_1, \ldots, M_{n-3}) \tag{2.1}$$

whenever $B, M_1, \ldots, M_{n-3} \geqslant 0$ and A is a symmetric matrix. For these and other facts about mixed discriminants, see [22, §2.3 and §4].

With these definitions in place, we have the following [23, Lemma 4.7].

Lemma 2.6 *Let $C_1, \ldots, C_{n-1} \in \mathcal{K}^n$ be of class C_+^2. Then*

$$dS_{C_1, \ldots, C_{n-1}} = D(D^2 h_{C_1}, \ldots, D^2 h_{C_{n-1}}) \, d\omega,$$

$$V(K, C_1, \ldots, C_{n-1}) = \frac{1}{n} \int h_K \, D(D^2 h_{C_1}, \ldots, D^2 h_{C_{n-1}}) \, d\omega$$

for any convex body K, where ω denotes the surface measure on S^{n-1}.

We now introduce the spectral interpretation of mixed volumes due to Hilbert. For simplicity, we will only consider the special case that is needed in this note; the same construction applies to general mixed volumes (cf. [22]).

Fix a body $K \in \mathcal{K}^n$ of class C_+^2 with the origin in its interior (so that $h_K > 0$). Then we define a measure μ_K on S^{n-1} as

$$d\mu_K := \frac{1}{n h_K} dS_{K, \ldots, K} = \frac{1}{n h_K} D(D^2 h_K, \ldots, D^2 h_K) \, d\omega,$$

and define the second-order differential operator $\mathscr{A}_K$ on S^{n-1} as

$$\mathscr{A}_K f := h_K \frac{D(D^2 f, D^2 h_K, \ldots, D^2 h_K)}{D(D^2 h_K, \ldots, D^2 h_K)}$$

for $f \in C^2(S^{n-1})$. The positivity of mixed discriminants of positive definite matrices implies that $\mathscr{A}_K$ is elliptic. Standard facts of elliptic regularity theory therefore imply the following, cf. [22, §3] or [14, Theorem 5.3]:

- $\mathscr{A}_K$ extends to a self-adjoint operator on $L^2(\mu_K)$ with $\mathrm{Dom}(\mathscr{A}_K) = H^2(S^{n-1})$.
- $\mathscr{A}_K$ has a discrete spectrum, that is, it has a countable sequence of eigenvalues $\lambda_1 > \lambda_2 \geqslant \lambda_3 \geqslant \cdots$ of tending to $-\infty$, and its eigenfunctions span $L^2(\mu_K)$.
- $\lambda_1 = 1$ is a simple eigenvalue, whose eigenspace is spanned by h_K.

These facts will be invoked in the sequel without further comment.

The point of this construction is that, by Lemmas 2.4 and 2.6, we evidently have

$$\langle f, \mathscr{A}_K g \rangle := \int f \mathscr{A}_K g \, d\mu_K = V(f, g, K, \ldots, K)$$

for any $f, g \in C^2(S^{n-1})$. Mixed volumes of this type may therefore be viewed as quadratic forms of the operator $\mathscr{A}_K$, which furnishes various geometric inequalities with a spectral interpretation as explained in Sect. 1.

When $K \in \mathcal{K}_s^n$ is symmetric, it is readily verified from the definitions that μ_K is an even measure, and that $\mathscr{A}_K$ leaves the spaces $L^2(\mu_K)_{\text{even}}$ and $L^2(\mu_K)_{\text{odd}}$ of even and odd functions on S^{n-1} invariant. As $L^2(\mu_K) = L^2(\mu_K)_{\text{even}} \oplus L^2(\mu_K)_{\text{odd}}$, it follows that when $K \in \mathcal{K}_s^n$ any $f \in L^2(\mu_K)_{\text{even}}$ can be expressed as a linear combination of the *even* eigenfunctions of $\mathscr{A}_K$; cf. [14, §5.1].

2.3 The Bochner Method

By Lemma 1.7, Minkowski's second inequality (1.6), and thus the Brunn-Minkowski inequality, is equivalent to the statement that the second largest eigenvalue of $\mathscr{A}_K$ satisfies $\lambda_2 \leq 0$. This idea was exploited by Hilbert to give a spectral proof of the Brunn-Minkowski inequality by means of an eigenvalue continuity argument.

A new proof of the above spectral condition was discovered by Shenfeld and the author in [22]. This proof is based on the elementary fact that the condition $\lambda_2 \leq 0$ would follow directly from the Lichnerowicz condition

$$\langle \mathscr{A}_K f, \mathscr{A}_K f \rangle \geq \langle f, \mathscr{A}_K f \rangle \quad \text{for all } f. \tag{2.2}$$

Indeed, if $\mathscr{A}_K f = \lambda f$, then (2.2) yields $\lambda^2 \geq \lambda$, i.e., $\lambda \geq 1$ or $\lambda \leq 0$. As $1 = \lambda_1 > \lambda_2$ by elliptic regularity theory, the conclusion $\lambda_2 \leq 0$ follows. While this is merely a reformulation of the problem, the beauty of (2.2) is that it is an immediate consequence of the following identity that admits a one-line proof.

Lemma 2.7 (Bochner Identity) *Let $K \in \mathcal{K}^n$ be of class C_+^2 and let $f \in C^2$. Then*

$$\langle \mathscr{A}_K f, \mathscr{A}_K f \rangle - \langle f, \mathscr{A}_K f \rangle =$$

$$\int \frac{h_K}{n} \left\{ \frac{\mathsf{D}(D^2 f, D^2 h_K, \ldots, D^2 h_K)^2}{\mathsf{D}(D^2 h_K, \ldots, D^2 h_K)} - \mathsf{D}(D^2 f, D^2 f, D^2 h_K, \ldots, D^2 h_K) \right\} d\omega. \tag{2.3}$$

Proof The identity is immediate from the definitions of $\mathscr{A}_K$ and μ_K, and as

$$\langle f, \mathscr{A}_K f \rangle = V(K, f, f, K, \ldots, K) = \frac{1}{n} \int h_K \, D(D^2 f, D^2 f, D^2 h_K, \ldots, D^2 h_K) \, d\omega$$

by Lemma 2.6 and as mixed volumes are symmetric in their arguments.   □

To deduce (2.2), it remains to recall that the integrand in (2.3) is nonnegative by the following special case of the mixed discriminant inequality (2.1):

$$D(A, B, \ldots, B)^2 \geqslant D(A, A, B, \ldots, B) \, D(B, \ldots, B), \tag{2.4}$$

In other words, the Bochner method reduces Minkowski's second inequality (1.6) to its linear-algebraic counterpart (2.4). This interpretation of the Bochner method will form the starting point for the main results of this note.

Remark 2.8 Lemma 2.7 is a trivial reformulation of the proof of [22, Lemma 3.1], as is explained in [22, §6.3]. Moreover, it is observed there that in the special case that K is the Euclidean ball, (2.3) is precisely the classical (integrated) Bochner formula on S^{n-1}. One may therefore naturally view the above approach as an analogue of the Bochner method of differential geometry.

The identity (2.3) was recently rediscovered by Milman [15]. A new insight of [15] is that (2.3) may in fact be viewed as a true Bochner formula in the sense of differential geometry for any body K of class C_+^2, by introducing a special centro-affine connection on ∂K. This interpretation does not appear to extend, however, to more general situations: for example, neither the more general identity that was used in [22] to prove the Alexandrov-Fenchel inequality, nor the "Bochner identities" of this note, are true Bochner formulas in the strictly formal sense, but should rather be viewed as a loose analogues of such a formula. The merits of taking a more liberal view on the Bochner method are illustrated by its diverse applications not only in convexity, but also in algebra and combinatorics [5, 6, 9, 22].

Remark 2.9 The Bochner method should not be confused with a different method to prove Brunn-Minkowski inequalities that was developed by Reilly [18] and considerably refined by Kolesnikov and Milman in [13, 14]. The basis for Reilly's method is an integrated form of the classical Bochner formula on $\mathbb{R}^n$ (or on a manifold), combined with the solution of a certain Neumann problem. This method appears to be unrelated to the Bochner method for the operator $\mathscr{A}_K$.

3   Proof of Theorem 1.4

The main step in the proof of Theorem 1.4 is the following analogue of (2.2).

Theorem 3.1 *Let $K \in \mathcal{K}_s^n$ be a zonoid of class C_+^2 and $f \in C^2(S^{n-1})_{\text{even}}$. Then*

$$\langle \mathscr{A}_K f, \mathscr{A}_K f \rangle \geqslant \frac{n-2}{n-1} \langle f, \mathscr{A}_K f \rangle + \frac{1}{n-1} \langle f, f \rangle. \tag{3.1}$$

Before we proceed, let us complete the proof of Theorem 1.4.

Proof of Theorem 1.4 Let $K \in \mathcal{K}_s^n$ be a zonoid of class C_+^2. As $\mathscr{A}_K$ is essentially self-adjoint on $C^2(S^{n-1})$, (3.1) extends directly to any even function $f \in \text{Dom}(\mathscr{A}_K)$. Thus if f is any even eigenfunction of $\mathscr{A}_K$ with eigenvalue λ, Theorem 3.1 yields

$$\lambda^2 \geqslant \frac{n-2}{n-1} \lambda + \frac{1}{n-1},$$

i.e., $\lambda \geqslant 1$ or $\lambda \leqslant -\frac{1}{n-1}$. But recall that the largest eigenvalue of $\mathscr{A}_K$ is $\lambda_1 = 1$ and its eigenspace is spanned by h_K. Thus any even eigenfunction f of $\mathscr{A}_K$ that is orthogonal to h_K must have eigenvalue $\lambda \leqslant -\frac{1}{n-1}$. In particular, as any $f \in L^2(\mu_K)_{\text{even}}$ is in the linear span of the even eigenfunctions of $\mathscr{A}_K$, we obtain

$$\langle f, \mathscr{A}_K f \rangle \leqslant -\frac{1}{n-1} \langle f, f \rangle \quad \text{whenever } f \in C^2(S^{n-1})_{\text{even}}, \ \langle f, h_K \rangle = 0.$$

For any $L \in \mathcal{K}_s^n$ of class C_+^2, we may now choose $f = h_L - \frac{\langle h_L, h_K \rangle}{\langle h_K, h_K \rangle} h_K$ and use

$$\langle h_L, \mathscr{A}_K h_L \rangle = V(L, L, K, \ldots, K), \qquad \langle h_L, h_L \rangle = \frac{1}{n} \int \frac{h_L^2}{h_K} dS_{K,\ldots,K},$$

$$\langle h_L, \mathscr{A}_K h_K \rangle = \langle h_L, h_K \rangle = V(L, K, \ldots, K)$$

to conclude the validity of (1.9) when K, L are of class C_+^2.

To conclude the proof, it suffices to show that for any $K, L \in \mathcal{K}_s^n$ such that K is a zonoid, there exist $K_n, L_n \in \mathcal{K}_s^n$ of class C_+^2 such that K_n is a zonoid and $K_n \to K$, $L_n \to L$ in the Hausdorff metric; the validity of (1.9) then follows by the continuity of mixed volumes and area measures. Both statements are classical; an approximation of L by C_+^2 bodies is given in [21, §3.4], while the approximation of K may be performed, for example, by choosing $h_{K_n} = \int h_{E_{n,u}} \eta(du)$ where η is the generating measure of K and $E_{n,u}$ are ellipsoids such that $E_{n,u} \to [-u, u]$. □

The remainder of this section is devoted to the proof of Theorem 3.1. In essence, the inequality (3.1) will follow from a "Bochner identity" in the spirit of (2.3). However, rather than reducing the validity of (1.9) to a linear algebraic analogue as was done in Sect. 2.3, the Bochner method will be used here to reduce (1.9) in dimension n to its validity in dimension $n - 1$. The conclusion then follows by induction. As will be explained below, the structure of the induction also provides an explanation for the appearance of the mysterious value $-\frac{1}{n-1}$ in Lemma 1.8.

3.1   The Induction Step

We begin with the following observation.

Lemma 3.2 *Let* $n \geqslant 3$, $K \in \mathcal{K}_s^n$ *be a zonoid, and* $f = h_M - h_{M'}$ *for* $M, M' \in \mathcal{K}_s^n$. *Assume that Theorem 1.4 has been proved in dimension* $n - 1$. *Then*

$$\frac{V([-u, u], f, K, \ldots, K)^2}{V([-u, u], K, \ldots, K)} \geqslant$$

$$\frac{n-2}{n-1} V([-u, u], f, f, K, \ldots, K) + \frac{1}{n(n-1)} \int \frac{f^2}{h_K} \, dS_{[-u,u], K, \ldots, K}$$

for every $u \in S^{n-1}$.

Proof Assume first that K is a zonoid of class C_+^2 and that $f \in C^2(S^{n-1})_{\text{even}}$. Then by Lemma 2.4, there exists a convex body $L \in \mathcal{K}_s^n$ of class C_+^2 and $a > 0$ such that $f = a(h_L - h_K)$. By expanding the squares, the inequality in the statement is readily seen to be equivalent to the inequality

$$\frac{V([-u, u], L, K, \ldots, K)^2}{V([-u, u], K, \ldots, K)} \geqslant$$

$$\frac{n-2}{n-1} V([-u, u], L, L, K, \ldots, K) + \frac{1}{n(n-1)} \int \frac{h_L^2}{h_K} \, dS_{[-u,u], K, \ldots, K}.$$

By Lemma 2.1, this is further equivalent to

$$\frac{V(\mathbf{P}_{u^\perp} L, \mathbf{P}_{u^\perp} K, \ldots, \mathbf{P}_{u^\perp} K)^2}{V(\mathbf{P}_{u^\perp} K, \ldots, \mathbf{P}_{u^\perp} K)} \geqslant$$

$$\frac{n-2}{n-1} V(\mathbf{P}_{u^\perp} L, \mathbf{P}_{u^\perp} L, \mathbf{P}_{u^\perp} K, \ldots, \mathbf{P}_{u^\perp} K) + \frac{1}{(n-1)^2} \int \frac{h_{\mathbf{P}_{u^\perp} L}^2}{h_{\mathbf{P}_{u^\perp} K}} \, dS_{\mathbf{P}_{u^\perp} K, \ldots, \mathbf{P}_{u^\perp} K},$$

where we used that $h_{\mathbf{P}_{u^\perp} L}(x) = h_L(x)$ for $x \in u^\perp$. But as $\mathbf{P}_{u^\perp} K$ is a zonoid, the latter inequality follows immediately from Theorem 1.4 in dimension $n - 1$. It remains to extend the conclusion to general K and $f = h_M - h_{M'}$ by approximating K, M, M' by C_+^2 bodies as in the proof of Theorem 1.4.   □

We are now ready to perform the induction step in the proof of Theorem 3.1.

Proposition 3.3 *Let* $n \geqslant 3$, *and assume that Theorem 1.4 has been proved in dimension* $n - 1$. *Then the conclusion of Theorem 3.1 holds in dimension* n.

Proof Let $n \geqslant 3$, $f \in C^2(S^{n-1})_{\text{even}}$, and $K \in \mathcal{K}_s^n$ be a zonoid of class C_+^2 with generating measure η. We may write

$$\langle \mathscr{A}_K f, \mathscr{A}_K f \rangle = \frac{1}{n} \int \int h_{[-u,u]} \frac{D(D^2 f, D^2 h_K, \ldots, D^2 h_K)^2}{D(D^2 h_K, \ldots, D^2 h_K)} \, d\omega \, \eta(du)$$

by the definitions of $\mathscr{A}_K$, μ_K and as $h_K = \int h_{[-u,u]} \, \eta(du)$. Now note that

$$\frac{1}{n} \int h_{[-u,u]} \frac{D(D^2 f, D^2 h_K, \ldots, D^2 h_K)^2}{D(D^2 h_K, \ldots, D^2 h_K)} \, d\omega \geqslant$$

$$\frac{\left(\frac{1}{n} \int h_{[-u,u]} D(D^2 f, D^2 h_K, \ldots, D^2 h_K) \, d\omega\right)^2}{\frac{1}{n} \int h_{[-u,u]} D(D^2 h_K, \ldots, D^2 h_K) \, d\omega} = \frac{V([-u, u], f, K, \ldots, K)^2}{V([-u, u], K, \ldots, K)}$$

for any u by Cauchy-Schwarz and Lemma 2.6. We therefore obtain

$$\langle \mathscr{A}_K f, \mathscr{A}_K f \rangle \geqslant \int \frac{V([-u, u], f, K, \ldots, K)^2}{V([-u, u], K, \ldots, K)} \, \eta(du)$$

$$\geqslant \int \left(\frac{n-2}{n-1} V([-u, u], f, f, K, \ldots, K)\right.$$

$$\left. + \frac{1}{n(n-1)} \int \frac{f^2}{h_K} \, dS_{[-u,u], K, \ldots, K}\right) \eta(du)$$

$$= \frac{n-2}{n-1} \langle f, \mathscr{A}_K f \rangle + \frac{1}{n-1} \langle f, f \rangle$$

using Lemma 3.2 and $h_K = \int h_{[-u,u]} \, \eta(du)$. $\qquad\square$

Remark 3.4 While we find it cleaner to formulate the proof of Proposition 3.3 in terms of inequalities, one may in principle interpret this proof as arising from a Bochner *identity* in the spirit of (2.3): indeed, combining the proofs of Lemma 3.2 and Proposition 3.3 yields for $f = a(h_L - h_K)$

$$\langle \mathscr{A}_K f, \mathscr{A}_K f \rangle - \frac{n-2}{n-1} \langle f, \mathscr{A}_K f \rangle - \frac{1}{n-1} \langle f, f \rangle =$$

$$\int \int \frac{h_{[-u,u]}}{h_K} \left(\mathscr{A}_K f - \frac{V([-u, u], f, K, \ldots, K)}{V([-u, u], K, \ldots, K)} h_K\right)^2 d\mu_K \, \eta(du) +$$

$$\frac{2a^2}{n} \int \left(\frac{V(\mathbf{P}_{u^\perp} L, \mathbf{P}_{u^\perp} K, \ldots, \mathbf{P}_{u^\perp} K)^2}{V(\mathbf{P}_{u^\perp} K, \ldots, \mathbf{P}_{u^\perp} K)}\right.$$

$$- \frac{n-2}{n-1} V(\mathbf{P}_{u^\perp} L, \mathbf{P}_{u^\perp} L, \mathbf{P}_{u^\perp} K, \ldots, \mathbf{P}_{u^\perp} K)$$

$$- \frac{1}{(n-1)^2} \int \frac{h^2_{\mathbf{P}_{u^\perp} L}}{h_{\mathbf{P}_{u^\perp} K}} \, dS_{\mathbf{P}_{u^\perp} K, \ldots, \mathbf{P}_{u^\perp} K} \Bigg) \eta(du),$$

where the two terms on the right-hand side are the deficits of the two inequalities used in the proof (the Cauchy-Schwarz inequality and (1.9) in dimension $n-1$, respectively). While it would be difficult to recognize this identity as a Bochner formula in the sense of differential geometry, it plays precisely the same role in the present proof as the Bochner identity (2.3) in Sect. 2.3.

Let us further note that an even eigenfunction $\mathscr{A}_K f = \lambda f$ yields equality in (3.1) if and only if $\lambda = 1$ or $\lambda = -\frac{1}{n-1}$. When this is the case, the right-hand side of the above Bochner identity must vanish. It then follows from the first term on the right that f must be proportional to h_K, so that $\lambda = 1$. In other words, when the zonoid K is of class C_+^2, any even eigenfunction that is orthogonal to h_K has eigenvalue strictly less than $-\frac{1}{n-1}$, and thus no nontrivial equality cases can arise in (1.9). However, nontrivial equality cases can arise when K is nonsmooth, which will be analyzed in Sect. 4 by a variation on the above argument.

Remark 3.5 At first sight, the formulation of the spectral condition of Lemma 1.8 is rather mysterious: what is the significance of the special value $-\frac{1}{n-1}$? The present proof provides one explanation for the appearance of this value: the constants in (3.1) in dimension n are precisely the same as those that appear in (1.9) in dimension $n-1$, so that the preservation of the sharp threshold $\lambda \leqslant -\frac{1}{n-1}$ by induction on the dimension n is explained by the quadratic relation (3.1).

3.2 The Induction Base

By Proposition 3.3 and induction on the dimension, the proof of Theorem 3.1 will be complete in any dimension $n \geqslant 3$ once we establish its validity in dimension $n = 2$. The latter is already known, however, by the results of [4] and Theorem 1.2. On the other hand, as we will presently explain, the $n = 2$ case may also be established directly by exactly the same method as was used in the proof of Proposition 3.3. This shows, in particular, that the Bochner method provides a unified explanation for the validity of Theorem 1.4 in every dimension.

Lemma 3.6 *The conclusion of Theorem 3.1 holds in dimension $n = 2$.*

Proof Let $f \in C^2(S^1)_{\text{even}}$, and let $K \in \mathcal{K}_s^2$ be a zonoid of class C_+^2. Applying the Cauchy-Schwarz inequality as in the proof of Proposition 3.3 yields

$$\langle \mathscr{A}_K f, \mathscr{A}_K f \rangle \geqslant \frac{1}{4} \int \frac{V([-u^\dagger, u^\dagger], f)^2}{V([-u^\dagger, u^\dagger], K)} \, S_K(du),$$

where we used Lemma 2.3 to compute the generating measure of a planar zonoid. However, as was observed in the proof of Lemma 2.3, we have

$$V([-u^\dagger, u^\dagger], K) = 2h_K(u), \qquad V([-u^\dagger, u^\dagger], f) = 2f(u)$$

(the latter follows as $f = a(h_L - h_K)$ for some $L \in \mathcal{K}_s^2$ by Lemma 2.4). Thus

$$\langle \mathscr{A}_K f, \mathscr{A}_K f \rangle \geqslant \frac{1}{2} \int \frac{f^2}{h_K} dS_K = \langle f, f \rangle,$$

concluding the proof. □

4 Proof of Theorem 1.5

As was already noted in Remark 3.4, we may expect in principle that one may deduce the equality cases of (1.9) by a careful analysis of the Bochner method. The immediate problem with this approach is that the most basic object that appears in the Bochner method—the Hilbert operator $\mathscr{A}_K$—is not even well defined unless K is of class C_+^2, and no nontrivial equality cases can arise in that setting. We will nonetheless pursue this strategy in the present section to settle the equality cases. This is possible, in essence, because it suffices for the purposes of characterizing equality to replace $\mathscr{A}_K f$ by $-\frac{1}{n-1} f$ in the Bochner identity, in which case the relevant formulas make sense also in nonsmooth situations.

We begin by making the latter idea precise in Sect. 4.1. We subsequently show in Sect. 4.2 what information on the equality cases may be extracted from the Bochner method. The proof of Theorem 1.5 will be completed in Sect. 4.3.

4.1 The Equality Condition

Before we proceed to the analysis of the equality cases, we state a slight generalization of (1.9) that will be needed in the sequel.

Lemma 4.1 Let $K \in \mathcal{K}_s^n$ be a zonoid. Then

$$\frac{V(f, K, \ldots, K)^2}{\text{Vol}(K)} \geqslant \frac{n-1}{n} V(f, f, K, \ldots, K) + \frac{1}{n^2} \int \frac{f^2}{h_K} dS_{K,\ldots,K}$$

holds whenever $f = h_L - h_M$ for some $L, M \in \mathcal{K}_s^n$.

Proof This follows from Theorem 1.4 as in the proof of Lemma 3.2. □

We can now obtain a basic reformulation of the equality condition in (1.9). The method is due to Alexandrov [1, pp. 80–81].

Lemma 4.2 *For any* $L \in \mathcal{K}_s^n$ *and any zonoid* $K \in \mathcal{K}_s^n$, *the following are equivalent:*

1. *Equality holds in* (1.9), *that is,*

$$\frac{V(L, K, \ldots, K)^2}{\mathrm{Vol}(K)} = \frac{n-1}{n} V(L, L, K, \ldots, K) + \frac{1}{n^2} \int \frac{h_L^2}{h_K} \, dS_{K,\ldots,K}.$$

2. *There exists* $a > 0$ *so that* $f = h_L - a h_K$ *satisfies*

$$h_K \, dS_{f,K,\ldots,K} = -\frac{1}{n-1} f \, dS_{K,\ldots,K}.$$

Proof We first prove that $2 \Rightarrow 1$. Integrating condition 2 yields $V(f, K, \ldots, K) = 0$, while multiplying condition 2 by $\frac{f}{h_K}$ and integrating yields

$$\frac{n-1}{n} V(f, f, K, \ldots, K) = -\frac{1}{n^2} \int \frac{f^2}{h_K} \, dS_{K,\ldots,K}.$$

We therefore obtain

$$\frac{V(f, K, \ldots, K)^2}{\mathrm{Vol}(K)} = \frac{n-1}{n} V(f, f, K, \ldots, K) + \frac{1}{n^2} \int \frac{f^2}{h_K} \, dS_{K,\ldots,K},$$

and condition 1 follows using $f = h_L - a h_K$ and expanding the squares.

We now prove the converse implication $1 \Rightarrow 2$. Let $g \in C^2(S^{n-1})_{\mathrm{even}}$ and define

$$\beta(t) := \frac{V(g_t, K, \ldots, K)^2}{\mathrm{Vol}(K)} - \frac{n-1}{n} V(g_t, g_t, K, \ldots, K) - \frac{1}{n^2} \int \frac{g_t^2}{h_K} \, dS_{K,\ldots,K}$$

where $g_t := h_L + tg$. Condition 1 implies $\beta(0) = 0$, while Lemma 4.1 implies $\beta(t) \geqslant 0$ for all t. Thus β is minimized at zero, so that $\beta'(0) = 0$ yields

$$\int g \, dS_{f,K,\ldots,K} = -\frac{1}{n-1} \int g \frac{f}{h_K} \, dS_{K,\ldots,K}$$

with $f = h_L - \frac{V(L,K,\ldots,K)}{\mathrm{Vol}(K)} h_K$. As $g \in C^2(S^{n-1})_{\mathrm{even}}$ is arbitrary and as $dS_{f,K,\ldots,K}$ and $\frac{f}{h_K} dS_{K,\ldots,K}$ are even measures, condition 2 follows. $\qquad\square$

It follows from the definition of the Hilbert operator $\mathscr{A}_K$ that when K, L are of class C_+^2, Lemma 4.2 states precisely that equality holds in (1.9) if and only if $\mathscr{A}_K f = -\frac{1}{n-1} f$ for $f = h_L - a h_K$. The point of Lemma 4.2 is that the same characterization can be formulated for nonsmooth bodies in the sense of measures. The latter will suffice to apply the Bochner method to study the equality cases.

4.2 The Bochner Method Revisited

Using Lemma 4.2, we can now essentially repeat the proof of Proposition 3.3 in the present setting to extract a necessary condition for equality in (1.9) from the Bochner method.

Lemma 4.3 *Let $K \in \mathcal{K}_s^n$ be a zonoid with generating measure η, and let $L \in \mathcal{K}_s^n$ be such that equality holds in (1.9). Then for every $u \in \operatorname{supp} \eta$, there exists $c(u) \geqslant 0$ such that $h_L(x) = c(u)h_K(x)$ for all $x \in \operatorname{supp} S_{K,...,K}$ with $|\langle u, x \rangle| > 0$.*

Proof Let $a > 0$ be such that $f = h_L - ah_K$ satisfies the second condition of Lemma 4.2, that is, $f \, dS_{K,...,K} = -(n-1) \, h_K \, dS_{f,K,...,K}$. Then we have

$$
\int \frac{f^2}{h_K} \, dS_{K,...,K} = \int \int \frac{f^2}{h_K^2} h_{[-u,u]} \, dS_{K,...,K} \, \eta(du)
$$

$$
\geqslant \int \frac{\left(\int \frac{f}{h_K} h_{[-u,u]} \, dS_{K,...,K} \right)^2}{\int h_{[-u,u]} \, dS_{K,...,K}} \, \eta(du)
$$

$$
= n(n-1)^2 \int \frac{V([-u,u], f, K, \ldots, K)^2}{V([-u,u], K, \ldots, K)} \, \eta(du)
$$

$$
\geqslant n(n-1)(n-2)V(f, f, K, \ldots, K) + (n-1) \int \frac{f^2}{h_K} \, dS_{K,...,K}
$$

$$
= \int \frac{f^2}{h_K} \, dS_{K,...,K}.
$$

Here we used $h_K = \int h_{[-u,u]} \, \eta(du)$ in the first line; the Cauchy-Schwarz inequality in the second line; the condition of Lemma 4.2 in the third line; Lemma 3.2 in the fourth line (or by the proof of Lemma 3.6 for $n = 2$); and the fifth line follows as

$$
V(f, f, K, \ldots, K) = \frac{1}{n} \int f \, dS_{f,K,...,K} = -\frac{1}{n(n-1)} \int \frac{f^2}{h_K} \, dS_{K,...,K}
$$

by the condition of Lemma 4.2.

Consequently, both inequalities used above must hold with equality. In particular, we have equality in the Cauchy-Schwarz inequality

$$
\int \frac{f^2}{h_K^2} h_{[-u,u]} \, dS_{K,...,K} = \frac{\left(\int \frac{f}{h_K} h_{[-u,u]} \, dS_{K,...,K} \right)^2}{\int h_{[-u,u]} \, dS_{K,...,K}}
$$

for every $u \in \operatorname{supp} \eta$. By the equality condition of the Cauchy-Schwarz inequality, this implies that for every $u \in \operatorname{supp} \eta$, there is a constant $c'(u)$ so that $f(x) = c'(u)h_K(x)$ for every $x \in \operatorname{supp} S_{K,...,K}$ with $h_{[-u,u]}(x) = |\langle u, x\rangle| > 0$. But as $f = h_L - ah_K$, the conclusion follows with $c(u) = a + c'(u)$. (Note that it must be the case that $c(u) \geqslant 0$ as h_K, h_L are positive functions.)   □

Remark 4.4 The proof of Lemma 4.3 actually provides more information than is expressed in its statement: not only do we get equality in Cauchy-Schwarz, but we also get equality in the application of Lemma 3.2. In particular, this implies that if equality holds in (1.9) for given K, L in dimension n, then the projections $\mathbf{P}_{u^\perp}K, \mathbf{P}_{u^\perp}L$ must also yield equality in (1.9) in dimension $n - 1$ for every $u \in \operatorname{supp} \eta$. It is a curious feature of the present problem that the latter information will not be needed to characterize the equality cases: the equality condition in Cauchy-Schwarz will already suffice to fully characterize the equality cases of (1.9).

4.3   Characterization of Equality

We are now ready to proceed to the proof of Theorem 1.5. The main difficulty is to show that the stated conditions are necessary for equality, which will be deduced from Lemma 4.3.

In the proof of the following result, we will encounter graphs that may have an uncountable number of vertices and edges. The standard properties of graphs that will be used in the proof—chiefly that a graph can be partitioned into its connected components—are valid at this level of generality; cf. [16, Chapter 2].

Proposition 4.5 *Let $K \in \mathcal{K}_s^n$ be a zonoid, and let $L \in \mathcal{K}_s^n$ be such that equality holds in* (1.9). *Then there exist* $1 \leqslant m \leqslant n$, $a_1, \ldots, a_m \geqslant 0$, *and zonoids* $C_1, \ldots, C_m$ *with* $\dim(C_1) + \cdots + \dim(C_m) = n$ *so that* $K = C_1 + \cdots + C_m$ *and*

$$h_L(x) = h_{a_1 C_1 + \cdots + a_m C_m}(x) \text{ for all } x \in \operatorname{supp} S_{K,...,K}.$$

Proof We define a graph (V, E) as follows:

- The vertices are $V = \operatorname{supp} \eta$, where η denotes the generating measure of K.
- There is an edge $\{u, v\} \in E$ between $u, v \in V$ if and only if there exists $x \in \operatorname{supp} S_{K,...,K}$ such that $|\langle u, x\rangle| > 0$ and $|\langle v, x\rangle| > 0$.

Denote by $V = \bigsqcup_{i \in I} V_i$ the partition of V into its connected components V_i.

For any edge $\{u, v\} \in E$, Lemma 4.3 implies that

$$c(u)h_K(x) = h_L(x) = c(v)h_K(x)$$

for some $x \in \operatorname{supp} S_{K,\ldots,K}$. As $h_K(x) > 0$, it follows that $c(u) = c(v)$. In particular, the value of $c(u)$ must be constant on each connected component. In the sequel, we will denote this value as $c(u) = a_i$ for $u \in V_i$.

Next, we make a key observation. $\qquad\square$

Claim For every $x \in \operatorname{supp} S_{K,\ldots,K}$, there exists $i \in I$ so that $x \perp V_j$ for all $j \neq i$.

Proof We can assume that $x \in \operatorname{supp} S_{K,\ldots,K}$ satisfies $|\langle u, x \rangle| > 0$ for some $i \in I$, $u \in V_i$, as otherwise the conclusion is trivial. But then we must have $|\langle v, x \rangle| = 0$ for all $j \neq i$, $v \in V_j$, as distinct connected components have no edge between them. $\qquad\square$

We also need the following.

Claim For every $u \in S^{n-1}$, there exists $x \in \operatorname{supp} S_{K,\ldots,K}$ so that $|\langle u, x \rangle| > 0$.

Proof If the conclusion were false, there would exist some $u \in S^{n-1}$ such that $0 = \int |\langle u, x \rangle| S_{K,\ldots,K}(dx) = 2\operatorname{Vol}(\mathbf{P}_{u^\perp} K)$ by Lemma 2.1. The latter is impossible as $K \in \mathcal{K}_s^n$ is assumed to have nonempty interior. $\qquad\square$

The above two claims imply that distinct V_i must lie in linearly independent subspaces $L_i = \operatorname{span} V_i$. Indeed, if this is not so, then there exists $z \in S^{n-1}$ so that

$$z = t_1 u_1 + \cdots + t_k u_k = s_1 v_1 + \cdots s_l v_l$$

for some $k, l \geq 1$, $i \in I$, $u_1, \ldots, u_k \in V_i$, $v_1, \ldots, v_l \in \bigcup_{j \neq i} V_j$, $t_1, \ldots, t_k, s_1, \ldots, s_l \neq 0$. By the second claim there exists $x \in \operatorname{supp} S_{K,\ldots,K}$ so that $|\langle z, x \rangle| > 0$. But by the first claim we must then have $x \perp v_1, \ldots, u_l$, which entails a contradiction. It follows, in particular, that there can be at most n connected components, so we can write $I = \{1, \ldots, m\}$ for some $1 \leq m \leq n$.

We now define zonoids $C_1, \ldots, C_m$ as

$$h_{C_i} = \int_{L_i} h_{[-u,u]} \, \eta(du).$$

As $L_1, \ldots, L_m$ are linearly independent and $\operatorname{supp} \eta = V \subseteq S^{n-1} \cap (L_1 \cup \cdots \cup L_m)$

$$h_{C_1} + \cdots + h_{C_m} = \int h_{[-u,u]} \, \eta(du) = h_K,$$

that is, $K = C_1 + \cdots + C_m$. Moreover, as $L_1, \ldots, L_m$ are linearly independent and K has nonempty interior, we must have $\dim(C_1) + \cdots + \dim(C_m) = n$.

Finally, let $x \in \operatorname{supp} S_{K,\ldots,K}$. By the first claim above, there exists $1 \leq i \leq m$ so that $h_{C_j}(x) = 0$ for all $j \neq i$. As this implies that $h_{C_i}(x) = h_K(x) > 0$, there must exist $u \in V_i$ so that $|\langle u, x \rangle| > 0$. Recalling that $c(u) = a_i$ for $u \in V_i$, we obtain

$$h_L(x) = a_i h_K(x) = a_i h_{C_i}(x) = a_1 h_{C_1}(x) + \cdots + a_m h_{C_m}(x)$$

by Lemma 4.3. As this holds for any $x \in \operatorname{supp} S_{K,\ldots,K}$, the proof is complete.

Before we complete the proof, we must verify the basic case of equality.

Lemma 4.6 *Suppose that $K = C_1 + \cdots + C_m$ for some convex bodies $C_1, \ldots, C_m$ such that $\dim(C_1) + \cdots + \dim(C_m) = n$, and that $L = a_1 C_1 + \cdots + a_m C_m$ for some $a_1, \ldots, a_m \geqslant 0$. Then equality holds in (1.9).*

Proof By Schneider [21, Theorem 5.1.8], the condition $\dim(C_1) + \cdots + \dim(C_m) = n$ implies that we have $V(C_{i_1}, \ldots, C_{i_n}) > 0$ if and only if each index $1 \leqslant j \leqslant m$ appears exactly $\dim(C_j)$ times among $(i_1, \ldots, i_n)$. Thus for any $b_1, \ldots, b_m \geqslant 0$

$$\mathrm{Vol}(b_1 C_1 + \cdots + b_m C_m) = \sum_{i_1,\ldots,i_n=1}^{m} b_{i_1} \cdots b_{i_n} V(C_{i_1}, \ldots, C_{i_n})$$

$$= \Gamma\, b_1^{\dim(C_1)} \cdots b_m^{\dim(C_m)}$$

for some constant Γ that depends only on $C_1, \ldots, C_m$. Therefore

$$0 = -\frac{\mathrm{Vol}(K)}{n^2} \frac{d^2}{dt^2} \log \mathrm{Vol}(e^{ta_1} C_1 + \cdots + e^{ta_m} C_m)\Big|_{t=0}$$

$$= \frac{V(L, K, \ldots, K)^2}{\mathrm{Vol}(K)} - \frac{n-1}{n} V(L, L, K, \ldots, K)$$

$$- \frac{1}{n^2} \int h_{a_1^2 C_1 + \cdots + a_m^2 C_m}\, dS_{K,\ldots,K}.$$

Now note that if $\int h_{C_i}\, dS_{C_{i_1},\ldots,C_{i_{n-1}}} > 0$, then using [21, Theorem 5.1.8] as above shows that $\int h_{C_j}\, dS_{C_{i_1},\ldots,C_{i_{n-1}}} = 0$ for all $j \neq i$. In particular, as we have $S_{K,\ldots,K} = \sum_{i_1,\ldots,i_{n-1}} S_{C_{i_1},\ldots,C_{i_{n-1}}}$, this implies that for every $x \in \mathrm{supp}\, S_{K,\ldots,K}$, there exists an index i so that $h_{C_j}(x) = 0$ for all $j \neq i$. It follows readily that

$$h_{a_1^2 C_1 + \cdots + a_m^2 C_m}(x) = \frac{h_L(x)^2}{h_K(x)} \quad \text{for all } x \in \mathrm{supp}\, S_{K,\ldots,K},$$

and the proof is complete.   □

We can now complete the proof of the necessity part of Theorem 1.5. In the proof, we use some nontrivial facts that do not appear elsewhere in this note.

Proof of Theorem 1.5 We first prove sufficiency. Suppose that $K = C_1 + \ldots + C_m$ for bodies $C_1, \ldots, C_m$ with $\dim(C_1) + \cdots + \dim(C_m) = n$, and that L and $L' := a_1 C_1 + \cdots + a_m C_m$ have the same supporting hyperplanes in all 1-extreme normal directions of K. The latter implies by Schneider [21, Theorem 4.5.3 and Lemma 7.6.15] that

$$h_L(x) = h_{L'}(x) \quad \text{for all } x \in \mathrm{supp}\, S_{M,K,\ldots,K} \tag{4.1}$$

for any convex body M. In particular, every term in (1.9) is unchanged if we replace L by L'. Thus equality holds in (1.9) by Lemma 4.6.

We now prove necessity. Suppose equality holds in (1.9). Then Proposition 4.5 provides $C_1, \ldots, C_m$ that satisfy all the required properties by construction except the last one: that is, what remains to be shown is that L and $L' := a_1 C_1 + \cdots + a_m C_m$ have the same supporting hyperplanes in all 1-extreme normal directions of K.

Let us write $f := h_L - h_{L'}$. By Proposition 4.5, we have $f = 0$ on supp $S_{K,\ldots,K}$. Moreover, as we clearly have $L' + C = (\max_k a_k) K$ for a convex body C, it follows that supp $S_{L',K,\ldots,K} \subseteq$ supp $S_{K,\ldots,K}$ and thus $f = 0$ on supp $S_{L',K,\ldots,K}$ as well. Substituting $h_L = h_{L'} + f$ into (1.9) and using that both L and L' yield equality in (1.9) (by assumption and by Lemma 4.6, respectively), we can readily compute

$$\mathsf{V}(f, K, \ldots, K) = 0, \qquad \mathsf{V}(f, f, K, \ldots, K) = 0.$$

Using that $f = h_L - h_{L'}$, this implies that we have equality

$$\mathsf{V}(L, L', K, \ldots, K)^2 = \mathsf{V}(L, L, K, \ldots, K)\,\mathsf{V}(L', L', K, \ldots, K)$$

in Minkowski's quadratic inequality. By the main result of [23], it follows that L and $aL' + v$ have the same supporting hyperplanes in all 1-extreme normal directions of K for some $a \geqslant 0$, $v \in \mathbb{R}^n$. But as L, L' are symmetric we must have $v = 0$, while $\mathsf{V}(f, K, \ldots, K) = 0$ and (4.1) imply $a = 1$. This concludes the proof. $\qquad\square$

5 Implications

As we recalled in Theorem 1.2, the validity of the local logarithmic Brunn-Minkowski inequality (1.9) for *all* $K \in \mathcal{K}_s^n$ is equivalent to the validity of the logarithmic Brunn-Minkowski and the logarithmic Minkowski inequalities. The proof of these facts is based on several recent deep results on uniqueness in the L^q-Minkowski problem for $q < 1$. While this equivalence does not hold for fixed $K \in \mathcal{K}_s^n$, it is explained in [15, §2.4] that the theory behind Theorem 1.2 still yields nontrivial implications when (1.9) is known to hold in a sufficiently rich sub-class of $\mathcal{K}_s^n$. The aim of the final section of this note is to investigate what conclusions may be drawn by combining these results with Theorems 1.4–1.5.

We begin with the proof of Corollary 1.6.

Proof of Corollary 1.6 By a routine approximation argument as in the proof of Theorem 1.4, it suffices to prove the validity of (1.8) for $K \in \mathcal{K}_s^n$ that are zonoids of class C_+^∞. Let us fix such a zonoid, and let $\mathcal{F} = \{(1 - t)K + tB : t \in [0, 1]\}$ where B is the Euclidean unit ball. Then every $K' \in \mathcal{F}$ is a zonoid of class C_+^∞. Moreover, it was observed in Remark 3.4 that every even eigenfunction of $\mathscr{A}_{K'}$ that is orthogonal to $h_{K'}$ has eigenvalue $\lambda < -\frac{1}{n-1}$. By the continuity of the eigenvalues of the Hilbert operator (cf. [14, Theorem 5.3]), there exists $\varepsilon > 0$ so that for every

$K' \in \mathcal{F}$, every even eigenfunction of $\mathscr{A}_{K'}$ that is orthogonal to $h_{K'}$ has eigenvalue $\lambda \leqslant -\frac{1}{n-1} - \varepsilon$. Thus there exists $p < 0$ so that condition (4) of [15, Theorem 2.1] holds for all $K' \in \mathcal{F}$. The conclusion now follows from the implication (4)$\Rightarrow$(3b) of [15, Theorem 2.1] (as the inequality in (3b) with $q = 0$ is precisely (1.8)).   □

The logarithmic Brunn-Minkowski conjecture is intimately connected to the uniqueness problem for cone volume measures; this was in fact the original motivation for the formulation of the conjecture [4]. Let us recall the definition.

Definition 5.1 The *cone volume measure* V_K of a convex body K is defined as

$$dV_K := \frac{1}{n} h_K \, dS_{K,\ldots,K}.$$

The basic question that arises here is whether the cone volume measure uniquely characterizes the convex body K. While this is not always the case, the question is closely connected to the equality cases of the logarithmic Minkowski inequality (1.8) in the case that $K, L \in \mathcal{K}_s^n$ are symmetric. For example, if $K, L \in \mathcal{K}_s^n$ satisfy $V_K = V_L$ (and thus *a fortiori* $\mathrm{Vol}(K) = \mathrm{Vol}(L)$ as $\mathrm{Vol}(K) = \int dV_K$), the validity of the logarithmic Brunn-Minkowski conjecture would yield

$$0 \leqslant \int h_K \log\left(\frac{h_L}{h_K}\right) dS_{K,\ldots,K} = \int h_L \log\left(\frac{h_L}{h_K}\right) dS_{L,\ldots,L} \leqslant 0$$

using (1.8) in the first inequality, $V_K = V_L$ in the equality, and (1.8) with the roles of K, L reversed in the second inequality. This would imply that V_K is uniquely determined by K whenever (1.8) does not admit nontrivial equality cases.

Unfortunately, even though we obtained a complete characterization of the equality cases of (1.9) when K is a zonoid, this information is lost in Corollary 1.6. The reason is that the proof of Corollary 1.6 required approximation of K by smooth bodies, which destroys the nontrivial equality cases. Nonetheless, for sufficiently smooth zonoids, uniqueness of cone volume measures follows by [15, Theorem 2.1]. Note that while the smoothness assumption on K is restrictive, the following statement requires neither that L is smooth nor that L is a zonoid.

Corollary 5.2 Let $K \in \mathcal{K}_s^n$ be a zonoid of class C_+^3. Then for any $L \in \mathcal{K}_s^n$, we have $V_K = V_L$ if and only if $K = L$.

Proof This follows from the implication (4)$\Rightarrow$(1) of [15, Theorem 2.1] by precisely the same argument as in the proof of Corollary 1.6.   □

Acknowledgments The results of this note were developed as a pedagogical example for the author's lectures at the Spring School on Convex Geometry and Random Matrices in High Dimension, Paris, June 2021. The author is grateful to M. Fradelizi, N. Gozlan, and O. Guédon for the invitation to present these lectures. The author also thanks K. Böröczky, E. Milman, and G. Paouris for helpful discussions, and E. Milman and the anonymous referee for helpful suggestions on the presentation. This work was supported in part by NSF grants DMS-1811735 and DMS-2054565, and by the Simons Collaboration on Algorithms & Geometry.

References

1. A.D. Alexandrov, *Selected Works. Part I* (Gordon and Breach Publishers, Amsterdam, 1996)
2. E.D. Bolker, A class of convex bodies. Trans. Am. Math. Soc. **145**, 323–345 (1969)
3. K.J. Böröczky, P. Kalantzopoulos, Log-Brunn-Minkowski inequality under symmetry. Trans. Amer. Math. Soc. **375**, 5987–6013 (2022)
4. K.J. Böröczky, E. Lutwak, D. Yang, G. Zhang, The log-Brunn-Minkowski inequality. Adv. Math. **231**(3–4), 1974–1997 (2012)
5. P. Brändén, J. Leake, Lorentzian polynomials on cones and the Heron-Rota-Welsh conjecture. Preprint (2021). arxiv:2110.00487
6. S.H. Chan, I. Pak, Log-concave poset inequalities. Preprint (2021). arxiv:2110.10740
7. S. Chen, Y. Huang, Q.-R. Li, J. Liu, The L^p-Brunn-Minkowski inequality for $p < 1$. Adv. Math. **368**, 107166 (2020)
8. A. Colesanti, G.V. Livshyts, A. Marsiglietti, On the stability of Brunn-Minkowski type inequalities. J. Funct. Anal. **273**(3), 1120–1139 (2017)
9. D. Cordero-Erausquin, B. Klartag, Q. Merigot, F. Santambrogio, One more proof of the Alexandrov-Fenchel inequality. C. R. Math. Acad. Sci. Paris **357**(8), 676–680 (2019)
10. R.J. Gardner, The Brunn-Minkowski inequality. Bull. Am. Math. Soc. **39**(3), 355–405 (2002)
11. Y. Gordon, M. Meyer, S. Reisner, Zonoids with minimal volume-product—a new proof. Proc. Am. Math. Soc. **104**(1), 273–276 (1988)
12. D. Hilbert, *Grundzüge einer allgemeinen Theorie der linearen Integralgleichungen.* (B. G. Teubner, 1912)
13. A.V. Kolesnikov, E. Milman, Poincaré and Brunn-Minkowski inequalities on the boundary of weighted Riemannian manifolds. Am. J. Math. **140**(5), 1147–1185 (2018)
14. A.V. Kolesnikov, E. Milman, Local L^p-Brunn-Minkowski inequalities for $p < 1$. Mem. Am. Math. Soc. **277**(1360), v+78 (2022)
15. E. Milman, Centro-affine differential geometry and the log-Minkowski problem. J. Eur. Math. Soc. (in press).
16. O. Ore, *Theory of Graphs.* American Mathematical Society Colloquium Publications, vol. XXXVIII (American Mathematical Society, Providence, 1962)
17. E. Putterman, Equivalence of the local and global versions of the L^p-Brunn-Minkowski inequality. J. Funct. Anal. **280**(9), Paper No. 108956, 20 (2021)
18. R.C. Reilly, Geometric applications of the solvability of Neumann problems on a Riemannian manifold. Arch. Ration. Mech. Anal. **75**(1), 23–29 (1980)
19. L. Rotem, A letter: the log-Brunn-Minkowski inequality for complex bodies. Preprint (2014). arxiv:1412.5321
20. C. Saroglou, Remarks on the conjectured log-Brunn-Minkowski inequality. Geom. Dedicata **177**, 353–365 (2015)
21. R. Schneider, *Convex Bodies: The Brunn-Minkowski Theory*, expanded edn. (Cambridge University Press, Cambridge, 2014)
22. Y. Shenfeld, R. van Handel, Mixed volumes and the Bochner method. Proc. Am. Math. Soc. **147**(12), 5385–5402 (2019)
23. Y. Shenfeld, R. Van Handel, The extremals of Minkowski's quadratic inequality. Duke Math. J. **171**(4), 957–1027 (2022)

Rapid Convergence of the Unadjusted Langevin Algorithm: Isoperimetry Suffices

Santosh S. Vempala and Andre Wibisono

Abstract We study the Unadjusted Langevin Algorithm (ULA) for sampling from a probability distribution $\nu = e^{-f}$ on $\mathbb{R}^n$. We prove a convergence guarantee in Kullback-Leibler (KL) divergence assuming ν satisfies a log-Sobolev inequality and the Hessian of f is bounded. Notably, we do not assume convexity or bounds on higher derivatives. We prove convergence guarantees in Rényi divergence of order $q > 1$ assuming the limit of ULA satisfies isoperimetry, namely either the log-Sobolev or Poincaré inequality. We also prove a bound on the bias of the limiting distribution of ULA assuming third-order smoothness of f, without requiring isoperimetry.

1 Introduction

Sampling is a fundamental algorithmic task. Many applications require sampling from probability distributions in high-dimensional spaces, and in modern applications the probability distributions are complicated and non-logconcave. While the setting of logconcave functions is well-studied, it is important to have efficient sampling algorithms with good convergence guarantees beyond the logconcavity assumption. There is a close interplay between sampling and optimization, either via optimization as a limit of sampling (annealing) [44, 67], or via sampling as optimization in the space of distributions [47, 77]. Motivated by the widespread use of

This work was supported in part by NSF awards CCF-1717349, DMS-1839323, CCF-2007443, and CCF-2106644.

S. S. Vempala (✉)
Georgia Institute of Technology, College of Computing, Atlanta, GA, USA
e-mail: vempala@gatech.edu

A. Wibisono
Yale University, Department of Computer Science, New Haven, CT, USA
e-mail: andre.wibisono@yale.edu

© The Author(s), under exclusive license to Springer Nature Switzerland AG 2023
R. Eldan et al. (eds.), *Geometric Aspects of Functional Analysis*, Lecture Notes
in Mathematics 2327, https://doi.org/10.1007/978-3-031-26300-2_15

non-convex optimization and sampling, there is resurgent interest in understanding non-logconcave sampling.

In this paper we study a simple algorithm, the Unadjusted Langevin Algorithm (ULA), for sampling from a target probability distribution $\nu = e^{-f}$ on $\mathbb{R}^n$. ULA is a discrete-time algorithm that starts from any $x_0 \in \mathbb{R}^n$ and applies the following update at each step:

$$x_{k+1} = x_k - \epsilon \nabla f(x_k) + \sqrt{2\epsilon}\, z_k$$

where $\epsilon > 0$ is step size and $z_k \sim \mathcal{N}(0, I)$ is an independent standard Gaussian random variable in $\mathbb{R}^n$. ULA requires oracle access to the gradient ∇f of the log density $f = -\log \nu$. In particular, ULA does not require knowledge of f, which makes it applicable in practice where we often only know ν up to a normalizing constant.

As the step size $\epsilon \to 0$, ULA recovers the Langevin dynamics, which is a continuous-time stochastic process in $\mathbb{R}^n$ that converges to ν. We recall the optimization interpretation of the Langevin dynamics for sampling as the gradient flow of the Kullback-Leibler (KL) divergence with respect to ν in the space of probability distributions with the Wasserstein metric [47]. When ν is strongly logconcave, the KL divergence is a strongly convex objective function, so the Langevin dynamics as gradient flow converges exponentially fast [5, 74]. From the classical theory of Markov chains and diffusion processes, there are several known conditions milder than logconcavity that are sufficient for rapid convergence *in continuous time*. These include isoperimetric inequalities such as Poincaré inequality or log-Sobolev inequality (LSI). Along the Langevin dynamics in continuous time, Poincaré inequality implies an exponential convergence rate in $L^2(\nu)$, while LSI—which is stronger—implies an exponential convergence rate in KL divergence (as well as in Rényi divergence).

However, in discrete time, sampling under Poincaré inequality or LSI is a more challenging problem. ULA is an inexact discretization of the Langevin dynamics, and it converges to a biased limit $\nu_\epsilon \neq \nu$. When ν is strongly logconcave and smooth, it is known how to control the bias and prove a convergence guarantee on KL divergence along ULA; see for example [19, 25, 26, 29]. When ν is strongly logconcave, there are many other sampling algorithms with provable rapid convergence; these include the ball walk and hit-and-run [48, 54–56] (which give truly polynomial algorithms), various discretizations of the overdamped or underdamped Langevin dynamics [9, 25, 26, 29, 30] (which have polynomial dependencies on smoothness parameters but low dependence on dimension), and more sophisticated methods such as the Hamiltonian Monte Carlo [17, 28, 50, 59, 60]. It is of great interest to extend these results to non-logconcave densities ν, where existing results require strong assumptions with bounds that grow exponentially with the dimension or other parameters [2, 20, 57, 61]. There are also recent works that analyze convergence of sampling using various techniques such as reflection coupling [32], kernel methods [38], and higher-order integrators [53], albeit still under some strong

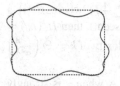 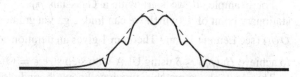

Fig. 1 Illustrations of non-logconcave distributions satisfying LSI or Poincaré inequality: the uniform distribution on a nonconvex set (left), and a small perturbation of a logconcave distribution, e.g., Gaussian (right)

conditions such as distant dissipativity, which is similar to strong logconcavity outside a bounded domain.

In this paper we study the convergence along ULA under minimal (and necessary) isoperimetric assumptions, namely, LSI and Poincaré inequality. These are sufficient for fast convergence in continuous time; moreover, in the case of logconcave distribution, the log-Sobolev and Poincaré constants can be bounded and lead to convergence guarantees for efficient sampling in discrete time. However, do they suffice on their own without the assumption of logconcavity?

We note that LSI and Poincaré inequality apply to a wider class of measures than logconcave distributions. In particular, LSI and Poincaré inequality are preserved under bounded perturbation and Lipschitz mapping (see Lemmas 16 and 19), whereas logconcavity would be destroyed. Given these properties, it is easy to exhibit examples of non-logconcave distributions satisfying LSI or Poincaré inequality. For example, we can take a small perturbation of a convex body to make it nonconvex but still satisfies isoperimetry; then the uniform probability distribution on the body (or a smooth approximation of it) is not logconcave but satisfies LSI or Poincaré inequality. Similarly, we can start with a strongly logconcave distribution such as a Gaussian, and subtract some small Gaussians from it; then the resulting (normalized) probability distribution is not logconcave, but it still satisfies LSI or Poincaré inequality as long as the Gaussians we subtract are small enough. See Fig. 1 for an illustration.

We measure the mode of convergence using KL divergence and Rényi divergence of order $q \geq 1$, which is stronger. Our first main result says that the only further assumption we need is smoothness, i.e., the gradient of f is Lipschitz (see Sect. 3.1). Here $H_\nu(\rho)$ is the KL divergence between ρ and ν. We say that $\nu = e^{-f}$ is L-smooth if ∇f is L-Lipschitz, or equivalently, $-LI \preceq \nabla^2 f(x) \preceq LI$ for all $x \in \mathbb{R}^n$.

Theorem 1 *Assume $\nu = e^{-f}$ satisfies log-Sobolev inequality with constant $\alpha > 0$ and is L-smooth. ULA with step size $0 < \epsilon \leq \frac{\alpha}{4L^2}$ satisfies*

$$H_\nu(\rho_k) \leq e^{-\alpha\epsilon k} H_\nu(\rho_0) + \frac{8\epsilon n L^2}{\alpha}.$$

In particular, for any $0 < \delta < 4n$, ULA with step size $\epsilon \leq \frac{\alpha\delta}{16L^2 n}$ reaches error $H_\nu(\rho_k) \leq \delta$ after $k \geq \frac{1}{\alpha\epsilon} \log \frac{2H_\nu(\rho_0)}{\delta}$ iterations.

For example, if we start with a Gaussian $\rho_0 = \mathcal{N}(x^*, \frac{1}{L}I)$ where x^* is a stationary point of f (which we can find, e.g., via gradient descent), then $H_\nu(\rho_0) = \tilde{O}(n)$ (see Lemma 1), and Theorem 1 gives an iteration complexity of $k = \tilde{\Theta}\left(\frac{L^2 n}{\alpha^2 \delta}\right)$ to achieve $H_\nu(\rho_k) \le \delta$ using ULA with step size $\epsilon = \Theta(\frac{\alpha\delta}{L^2 n})$.

The result above matches previous known bounds for ULA when ν is strongly logconcave [19, 25, 26, 29]. Our result complements the work of Ma et al. [57] who study the underdamped version of the Langevin dynamics under LSI and show an iteration complexity for the discrete-time algorithm that has better dependence on the dimension ($\sqrt{\frac{n}{\delta}}$ in place of $\frac{n}{\delta}$ above for ULA), but under an additional smoothness assumption (f has bounded third derivatives) and with higher polynomial dependence on other parameters. Our result also complements the work of Mangoubi and Vishnoi [61] who study the Metropolis-adjusted version of ULA (MALA) for non-logconcave ν and show a $\log(\frac{1}{\delta})$ iteration complexity from a warm start, under the additional assumption that f has bounded third and fourth derivatives in an appropriate ∞-norm.

We note that in general some isoperimetry condition is needed for rapid mixing of Markov chains (such as the Langevin dynamics and ULA), otherwise there are bad regions in the state space from which the chains take arbitrarily long to escape. Smoothness or bounded Hessian is a common assumption that seems to be needed for the analysis of discrete-time algorithms (such as gradient descent or ULA above).

In the second part of this paper, we study the convergence of Rényi divergence of order $q > 1$ along ULA. Rényi divergence is a family of generalizations of KL divergence [12, 68, 71], which becomes stronger as the order q increases. There are physical and operational interpretations of Rényi divergence [3, 40]. Rényi divergence has been useful in many applications, including for the exponential mechanism in differential privacy [1, 13, 31, 64], lattice-based cryptography [4], information-theoretic encryption [45], variational inference [52], machine learning [41, 62], information theory and statistics [24, 65], and black hole physics [27].

Our second main result proves a convergence bound for the Rényi divergence of order $q > 1$. While this is a stronger measure of convergence than KL divergence, the situation here is more complicated. First, we can only hope to converge to the target for finite q for any step-size ϵ (as we illustrate with an example). Second, it is unclear how to bound the Rényi divergence between the biased limit ν_ϵ and ν. We first show the convergence of Rényi divergence along Langevin dynamics in continuous time under LSI; see Theorem 2 in Sect. 4.2. Here $R_{q,\nu}(\rho)$ is the Rényi divergence of order q between ρ and ν.

Theorem 2 *Suppose ν satisfies LSI with constant $\alpha > 0$. Let $q \ge 1$. Along the Langevin dynamics,*

$$R_{q,\nu}(\rho_t) \le e^{-\frac{2\alpha t}{q}} R_{q,\nu}(\rho_0).$$

We also have the following convergence of Rényi divergence along Langevin dynamics under Poincaré inequality; see Theorem 3 in Sect. 6.1.

Theorem 3 *Suppose ν satisfies Poincaré inequality with constant $\alpha > 0$. Let $q \geq 2$. Along the Langevin dynamics,*

$$R_{q,\nu}(\rho_t) \leq \begin{cases} R_{q,\nu}(\rho_0) - \frac{2\alpha t}{q} & \text{if } R_{q,\nu}(\rho_0) \geq 1 \text{ and as long as } R_{q,\nu}(\rho_t) \geq 1, \\ e^{-\frac{2\alpha t}{q}} R_{q,\nu}(\rho_0) & \text{if } R_{q,\nu}(\rho_0) \leq 1. \end{cases}$$

The reader will notice that under Poincaré inequality, compared to LSI, the convergence is slower in the beginning before it becomes exponential. For a reasonable starting distribution (such as a Gaussian centered at a stationary point), this leads to an extra factor of n compared to the convergence under LSI.

We then turn to the discrete-time algorithm and show that ULA converges in Rényi divergence to the biased limit ν_ϵ under the assumption that ν_ϵ itself satisfies either LSI or Poincaré inequality. We combine this with a decomposition result on Rényi divergence to derive a convergence guarantee in Rényi divergence to ν; see Theorem 5 in Sect. 5.3 and Theorem 6 in Sect. 6.3.

Finally, we show some properties on the biased limit of ULA. Previously, from the convergence analysis of ULA in Theorem 1, we could deduce a bound on the bias of ULA in KL divergence under LSI and smoothness. Here we provide a direct bound on the bias of ULA in relative Fisher information assuming third-order smoothness, without isoperimetry; see Theorem 7. We also show the biased limit satisfies LSI if the original target is smooth and strongly log-concave; see Theorem 8.

In what follows, we review KL divergence and its properties along the Langevin dynamics in Sect. 2, and prove a convergence guarantee for KL divergence along ULA under LSI in Sect. 3. We provide a review of Rényi divergence and its properties along the Langevin dynamics in Sect. 4. We then prove the convergence guarantee for Rényi divergence along ULA under LSI in Sect. 5, and under Poincaré inequality in Sect. 6. We show properties on the biased limit of ULA in Sect. 7. We provide all proofs and details in Sect. 8. We conclude with a discussion in Sect. 9, including subsequent work that used some of the analysis techniques from this paper.

2 Review of KL Divergence Along Langevin Dynamics

In this section we review the definition of Kullback-Leibler (KL) divergence, log-Sobolev inequality, and the convergence of KL divergence along the Langevin dynamics in continuous time under log-Sobolev inequality. See section "Review on Notation and Basic Properties" in Appendix for a review on notation.

2.1 KL Divergence

Let ρ, ν be probability distributions on $\mathbb{R}^n$, represented via their probability density functions with respect to the Lebesgue measure on $\mathbb{R}^n$. We assume ρ, ν have full support and smooth densities.

Recall the **Kullback-Leibler (KL) divergence** of ρ with respect to ν is

$$H_\nu(\rho) = \int_{\mathbb{R}^n} \rho(x) \log \frac{\rho(x)}{\nu(x)} \, dx. \tag{1}$$

KL divergence is the relative form of *Shannon entropy* $H(\rho) = -\int_{\mathbb{R}^n} \rho(x) \log \rho(x) \, dx$. Whereas Shannon entropy can be positive or negative, KL divergence is nonnegative and minimized at ν: $H_\nu(\rho) \geq 0$ for all ρ, and $H_\nu(\rho) = 0$ if and only if $\rho = \nu$. Therefore, KL divergence serves as a measure of (albeit asymmetric) "distance" of a probability distribution ρ from a base distribution ν. KL divergence is a relatively strong measure of distance; for example, Pinsker's inequality implies that KL divergence controls total variation distance. Furthermore, under log-Sobolev (or Talagrand) inequality, KL divergence also controls the quadratic Wasserstein W_2 distance, as we review below.

We say $\nu = e^{-f}$ is L-**smooth** if f has bounded Hessian: $-LI \preceq \nabla^2 f(x) \preceq LI$ for all $x \in \mathbb{R}^n$.

Lemma 1 *Suppose* $\nu = e^{-f}$ *is L-smooth. Let* $\rho = \mathcal{N}(x^*, \frac{1}{L}I)$ *where x^* is a stationary point of f. Then* $H_\nu(\rho) \leq f(x^*) + \frac{n}{2} \log \frac{L}{2\pi}$.

We provide the proof of Lemma 1 in Sect. 8.1.1.

2.2 Log-Sobolev Inequality

Recall we say ν satisfies the **log-Sobolev inequality (LSI)** with a constant $\alpha > 0$ if for all smooth function $g \colon \mathbb{R}^n \to \mathbb{R}$ with $\mathbb{E}_\nu[g^2] < \infty$,

$$\mathbb{E}_\nu[g^2 \log g^2] - \mathbb{E}_\nu[g^2] \log \mathbb{E}_\nu[g^2] \leq \frac{2}{\alpha} \mathbb{E}_\nu[\|\nabla g\|^2]. \tag{2}$$

Recall the **relative Fisher information** of ρ with respect to ν is

$$J_\nu(\rho) = \int_{\mathbb{R}^n} \rho(x) \left\| \nabla \log \frac{\rho(x)}{\nu(x)} \right\|^2 dx. \tag{3}$$

LSI is equivalent to the following relation between KL divergence and Fisher information for all ρ:

$$H_\nu(\rho) \leq \frac{1}{2\alpha} J_\nu(\rho). \tag{4}$$

Indeed, to obtain (4) we choose $g^2 = \frac{\rho}{\nu}$ in (2); conversely, to obtain (2) we choose $\rho = \frac{g^2 \nu}{\mathbb{E}_\nu[g^2]}$ in (4).

LSI is a strong isoperimetry statement and implies, among others, concentration of measure and sub-Gaussian tail property [49]. LSI was first shown by Gross [39] for the case of Gaussian ν. It was extended by Bakry and Émery [5] to strongly log-concave ν; namely, when $f = -\log \nu$ is α-strongly convex, then ν satisfies LSI with constant α. However, LSI applies more generally. For example, the classical perturbation result by Holley and Stroock [43] states that LSI is stable under bounded perturbation. Furthermore, LSI is preserved under a Lipschitz mapping. In one dimension, there is an exact characterization of when a probability distribution on $\mathbb{R}$ satisfies LSI [10]. Moreover, LSI satisfies a tensorization property [49]: If ν_1, ν_2 satisfy LSI with constants $\alpha_1, \alpha_2 > 0$, respectively, then $\nu_1 \otimes \nu_2$ satisfies LSI with constant $\min\{\alpha_1, \alpha_2\} > 0$. Thus, there are many examples of non-logconcave distributions ν on $\mathbb{R}^n$ satisfying LSI (with a constant independent of dimension). There are also Lyapunov function criteria and exponential integrability conditions that can be used to verify when a probability distribution satisfies LSI; see for example [8, 15, 16, 63, 75].

2.2.1 Talagrand Inequality

Recall the **Wasserstein distance** between ρ and ν is

$$W_2(\rho, \nu) = \inf_\Pi \mathbb{E}_\Pi[\|X - Y\|^2]^{\frac{1}{2}} \tag{5}$$

where the infimum is over joint distributions Π of (X, Y) with the correct marginals $X \sim \rho, Y \sim \nu$.

Recall we say ν satisfies **Talagrand inequality** with a constant $\alpha > 0$ if for all ρ:

$$\frac{\alpha}{2} W_2(\rho, \nu)^2 \leq H_\nu(\rho). \tag{6}$$

Talagrand's inequality implies concentration of measure of Gaussian type. It was first studied by Talagrand [70] for Gaussian ν, and extended by Otto and Villani [66] to all ν satisfying LSI; namely, if ν satisfies LSI with constant $\alpha > 0$, then ν also satisfies Talagrand's inequality with the same constant [66, Theorem 1]. Therefore, under LSI, KL divergence controls the Wasserstein distance. Moreover, when ν is log-concave, LSI and Talagrand's inequality are equivalent [66, Corollary 3.1].

We recall the geometric interpretation of LSI and Talagrand's inequality from [66]. In the space of probability distributions with the Riemannian metric defined by the Wasserstein W_2 distance, the relative Fisher information (3) is the squared norm of the gradient of KL divergence (1). Therefore, LSI (4) is the gradient dominated condition (also known as the Polyak-Łojaciewicz (PL) inequality) for KL divergence. On the other hand, Talagrand's inequality (6) is the quadratic growth condition for KL divergence. In general, the gradient dominated condition implies the quadratic growth condition [66, Proposition 1']; therefore, LSI implies Talagrand's inequality.

2.3 Langevin Dynamics

The **Langevin dynamics** for target distribution $\nu = e^{-f}$ is a continuous-time stochastic process $(X_t)_{t \geq 0}$ in $\mathbb{R}^n$ that evolves following the stochastic differential equation:

$$dX_t = -\nabla f(X_t)\, dt + \sqrt{2}\, dW_t \tag{7}$$

where $(W_t)_{t \geq 0}$ is the standard Brownian motion in $\mathbb{R}^n$ with $W_0 = 0$.

If $(X_t)_{t \geq 0}$ evolves following the Langevin dynamics (7), then their probability density function $(\rho_t)_{t \geq 0}$ evolves following the **Fokker-Planck equation**:

$$\frac{\partial \rho_t}{\partial t} = \nabla \cdot (\rho_t \nabla f) + \Delta \rho_t = \nabla \cdot \left(\rho_t \nabla \log \frac{\rho_t}{\nu} \right). \tag{8}$$

Here $\nabla \cdot$ is the divergence and Δ is the Laplacian operator. We provide a derivation in section "Derivation of the Fokker-Planck Equation" in Appendix. From (8), if $\rho_t = \nu$, then $\frac{\partial \rho_t}{\partial t} = 0$, so ν is the stationary distribution for the Langevin dynamics (7). Moreover, the Langevin dynamics brings any distribution $X_t \sim \rho_t$ closer to the target distribution ν, as the following lemma shows.

Lemma 2 *Along the Langevin dynamics (7) (or equivalently, the Fokker-Planck equation (8)),*

$$\frac{d}{dt} H_\nu(\rho_t) = -J_\nu(\rho_t). \tag{9}$$

We provide the proof of Lemma 2 in Sect. 8.1.2. Since $J_\nu(\rho) \geq 0$, the identity (9) shows that KL divergence with respect to ν is decreasing along the Langevin dynamics, so indeed the distribution ρ_t converges to ν.

2.3.1 Exponential Convergence of KL Divergence Along Langevin Dynamics Under LSI

When ν satisfies LSI, KL divergence converges exponentially fast along the Langevin dynamics.

Theorem 4 *Suppose ν satisfies LSI with constant $\alpha > 0$. Along the Langevin dynamics (7),*

$$H_\nu(\rho_t) \leq e^{-2\alpha t} H_\nu(\rho_0). \tag{10}$$

Furthermore, $W_2(\rho_t, \nu) \leq \sqrt{\frac{2}{\alpha} H_\nu(\rho_0)} \, e^{-\alpha t}$.

We provide the proof of Theorem 4 in Sect. 8.1.3. We also recall the optimization interpretation of Langevin dynamics as the gradient flow of KL divergence in the space of distributions with the Wasserstein metric [47, 66, 74]. Then the exponential convergence rate in Theorem 4 is a manifestation of the general fact that gradient flow converges exponentially fast under gradient domination condition. This provides a justification for using the Langevin dynamics for sampling from ν, as a natural steepest descent flow that minimizes the KL divergence H_ν.

3 Unadjusted Langevin Algorithm

In this section we study the behavior of KL divergence along the Unadjusted Langevin Algorithm (ULA) in discrete time under log-Sobolev inequality assumption.

Suppose we wish to sample from a smooth target probability distribution $\nu = e^{-f}$ in $\mathbb{R}^n$. The **Unadjusted Langevin Algorithm (ULA)** with step size $\epsilon > 0$ is the discrete-time algorithm

$$x_{k+1} = x_k - \epsilon \nabla f(x_k) + \sqrt{2\epsilon} \, z_k \tag{11}$$

where $z_k \sim \mathcal{N}(0, I)$ is an independent standard Gaussian random variable in $\mathbb{R}^n$. Let ρ_k denote the probability distribution of x_k that evolves following ULA.

As $\epsilon \to 0$, ULA recovers the Langevin dynamics (7) in continuous-time. However, for fixed $\epsilon > 0$, ULA converges to a biased limiting distribution $\nu_\epsilon \neq \nu$. Therefore, KL divergence $H_\nu(\rho_k)$ does not tend to 0 along ULA, as it has an asymptotic bias $H_\nu(\nu_\epsilon) > 0$.

Example 1 Let $\nu = \mathcal{N}(0, \frac{1}{\alpha}I)$. The ULA iteration is $x_{k+1} = (1 - \epsilon\alpha)x_k + \sqrt{2\epsilon} z_k$, $z_k \sim \mathcal{N}(0, I)$. For $0 < \epsilon < \frac{2}{\alpha}$, the limit is $\nu_\epsilon = \mathcal{N}\left(0, \frac{1}{\alpha(1 - \frac{\epsilon\alpha}{2})}\right)$, and the bias is $H_\nu(\nu_\epsilon) = \frac{n}{2}\left(\frac{\epsilon\alpha}{2(1-\frac{\epsilon\alpha}{2})} + \log\left(1 - \frac{\epsilon\alpha}{2}\right)\right)$. In particular, $H_\nu(\nu_\epsilon) \le \frac{n\epsilon^2\alpha^2}{16(1-\frac{\epsilon\alpha}{2})^2} = O(\epsilon^2)$.

3.1 Convergence of KL Divergence Along ULA Under LSI

When the true target distribution ν satisfies LSI and a smoothness condition, we can prove a convergence guarantee in KL divergence along ULA. Recall we say $\nu = e^{-f}$ is L-smooth, $0 < L < \infty$, if $-LI \preceq \nabla^2 f(x) \preceq LI$ for all $x \in \mathbb{R}^n$.

A key part in our analysis is the following lemma which bounds the decrease in KL divergence along one iteration of ULA. Here $x_{k+1} \sim \rho_{k+1}$ is the output of one step of ULA (11) from $x_k \sim \rho_k$.

Lemma 3 *Suppose ν satisfies LSI with constant $\alpha > 0$ and is L-smooth. If $0 < \epsilon \le \frac{\alpha}{4L^2}$, then along each step of ULA (11),*

$$H_\nu(\rho_{k+1}) \le e^{-\alpha\epsilon} H_\nu(\rho_k) + 6\epsilon^2 nL^2. \tag{12}$$

We provide the proof of Lemma 3 in Sect. 8.2.1. The proof of Lemma 3 compares the evolution of KL divergence along one step of ULA with the evolution along the Langevin dynamics in continuous time (which converges exponentially fast under LSI), and bounds the discretization error; see Fig. 2 for an illustration. This high-level comparison technique has been used in many papers. Our proof structure is similar to that of Cheng and Bartlett [19], whose analysis needs ν to be strongly log-concave.

With Lemma 3, we can prove our main result on the convergence rate of ULA under LSI.

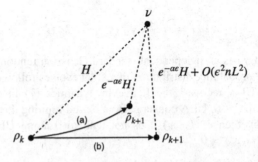

Fig. 2 An illustration for the proof of Lemma 3. In each iteration, we compare the evolution of (a) the continuous-time Langevin dynamics for time ϵ, and (b) one step of ULA. If the current KL divergence is $H \equiv H_\nu(\rho_k)$, then after the Langevin dynamics (a) the KL divergence is $H_\nu(\tilde{\rho}_{k+1}) \le e^{-\alpha\epsilon} H$, and we show that after ULA (b) the KL divergence is $H_\nu(\rho_{k+1}) \le e^{-\alpha\epsilon} H + O(\epsilon^2 nL^2)$

Theorem 1 *Suppose ν satisfies LSI with constant $\alpha > 0$ and is L-smooth. For any $x_0 \sim \rho_0$ with $H_\nu(\rho_0) < \infty$, the iterates $x_k \sim \rho_k$ of ULA (11) with step size $0 < \epsilon \leq \frac{\alpha}{4L^2}$ satisfy*

$$H_\nu(\rho_k) \leq e^{-\alpha \epsilon k} H_\nu(\rho_0) + \frac{8\epsilon n L^2}{\alpha}. \tag{13}$$

Thus, for any $\delta > 0$, to achieve $H_\nu(\rho_k) < \delta$, it suffices to run ULA with step size $\epsilon \leq \frac{\alpha}{4L^2} \min\{1, \frac{\delta}{4n}\}$ for $k \geq \frac{1}{\alpha \epsilon} \log \frac{2H_\nu(\rho_0)}{\delta}$ iterations.

We provide the proof of Theorem 1 in Sect. 8.2.2.

In particular, suppose $\delta < 4n$ and we choose the largest permissible step size $\epsilon = \Theta\left(\frac{\alpha \delta}{L^2 n}\right)$. Suppose we start with a Gaussian $\rho_0 = \mathcal{N}(x^*, \frac{1}{L} I)$, where x^* is a stationary point of f (which we can find, e.g., via gradient descent), so $H_\nu(\rho_0) \leq f(x^*) + \frac{n}{2} \log \frac{L}{2\pi} = \tilde{O}(n)$ by Lemma 1. Therefore, Theorem 1 states that to achieve $H_\nu(\rho_k) \leq \delta$, ULA has iteration complexity $k = \tilde{\Theta}\left(\frac{L^2 n}{\alpha^2 \delta}\right)$. Since LSI implies Talagrand's inequality, Theorem 1 also yields a convergence guarantee in Wasserstein distance.

As $k \to \infty$, Theorem 1 implies the following bound on the bias between ν_ϵ and ν under LSI. However, note that the bound in Corollary 1 is $H_\nu(\nu_\epsilon) = O(\epsilon)$, while from Example 1 we see that $H_\nu(\nu_\epsilon) = O(\epsilon^2)$ in the Gaussian case.

Corollary 1 *Suppose ν satisfies LSI with constant $\alpha > 0$ and is L-smooth. For $0 < \epsilon \leq \frac{\alpha}{4L^2}$, the biased limit ν_ϵ of ULA with step size ϵ satisfies $H_\nu(\nu_\epsilon) \leq \frac{8nL^2\epsilon}{\alpha}$ and $W_2(\nu, \nu_\epsilon)^2 \leq \frac{16nL^2\epsilon}{\alpha^2}$.*

Remark 1 If f satisfies a third-order smoothness condition (without isoperimetry), then we can show a bound on the bias in relative Fisher information; see Sect. 7.1.

4 Review of Rényi Divergence Along Langevin Dynamics

In this section we review the definition of Rényi divergence and the exponential convergence of Rényi divergence along the Langevin dynamics under LSI.

4.1 Rényi Divergence

Rényi divergence [68] is a family of generalizations of KL divergence. We refer to [12, 71] for basic properties of Rényi divergence.

For $q > 0$, $q \neq 1$, the **Rényi divergence** of order q of a probability distribution ρ with respect to ν is

$$R_{q,\nu}(\rho) = \frac{1}{q-1} \log F_{q,\nu}(\rho) \tag{14}$$

where

$$F_{q,\nu}(\rho) = \mathbb{E}_\nu \left[\left(\frac{\rho}{\nu} \right)^q \right] = \int_{\mathbb{R}^n} \nu(x) \frac{\rho(x)^q}{\nu(x)^q} \, dx = \int_{\mathbb{R}^n} \frac{\rho(x)^q}{\nu(x)^{q-1}} dx. \tag{15}$$

Rényi divergence is the relative form of *Rényi entropy* [68]: $H_q(\rho) = \frac{1}{q-1} \log \int \rho(x)^q \, dx$. The case $q = 1$ is defined via limit, and recovers the KL divergence (1):

$$R_{1,\nu}(\rho) = \lim_{q \to 1} R_{q,\nu}(\rho) = \mathbb{E}_\nu \left[\frac{\rho}{\nu} \log \frac{\rho}{\nu} \right] = \mathbb{E}_\rho \left[\log \frac{\rho}{\nu} \right] = H_\nu(\rho). \tag{16}$$

Rényi divergence has the property that $R_{q,\nu}(\rho) \geq 0$ for all ρ, and $R_{q,\nu}(\rho) = 0$ if and only if $\rho = \nu$. Furthermore, the map $q \mapsto R_{q,\nu}(\rho)$ is increasing (see Sect. 8.3.1). Therefore, Rényi divergence provides an alternative measure of "distance" of ρ from ν, which becomes stronger as q increases. In particular, $R_{\infty,\nu}(\rho) = \log \left\| \frac{\rho}{\nu} \right\|_\infty = \log \sup_x \frac{\rho(x)}{\nu(x)}$ is finite if and only if ρ is *warm* relative to ν. It is possible that $R_{q,\nu}(\rho) = \infty$ for large enough q, as the following example shows.

Example 2 Let $\rho = \mathcal{N}(0, \sigma^2 I)$ and $\nu = \mathcal{N}(0, \lambda^2 I)$. If $\sigma^2 > \lambda^2$ and $q \geq \frac{\sigma^2}{\sigma^2 - \lambda^2}$, then $R_{q,\nu}(\rho) = \infty$. Otherwise, $R_{q,\nu}(\rho) = \frac{n}{2} \log \frac{\lambda^2}{\sigma^2} - \frac{n}{2(q-1)} \log \left(q - (q-1) \frac{\sigma^2}{\lambda^2} \right)$.

Analogous to Lemma 1, we have the following estimate of the Rényi divergence of a Gaussian.

Lemma 4 *Suppose* $\nu = e^{-f}$ *is L-smooth. Let* $\rho = \mathcal{N}(x^*, \frac{1}{L}I)$ *where* x^* *is a stationary point of* f. *Then for all* $q \geq 1$, $R_{q,\nu}(\rho) \leq f(x^*) + \frac{n}{2} \log \frac{L}{2\pi}$.

We provide the proof of Lemma 4 in Sect. 8.3.2.

4.1.1 Log-Sobolev Inequality

For $q > 0$, we define the **Rényi information** of order q of ρ with respect to ν as

$$G_{q,\nu}(\rho) = \mathbb{E}_\nu \left[\left(\frac{\rho}{\nu} \right)^q \left\| \nabla \log \frac{\rho}{\nu} \right\|^2 \right]$$

$$= \mathbb{E}_\nu \left[\left(\frac{\rho}{\nu} \right)^{q-2} \left\| \nabla \frac{\rho}{\nu} \right\|^2 \right] = \frac{4}{q^2} \mathbb{E}_\nu \left[\left\| \nabla \left(\frac{\rho}{\nu} \right)^{\frac{q}{2}} \right\|^2 \right]. \tag{17}$$

The case $q = 1$ recovers relative Fisher information (3): $G_{1,\nu}(\rho) = \mathbb{E}_\nu\left[\frac{\rho}{\nu}\left\|\nabla\log\frac{\rho}{\nu}\right\|^2\right] = J_\nu(\rho)$. We have the following relation under log-Sobolev inequality. Note that the case $q = 1$ recovers LSI in the form (4) involving KL divergence and relative Fisher information.

Lemma 5 *Suppose ν satisfies LSI with constant $\alpha > 0$. Let $q \geq 1$. For all ρ,*

$$\frac{G_{q,\nu}(\rho)}{F_{q,\nu}(\rho)} \geq \frac{2\alpha}{q^2}R_{q,\nu}(\rho).$$ (18)

We provide the proof of Lemma 5 in Sect. 8.3.3.

4.2 Langevin Dynamics

Along the Langevin dynamics (7) for ν, we can compute the rate of change of the Rényi divergence.

Lemma 6 *For all $q > 0$, along the Langevin dynamics (7),*

$$\frac{d}{dt}R_{q,\nu}(\rho_t) = -q\frac{G_{q,\nu}(\rho_t)}{F_{q,\nu}(\rho_t)}.$$ (19)

We provide the proof of Lemma 6 in Sect. 8.3.4. In particular, $\frac{d}{dt}R_{q,\nu}(\rho_t) \leq 0$, so Rényi divergence is always decreasing along the Langevin dynamics. Furthermore, analogous to how the Langevin dynamics is the gradient flow of KL divergence under the Wasserstein metric, one can also show that the Langevin dynamics is the gradient flow of Rényi divergence with respect to a suitably defined metric (which depends on the target distribution ν) on the space of distributions; see [14].

4.2.1 Convergence of Rényi Divergence Along Langevin Dynamics Under LSI

When ν satisfies LSI, Rényi divergence converges exponentially fast along the Langevin dynamics. Note the case $q = 1$ recovers the exponential convergence rate of KL divergence from Theorem 4.

Theorem 2 *Suppose ν satisfies LSI with constant $\alpha > 0$. Let $q \geq 1$. Along the Langevin dynamics (7),*

$$R_{q,\nu}(\rho_t) \leq e^{-\frac{2\alpha t}{q}}R_{q,\nu}(\rho_0).$$ (20)

We provide the proof of Theorem 2 in Sect. 8.3.5. Theorem 2 shows that if the initial Rényi divergence is finite, then it converges exponentially fast. However, even if initially the Rényi divergence of some order is infinite, it will be eventually finite along the Langevin dynamics, after which time Theorem 2 applies. This is because when ν satisfies LSI, the Langevin dynamics satisfies a *hypercontractivity* property [11, 39, 74]; see Sect. 8.3.6. Furthermore, as shown in [14], we can combine the exponential convergence rate above with the hypercontractivity property to improve the exponential rate to be 2α, independent of q, at the cost of some initial waiting time; here we leave the rate as above for simplicity.

Remark 2 When ν satisfies Poincaré inequality, we can still prove the convergence of Rényi divergence along the Langevin dynamics. However, in this case, Rényi divergence initially decreases linearly, then exponentially once it is less than 1. See Sect. 6.1.

5   Rényi Divergence Along ULA

In this section we prove a convergence guarantee for Rényi divergence along ULA under the assumption that the biased limit satisfies LSI.

As before, let $\nu = e^{-f}$, and let ν_ϵ denote the biased limit of ULA (11) with step size $\epsilon > 0$. We first note that the asymptotic bias $R_{q,\nu}(\nu_\epsilon)$ may be infinite for large enough q.

Example 3 As in Examples 1 and 2, let $\nu = \mathcal{N}(0, \frac{1}{\alpha}I)$, so $\nu_\epsilon = \mathcal{N}\left(0, \frac{1}{\alpha(1-\frac{\epsilon\alpha}{2})}\right)$. The bias is

$$R_{q,\nu}(\nu_\epsilon) = \begin{cases} \frac{n}{2(q-1)}\left(q\log\left(1-\frac{\epsilon\alpha}{2}\right) - \log\left(1-\frac{q\epsilon\alpha}{2}\right)\right) & \text{if } 1 < q < \frac{2}{\epsilon\alpha}, \\ \infty & \text{if } q \geq \frac{2}{\epsilon\alpha}. \end{cases}$$

For $1 < q < \frac{2}{\epsilon\alpha}$, we can bound $R_{q,\nu}(\nu_\epsilon) \leq \frac{n\alpha^2 q^2\epsilon^2}{8(q-1)(1-\frac{q\epsilon\alpha}{2})}$.

Thus, for each fixed $q > 1$, there is an asymptotic bias $R_{q,\nu}(\nu_\epsilon)$ which is finite for small $\epsilon > 0$. In Example 3, we have $R_{q,\nu}(\nu_\epsilon) = O(\epsilon^2)$.

5.1   Decomposition of Rényi Divergence

For order $q > 1$, we have the following decomposition of Rényi divergence.

Lemma 7 *Let* $q > 1$. *For all probability distribution* ρ,

$$R_{q,\nu}(\rho) \leq \left(\frac{q - \frac{1}{2}}{q - 1}\right) R_{2q,\nu_\epsilon}(\rho) + R_{2q-1,\nu}(\nu_\epsilon). \tag{21}$$

We provide the proof of Lemma 7 in Sect. 8.4.1. The first term in the bound above is the Rényi divergence with respect to the biased limit, which converges exponentially fast under LSI assumption (see Lemma 8). The second term in (21) is the asymptotic bias in Rényi divergence.

5.2 Rapid Convergence of Rényi Divergence to Biased Limit Under LSI

We show that Rényi divergence with respect to the biased limit ν_ϵ converges exponentially fast along ULA, assuming ν_ϵ itself satisfies LSI.

Assumption 1 *The probability distribution* ν_ϵ *satisfies LSI with a constant* $\beta \equiv \beta_\epsilon > 0$.

We can verify Assumption 1 in the Gaussian case. We can also verify Assumption 1 when ν is smooth and strongly log-concave; see Sect. 7.2. However, it is unclear how to verify Assumption 1 in general. One might hope to prove that if ν satisfies LSI, then Assumption 1 holds.

Example 4 Let $\nu = \mathcal{N}(0, \frac{1}{\alpha}I)$, so $\nu_\epsilon = \mathcal{N}\left(0, \frac{1}{\alpha(1-\frac{\epsilon\alpha}{2})}I\right)$, which is strongly log-concave (and hence satisfies LSI) with parameter $\beta = \alpha\left(1 - \frac{\epsilon\alpha}{2}\right)$. In particular, $\beta \geq \frac{\alpha}{2}$ for $\epsilon \leq \frac{1}{\alpha}$.

Under Assumption 1, we can prove an exponential convergence rate to the biased limit ν_ϵ.

Lemma 8 *Assume Assumption 1. Suppose* $\nu = e^{-f}$ *is* L-*smooth, and let* $0 < \epsilon \leq \min\left\{\frac{1}{3L}, \frac{1}{9\beta}\right\}$. *For* $q \geq 1$, *along ULA* (11),

$$R_{q,\nu_\epsilon}(\rho_k) \leq e^{-\frac{\beta\epsilon k}{q}} R_{q,\nu_\epsilon}(\rho_0). \tag{22}$$

We provide the proof of Lemma 8 in Sect. 8.4.2. In the proof of Lemma 8, we decompose each step of ULA as a sequence of two operations; see Fig. 3 for an illustration. In the first part, we take a gradient step; this is a deterministic bijective map, so it preserves Rényi divergence. In the second part, we add an independent Gaussian; this is the result of evolution along the heat flow, and we can derive a formula on the decrease in Rényi divergence (which is similar to the formula (19) along the Langevin dynamics; see Sect. 8.4.2 for detail).

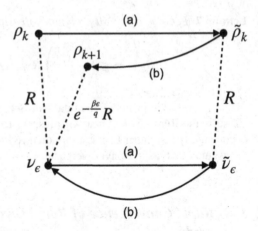

Fig. 3 An illustration for the proof of Lemma 8. We decompose each step of ULA into two operations: (a) a deterministic gradient step, and (b) an evolution along the heat flow. If the current Rényi divergence is $R \equiv R_{q,\nu_\epsilon}(\rho_k)$, then the gradient step (a) does not change the Rényi divergence: $R_{q,\tilde{\nu}_\epsilon}(\tilde{\rho}_k) = R$, while the heat flow (b) decreases the Rényi divergence: $R_{q,\nu_\epsilon}(\rho_{k+1}) \leq e^{-\alpha\epsilon} R$

5.3 Convergence of Rényi Divergence Along ULA Under LSI

We combine Lemmas 7 and 8 to obtain the following characterization of the convergence of Rényi divergence along ULA under LSI.

Theorem 5 *Assume Assumption 1. Suppose $v = e^{-f}$ is L-smooth, and let $0 < \epsilon \leq \min\left\{\frac{1}{3L}, \frac{1}{9\beta}\right\}$. Let $q > 1$, and suppose $R_{2q,\nu_\epsilon}(\rho_0) < \infty$. Then along ULA (11),*

$$R_{q,v}(\rho_k) \leq \left(\frac{q - \frac{1}{2}}{q - 1}\right) R_{2q,\nu_\epsilon}(\rho_0) e^{-\frac{\beta\epsilon k}{2q}} + R_{2q-1,v}(\nu_\epsilon). \tag{23}$$

We provide the proof of Theorem 5 in Sect. 8.4.3. For $\delta > 0$, let $\gamma_q(\delta) = \sup\{\epsilon > 0 : R_{q,v}(\nu_\epsilon) \leq \delta\}$. Theorem 5 states that to achieve $R_{q,v}(\rho_k) \leq \delta$, it suffices to run ULA with step size $\epsilon = \Theta\left(\min\left\{\frac{1}{L}, \gamma_{2q-1}\left(\frac{\delta}{2}\right)\right\}\right)$ for $k = \Theta\left(\frac{1}{\beta\epsilon} \log \frac{R_{2q,\nu_\epsilon}(\rho_0)}{\delta}\right)$ iterations. Suppose δ is small enough that $\gamma_{2q-1}\left(\frac{\delta}{2}\right) < \frac{1}{L}$. Note that ν_ϵ is $\frac{1}{2\epsilon}$-smooth, so by choosing ρ_0 to be a Gaussian with covariance $2\epsilon I$, we have $R_{2q,\nu_\epsilon}(\rho_0) = \tilde{O}(n)$ by Lemma 4. Therefore, Theorem 5 yields an iteration complexity of $k = \tilde{\Theta}\left(\frac{1}{\beta\gamma_{2q-1}\left(\frac{\delta}{2}\right)}\right)$.

For example, if $R_{q,v}(\nu_\epsilon) = O(\epsilon)$, then $\gamma_q(\delta) = \Omega(\delta)$, so the iteration complexity is $k = \tilde{\Theta}\left(\frac{1}{\beta\delta}\right)$ with step size $\epsilon = \Theta(\delta)$. On the other hand, if $R_{q,v}(\epsilon) = O(\epsilon^2)$, as in Example 3, then $\gamma_q(\delta) = \Omega(\sqrt{\delta})$, so the iteration complexity is $k = \tilde{\Theta}\left(\frac{1}{\beta\sqrt{\delta}}\right)$ with step size $\epsilon = \Theta(\sqrt{\delta})$.

Remark 3 Our result for Rényi divergence above involves the asymptotic bias, which we do not bound. Another approach to analyze ULA in Rényi divergence was proposed in [35] (and improved in [34]), albeit with a bound that does not provide an estimate of the Rényi bias. The work of [22] extended our one-step interpolation technique to show the convergence of ULA in Rényi divergence under LSI and smoothness, and provides an estimate on the Rényi bias.

6 Poincaré Inequality

In this section we review the definition of Poincaré inequality and prove convergence guarantees for the Rényi divergence along the Langevin dynamics and ULA. As before, let ρ, ν be smooth probability distributions on $\mathbb{R}^n$.

Recall we say ν satisfies **Poincaré inequality (PI)** with a constant $\alpha > 0$ if for all smooth function $g \colon \mathbb{R}^n \to \mathbb{R}$,

$$\text{Var}_\nu(g) \leq \frac{1}{\alpha} \mathbb{E}_\nu[\|\nabla g\|^2] \tag{24}$$

where $\text{Var}_\nu(g) = \mathbb{E}_\nu[g^2] - \mathbb{E}_\nu[g]^2$ is the variance of g under ν. Poincaré inequality is an isoperimetric-type statement, but it is weaker than LSI. It is known that LSI implies PI with the same constant; in fact, PI is a linearization of LSI (4), i.e., when $\rho = (1 + \eta g)\nu$ as $\eta \to 0$ [69, 74]. Furthermore, it is also known that Talagrand's inequality implies PI with the same constant, and in fact PI is also a linearization of Talagrand's inequality [66]. Poincaré inequality is better behaved than LSI [16], and there are various Lyapunov function criteria and integrability conditions that can be used to verify when a probability distribution satisfies Poincaré inequality; see for example [6, 23, 63].

Under Poincaré inequality, we can prove the following bound on Rényi divergence, which is analogous to Lemma 5 under LSI. When $R_{q,\nu}(\rho)$ is small, the two bounds are approximately equivalent.

Lemma 9 *Suppose ν satisfies Poincaré inequality with constant $\alpha > 0$. Let $q \geq 2$. For all ρ,*

$$\frac{G_{q,\nu}(\rho)}{F_{q,\nu}(\rho)} \geq \frac{4\alpha}{q^2}\left(1 - e^{-R_{q,\nu}(\rho)}\right).$$

We provide the proof of Lemma 9 in Sect. 8.5.1.

6.1 Convergence of Rényi Divergence Along Langevin Dynamics Under Poincaré

When ν satisfies Poincaré inequality, Rényi divergence converges along the Langevin dynamics. The convergence is initially linear, then becomes exponential once the Rényi divergence falls below a constant.

Theorem 3 *Suppose ν satisfies Poincaré inequality with constant $\alpha > 0$. Let $q \geq 2$.*
Along the Langevin dynamics (7),

$$R_{q,\nu}(\rho_t) \leq \begin{cases} R_{q,\nu}(\rho_0) - \frac{2\alpha t}{q} & \text{if } R_{q,\nu}(\rho_0) \geq 1 \text{ and as long as } R_{q,\nu}(\rho_t) \geq 1, \\ e^{-\frac{2\alpha t}{q}} R_{q,\nu}(\rho_0) & \text{if } R_{q,\nu}(\rho_0) \leq 1. \end{cases}$$

We provide the proof of Theorem 3 in Sect. 8.5.2. Theorem 3 states that starting from $R_{q,\nu}(\rho_0) \geq 1$, the Langevin dynamics reaches $R_{q,\nu}(\rho_t) \leq \delta$ in $t \leq O\left(\frac{q}{\alpha}\left(R_{q,\nu}(\rho_0) + \log\frac{1}{\delta}\right)\right)$ time.

6.2 Convergence of Rényi Divergence to Biased Limit Under Poincaré

We show that Rényi divergence with respect to the biased limit ν_ϵ converges exponentially fast along ULA, assuming ν_ϵ satisfies Poincaré inequality.

Assumption 2 *The distribution ν_ϵ satisfies Poincaré inequality with a constant $\beta \equiv \beta_\epsilon > 0$.*

We can verify Assumption 2 in the Gaussian case, and when ν is smooth and strongly log-concave; see Sect. 7.2. However, it is unclear how to verify Assumption 2 in general. One might hope to prove that if ν satisfies Poincaré, then Assumption 2 holds.

Analogous to Lemma 8, we have the following convergence to the biased limit in discrete time, at a rate which matches the continuous-time convergence in Theorem 6.

Lemma 10 *Assume Assumption 2. Suppose $\nu = e^{-f}$ is L-smooth, and let $0 < \epsilon \leq \min\left\{\frac{1}{3L}, \frac{1}{9\beta}\right\}$. For $q \geq 2$, along ULA (11),*

$$R_{q,\nu_\epsilon}(\rho_k) \leq \begin{cases} R_{q,\nu_\epsilon}(\rho_0) - \frac{\beta\epsilon k}{q} & \text{if } R_{q,\nu_\epsilon}(\rho_0) \geq 1 \text{ and as long as } R_{q,\nu_\epsilon}(\rho_k) \geq 1, \\ e^{-\frac{\beta\epsilon k}{q}} R_{q,\nu_\epsilon}(\rho_0) & \text{if } R_{q,\nu_\epsilon}(\rho_0) \leq 1. \end{cases}$$

$$(25)$$

We provide the proof of Lemma 10 in Sect. 8.5.3. Lemma 10 states that starting from $R_{q,\nu_\epsilon}(\rho_0) \geq 1$, ULA reaches $R_{q,\nu_\epsilon}(\rho_k) \leq \delta$ in $k \leq O\left(\frac{q}{\epsilon\beta}\left(R_{q,\nu_\epsilon}(\rho_0) + \log\frac{1}{\delta}\right)\right)$ iterations.

6.3 Convergence of Rényi Divergence Along ULA Under Poincaré

We combine Lemmas 7 and 10 to obtain the following characterization of the convergence of Rényi divergence along ULA to the true target distribution under Poincaré inequality.

Theorem 6 *Assume Assumption 2. Suppose $v = e^{-f}$ is L-smooth, and let $0 < \epsilon \le \min\left\{\frac{1}{3L}, \frac{1}{9\beta}\right\}$. Let $q > 1$, and suppose $1 \le R_{2q,v_\epsilon}(\rho_0) < \infty$. Then along ULA (11), for $k \ge k_0 := \frac{2q}{\beta\epsilon}(R_{2q,v_\epsilon}(\rho_0) - 1)$,*

$$R_{q,v}(\rho_k) \le \left(\frac{q - \frac{1}{2}}{q - 1}\right) e^{-\frac{\beta\epsilon(k-k_0)}{2q}} + R_{2q-1,v}(v_\epsilon). \tag{26}$$

We provide the proof of Theorem 6 in Sect. 8.5.4.

For $\delta > 0$, recall $\gamma_q(\delta) = \sup\{\epsilon > 0: R_{q,v}(v_\epsilon) \le \delta\}$. Theorem 6 states that to achieve $R_{q,v}(\rho_k) \le \delta$, it suffices to run ULA with step size $\epsilon = \Theta\left(\min\left\{\frac{1}{L}, \gamma_{2q-1}\left(\frac{\delta}{2}\right)\right\}\right)$ for $k = \Theta\left(\frac{1}{\beta\epsilon}\left(R_{2q,v_\epsilon}(\rho_0) + \log\frac{1}{\delta}\right)\right)$ iterations. Suppose δ is small enough that $\gamma_{2q-1}\left(\frac{\delta}{2}\right) < \frac{1}{L}$. Note that v_ϵ is $\frac{1}{2\epsilon}$-smooth, so by choosing ρ_0 to be a Gaussian with covariance $2\epsilon I$, we have $R_{2q,v_\epsilon}(\rho_0) = \tilde{O}(n)$ by Lemma 4. Therefore, Theorem 6 yields an iteration complexity of $k = \tilde{\Theta}\left(\frac{n}{\beta\gamma_{2q-1}\left(\frac{\delta}{2}\right)}\right)$. Note the additional dependence on dimension, compared to the LSI case in Sect. 5.3.

For example, if $R_{q,v}(v_\epsilon) = O(\epsilon)$, then $\gamma_q(\delta) = \Omega(\delta)$, so the iteration complexity is $k = \tilde{\Theta}\left(\frac{n}{\beta\delta}\right)$ with step size $\epsilon = \Theta(\delta)$. On the other hand, if $R_{q,v}(v_\epsilon) = O(\epsilon^2)$, as in Example 3, then $\gamma_q(\delta) = \Omega(\sqrt{\delta})$, so the iteration complexity is $k = \tilde{\Theta}\left(\frac{n}{\beta\sqrt{\delta}}\right)$ with step size $\epsilon = \Theta(\sqrt{\delta})$.

7 Properties of Biased Limit

7.1 Bound on Bias Under Third-Order Smoothness

Let v_ϵ be the biased limit of ULA with step size $\epsilon > 0$. Let $\mu_\epsilon = (I - \epsilon\nabla f)_\# v_\epsilon$, so v_ϵ satisfies

$$v_\epsilon = \mu_\epsilon * \mathcal{N}(0, 2\epsilon I).$$

We will bound the relative Fisher information $J_\nu(\nu_\epsilon)$ under third-order smoothness. We say f is (L, M)-**smooth** if f is L-smooth (∇f is L-Lipschitz), and $\nabla^2 f$ is M-Lipschitz, or $\|\nabla^3 f\|_{\mathrm{op}} \leq M$. We provide the proof of Theorem 7 in Sects. 8.6.4 and 8.6.5.

Theorem 7 *We have the following:*

1. *If f is (L, M)-smooth and $\epsilon \leq \frac{1}{2L}$, then:*

$$J_\nu(\nu_\epsilon) \leq 2\epsilon n \left(L^2 + 2\sqrt{n}LM + 3nM^2 \right).$$

2. *For any f and $\epsilon > 0$ (such that ν_ϵ exists and the quantities below are defined):*

$$J_\nu(\nu_\epsilon) \geq \frac{\epsilon^2}{4} \frac{\left(\mathbb{E}_{\nu_\epsilon}[\|\nabla f\|^2] \right)^2}{\mathrm{Var}_{\nu_\epsilon}(X)}.$$

Note the dependence on ϵ in the upper bound above is $O(\epsilon)$, while the lower bound is $\Omega(\epsilon^2)$.

Example 5 Recall that if $\nu = \mathcal{N}(0, \frac{1}{\alpha}I)$, then $\nu_\epsilon = \mathcal{N}(0, \frac{1}{\alpha(1-\frac{\epsilon\alpha}{2})}I)$ for $\epsilon < \frac{2}{\alpha}$. Then[1]

$$J_\nu(\nu_\epsilon) = \frac{\epsilon^2}{4} \frac{n\alpha^2}{(1 - \frac{\epsilon\alpha}{2})^2} = \Theta(\epsilon^2 n\alpha^2). \tag{27}$$

Note the lower bound in Theorem 7 has the correct order of ϵ, but the upper bound is not tight.

Remark 4 Recall from Theorem 1 that under LSI and L-smoothness we have $H_\nu(\nu_\epsilon) \leq O(\frac{\epsilon n L^2}{\alpha})$. Under α-LSI, the upper bound in Theorem 7 implies $H_\nu(\nu_\epsilon) \leq O(\frac{\epsilon n}{\alpha}(L^2 + \sqrt{n}LM + nM^2))$; this has the same first term as in the bound from Theorem 1, but has an additional dependence on third-order smoothness.

We note that in general, convergence in relative Fisher information does not necessarily imply convergence of the underlying distributions in total variation; see for example [7, Proposition 1]. We also note that by examining the proof of the upper bound in Theorem 7, we can conclude that $J_\nu(\nu_\epsilon) \leq \epsilon n L^2$ assuming f is L-smooth and $\Delta\Delta f \geq 0$.

---

[1] Recall for $\nu = \mathcal{N}(0, \frac{1}{\alpha}I)$ and $\rho = \mathcal{N}(0, \frac{1}{\beta}I)$ on $\mathbb{R}^n$, the relative Fisher information is $J_\nu(\rho) = \frac{n}{\beta}(\beta - \alpha)^2$.

7.2 Isoperimetry of Biased Limit Under Strong Log-Concavity and Smoothness

If ν is smooth and strongly log-concave, then the biased limit ν_ϵ satisfies LSI (hence also Poincaré), so Assumptions 1 and 2 are satisfied. We provide the proof of Theorem 8 in Sect. 8.6.6.

Theorem 8 *If ν is α-strongly log-concave and L-smooth, and $\epsilon \leq \frac{1}{L}$, then ν_ϵ is β-LSI with $\beta \geq \frac{\alpha}{2}$.*

With Theorem 8, we know that for target distributions which are smooth and strongly log-concave, we have convergence of ULA in Rényi divergence to the biased limit, as in Theorem 5. However, the final bound is in terms of the bias in Rényi divergence, which we do not bound. (Under third-order smoothness, we can bound it in relative Fisher information as in Theorem 7, but it does not bound the Rényi divergence.) The work of [22] extends our interpolation technique to show the convergence in Rényi divergence under LSI as well as a general family of isoperimetric inequalities, and proves a bound on the Rényi bias under LSI and smoothness.

8 Proofs and Details

8.1 Proofs for Sect. 2: KL Divergence Along Langevin Dynamics

8.1.1 Proof of Lemma 1

Proof of Lemma 1 Since f is L-smooth and $\nabla f(x^*) = 0$, we have the bound

$$f(x) \leq f(x^*) + \langle \nabla f(x^*), x - x^* \rangle + \frac{L}{2}\|x - x^*\|^2 = f(x^*) + \frac{L}{2}\|x - x^*\|^2.$$

Let $X \sim \rho = \mathcal{N}(x^*, \frac{1}{L}I)$. Then

$$\mathbb{E}_\rho[f(X)] \leq f(x^*) + \frac{L}{2}\text{Var}_\rho(X) = f(x^*) + \frac{n}{2}.$$

Recall the entropy of ρ is $H(\rho) = -\mathbb{E}_\rho[\log \rho(X)] = \frac{n}{2}\log\frac{2\pi e}{L}$. Therefore, the KL divergence is

$$H_\nu(\rho) = \int \rho (\log \rho + f)\, dx = -H(\rho) + \mathbb{E}_\rho[f] \leq f(x^*) + \frac{n}{2}\log\frac{L}{2\pi}.$$

$\square$

8.1.2   Proof of Lemma 2

Proof of Lemma 2 Recall the time derivative of KL divergence along any flow is given by

$$\frac{d}{dt} H_\nu(\rho_t) = \frac{d}{dt} \int_{\mathbb{R}^n} \rho_t \log \frac{\rho_t}{\nu} \, dx = \int_{\mathbb{R}^n} \frac{\partial \rho_t}{\partial t} \log \frac{\rho_t}{\nu} \, dx$$

since the second part of the chain rule is zero: $\int \rho_t \frac{\partial}{\partial t} \log \frac{\rho_t}{\nu} \, dx = \int \frac{\partial \rho_t}{\partial t} \, dx = \frac{d}{dt} \int \rho_t \, dx = 0$. Therefore, along the Fokker-Planck equation (8) for the Langevin dynamics (7),

$$\frac{d}{dt} H_\nu(\rho_t) = \int \nabla \cdot \left(\rho_t \nabla \log \frac{\rho_t}{\nu} \right) \log \frac{\rho_t}{\nu} \, dx$$

$$= - \int \rho_t \left\| \nabla \log \frac{\rho_t}{\nu} \right\|^2 dx$$

$$= -J_\nu(\rho_t)$$

where in the second equality we have applied integration by parts.   □

8.1.3   Proof of Theorem 4

Proof of Theorem 4 From Lemma 2 and the LSI assumption (4),

$$\frac{d}{dt} H_\nu(\rho_t) = -J_\nu(\rho_t) \le -2\alpha H_\nu(\rho_t).$$

Integrating implies the desired bound $H_\nu(\rho_t) \le e^{-2\alpha t} H_\nu(\rho_0)$.

Furthermore, since ν satisfies LSI with constant α, it also satisfies Talagrand's inequality (6) with constant α [66, Theorem 1]. Therefore, $W_2(\rho_t, \nu)^2 \le \frac{2}{\alpha} H_\nu(\rho_t) \le \frac{2}{\alpha} e^{-2\alpha t} H_\nu(\rho_0)$, as desired.   □

8.2   Proofs for Sect. 3: Unadjusted Langevin Algorithm

8.2.1   Proof of Lemma 3

We will use the following auxiliary results.

Lemma 11 *Assume $\nu = e^{-f}$ is L-smooth. Then*

$$\mathbb{E}_\nu[\|\nabla f\|^2] \le nL.$$

Proof Since $v = e^{-f}$, by integration by parts we can write

$$\mathbb{E}_v[\|\nabla f\|^2] = \mathbb{E}_v[\Delta f].$$

Since v is L-smooth, $\nabla^2 f(x) \preceq L I$, so $\Delta f(x) \leq nL$ for all $x \in \mathbb{R}^n$. Therefore, $\mathbb{E}_v[\|\nabla f\|^2] = \mathbb{E}_v[\Delta f] \leq nL$, as desired.   □

Lemma 12 *Suppose v satisfies Talagrand's inequality with constant $\alpha > 0$ and is L-smooth. For any ρ,*

$$\mathbb{E}_\rho[\|\nabla f\|^2] \leq \frac{4L^2}{\alpha} H_v(\rho) + 2nL.$$

Proof Let $x \sim \rho$ and $x^* \sim v$ with an optimal coupling (x, x^*) so that $\mathbb{E}[\|x - x^*\|^2] = W_2(\rho, v)^2$. Since $v = e^{-f}$ is L-smooth, ∇f is L-Lipschitz. By triangle inequality,

$$\|\nabla f(x)\| \leq \|\nabla f(x) - \nabla f(x^*)\| + \|\nabla f(x^*)\|$$
$$\leq L\|x - x^*\| + \|\nabla f(x^*)\|.$$

Squaring, using $(a + b)^2 \leq 2a^2 + 2b^2$, and taking expectation, we get

$$\mathbb{E}_\rho[\|\nabla f(x)\|^2] \leq 2L^2 \mathbb{E}[\|x - x^*\|^2] + 2\mathbb{E}_v[\|\nabla f(x^*)\|^2]$$
$$= 2L^2 W_2(\rho, v)^2 + 2\mathbb{E}_v[\|\nabla f(x^*)\|^2].$$

By Talagrand's inequality (6), $W_2(\rho, v)^2 \leq \frac{2}{\alpha} H_v(\rho)$. By Lemma 11 we have $\mathbb{E}_v[\|\nabla f(x^*)\|^2] \leq nL$. Plugging these to the bound above gives the desired result.   □

We are now ready to prove Lemma 3.

Proof of Lemma 3 For simplicity suppose $k = 0$, so we start at $x_0 \sim \rho_0$. We write one step of ULA

$$x_0 \mapsto x_0 - \epsilon \nabla f(x_0) + \sqrt{2\epsilon} z_0$$

as the output at time ϵ of the stochastic differential equation

$$dx_t = -\nabla f(x_0) \, dt + \sqrt{2} \, dW_t \tag{28}$$

where W_t is the standard Brownian motion in $\mathbb{R}^n$ starting at $W_0 = 0$. Indeed, the solution to (28) at time $t = \epsilon$ is

$$x_\epsilon = x_0 - \epsilon \nabla f(x_0) + \sqrt{2} \, W_\epsilon$$

$$\stackrel{d}{=} x_0 - \epsilon \nabla f(x_0) + \sqrt{2\epsilon} \, z_0 \tag{29}$$

where $z_0 \sim \mathcal{N}(0, I)$, which is identical to the ULA update.

We derive the continuity equation corresponding to (28) as follows. For each $t > 0$, let $\rho_{0t}(x_0, x_t)$ denote the joint distribution of (x_0, x_t), which we write in terms of the conditionals and marginals as

$$\rho_{0t}(x_0, x_t) = \rho_0(x_0)\rho_{t|0}(x_t \mid x_0) = \rho_t(x_t)\rho_{0|t}(x_0 \mid x_t).$$

Conditioning on x_0, the drift vector field $-\nabla f(x_0)$ is a constant, so the Fokker-Planck formula for the conditional density $\rho_{t|0}(x_t \mid x_0)$ is

$$\frac{\partial \rho_{t|0}(x_t \mid x_0)}{\partial t} = \nabla \cdot \left(\rho_{t|0}(x_t \mid x_0) \nabla f(x_0) \right) + \Delta \rho_{t|0}(x_t \mid x_0). \tag{30}$$

To derive the evolution of ρ_t, we take expectation over $x_0 \sim \rho_0$. Multiplying both sides of (30) by $\rho_0(x_0)$ and integrating over x_0, we obtain

$$\frac{\partial \rho_t(x)}{\partial t} = \int_{\mathbb{R}^n} \frac{\partial \rho_{t|0}(x \mid x_0)}{\partial t} \rho_0(x_0) \, dx_0$$

$$= \int_{\mathbb{R}^n} \left(\nabla \cdot \left(\rho_{t|0}(x \mid x_0) \nabla f(x_0) \right) + \Delta \rho_{t|0}(x \mid x_0) \right) \rho_0(x_0) \, dx_0$$

$$= \int_{\mathbb{R}^n} \left(\nabla \cdot \left(\rho_{t,0}(x, x_0) \nabla f(x_0) \right) + \Delta \rho_{t,0}(x, x_0) \right) dx_0$$

$$= \nabla \cdot \left(\rho_t(x) \int_{\mathbb{R}^n} \rho_{0|t}(x_0 \mid x) \nabla f(x_0) \, dx_0 \right) + \Delta \rho_t(x)$$

$$= \nabla \cdot \left(\rho_t(x) \mathbb{E}_{\rho_{0|t}}[\nabla f(x_0) \mid x_t = x] \right) + \Delta \rho_t(x). \tag{31}$$

Observe that the difference between the Fokker-Planck equations (31) for ULA and (8) for Langevin dynamics is in the first term, that the drift is now the conditional expectation $\mathbb{E}_{\rho_{0|t}}[\nabla f(x_0) \mid x_t = x]$, rather than the true gradient $\nabla f(x)$.

Recall the time derivative of relative entropy along any flow is given by

$$\frac{d}{dt} H_\nu(\rho_t) = \frac{d}{dt} \int_{\mathbb{R}^n} \rho_t \log \frac{\rho_t}{\nu} \, dx = \int_{\mathbb{R}^n} \frac{\partial \rho_t}{\partial t} \log \frac{\rho_t}{\nu} \, dx$$

since the second part of the chain rule is zero: $\int \rho_t \frac{\partial}{\partial t} \log \frac{\rho_t}{\nu} \, dx = \int \frac{\partial \rho_t}{\partial t} \, dx = \frac{d}{dt} \int \rho_t \, dx = 0$.

Therefore, the time derivative of relative entropy for ULA, using the Fokker-Planck equation (31) and integrating by parts, is given by:

$$
\begin{aligned}
\frac{d}{dt} H_v(\rho_t) &= \int_{\mathbb{R}^n} \left(\nabla \cdot \left(\rho_t(x) \mathbb{E}_{\rho_{0|t}}[\nabla f(x_0) \mid x_t = x] \right) + \Delta \rho_t(x) \right) \log \frac{\rho_t(x)}{v(x)} \, dx \\
&= \int_{\mathbb{R}^n} \left(\nabla \cdot \left(\rho_t(x) \left(\nabla \log \frac{\rho_t(x)}{v(x)} + \mathbb{E}_{\rho_{0|t}}[\nabla f(x_0) \mid x_t = x] \right. \right. \right. \\
&\quad \left. \left. \left. - \nabla f(x) \right) \right) \right) \log \frac{\rho_t(x)}{v(x)} \, dx \\
&= -\int_{\mathbb{R}^n} \rho_t(x) \left\langle \nabla \log \frac{\rho_t(x)}{v(x)} + \mathbb{E}_{\rho_{0|t}}[\nabla f(x_0) \mid x_t = x] \right. \\
&\quad \left. - \nabla f(x), \, \nabla \log \frac{\rho_t(x)}{v(x)} \right\rangle dx \\
&= -\int_{\mathbb{R}^n} \rho_t(x) \left\| \nabla \log \frac{\rho_t}{v} \right\|^2 dx \\
&\quad + \int_{\mathbb{R}^n} \rho_t(x) \left\langle \nabla f(x) - \mathbb{E}_{\rho_{0|t}}[\nabla f(x_0) \mid x_t = x], \, \nabla \log \frac{\rho_t(x)}{v(x)} \right\rangle dx \\
&= -J_v(\rho_t) + \int_{\mathbb{R}^n \times \mathbb{R}^n} \rho_{0t}(x_0, x) \left\langle \nabla f(x) - \nabla f(x_0), \, \nabla \log \frac{\rho_t(x)}{v(x)} \right\rangle dx_0 \, dx \\
&= -J_v(\rho_t) + \mathbb{E}_{\rho_{0t}} \left[\left\langle \nabla f(x_t) - \nabla f(x_0), \, \nabla \log \frac{\rho_t(x_t)}{v(x_t)} \right\rangle \right] \quad (32)
\end{aligned}
$$

where in the last step we have renamed x as x_t. The first term in (32) is the same as in the Langevin dynamics. The second term in (32) is the discretization error, which we can bound as follows. Using $\langle a, b \rangle \leq \|a\|^2 + \frac{1}{4}\|b\|^2$ and since ∇f is L-Lipschitz,

$$
\begin{aligned}
\mathbb{E}_{\rho_{0t}} \left[\left\langle \nabla f(x_t) - \nabla f(x_0), \, \nabla \log \frac{\rho_t(x_t)}{v(x_t)} \right\rangle \right] &\leq \mathbb{E}_{\rho_{0t}}[\|\nabla f(x_t) - \nabla f(x_0)\|^2] \\
&\quad + \frac{1}{4} \mathbb{E}_{\rho_{0t}} \left[\left\| \nabla \log \frac{\rho_t(x_t)}{v(x_t)} \right\|^2 \right] \\
&= \mathbb{E}_{\rho_{0t}}[\|\nabla f(x_t) - \nabla f(x_0)\|^2] \\
&\quad + \frac{1}{4} J_v(\rho_t) \\
&\leq L^2 \mathbb{E}_{\rho_{0t}}[\|x_t - x_0\|^2] + \frac{1}{4} J_v(\rho_t) \\
&\quad (33)
\end{aligned}
$$

Recall from (29) the solution of ULA is $x_t \overset{d}{=} x_0 - t\nabla f(x_0) + \sqrt{2t}\, z_0$, where $z_0 \sim \mathcal{N}(0, I)$ is independent of x_0. Then

$$
\begin{aligned}
\mathbb{E}_{\rho_{0t}}[\|x_t - x_0\|^2] &= \mathbb{E}_{\rho_{0t}}[\| - t\nabla f(x_0) + \sqrt{2t}\, z_0\|^2] \\
&= t^2 \mathbb{E}_{\rho_0}[\|\nabla f(x_0)\|^2] + 2tn \\
&\leq \frac{4t^2 L^2}{\alpha} H_\nu(\rho_0) + 2t^2 nL + 2tn
\end{aligned}
$$

where in the last inequality we have used Lemma 12. This bounds the discretization error by

$$
\mathbb{E}_{\rho_{0t}}\left[\left\langle \nabla f(x_t) - \nabla f(x_0), \nabla \log \frac{\rho_t(x_t)}{\nu(x_t)}\right\rangle\right] \leq \frac{4t^2 L^4}{\alpha} H_\nu(\rho_0) + 2t^2 nL^3
$$

$$
+ 2tnL^2 + \frac{1}{4} J_\nu(\rho_t).
$$

Therefore, from (32), the time derivative of KL divergence along ULA is bounded by

$$
\frac{d}{dt} H_\nu(\rho_t) \leq -\frac{3}{4} J_\nu(\rho_t) + \frac{4t^2 L^4}{\alpha} H_\nu(\rho_0) + 2t^2 nL^3 + 2tnL^2.
$$

Then by the LSI (4) assumption,

$$
\frac{d}{dt} H_\nu(\rho_t) \leq -\frac{3\alpha}{2} H_\nu(\rho_t) + \frac{4t^2 L^4}{\alpha} H_\nu(\rho_0) + 2t^2 nL^3 + 2tnL^2.
$$

We wish to integrate the inequality above for $0 \leq t \leq \epsilon$. Using $t \leq \epsilon$ and since $\epsilon \leq \frac{1}{2L}$, we simplify the above to

$$
\frac{d}{dt} H_\nu(\rho_t) \leq -\frac{3\alpha}{2} H_\nu(\rho_t) + \frac{4\epsilon^2 L^4}{\alpha} H_\nu(\rho_0) + 2\epsilon^2 nL^3 + 2\epsilon nL^2
$$

$$
\leq -\frac{3\alpha}{2} H_\nu(\rho_t) + \frac{4\epsilon^2 L^4}{\alpha} H_\nu(\rho_0) + 3\epsilon nL^2.
$$

Multiplying both sides by $e^{\frac{3\alpha}{2}t}$, we can write the above as

$$
\frac{d}{dt}\left(e^{\frac{3\alpha}{2}t} H_\nu(\rho_t)\right) \leq e^{\frac{3\alpha}{2}t}\left(\frac{4\epsilon^2 L^4}{\alpha} H_\nu(\rho_0) + 3\epsilon nL^2\right).
$$

Integrating from $t = 0$ to $t = \epsilon$ gives

$$e^{\frac{3}{2}\alpha\epsilon} H_\nu(\rho_\epsilon) - H_\nu(\rho_0) \le \frac{2(e^{\frac{3}{2}\alpha\epsilon} - 1)}{3\alpha} \left(\frac{4\epsilon^2 L^4}{\alpha} H_\nu(\rho_0) + 3\epsilon n L^2 \right)$$

$$\le 2\epsilon \left(\frac{4\epsilon^2 L^4}{\alpha} H_\nu(\rho_0) + 3\epsilon n L^2 \right)$$

where in the last step we have used the inequality $e^c \le 1 + 2c$ for $0 < c = \frac{3}{2}\alpha\epsilon \le 1$, which holds because $0 < \epsilon \le \frac{2}{3\alpha}$. Rearranging, the inequality above gives

$$H_\nu(\rho_\epsilon) \le e^{-\frac{3}{2}\alpha\epsilon} \left(1 + \frac{8\epsilon^3 L^4}{\alpha} \right) H_\nu(\rho_0) + e^{-\frac{3}{2}\alpha\epsilon} 6\epsilon^2 n L^2.$$

Since $1 + \frac{8\epsilon^3 L^4}{\alpha} \le 1 + \frac{\alpha\epsilon}{2} \le e^{\frac{1}{2}\alpha\epsilon}$ for $\epsilon \le \frac{\alpha}{4L^2}$, and using $e^{-\frac{3}{2}\alpha\epsilon} \le 1$, we conclude that

$$H_\nu(\rho_\epsilon) \le e^{-\alpha\epsilon} H_\nu(\rho_0) + 6\epsilon^2 n L^2.$$

This is the desired inequality, after renaming $\rho_0 \equiv \rho_k$ and $\rho_\epsilon \equiv \rho_{k+1}$. Note that the conditions $\epsilon \le \frac{1}{2L}$ and $\epsilon \le \frac{2}{3\alpha}$ above are also implied by the assumption $\epsilon \le \frac{\alpha}{4L^2}$ since $\alpha \le L$. □

8.2.2 Proof of Theorem 1

Proof of Theorem 1 Applying the recursion (12) from Lemma 3, we obtain

$$H_\nu(\rho_k) \le e^{-\alpha\epsilon k} H_\nu(\rho_0) + \frac{6\epsilon^2 n L^2}{1 - e^{-\alpha\epsilon}} \le e^{-\alpha\epsilon k} H_\nu(\rho_0) + \frac{8\epsilon n L^2}{\alpha}$$

where in the last step we have used the inequality $1 - e^{-c} \ge \frac{3}{4}c$ for $0 < c = \alpha\epsilon \le \frac{1}{4}$, which holds since $\epsilon \le \frac{\alpha}{4L^2} \le \frac{1}{4\alpha}$.

Given $\delta > 0$, if we further assume $\epsilon \le \frac{\delta\alpha}{16nL^2}$, then the above implies $H_\nu(\rho_k) \le e^{-\alpha\epsilon k} H_\nu(\rho_0) + \frac{\delta}{2}$. This means for $k \ge \frac{1}{\alpha\epsilon} \log \frac{2H_\nu(\rho_0)}{\delta}$, we have $H_\nu(\rho_k) \le \frac{\delta}{2} + \frac{\delta}{2} = \delta$, as desired. □

8.3   Details for Sect. 4: Rényi Divergence Along Langevin Dynamics

8.3.1   Properties of Rényi Divergence

We recall that Rényi divergence is increasing in the order.

Lemma 13 *For any probability distributions* ρ, ν, *the map* $q \mapsto R_{q,\nu}(\rho)$ *is increasing for* $q > 0$.

Proof Let $0 < q \leq r$. We will show that $R_{q,\nu}(\rho) \leq R_{r,\nu}(\rho)$.

First suppose $q > 1$. We write $F_{q,\nu}(\rho)$ as an expectation over ρ and use power mean inequality:

$$F_{q,\nu}(\rho) = \mathbb{E}_\nu\left[\left(\frac{\rho}{\nu}\right)^q\right] = \mathbb{E}_\rho\left[\left(\frac{\rho}{\nu}\right)^{q-1}\right] \leq \mathbb{E}_\rho\left[\left(\frac{\rho}{\nu}\right)^{r-1}\right]^{\frac{q-1}{r-1}}$$

$$= \mathbb{E}_\nu\left[\left(\frac{\rho}{\nu}\right)^r\right]^{\frac{q-1}{r-1}} = F_{r,\nu}(\rho)^{\frac{q-1}{r-1}}.$$

Taking logarithm and dividing by $q - 1 > 0$ gives

$$R_{q,\nu}(\rho) = \frac{1}{q-1}\log F_{q,\nu}(\rho) \leq \frac{1}{r-1}\log F_{r,\nu}(\rho) = R_{r,\nu}(\rho).$$

The case $q = 1$ follows by taking limit $q \to 1$.

Now suppose $q \leq r < 1$, so $1 - q \geq 1 - r > 0$. We again write $F_{q,\nu}(\rho)$ as an expectation over ρ and use power mean inequality:

$$F_{q,\nu}(\rho) = \mathbb{E}_\nu\left[\left(\frac{\rho}{\nu}\right)^q\right] = \mathbb{E}_\rho\left[\left(\frac{\nu}{\rho}\right)^{1-q}\right] \geq \mathbb{E}_\rho\left[\left(\frac{\nu}{\rho}\right)^{1-r}\right]^{\frac{1-q}{1-r}}$$

$$= \mathbb{E}_\nu\left[\left(\frac{\rho}{\nu}\right)^r\right]^{\frac{1-q}{1-r}} = F_{r,\nu}(\rho)^{\frac{1-q}{1-r}}.$$

Taking logarithm and dividing by $q - 1 < 0$ (which flips the inequality) gives

$$R_{q,\nu}(\rho) = \frac{1}{q-1}\log F_{q,\nu}(\rho) \leq \frac{1}{r-1}\log F_{r,\nu}(\rho) = R_{r,\nu}(\rho).$$

The case $q < 1 \leq r$ follows since $R_{q,\nu}(\rho) \leq R_{1,\nu}(\rho) \leq R_{r,\nu}(\rho)$. $\square$

8.3.2 Proof of Lemma 4

Proof of Lemma 4 Since f is L-smooth and x^* is a stationary point of f, for all $x \in \mathbb{R}^n$ we have

$$f(x) \le f(x^*) + \langle \nabla f(x^*), x - x^* \rangle + \frac{L}{2}\|x - x^*\|^2 = f(x^*) + \frac{L}{2}\|x - x^*\|^2.$$

Let $q > 1$. Then for $\rho = \mathcal{N}(x^*, \sigma^2 I)$ with $\frac{q}{\sigma^2} > (q-1)L$,

$$
\begin{aligned}
F_{q,\nu}(\rho) &= \int_{\mathbb{R}^n} \frac{\rho(x)^q}{\nu(x)^{q-1}} dx \\
&= \frac{1}{(2\pi\sigma^2)^{\frac{nq}{2}}} \int_{\mathbb{R}^n} e^{-\frac{q}{2\sigma^2}\|x-x^*\|^2 + (q-1)f(x)} dx \\
&\le \frac{1}{(2\pi\sigma^2)^{\frac{nq}{2}}} \int_{\mathbb{R}^n} e^{-\frac{q}{2\sigma^2}\|x-x^*\|^2 + (q-1)f(x^*) + \frac{(q-1)L}{2}\|x-x^*\|^2} dx \\
&= \frac{e^{(q-1)f(x^*)}}{(2\pi\sigma^2)^{\frac{nq}{2}}} \int_{\mathbb{R}^n} e^{-\frac{1}{2}\left(\frac{q}{\sigma^2} - (q-1)L\right)\|x-x^*\|^2} dx \\
&= \frac{e^{(q-1)f(x^*)}}{(2\pi\sigma^2)^{\frac{nq}{2}}} \left(\frac{2\pi}{\frac{q}{\sigma^2} - (q-1)L}\right)^{\frac{n}{2}} \\
&= \frac{e^{(q-1)f(x^*)}}{(2\pi)^{\frac{n}{2}(q-1)}(\sigma^2)^{\frac{nq}{2}}} \frac{1}{\left(\frac{q}{\sigma^2} - (q-1)L\right)^{\frac{n}{2}}}.
\end{aligned}
$$

Therefore,

$$R_{q,\nu}(\rho) = \frac{1}{q-1}\log F_{q,\nu}(\rho) \le f(x^*) - \frac{n}{2}\log 2\pi$$

$$- \frac{n}{2(q-1)}\log \sigma^{2q}\left(\frac{q}{\sigma^2} - (q-1)L\right).$$

In particular, if $\sigma^2 = \frac{1}{L}$, then $\frac{q}{\sigma^2} - (q-1)L = L > 0$, and the bound above becomes

$$R_{q,\nu}(\rho) \le f(x^*) + \frac{n}{2}\log \frac{L}{2\pi}.$$

The case $q = 1$ follows from Lemma 1. $\qquad\qquad\square$

8.3.3 Proof of Lemma 5

Proof of Lemma 5 We plug in $h^2 = \left(\frac{\rho}{\nu}\right)^q$ to the LSI definition (2) to obtain

$$\frac{q^2}{2\alpha} G_{q,\nu}(\rho) \geq q \mathbb{E}_\nu \left[\left(\frac{\rho}{\nu}\right)^q \log \frac{\rho}{\nu} \right] - F_{q,\nu}(\rho) \log F_{q,\nu}(\rho) \qquad (34)$$

$$= q \frac{\partial}{\partial q} F_{q,\nu}(\rho) - F_{q,\nu}(\rho) \log F_{q,\nu}(\rho).$$

Therefore,

$$\frac{q^2}{2\alpha} \frac{G_{q,\nu}(\rho)}{F_{q,\nu}(\rho)} \geq q \frac{\partial}{\partial q} \log F_{q,\nu}(\rho) - \log F_{q,\nu}(\rho)$$

$$= q \frac{\partial}{\partial q} \left((q-1) R_{q,\nu}(\rho) \right) - (q-1) R_{q,\nu}(\rho)$$

$$= q R_{q,\nu}(\rho) + q(q-1) \frac{\partial}{\partial q} R_{q,\nu}(\rho) - (q-1) R_{q,\nu}(\rho)$$

$$= R_{q,\nu}(\rho) + q(q-1) \frac{\partial}{\partial q} R_{q,\nu}(\rho)$$

$$\geq R_{q,\nu}(\rho)$$

where in the last inequality we have used $q \geq 1$ and $\frac{\partial}{\partial q} R_{q,\nu}(\rho) \geq 0$ since $q \mapsto R_{q,\nu}(\rho)$ is increasing by Lemma 13.   □

8.3.4 Proof of Lemma 6

Proof of Lemma 6 Let $q > 0$, $q \neq 1$. By the Fokker-Planck formula (8) and integration by parts,

$$\frac{d}{dt} F_{q,\nu}(\rho_t) = \int_{\mathbb{R}^n} \nu \frac{\frac{\partial}{\partial t}(\rho_t^q)}{\nu^q} dx$$

$$= q \int_{\mathbb{R}^n} \frac{\rho_t^{q-1}}{\nu^{q-1}} \frac{\partial \rho_t}{\partial t} dx$$

$$= q \int_{\mathbb{R}^n} \left(\frac{\rho_t}{\nu}\right)^{q-1} \nabla \cdot \left(\rho_t \nabla \log \frac{\rho_t}{\nu} \right) dx$$

$$= -q \int_{\mathbb{R}^n} \rho_t \left\langle \nabla \left(\frac{\rho_t}{\nu}\right)^{q-1}, \nabla \log \frac{\rho_t}{\nu} \right\rangle dx$$

$$= -q(q-1) \int_{\mathbb{R}^n} \rho_t \left\langle \left(\frac{\rho_t}{\nu}\right)^{q-2} \nabla \frac{\rho_t}{\nu}, \left(\frac{\rho_t}{\nu}\right)^{-1} \nabla \frac{\rho_t}{\nu} \right\rangle dx$$

$$= -q(q-1)\mathbb{E}_\nu \left[\left(\frac{\rho_t}{\nu}\right)^{q-2} \left\| \nabla \frac{\rho_t}{\nu} \right\|^2 \right]$$

$$= -q(q-1)G_{q,\nu}(\rho_t). \tag{35}$$

Therefore,

$$\frac{d}{dt} R_{q,\nu}(\rho_t) = \frac{1}{q-1} \frac{\frac{d}{dt} F_{q,\nu}(\rho_t)}{F_{q,\nu}(\rho_t)} = -q \frac{G_{q,\nu}(\rho_t)}{F_{q,\nu}(\rho_t)}.$$

For $q = 1$, we have $R_{1,\nu}(\rho_t) = H_\nu(\rho_t)$, $G_{1,\nu}(\rho_t) = J_\nu(\rho_t)$, and $F_{1,\nu}(\rho_t) = 1$, and the claim (19) follows from Lemma 2. □

8.3.5 Proof of Theorem 2

Proof of Theorem 2 By Lemmas 5 and 6,

$$\frac{d}{dt} R_{q,\nu}(\rho_t) = -q \frac{G_{q,\nu}(\rho_t)}{F_{q,\nu}(\rho_t)} \leq -\frac{2\alpha}{q} R_{q,\nu}(\rho_t).$$

Integrating gives

$$R_{q,\nu}(\rho_t) \leq e^{-\frac{2\alpha}{q}t} R_{q,\nu}(\rho_0)$$

as desired. □

8.3.6 Hypercontractivity

Lemma 14 *Suppose ν satisfies LSI with constant $\alpha > 0$. Let $q_0 > 1$, and suppose $R_{q_0,\nu}(\rho_0) < \infty$. Define $q_t = 1 + e^{2\alpha t}(q_0 - 1)$. Along the Langevin dynamics (7), for all $t \geq 0$,*

$$\left(1 - \frac{1}{q_t}\right) R_{q_t,\nu}(\rho_t) \leq \left(1 - \frac{1}{q_0}\right) R_{q_0,\nu}(\rho_0). \tag{36}$$

In particular, for any $q \geq q_0$, we have $R_{q,\nu}(\rho_t) \leq R_{q_0,\nu}(\rho_0) < \infty$ for all $t \geq \frac{1}{2\alpha} \log \frac{q-1}{q_0-1}$.

Proof We will show $\frac{d}{dt} \left\{ \left(1 - \frac{1}{q_t}\right) R_{q_t,\nu}(\rho_t) \right\} \leq 0$, which implies the desired relation (36). Since $q_t = 1 + e^{2\alpha t}(q_0 - 1)$, we have $\dot{q}_t = \frac{d}{dt} q_t = 2\alpha(q_t - 1)$.

Note that

$$\frac{d}{dt} R_{q_t,\nu}(\rho_t) = \frac{d}{dt} \left(\frac{\log F_{q_t,\nu}(\rho_t)}{q_t - 1} \right)$$

$$\overset{(35)}{=} -\frac{\dot{q}_t \log F_{q_t,\nu}(\rho_t)}{(q_t - 1)^2} + \frac{\dot{q}_t \mathbb{E}_\nu \left[\left(\frac{\rho_t}{\nu}\right)^{q_t} \log \frac{\rho_t}{\nu} \right] - q_t(q_t - 1)G_{q_t,\nu}(\rho_t)}{(q_t - 1)F_{q_t,\nu}(\rho_t)}$$

$$= -2\alpha R_{q_t,\nu}(\rho_t) + 2\alpha \frac{\mathbb{E}_\nu \left[\left(\frac{\rho_t}{\nu}\right)^{q_t} \log \frac{\rho_t}{\nu} \right]}{F_{q_t,\nu}(\rho_t)} - q_t \frac{G_{q_t,\nu}(\rho_t)}{F_{q_t,\nu}(\rho_t)}.$$

In the second equality above we have used our earlier calculation (35) which holds for fixed q. Then by LSI in the form (34), we have

$$\frac{d}{dt} R_{q_t,\nu}(\rho_t) \leq -2\alpha R_{q_t,\nu}(\rho_t) + 2\alpha \left(\frac{q_t}{2\alpha} \frac{G_{q_t,\nu}(\rho_t)}{F_{q_t,\nu}(\rho_t)} + \frac{1}{q_t} \log F_{q_t,\nu}(\rho_t) \right)$$

$$- q_t \frac{G_{q_t,\nu}(\rho_t)}{F_{q_t,\nu}(\rho_t)}$$

$$= -2\alpha R_{q_t,\nu}(\rho_t) + 2\alpha \left(1 - \frac{1}{q_t} \right) R_{q_t,\nu}(\rho_t)$$

$$= -\frac{2\alpha}{q_t} R_{q_t,\nu}(\rho_t).$$

Therefore,

$$\frac{d}{dt} \left\{ \left(1 - \frac{1}{q_t} \right) R_{q_t,\nu}(\rho_t) \right\} = \frac{\dot{q}_t}{q_t^2} R_{q_t,\nu}(\rho_t) + \left(1 - \frac{1}{q_t} \right) \frac{d}{dt} R_{q_t,\nu}(\rho_t)$$

$$\leq \frac{2\alpha(q_t - 1)}{q_t^2} R_{q_t,\nu}(\rho_t) - \left(1 - \frac{1}{q_t} \right) \frac{2\alpha}{q_t} R_{q_t,\nu}(\rho_t)$$

$$= 0,$$

as desired.

Now given $q \geq q_0$, let $t_0 = \frac{1}{2\alpha} \log \frac{q-1}{q_0-1}$ so $q_{t_0} = q$. Then $R_{q,\nu}(\rho_{t_0}) \leq \frac{q}{(q-1)} \frac{(q_0-1)}{q_0} R_{q_0,\nu}(\rho_0) \leq R_{q_0,\nu}(\rho_0) < \infty$. For $t > t_0$, by applying Theorem 2 starting from ρ_{t_0}, we obtain $R_{q,\nu}(\rho_t) \leq e^{-\frac{2\alpha}{q}(t-t_0)} R_{q,\nu}(\rho_{t_0}) \leq R_{q,\nu}(\rho_{t_0}) \leq R_{q_0,\nu}(\rho_0) < \infty$. $\qquad\square$

By combining Theorem 2 and Lemma 14, we obtain the following characterization of the behavior of Renyi divergence along the Langevin dynamics under LSI.

Corollary 2 *Suppose ν satisfies LSI with constant $\alpha > 0$. Suppose ρ_0 satisfies $R_{q_0,\nu}(\rho_0) < \infty$ for some $q_0 > 1$. Along the Langevin dynamics (7), for all $q \geq q_0$*

and $t \geq t_0 := \frac{1}{2\alpha} \log \frac{q-1}{q_0-1}$,

$$R_{q,\nu}(\rho_t) \leq e^{-\frac{2\alpha}{q}(t-t_0)} R_{q_0,\nu}(\rho_0). \tag{37}$$

Proof By Lemma 14, at $t = t_0$ we have $R_{q,\nu}(\rho_{t_0}) \leq R_{q_0,\nu}(\rho_0)$. For $t > t_0$, by applying Theorem 2 starting from ρ_{t_0}, we have $R_{q,\nu}(\rho_t) \leq e^{-\frac{2\alpha}{q}(t-t_0)} R_{q,\nu}(\rho_{t_0}) \leq e^{-\frac{2\alpha}{q}(t-t_0)} R_{q_0,\nu}(\rho_0)$. □

8.4 Proofs for Sect. 5: Rényi Divergence Along ULA

8.4.1 Proof of Lemma 7

Proof of Lemma 7 By Cauchy-Schwarz inequality,

$$
\begin{aligned}
F_{q,\nu}(\rho) &= \int \frac{\rho^q}{\nu^{q-1}} \, dx \\
&= \int \nu_\epsilon \left(\frac{\rho}{\nu_\epsilon}\right)^q \left(\frac{\nu_\epsilon}{\nu}\right)^{q-1} dx \\
&\leq \left(\int \nu_\epsilon \left(\frac{\rho}{\nu_\epsilon}\right)^{2q} dx\right)^{\frac{1}{2}} \left(\int \nu_\epsilon \left(\frac{\nu_\epsilon}{\nu}\right)^{2(q-1)} dx\right)^{\frac{1}{2}} \\
&= F_{2q,\nu_\epsilon}(\rho)^{\frac{1}{2}} F_{2q-1,\nu}(\nu_\epsilon)^{\frac{1}{2}}.
\end{aligned}
$$

Taking logarithm gives

$$(q-1)R_{q,\nu}(\rho) \leq \frac{(2q-1)}{2} R_{2q,\nu_\epsilon}(\rho) + \frac{(2q-2)}{2} R_{2q-1,\nu}(\nu_\epsilon).$$

Dividing both sides by $q - 1 > 0$ gives the desired inequality (21). □

8.4.2 Proof of Lemma 8

We will use the following auxiliary results. Recall that given a map $T : \mathbb{R}^n \to \mathbb{R}^n$ and a probability distribution ρ, the pushforward $T_{\#}\rho$ is the distribution of $T(x)$ when $x \sim \rho$.

Lemma 15 *Let* $T : \mathbb{R}^n \to \mathbb{R}^n$ *be a differentiable bijective map. For any probability distributions* ρ, ν, *and for all* $q > 0$,

$$R_{q,T_{\#}\nu}(T_{\#}\rho) = R_{q,\nu}(\rho).$$

Proof Let $\tilde{\rho} = T_{\#}\rho$ and $\tilde{v} = T_{\#}v$. By the change of variable formula,

$$\rho(x) = \det(\nabla T(x))\,\tilde{\rho}(T(x)),$$

$$v(x) = \det(\nabla T(x))\,\tilde{v}(T(x)).$$

Since T is differentiable and bijective, $\det(\nabla T(x)) \neq 0$. Therefore,

$$\frac{\tilde{\rho}(T(x))}{\tilde{v}(T(x))} = \frac{\rho(x)}{v(x)}.$$

Now let $X \sim v$, so $T(X) \sim \tilde{v}$. Then for all $q > 0$.

$$F_{q,\tilde{v}}(\tilde{\rho}) = \mathbb{E}_{\tilde{v}}\left[\left(\frac{\tilde{\rho}}{\tilde{v}}\right)^q\right] = \mathbb{E}_{X\sim v}\left[\left(\frac{\tilde{\rho}(T(X))}{\tilde{v}(T(X))}\right)^q\right]$$

$$= \mathbb{E}_{X\sim v}\left[\left(\frac{\rho(X)}{v(X)}\right)^q\right] = F_{q,v}(\rho).$$

Suppose $q \neq 1$. Taking logarithm on both sides and dividing by $q - 1 \neq 0$ yields $R_{q,\tilde{v}}(\tilde{\rho}) = R_{q,v}(\rho)$, as desired. The case $q = 1$ follows from taking limit $q \to 1$, or by an analogous direct argument:

$$H_{\tilde{v}}(\tilde{\rho}) = \mathbb{E}_{\tilde{v}}\left[\frac{\tilde{\rho}}{\tilde{v}}\log\frac{\tilde{\rho}}{\tilde{v}}\right] = \mathbb{E}_{X\sim v}\left[\frac{\tilde{\rho}(T(X))}{\tilde{v}(T(X))}\log\frac{\tilde{\rho}(T(X))}{\tilde{v}(T(X))}\right]$$

$$= \mathbb{E}_{X\sim v}\left[\frac{\rho(X)}{v(X)}\log\frac{\rho(X)}{v(X)}\right] = H_v(\rho).$$

$\square$

We have the following standard result on how the LSI constant changes under a Lipschitz mapping. We recall that $T\colon \mathbb{R}^n \to \mathbb{R}^n$ is L-Lipschitz if $\|T(x) - T(y)\| \leq L\|x - y\|$ for all $x, y \in \mathbb{R}^n$. For completeness, we provide the proof of Lemma 16 in Appendix.

Lemma 16 *Suppose a probability distribution v satisfies LSI with constant $\alpha > 0$. Let $T\colon \mathbb{R}^n \to \mathbb{R}^n$ be a differentiable L-Lipschitz map. Then $\tilde{v} = T_{\#}v$ satisfies LSI with constant α/L^2.*

We also recall the following result on how the LSI constant changes along Gaussian convolution. We provide the proof of Lemma 17 in Appendix.

Lemma 17 *Suppose a probability distribution v satisfies LSI with constant $\alpha > 0$. For $t > 0$, the probability distribution $\tilde{v}_t = v * \mathcal{N}(0, 2t\,I)$ satisfies LSI with constant $\left(\frac{1}{\alpha} + 2t\right)^{-1}$.*

We now derive a formula for the decrease of Rényi divergence along simultaneous heat flow. We note the resulting formula (38) is similar to the formula (19) for the decrease of Rényi divergence along the Langevin dynamics. A generalization of the following formula is also useful for analyzing a proximal sampling algorithm under isoperimetry [18].

Lemma 18 *For any probability distributions ρ_0, ν_0, and for any $t \geq 0$, let $\rho_t = \rho_0 * \mathcal{N}(0, 2tI)$ and $\nu_t = \nu_0 * \mathcal{N}(0, 2tI)$. Then for all $q > 0$,*

$$\frac{d}{dt} R_{q,\nu_t}(\rho_t) = -q \frac{G_{q,\nu_t}(\rho_t)}{F_{q,\nu_t}(\rho_t)}. \tag{38}$$

Proof By definition, ρ_t and ν_t evolve following the simultaneous heat flow:

$$\frac{\partial \rho_t}{\partial t} = \Delta \rho_t, \qquad \frac{\partial \nu_t}{\partial t} = \Delta \nu_t. \tag{39}$$

We will use the following identity for any smooth function $h \colon \mathbb{R}^n \to \mathbb{R}$,

$$\Delta(h^q) = \nabla \cdot \left(q h^{q-1} \nabla h \right) = q(q-1)h^{q-2}\|\nabla h\|^2 + q h^{q-1} \Delta h.$$

We will also use the integration by parts formula (A.1). Then along the simultaneous heat flow (39),

$$\frac{d}{dt} F_{q,\nu_t}(\rho_t) = \frac{d}{dt} \int \frac{\rho_t^q}{\nu_t^{q-1}} dx$$

$$= \int q \left(\frac{\rho_t}{\nu_t} \right)^{q-1} \frac{\partial \rho_t}{\partial t} dx - \int (q-1) \left(\frac{\rho_t}{\nu_t} \right)^q \frac{\partial \nu_t}{\partial t} dx$$

$$= q \int \left(\frac{\rho_t}{\nu_t} \right)^{q-1} \Delta \rho_t \, dx - (q-1) \int \left(\frac{\rho_t}{\nu_t} \right)^q \Delta \nu_t \, dx$$

$$= q \int \Delta \left(\left(\frac{\rho_t}{\nu_t} \right)^{q-1} \right) \rho_t \, dx - (q-1) \int \Delta \left(\left(\frac{\rho_t}{\nu_t} \right)^q \right) \nu_t \, dx$$

$$= q \int \left((q-1)(q-2) \left(\frac{\rho_t}{\nu_t} \right)^{q-3} \left\| \nabla \frac{\rho_t}{\nu_t} \right\|^2 \right.$$

$$\left. + (q-1) \left(\frac{\rho_t}{\nu_t} \right)^{q-2} \Delta \frac{\rho_t}{\nu_t} \right) \rho_t \, dx$$

$$- (q-1) \int \left(q(q-1) \left(\frac{\rho_t}{\nu_t} \right)^{q-2} \left\| \nabla \frac{\rho_t}{\nu_t} \right\|^2 \right.$$

$$\left. + q \left(\frac{\rho_t}{\nu_t} \right)^{q-1} \Delta \frac{\rho_t}{\nu_t} \right) \nu_t \, dx$$

$$= -q(q-1) \int v_t \left(\frac{\rho_t}{v_t}\right)^{q-2} \left\|\nabla \frac{\rho_t}{v_t}\right\|^2 dx$$

$$= -q(q-1)G_{q,v_t}(\rho_t). \tag{40}$$

Note that the identity (40) above is analogous to the identity (35) along the Langevin dynamics. Therefore, for $q \neq 1$,

$$\frac{d}{dt}R_{q,v_t}(\rho_t) = \frac{1}{q-1}\frac{\frac{d}{dt}F_{q,v_t}(\rho_t)}{F_{q,v_t}(\rho_t)} = -q\frac{G_{q,v_t}(\rho_t)}{F_{q,v_t}(\rho_t)},$$

as desired.

The case $q = 1$ follows from taking limit $q \to 1$, or by an analogous direct calculation. We will use the following identity for $h: \mathbb{R}^n \to \mathbb{R}_{>0}$,

$$\Delta \log h = \nabla \cdot \left(\frac{\nabla h}{h}\right) = \frac{\Delta h}{h} - \|\nabla \log h\|^2.$$

Then along the simultaneous heat flow (39),

$$\frac{d}{dt}H_{v_t}(\rho_t) = \frac{d}{dt}\int \rho_t \log \frac{\rho_t}{v_t} dx$$

$$= \int \frac{\partial \rho_t}{\partial t} \log \frac{\rho_t}{v_t} dx + \int \rho_t \frac{v_t}{\rho_t}\frac{\partial}{\partial t}\left(\frac{\rho_t}{v_t}\right) dx$$

$$= \int \Delta \rho_t \log \frac{\rho_t}{v_t} dx + \int v_t \left(\frac{1}{v_t}\frac{\partial \rho_t}{\partial t} dx - \frac{\rho_t}{v_t^2}\frac{\partial v_t}{\partial t}\right) dx$$

$$= \int \rho_t \Delta \log \frac{\rho_t}{v_t} dx - \int \frac{\rho_t}{v_t}\Delta v_t dx$$

$$= \int \rho_t \left(\frac{v_t}{\rho_t}\Delta\left(\frac{\rho_t}{v_t}\right) - \left\|\nabla \log \frac{\rho_t}{v_t}\right\|^2\right) dx - \int \frac{\rho_t}{v_t}\Delta v_t dx$$

$$= -J_{v_t}(\rho_t),$$

as desired. Note that this is also analogous to the identity (9) along the Langevin dynamics. □

We are now ready to prove Lemma 8.

Proof of Lemma 8 We will prove that along each step of ULA (11) from $x_k \sim \rho_k$ to $x_{k+1} \sim \rho_{k+1}$, the Rényi divergence with respect to v_ϵ decreases by a constant factor:

$$R_{q,\nu_\epsilon}(\rho_{k+1}) \le e^{-\frac{\beta\epsilon}{q}} R_{q,\nu_\epsilon}(\rho_k). \tag{41}$$

Iterating the bound above yields the desired claim (22).

We decompose each step of ULA (11) into a sequence of two steps:

$$\tilde{\rho}_k = (I - \epsilon\nabla f)_\# \rho_k, \tag{42a}$$

$$\rho_{k+1} = \tilde{\rho}_k * \mathcal{N}(0, 2\epsilon I). \tag{42b}$$

In the first step (42a), we apply a smooth deterministic map $T(x) = x - \epsilon\nabla f(x)$. Since ∇f is L-Lipschitz and $\epsilon < \frac{1}{L}$, T is a bijection. Then by Lemma 15,

$$R_{q,\nu_\epsilon}(\rho_k) = R_{q,\tilde{\nu}_\epsilon}(\tilde{\rho}_k) \tag{43}$$

where $\tilde{\nu}_\epsilon = (I - \epsilon\nabla f)_\# \nu_\epsilon$. Recall by Assumption 1 that ν_ϵ satisfies LSI with constant β. Since the map $T(x) = x - \epsilon\nabla f(x)$ is $(1+\epsilon L)$-Lipschitz, by Lemma 16 we know that $\tilde{\nu}_\epsilon$ satisfies LSI with constant $\frac{\beta}{(1+\epsilon L)^2}$.

In the second step (42b), we convolve with a Gaussian distribution, which is the result of evolving along the heat flow at time ϵ. For $0 \le t \le \epsilon$, let $\tilde{\rho}_{k,t} = \tilde{\rho}_k * \mathcal{N}(0, 2tI)$ and $\tilde{\nu}_{\epsilon,t} = \tilde{\nu}_\epsilon * \mathcal{N}(0, 2tI)$, so $\tilde{\rho}_{k,\epsilon} = \tilde{\rho}_{k+1}$ and $\tilde{\nu}_{\epsilon,\epsilon} = \nu_\epsilon$. By Lemma 18,

$$\frac{d}{dt} R_{q,\tilde{\nu}_{\epsilon,t}}(\tilde{\rho}_{k,t}) = -q \frac{G_{q,\tilde{\nu}_{\epsilon,t}}(\tilde{\rho}_{k,t})}{F_{q,\tilde{\nu}_{\epsilon,t}}(\tilde{\rho}_{k,t})}.$$

Since $\tilde{\nu}_\epsilon$ satisfies LSI with constant $\frac{\beta}{(1+\epsilon L)^2}$, by Lemma 17 we know that $\tilde{\nu}_{\epsilon,t}$ satisfies LSI with constant $\left(\frac{(1+\epsilon L)^2}{\beta} + 2t\right)^{-1} \ge \left(\frac{(1+\epsilon L)^2}{\beta} + 2\epsilon\right)^{-1}$ for $0 \le t \le \epsilon$. In particular, since $\epsilon \le \min\{\frac{1}{3L}, \frac{1}{9\beta}\}$, the LSI constant is $\left(\frac{(1+\epsilon L)^2}{\beta} + 2\epsilon\right)^{-1} \ge \left(\frac{16}{9\beta} + \frac{2}{9\beta}\right)^{-1} = \frac{\beta}{2}$. Then by Lemma 5,

$$\frac{d}{dt} R_{q,\tilde{\nu}_{\epsilon,t}}(\tilde{\rho}_{k,t}) = -q \frac{G_{q,\tilde{\nu}_{\epsilon,t}}(\tilde{\rho}_{k,t})}{F_{q,\tilde{\nu}_{\epsilon,t}}(\tilde{\rho}_{k,t})} \le -\frac{\beta}{q} R_{q,\tilde{\nu}_{\epsilon,t}}(\tilde{\rho}_{\epsilon,t}).$$

Integrating over $0 \le t \le \epsilon$ gives

$$R_{q,\nu_\epsilon}(\rho_{k+1}) = R_{q,\tilde{\nu}_{\epsilon,\epsilon}}(\tilde{\rho}_{k,\epsilon}) \le e^{-\frac{\beta\epsilon}{q}} R_{q,\tilde{\nu}_\epsilon}(\tilde{\rho}_k). \tag{44}$$

Combining (43) and (44) gives the desired inequality (41). $\square$

8.4.3   Proof of Theorem 5

Proof of Theorem 5 This follows directly from Lemmas 7 and 8.   □

8.5   Details for Sect. 6: Poincaré Inequality

8.5.1   Proof of Lemma 9

Proof of Lemma 9 We plug in $g^2 = \left(\frac{\rho}{\nu}\right)^q$ to Poincaré inequality (24) and use the monotonicity condition from Lemma 13 to obtain

$$\frac{q^2}{4\alpha} G_{q,\nu}(\rho) \geq F_{q,\nu}(\rho) - F_{\frac{q}{2},\nu}(\rho)^2$$

$$= e^{(q-1)R_{q,\nu}(\rho)} - e^{(q-2)R_{\frac{q}{2},\nu}(\rho)}$$

$$\geq e^{(q-1)R_{q,\nu}(\rho)} - e^{(q-2)R_{q,\nu}(\rho)}$$

$$= F_{q,\nu}(\rho)\left(1 - e^{-R_{q,\nu}(\rho)}\right).$$

Dividing both sides by $F_{q,\nu}(\rho)$ and rearranging yields the desired inequality.   □

8.5.2   Proof of Theorem 3

Proof of Theorem 3 By Lemmas 6 and 9,

$$\frac{d}{dt} R_{q,\nu}(\rho_t) = -q \frac{G_{q,\nu}(\rho_t)}{F_{q,\nu}(\rho_t)} \leq -\frac{4\alpha}{q}\left(1 - e^{-R_{q,\nu}(\rho_t)}\right).$$

We now consider two possibilities:

1. If $R_{q,\nu}(\rho_0) \geq 1$, then as long as $R_{q,\nu}(\rho_t) \geq 1$, we have $1 - e^{-R_{q,\nu}(\rho_t)} \geq 1 - e^{-1} > \frac{1}{2}$, so $\frac{d}{dt} R_{q,\nu}(\rho_t) \leq -\frac{2\alpha}{q}$, which implies $R_{q,\nu}(\rho_t) \leq R_{q,\nu}(\rho_0) - \frac{2\alpha t}{q}$.

2. If $R_{q,\nu}(\rho_0) \leq 1$, then $R_{q,\nu}(\rho_t) \leq 1$, and thus $\frac{1-e^{-R_{q,\nu}(\rho_t)}}{R_{q,\nu}(\rho_t)} \geq \frac{1}{1+R_{q,\nu}(\rho_t)} \geq \frac{1}{2}$. Thus, in this case $\frac{d}{dt} R_{q,\nu}(\rho_t) \leq -\frac{2\alpha}{q} R_{q,\nu}(\rho_t)$, and integrating gives $R_{q,\nu}(\rho_t) \leq e^{-\frac{2\alpha t}{q}} R_{q,\nu}(\rho_0)$, as desired.

                                                                                   □

8.5.3   Proof of Lemma 10

We will use the following auxiliary results, which are analogous to Lemmas 16 and 17. We provide the proof of Lemma 19 in Appendix, and the proof of Lemma 20 in Appendix.

Lemma 19 *Suppose a probability distribution ν satisfies Poincaré inequality with constant $\alpha > 0$. Let $T : \mathbb{R}^n \to \mathbb{R}^n$ be a differentiable L-Lipschitz map. Then $\tilde{\nu} = T_\# \nu$ satisfies Poincaré inequality with constant α / L^2.*

Lemma 20 *Suppose a probability distribution ν satisfies Poincaré inequality with constant $\alpha > 0$. For $t > 0$, the probability distribution $\tilde{\nu}_t = \nu * \mathcal{N}(0, 2tI)$ satisfies Poincaré inequality with constant $\left(\frac{1}{\alpha} + 2t\right)^{-1}$.*

We are now ready to prove Lemma 10.

Proof of Lemma 10 Following the proof of Lemma 8, we decompose each step of ULA (11) into two steps:

$$\tilde{\rho}_k = (I - \epsilon \nabla f)_\# \rho_k, \tag{45a}$$

$$\rho_{k+1} = \tilde{\rho}_k * \mathcal{N}(0, 2\epsilon I). \tag{45b}$$

The first step (45a) is a deterministic bijective map, so it preserves Rényi divergence by Lemma 15: $R_{q,\nu_\epsilon}(\rho_k) = R_{q,\tilde{\nu}_\epsilon}(\tilde{\rho}_k)$, where $\tilde{\nu}_\epsilon = (I - \epsilon \nabla f)_\# \nu_\epsilon$. Recall by Assumption 2 that ν_ϵ satisfies Poincaré inequality with constant β. Since the map $T(x) = x - \epsilon \nabla f(x)$ is $(1 + \epsilon L)$-Lipschitz, by Lemma 19 we know that $\tilde{\nu}_\epsilon$ satisfies Poincaré inequality with constant $\frac{\beta}{(1+\epsilon L)^2}$.

The second step (45b) is convolution with a Gaussian distribution, which is the result of evolving along the heat flow at time ϵ. For $0 \le t \le \epsilon$, let $\tilde{\rho}_{k,t} = \tilde{\rho}_k * \mathcal{N}(0, 2tI)$ and $\tilde{\nu}_{\epsilon,t} = \tilde{\nu}_\epsilon * \mathcal{N}(0, 2tI)$, so $\tilde{\rho}_{k,\epsilon} = \tilde{\rho}_{k+1}$ and $\tilde{\nu}_{\epsilon,\epsilon} = \nu_\epsilon$. Since $\tilde{\nu}_\epsilon$ satisfies Poincaré inequality with constant $\frac{\beta}{(1+\epsilon L)^2}$, by Lemma 20 we know that $\tilde{\nu}_{\epsilon,t}$ satisfies Poincaré inequality with constant $\left(\frac{(1+\epsilon L)^2}{\beta} + 2t\right)^{-1} \ge \left(\frac{(1+\epsilon L)^2}{\beta} + 2\epsilon\right)^{-1}$ for $0 \le t \le \epsilon$. In particular, since $\epsilon \le \min\{\frac{1}{3L}, \frac{1}{9\beta}\}$, the Poincaré constant is $\left(\frac{(1+\epsilon L)^2}{\beta} + 2\epsilon\right)^{-1} \ge \left(\frac{16}{9\beta} + \frac{2}{9\beta}\right)^{-1} = \frac{\beta}{2}$. Then by Lemmas 18 and 9,

$$\frac{d}{dt} R_{q,\tilde{\nu}_{\epsilon,t}}(\tilde{\rho}_{k,t}) = -q \frac{G_{q,\tilde{\nu}_{\epsilon,t}}(\tilde{\rho}_{k,t})}{F_{q,\tilde{\nu}_{\epsilon,t}}(\tilde{\rho}_{k,t})} \le -\frac{2\beta}{q} \left(1 - e^{-R_{q,\tilde{\nu}_{\epsilon,t}}(\tilde{\rho}_{k,t})}\right).$$

We now consider two possibilities, as in Theorem 3:

1. If $R_{q,\nu_\epsilon}(\rho_k) = R_{q,\tilde{\nu}_{\epsilon,0}}(\tilde{\rho}_{k,0}) \ge 1$, then as long as $R_{q,\nu_\epsilon}(\rho_{k+1}) = R_{q,\tilde{\nu}_{\epsilon,\epsilon}}(\tilde{\rho}_{k,\epsilon}) \ge 1$, we have $1 - e^{-R_{q,\tilde{\nu}_{\epsilon,t}}(\tilde{\rho}_{k,t})} \ge 1 - e^{-1} > \frac{1}{2}$, so $\frac{d}{dt} R_{q,\tilde{\nu}_{\epsilon,t}}(\tilde{\rho}_{k,t}) \le -\frac{\beta}{q}$, which implies $R_{q,\nu_\epsilon}(\rho_{k+1}) \le R_{q,\nu_\epsilon}(\rho_k) - \frac{\beta\epsilon}{q}$. Iterating this step, we have that $R_{q,\nu_\epsilon}(\rho_k) \le R_{q,\nu_\epsilon}(\rho_0) - \frac{\beta\epsilon k}{q}$ if $R_{q,\nu_\epsilon}(\rho_0) \ge 1$ and as long as $R_{q,\nu_\epsilon}(\rho_k) \ge 1$.

2. If $R_{q,\nu_\epsilon}(\rho_k) = R_{q,\tilde{\nu}_{\epsilon,0}}(\tilde{\rho}_{k,0}) \leq 1$, then $R_{q,\tilde{\nu}_{\epsilon,t}}(\tilde{\rho}_{k,t}) \leq 1$, and thus $\frac{1-e^{-R_{q,\tilde{\nu}_{\epsilon,t}}(\tilde{\rho}_{k,t})}}{R_{q,\tilde{\nu}_{\epsilon,t}}(\tilde{\rho}_{k,t})} \geq \frac{1}{1+R_{q,\tilde{\nu}_{\epsilon,t}}(\tilde{\rho}_{k,t})} \geq \frac{1}{2}$. Thus, in this case $\frac{d}{dt} R_{q,\tilde{\nu}_{\epsilon,t}}(\tilde{\rho}_{k,t}) \leq -\frac{\beta}{q} R_{q,\tilde{\nu}_{\epsilon,t}}(\tilde{\rho}_{k,t})$. Integrating over $0 \leq t \leq \epsilon$ gives $R_{q,\nu_\epsilon}(\rho_{k+1}) = R_{q,\tilde{\nu}_{\epsilon,\epsilon}}(\tilde{\rho}_{k,\epsilon}) \leq e^{-\frac{\beta\epsilon}{q}} R_{q,\tilde{\nu}_{\epsilon,0}}(\tilde{\rho}_{k,0}) = e^{-\frac{\beta\epsilon}{q}} R_{q,\nu_\epsilon}(\rho_k)$. Iterating this step gives $R_{q,\nu_\epsilon}(\rho_k) \leq e^{-\frac{\beta\epsilon k}{q}} R_{q,\nu_\epsilon}(\rho_0)$ if $R_{q,\nu_\epsilon}(\rho_0) \leq 1$, as desired.

$\square$

8.5.4 Proof of Theorem 6

Proof of Theorem 6 By Lemma 10 (which applies since $2q > 2$), after k_0 iterations we have $R_{2q,\nu_\epsilon}(\rho_{k_0}) \leq 1$. Applying the second case of Lemma 10 starting from k_0 gives $R_{2q,\nu_\epsilon}(\rho_k) \leq e^{-\frac{\beta\epsilon(k-k_0)}{2q}} R_{2q,\nu_\epsilon}(\rho_{k_0}) \leq e^{-\frac{\beta\epsilon(k-k_0)}{2q}}$. Then by Lemma 7,

$$R_{q,\nu}(\rho_k) \leq \left(\frac{q-\frac{1}{2}}{q-1}\right) R_{2q,\nu_\epsilon}(\rho_k) + R_{2q-1,\nu}(\nu_\epsilon)$$

$$\leq \left(\frac{q-\frac{1}{2}}{q-1}\right) e^{-\frac{\beta\epsilon(k-k_0)}{2q}} + R_{2q-1,\nu}(\nu_\epsilon)$$

as desired.

$\square$

8.6 Proofs for Sect. 7: Properties of Biased Limit

8.6.1 Bounding Relative Fisher Information

Let $H(\rho) = -\mathbb{E}_\rho[\log \rho]$ be Shannon entropy, $J(\rho) = \mathbb{E}_\rho[\|\nabla \log \rho\|^2]$ be the Fisher information, and $K(\rho) = \mathbb{E}_\rho[\|\nabla^2 \log \rho\|^2_{\mathrm{HS}}]$ be the second-order Fisher information. We can write relative entropy as

$$H_\nu(\rho) = \mathbb{E}_\rho\left[\log\frac{\rho}{\nu}\right] = -H(\rho) + \mathbb{E}_\rho[f]$$

and we can write relative Fisher information as

$$J_\nu(\nu_\epsilon) = \mathbb{E}_{\nu_\epsilon}\left[\left\|\nabla \log \frac{\nu_\epsilon}{\nu}\right\|^2\right] = J(\nu_\epsilon) + 2\mathbb{E}_{\nu_\epsilon}[\langle\nabla \log \nu_\epsilon, \nabla f\rangle] + \mathbb{E}_{\nu_\epsilon}[\|\nabla f\|^2]$$

$$= J(\nu_\epsilon) + \mathbb{E}_{\nu_\epsilon}[\|\nabla f\|^2 - 2\Delta f] \tag{46}$$

where the last step follows from integration by parts.

We first prove the following, which only requires second-order smoothness.

Lemma 21 *Assume f is L-smooth $(-LI \preceq \nabla^2 f \preceq LI)$ and $\epsilon \leq \frac{1}{2L}$. Then*

$$J_v(\nu_\epsilon) \leq \mathbb{E}_{\nu_\epsilon}[\|\nabla f\|^2 - \Delta f] + \epsilon n L^2.$$

Proof We examine how entropy changes from ν_ϵ to μ_ϵ and back, which will give us an estimate on the Fisher information. By the change-of variable formula for $\mu_\epsilon = (I - \epsilon \nabla f)_{\#} \nu_\epsilon$, we have

$$\log \nu_\epsilon(x) = \log \det(I - \epsilon \nabla^2 f(x)) + \log \mu_\epsilon(x - \epsilon \nabla f(x)). \tag{47}$$

By taking expectation over $x \sim \nu_\epsilon$ (equivalently, $x - \epsilon \nabla f(x) \sim \mu_\epsilon$), we get

$$H(\nu_\epsilon) = H(\mu_\epsilon) - \mathbb{E}_{\nu_\epsilon}[\log \det(I - \epsilon \nabla^2 f)]. \tag{48}$$

On the other hand, recall that along the heat flow $\rho_t = \rho_0 * \mathcal{N}(0, 2tI)$, we have the relations

$$\frac{d}{dt} H(\rho_t) = J(\rho_t),$$

$$\frac{d}{dt} J(\rho_t) = -K(\rho_t) \leq 0.$$

See for example [73]. Thus, $\nu_\epsilon = \mu_\epsilon * \mathcal{N}(0, 2\epsilon I)$ satisfies

$$H(\nu_\epsilon) = H(\mu_\epsilon) + \int_0^\epsilon J(\rho_t)\, dt \geq H(\mu_\epsilon) + \epsilon J(\nu_\epsilon) \tag{49}$$

where $\rho_t = \rho_0 * \mathcal{N}(0, 2tI)$ is the heat flow from $\rho_0 = \mu_\epsilon$ to $\rho_\epsilon = \nu_\epsilon$, and the last inequality holds since $t \mapsto J(\rho_t)$ is decreasing. Combining (48) and (49), we get

$$\epsilon J(\nu_\epsilon) \leq H(\nu_\epsilon) - H(\mu_\epsilon) = -\mathbb{E}_{\nu_\epsilon}[\log \det(I - \epsilon \nabla^2 f)]. \tag{50}$$

Let $\lambda_1, \ldots, \lambda_n$ be the eigenvalues of $\nabla^2 f$. Since f is L-smooth, $|\lambda_i| \leq L$. Using the inequality $\log(1 - \epsilon \lambda_i) \geq -\epsilon \lambda_i - \epsilon^2 \lambda_i^2$, which holds since $\epsilon |\lambda_i| \leq \frac{1}{2}$ from $\epsilon \leq \frac{1}{2L}$, we have

$$-\mathbb{E}_{\nu_\epsilon}[\log \det(I - \epsilon \nabla^2 f)] = \sum_{i=1}^n \mathbb{E}_{\nu_\epsilon}[-\log(1 - \epsilon \lambda_i)]$$

$$\leq \sum_{i=1}^n \mathbb{E}_{\nu_\epsilon}[\epsilon \lambda_i + \epsilon^2 \lambda_i^2]$$

$$= \epsilon \mathbb{E}_{\nu_\epsilon}[\Delta f] + \epsilon^2 \, \mathbb{E}_{\nu_\epsilon}[\|\nabla^2 f\|_{HS}^2]$$

$$\leq \epsilon \mathbb{E}_{\nu_\epsilon}[\Delta f] + \epsilon^2 n L^2.$$

Plugging this to (50) gives

$$J(\nu_\epsilon) \leq -\frac{1}{\epsilon} \mathbb{E}_{\nu_\epsilon}[\log \det(I - \epsilon \nabla^2 f)] \leq \mathbb{E}_{\nu_\epsilon}[\Delta f] + \epsilon n L^2. \tag{51}$$

Therefore, we can bound the relative Fisher information (46):

$$J_\nu(\nu_\epsilon) = J(\nu_\epsilon) + \mathbb{E}_{\nu_\epsilon}[\|\nabla f\|^2 - 2\Delta f] \leq \mathbb{E}_{\nu_\epsilon}[\|\nabla f\|^2 - \Delta f] + \epsilon n L^2.$$

□

8.6.2 Bounding the Expected Value

Recall that for $\nu = e^{-f}$, we have $\mathbb{E}_\nu[\|\nabla f\|^2 - \Delta f] = 0$. Under third-order smoothness, we will prove $\mathbb{E}_{\nu_\epsilon}[\|\nabla f\|^2 - \Delta f] = O(\epsilon)$.

Lemma 22 *Assume f is (L, M)-smooth. Then for $\epsilon \leq \frac{1}{2L}$,*

$$\mathbb{E}_{\nu_\epsilon}[\|\nabla f\|^2 - \Delta f] \leq \epsilon n \left(L^2 + M \sqrt{J(\mu_\epsilon)} \right). \tag{52}$$

Proof We examine how the expected value of f changes from ν_ϵ to μ_ϵ and back, which will give us an estimate on the desired quantity.

Let $x \sim \nu_\epsilon$ and $y = x - \epsilon \nabla f(x) \sim \mu_\epsilon$, so $x' = y + \sqrt{2\epsilon}\, Z \sim \nu_\epsilon$ where $Z \sim \mathcal{N}(0, I)$ is independent. Since f is L-smooth, we have the bound:

$$f(y) \leq f(x) - \epsilon \left(1 - \frac{\epsilon L}{2} \right) \|\nabla f(x)\|^2.$$

Taking expectation over $x \sim \nu_\epsilon$ (equivalently, $y \sim \mu_\epsilon$) yields

$$\mathbb{E}_{\mu_\epsilon}[f] \leq \mathbb{E}_{\nu_\epsilon}[f] - \epsilon \left(1 - \frac{\epsilon L}{2} \right) \mathbb{E}_{\nu_\epsilon}[\|\nabla f\|^2].$$

On the other hand, let $\rho_t = \rho_0 * \mathcal{N}(0, 2tI)$ be the heat flow from $\rho_0 = \mu_\epsilon$ to $\rho_\epsilon = \nu_\epsilon$, and recall that along the heat flow, $\frac{d}{dt}\mathbb{E}_{\rho_t}[f] = \mathbb{E}_{\rho_t}[\Delta f]$. Then

$$\mathbb{E}_{\nu_\epsilon}[f] = \mathbb{E}_{\mu_\epsilon}[f] + \int_0^\epsilon \mathbb{E}_{\rho_t}[\Delta f]\, dt.$$

Combining the two relations above,

$$\epsilon \left(1 - \frac{\epsilon L}{2}\right) \mathbb{E}_{\nu_\epsilon}[\|\nabla f(x)\|^2] \le \mathbb{E}_{\nu_\epsilon}[f] - \mathbb{E}_{\mu_\epsilon}[f] = \int_0^\epsilon \mathbb{E}_{\rho_t}[\Delta f]\,dt.$$

Therefore,

$$
\begin{aligned}
\epsilon \left(1 - \frac{\epsilon L}{2}\right) \mathbb{E}_{\nu_\epsilon}[\|\nabla f\|^2 - \Delta f] &\le \int_0^\epsilon \mathbb{E}_{\rho_t}[\Delta f]\,dt - \epsilon \left(1 - \frac{\epsilon L}{2}\right) \mathbb{E}_{\nu_\epsilon}[\Delta f] \\
&= \int_0^\epsilon (\mathbb{E}_{\rho_t}[\Delta f] - \mathbb{E}_{\rho_\epsilon}[\Delta f])\,dt + \frac{\epsilon^2 L}{2}\mathbb{E}_{\nu_\epsilon}[\Delta f] \\
&\le \int_0^\epsilon (\mathbb{E}_{\rho_t}[\Delta f] - \mathbb{E}_{\rho_\epsilon}[\Delta f])\,dt + \frac{\epsilon^2 n L^2}{2}.
\end{aligned}
$$

$$(53)$$

Since ρ_t evolves following the heat flow, by Lemma 23 we have for any $0 \le t \le \epsilon$:

$$W_2(\rho_t, \rho_\epsilon)^2 \le (\epsilon - t)^2 J(\rho_t) \le (\epsilon - t)^2 J(\rho_0) = (\epsilon - t)^2 J(\mu_\epsilon)$$

where the second inequality above follows from the fact that Fisher information is decreasing along heat flow.

Since we assume $\nabla^2 f$ is M-Lipschitz, the Laplacian $\Delta f = \mathrm{Tr}(\nabla^2 f)$ is (nM)-Lipschitz. Then by the dual formulation of W_1 distance,[2]

$$\mathbb{E}_{\rho_t}[\Delta f] - \mathbb{E}_{\rho_\epsilon}[\Delta f] \le nM\, W_1(\rho_t, \rho_\epsilon) \le nM\, W_2(\rho_t, \rho_\epsilon) \le (\epsilon - t)\, nM\sqrt{J(\mu_\epsilon)}.$$

Integrating over $0 \le t \le \epsilon$ gives

$$\int_0^\epsilon (\mathbb{E}_{\rho_t}[\Delta f] - \mathbb{E}_{\rho_\epsilon}[\Delta f])\,dt \le \frac{\epsilon^2}{2} nM\sqrt{J(\mu_\epsilon)}.$$

Plugging this to (53) gives

$$\epsilon \left(1 - \frac{\epsilon L}{2}\right) \mathbb{E}_{\nu_\epsilon}[\|\nabla f\|^2 - \Delta f] \le \frac{\epsilon^2}{2} nM\sqrt{J(\mu_\epsilon)} + \frac{\epsilon^2 n L^2}{2}.$$

---

[2] Recall $W_1(\rho, \nu) = \sup\{\mathbb{E}_\rho[g] - \mathbb{E}_\nu[g]: g \text{ is } 1\text{-Lipschitz}\}$.

Since $1 - \frac{\epsilon L}{2} \geq \frac{3}{4} > \frac{1}{2}$ for $\epsilon \leq \frac{1}{2L}$, this also implies

$$\frac{\epsilon}{2} \mathbb{E}_{\nu_\epsilon}[\|\nabla f\|^2 - \Delta f] \leq \frac{\epsilon^2}{2} nM\sqrt{J(\mu_\epsilon)} + \frac{\epsilon^2 n L^2}{2}.$$

Dividing by $\frac{\epsilon}{2}$ gives the claim.   □

Remark 5 Observe from (53) that if $\Delta\Delta f \geq 0$, then $\mathbb{E}_{\rho_t}[\Delta f] - \mathbb{E}_{\rho_\epsilon}[\Delta f] \leq 0$, so $\mathbb{E}_{\nu_\epsilon}[\|\nabla f\|^2 - \Delta f] \leq \epsilon n L^2$. Plugging this to Lemma 21, we obtain the bound $J_\nu(\nu_\epsilon) \leq 2\epsilon n L^2$ assuming f is L-smooth and $\Delta\Delta f \geq 0$.

In the proof above, we use the following lemma on the distance along the heat flow. Note that a simple coupling argument gives $W_2(\rho_\epsilon, \rho_0)^2 \leq O(\epsilon)$, rather than $O(\epsilon^2)$ below (when $J(\rho_0) < \infty$).

Lemma 23 *For any probability distribution ρ_0 and for any $\epsilon > 0$, let $\rho_\epsilon = \rho_0 * \mathcal{N}(0, 2\epsilon I)$. Then*

$$W_2(\rho_\epsilon, \rho_0)^2 \leq \epsilon^2 J(\rho_0).$$

Proof By definition, ρ_ϵ evolves following the heat flow $\frac{\partial \rho_t}{\partial t} = \Delta\rho_t$ from time $t = 0$ to time $t = \epsilon$. Fix $\epsilon > 0$, and let us rescale time to be from 0 to 1: Let $\tilde{\rho}_\tau = \rho_{\tau\epsilon}$, so $\tilde{\rho}_0 = \rho_0$ and $\tilde{\rho}_1 = \rho_\epsilon$. Then $\tilde{\rho}_\tau$ evolves following a rescaled heat flow:

$$\frac{\partial \tilde{\rho}_\tau}{\partial \tau} = \frac{\partial \rho_{\tau\epsilon}}{\partial \tau} = \epsilon \Delta\rho_{\tau\epsilon} = \epsilon \Delta\tilde{\rho}_\tau = \epsilon \nabla \cdot (\tilde{\rho}_\tau \nabla \log \tilde{\rho}_\tau) \qquad (54)$$

Since $(\tilde{\rho}_\tau)_{0 \leq \tau \leq 1}$ connects $\tilde{\rho}_0 = \rho_0$ to $\tilde{\rho}_1 = \rho_\epsilon$, its length must exceed the W_2 distance:

$$W_2(\rho_\epsilon, \rho_0)^2 \leq \int_0^1 \mathbb{E}_{\tilde{\rho}_\tau}[\|\epsilon \nabla \log \tilde{\rho}_\tau\|^2] \, d\tau = \epsilon^2 \int_0^1 J(\tilde{\rho}_\tau) \, d\tau \leq \epsilon^2 J(\rho_0).$$

In the last step we have used the fact that Fisher information is decreasing along the heat flow: $J(\tilde{\rho}_\tau) \leq J(\rho_0)$.   □

8.6.3 Bounding the Fisher Information

Lemma 24

1. *If f is L-smooth and $\epsilon \leq \frac{1}{2L}$, then*

$$J(\nu_\epsilon) \leq \frac{3}{2} nL.$$

2. *If f is (L, M)-smooth and $\epsilon \leq \frac{1}{2L}$, then*

$$J(\mu_\epsilon) \leq 12n(L + 3nM^2).$$

Proof First, since f is L-smooth and $\epsilon \leq \frac{1}{2L}$, from (51) we can bound

$$J(\nu_\epsilon) \leq \mathbb{E}_{\nu_\epsilon}[\Delta f] + \epsilon \, \mathbb{E}_{\nu_\epsilon}[\|\nabla^2 f\|_{\mathrm{HS}}^2] \leq nL + \epsilon n L^2 \leq \frac{3}{2}nL. \tag{55}$$

Second, by taking gradient in the formula (47) for $\mu_\epsilon = (I - \epsilon \nabla f)_{\#}\nu_\epsilon$, we get

$$\nabla \log \nu_\epsilon(x) = -\epsilon \nabla^3 f(x) A(x)^{-1} + A(x)\nabla \log \mu_\epsilon(x - \epsilon \nabla f(x))$$

or equivalently,

$$\nabla \log \mu_\epsilon(x - \epsilon \nabla f(x)) = A(x)^{-1}\nabla \log \nu_\epsilon(x) + \epsilon A(x)^{-1}\nabla^3 f(x) A(x)^{-1} \tag{56}$$

where

$$A(x) = I - \epsilon \nabla^2 f(x).$$

satisfies $\frac{1}{2}I \preceq A(x) \preceq \frac{3}{2}I$ since $-LI \preceq \nabla^2 f(x) \preceq LI$ and $\epsilon \leq \frac{1}{2L}$. In particular,

$$\frac{2}{3}I \preceq A(x)^{-1} \preceq 2I.$$

Therefore, the first term in (56) we can bound as

$$\|A(x)^{-1}\nabla \log \nu_\epsilon(x)\|_2 \leq 2\|\nabla \log \nu_\epsilon(x)\|_2.$$

For the second term, using the assumption $\|\nabla^3 f(x)\|_{\mathrm{op}} \leq M$ and Lemma 25, we have

$$\|A(x)^{-1}\nabla^3 f(x) A(x)^{-1}\|_2 \leq 2\|\nabla^3 f(x) A(x)^{-1}\|_2 \leq 4nM.$$

Therefore, from (56), we get

$$\|\nabla \log \mu_\epsilon(x - \epsilon \nabla f(x))\|_2 \leq 2\|\nabla \log \nu_\epsilon(x)\|_2 + 4nM.$$

This implies

$$\|\nabla \log \mu_\epsilon(x - \epsilon \nabla f(x))\|_2^2 \leq 8\|\nabla \log \nu_\epsilon(x)\|_2^2 + 32n^2M^2.$$

Taking expectation over $x \sim \nu_\epsilon$ (equivalently, $x - \epsilon \nabla f(x) \sim \mu_\epsilon$), we conclude that

$$J(\mu_\epsilon) \le 8J(\nu_\epsilon) + 32n^2 M^2 \le 12nL + 32n^2 M^2 \le 12n(L + 3nM^2).$$

$\square$

In the above, we use the following bound from smoothness.

Lemma 25 *Let* $T \in \mathbb{R}^{n \times n \times n}$ *be a 3-tensor with* $\|T\|_{\mathrm{op}} \le M$. *For any symmetric matrix* $B \in \mathbb{R}^{n \times n}$ *with* $\|B\|_{\mathrm{op}} \le \beta$, *the vector* $TB \in \mathbb{R}^n$ *satisfies* $\|TB\|_2 \le n\beta M$.

Proof Since $\|T\|_{\mathrm{op}} \le M$, for any $u, v, w \in \mathbb{R}^n$ with $\|u\| = \|v\| = \|w\| = 1$, $|T[u, v, w]| \le M$. In particular, for any $u \in \mathbb{R}^n$ with $\|u\| = 1$, $p = T[u, u] \in \mathbb{R}^n$ satisfies $\|p\| \le M$. We eigendecompose $B = \sum_{i=1}^{n} \lambda_i u_i u_i^\top$ with eigenvectors $u_1, \ldots, u_n \in \mathbb{R}^n$ and eigenvalues $\lambda_1, \ldots, \lambda_n \in \mathbb{R}$ with $\|u_i\| = 1$, $|\lambda_i| \le \beta$. Then

$$\|TB\|_2 = \left\| T \sum_{i=1}^{n} \lambda_i u_i u_i^\top \right\|_2 \le \sum_{i=1}^{n} |\lambda_i| \cdot \|T[u_i, u_i]\|_2 \le \sum_{i=1}^{n} \beta M = n\beta M.$$

$\square$

8.6.4   Proof of Upper Bound in Theorem 7

Proof of Upper Bound in Theorem 7 By combining Lemmas 21, 22, and 24:

$$\begin{aligned}
J_\nu(\nu_\epsilon) &\le \mathbb{E}_{\nu_\epsilon}[\|\nabla f\|^2 - \Delta f] + \epsilon n L^2 \\
&\le \epsilon n \left(2L^2 + M\sqrt{J(\mu_\epsilon)} \right) \\
&\le \epsilon n \left(2L^2 + M\sqrt{12n(L + 3nM^2)} \right) \\
&\le \epsilon n \left(2L^2 + M(4\sqrt{nL} + 6nM) \right) \\
&\le 2\epsilon n \left(L^2 + 2\sqrt{nL}M + 3nM^2 \right).
\end{aligned}$$

$\square$

8.6.5   Proof of Lower Bound in Theorem 7

For the lower bound, we first prove the following properties. Observe that for $\nu \propto e^{-f}$, $\mathbb{E}_\nu[\nabla f] = 0$ and $\mathbb{E}_\nu[\langle x, \nabla f(x)\rangle] = n$. We show similar properties still hold for the biased limit.

Lemma 26 *For any f and $\epsilon > 0$, the biased limit ν_ϵ satisfies:*

1. $\mathbb{E}_{\nu_\epsilon}[\nabla f] = 0$.
2. $\mathbb{E}_{\nu_\epsilon}[\langle x, \nabla f(x) \rangle] = n + \frac{\epsilon}{2}\mathbb{E}_{\nu_\epsilon}[\|\nabla f\|^2]$.

Proof Let $x \sim \nu_\epsilon$, $y = x - \epsilon \nabla f(x) \sim \mu_\epsilon$, and $x' = y + \sqrt{2\epsilon}z \sim \nu_\epsilon$ where $z \sim \mathcal{N}(0, I)$ is independent. Then

$$x' = x - \epsilon \nabla f(x) + \sqrt{2\epsilon}z.$$

By taking expectation over $x \sim \nu_\epsilon$ (so $x' \sim \nu_\epsilon$), we get:

$$\mathbb{E}_{\nu_\epsilon}[x'] = \mathbb{E}_{\nu_\epsilon}[x] - \mathbb{E}_{\nu_\epsilon}[\nabla f(x)]$$

which implies $\mathbb{E}_{\nu_\epsilon}[\nabla f] = 0$.

Next, by taking covariance, we get:

$$\begin{aligned}
\mathrm{Cov}_{\nu_\epsilon}(x') &= \mathrm{Cov}_{\nu_\epsilon}(x - \epsilon \nabla f(x)) + 2\epsilon I \\
&= \mathrm{Cov}_{\nu_\epsilon}(x) - \epsilon \mathrm{Cov}_{\nu_\epsilon}(x, \nabla f(x)) - \epsilon \mathrm{Cov}_{\nu_\epsilon}(\nabla f(x), x) \\
&\quad + \epsilon^2 \mathrm{Cov}_{\nu_\epsilon}(\nabla f(x)) + 2\epsilon I
\end{aligned}$$

so

$$\mathrm{Cov}_{\nu_\epsilon}(x, \nabla f(x)) + \mathrm{Cov}_{\nu_\epsilon}(\nabla f(x), x) = \epsilon \mathrm{Cov}_{\nu_\epsilon}(\nabla f(x)) + 2I.$$

Since $\mathbb{E}_{\nu_\epsilon}[\nabla f] = 0$, this means

$$\mathbb{E}_{\nu_\epsilon}[x\, \nabla f(x)^\top] + \mathbb{E}_{\nu_\epsilon}[\nabla f(x)\, x^\top] = \epsilon \mathbb{E}_{\nu_\epsilon}[\nabla f(x)\, \nabla f(x)^\top] + 2I.$$

Taking trace and dividing by 2 gives

$$\mathbb{E}_{\nu_\epsilon}[\langle x, \nabla f(x) \rangle] = n + \frac{\epsilon}{2}\mathbb{E}_{\nu_\epsilon}[\|\nabla f(x)\|^2].$$

$\square$

We are now ready to prove the lower bound.

Proof of Lower Bound in Theorem 7 From Lemma 26, using the identity $n = \mathbb{E}_{\nu_\epsilon}[\langle x, -\nabla \log \nu_\epsilon \rangle]$ and Cauchy-Schwarz inequality, we can derive the bound:

$$\begin{aligned}
\frac{\epsilon}{2}\mathbb{E}_{\nu_\epsilon}[\|\nabla f(x)\|^2] &= \mathbb{E}_{\nu_\epsilon}[\langle x, \nabla f(x) \rangle] - n \\
&= \mathbb{E}_{\nu_\epsilon}\left[\left\langle x, \nabla \log \frac{\nu_\epsilon}{\nu} \right\rangle\right]
\end{aligned}$$

$$= \mathbb{E}_{\nu_\epsilon}\left[\left\langle x - \mathbb{E}_{\nu_\epsilon}[x], \nabla \log \frac{\nu_\epsilon}{\nu}\right\rangle\right]$$

$$\leq \sqrt{\mathrm{Var}_{\nu_\epsilon}(x)} \cdot \sqrt{J_\nu(\nu_\epsilon)}.$$

Rearranging gives us the desired result. In the second step above we can subtract $\mathbb{E}_{\nu_\epsilon}[x]$ because for any $c \in \mathbb{R}^n$, by Lemma 26,

$$\mathbb{E}_{\nu_\epsilon}\left[\left\langle c, \nabla \log \frac{\nu_\epsilon}{\nu}\right\rangle\right] = \langle c, \mathbb{E}_{\nu_\epsilon}[\nabla \log \nu_\epsilon] + \mathbb{E}_{\nu_\epsilon}[\nabla f]\rangle = 0.$$

$\square$

8.6.6 Proof of Theorem 8

Proof of Theorem 8 Suppose we run ULA from $x_0 \sim \rho_0$ to obtain $x_k \sim \rho_k$, so $\rho_k \to \nu_\epsilon$ as $k \to \infty$. Let α_k denote the LSI constant of ρ_k, i.e. the largest constant $\tilde{\alpha} > 0$ such that (4) holds. Since $0 \leq \epsilon \leq \frac{1}{L}$ and f is α-strongly convex, the map $x \mapsto x - \epsilon\nabla f(x)$ is $(1 - \epsilon\alpha)$-Lipschitz. Since $x_k \sim \rho_k$ is α_k-LSI, by Lemma 16, the distribution of $x_k - \epsilon\nabla f(x_k)$ satisfies LSI with constant $\alpha_k/(1 - \epsilon\alpha)^2$. Then by Lemma 17, $x_{k+1} = x_k - \epsilon\nabla f(x_k) + \sqrt{2\epsilon}\, z_k \sim \rho_{k+1}$ satisfies α_{k+1}-LSI with

$$\frac{1}{\alpha_{k+1}} \leq \frac{(1 - \epsilon\alpha)^2}{\alpha_k} + 2\epsilon.$$

Suppose we start $\alpha_0 \geq \frac{\alpha}{2}$. We claim that $\alpha_k \geq \frac{\alpha}{2}$ for all $k \geq 0$. Indeed, if $\frac{1}{\alpha_k} \leq \frac{2}{\alpha}$, then since $\epsilon \leq \frac{1}{L} \leq \frac{1}{\alpha}$, we have

$$\frac{1}{\alpha_{k+1}} \leq \frac{(1 - \epsilon\alpha)^2}{\alpha/2} + 2\epsilon = \frac{2}{\alpha} - 2\epsilon(1 - \epsilon\alpha) \leq \frac{2}{\alpha}.$$

Thus by induction, $\alpha_k \geq \frac{\alpha}{2}$ for all $k \geq 0$. Taking the limit $k \to \infty$, this shows that $\nu_\epsilon = \lim_{k\to\infty} \rho_k$ also satisfies LSI with constant $\beta \geq \frac{\alpha}{2}$. $\square$

9 Discussion

In this paper we proved convergence guarantees on KL divergence and Rényi divergence along ULA under isoperimetric assumptions and bounded Hessian, without assuming convexity or bounds on higher derivatives. In particular, under LSI and bounded Hessian, we prove a complexity guarantee of $O(\frac{\kappa^2 n}{\delta})$ to achieve $H_\nu(\rho_k) \leq \delta$, where $\kappa := L/\alpha$ is the condition number. We note the dependence on κ may not be tight. In particular, the asymptotic bias in KL divergence from our result scales linearly with step size, while from the Gaussian example we see it should scale quadratically with step size. We can achieve a smaller bias using a different algorithm, e.g. the underdamped Langevin algorithm [57] or the proximal Langevin algorithm [78]. However, it remains open whether we can provide a better analysis of ULA under LSI and smoothness that yields the optimal bias.

Our convergence results for ULA in Rényi divergence hold assuming the biased limit satisfies isoperimetry (Assumptions 1 and 2), which we can verify assuming strong log-concavity and smoothness of the target distribution. It would be interesting to verify when Assumptions 1 and 2 hold more generally, whether they can be relaxed, or if they follow from assuming isoperimetry and smoothness for the target density.

Another intriguing question is whether there is an affine-invariant version of the Langevin dynamics. This might lead to a sampling algorithm with logarithmic dependence on smoothness parameters, rather than the current polynomial dependence. There are some approaches that achieve affine invariance in continuous time, for example via interacting Langevin dynamics [36] or the Newton Langevin dynamics [21]; however, the discretization analysis remains a challenge.

Since the publication of the conference version of this paper [72], some of our techniques and results have been generalized. The one-step interpolation technique that we use in Lemma 3 proves to be useful for analyzing ULA or its variants under various assumptions. It has been extended to analyze ULA for sampling on manifolds, for example, on a product of spheres with applications to solving semidefinite programms [51]; and to sampling on Riemannian manifolds [37, 76]. It has also been used to analyze ULA for sampling from distributions with sub-Gaussian tail growth and Hölder-continuous gradient [33]; for sampling from heavy-tailed distributions [42]; for sampling in Rényi divergence under a family of isoperimetric inequalities interpolating between LSI and Poincaré [22]; and for sampling from non-log-concave distributions with convergence in Fisher information [7]. The interpolation technique has also been useful for analyzing other sampling algorithms, e.g. the Proximal Langevin Algorithm (which uses the proximal method for f rather than gradient descent), which yields a smaller (and tight) asymptotic bias [78]. It has also been used to analyze the Mirror Langevin Algorithm (which uses Hessian metric and discretizes in the dual space) for sampling under mirror isoperimetric inequalities [46]. Further, the calculation along simultaneous heat flow (Lemma 18) is also useful for analyzing the convergence of a new proximal sampler algorithm under isoperimetry [18].

Appendix

Review on Notation and Basic Properties

Throughout, we represent a probability distribution ρ on $\mathbb{R}^n$ via its probability density function with respect to the Lebesgue measure, so $\rho \colon \mathbb{R}^n \to \mathbb{R}$ with $\int_{\mathbb{R}^n} \rho(x)dx = 1$. We typically assume ρ has full support and smooth density, so $\rho(x) > 0$ and $x \mapsto \rho(x)$ is differentiable. Given a function $f \colon \mathbb{R}^n \to \mathbb{R}$, we denote the expected value of f under ρ by

$$\mathbb{E}_\rho[f] = \int_{\mathbb{R}^n} f(x)\rho(x)\,dx.$$

We use the Euclidean inner product $\langle x, y \rangle = \sum_{i=1}^n x_i y_i$ for $x = (x_i)_{1 \le i \le n}$, $y = (y_i)_{1 \le i \le n} \in \mathbb{R}^n$. For symmetric matrices $A, B \in \mathbb{R}^{n \times n}$, let $A \preceq B$ denote that $B - A$ is positive semidefinite. For $\mu \in \mathbb{R}^n$, $\Sigma \succ 0$, let $\mathcal{N}(\mu, \Sigma)$ denote the Gaussian distribution on $\mathbb{R}^n$ with mean μ and covariance matrix Σ.

Given a smooth function $f \colon \mathbb{R}^n \to \mathbb{R}$, its **gradient** $\nabla f \colon \mathbb{R}^n \to \mathbb{R}^n$ is the vector of partial derivatives:

$$\nabla f(x) = \left(\frac{\partial f(x)}{\partial x_1}, \ldots, \frac{\partial f(x)}{\partial x_n} \right).$$

The **Hessian** $\nabla^2 f \colon \mathbb{R}^n \to \mathbb{R}^{n \times n}$ is the matrix of second partial derivatives:

$$\nabla^2 f(x) = \left(\frac{\partial^2 f(x)}{\partial x_i x_j} \right)_{1 \le i, j \le n}.$$

The **Laplacian** $\Delta f \colon \mathbb{R}^n \to \mathbb{R}$ is the trace of its Hessian:

$$\Delta f(x) = \mathrm{Tr}(\nabla^2 f(x)) = \sum_{i=1}^n \frac{\partial^2 f(x)}{\partial x_i^2}.$$

Given a smooth vector field $v = (v_1, \ldots, v_n) \colon \mathbb{R}^n \to \mathbb{R}^n$, its **divergence** $\nabla \cdot v \colon \mathbb{R}^n \to \mathbb{R}$ is

$$(\nabla \cdot v)(x) = \sum_{i=1}^n \frac{\partial v_i(x)}{\partial x_i}.$$

In particular, the divergence of gradient is the Laplacian:

$$(\nabla \cdot \nabla f)(x) = \sum_{i=1}^n \frac{\partial^2 f(x)}{\partial x_i^2} = \Delta f(x).$$

For any function $f: \mathbb{R}^n \to \mathbb{R}$ and vector field $v: \mathbb{R}^n \to \mathbb{R}^n$ with sufficiently fast decay at infinity, we have the following **integration by parts** formula:

$$\int_{\mathbb{R}^n} \langle v(x), \nabla f(x) \rangle dx = - \int_{\mathbb{R}^n} f(x)(\nabla \cdot v)(x) dx.$$

Furthermore, for any two functions $f, g: \mathbb{R}^n \to \mathbb{R}$,

$$\int_{\mathbb{R}^n} f(x) \Delta g(x) dx = - \int_{\mathbb{R}^n} \langle \nabla f(x), \nabla g(x) \rangle dx = \int_{\mathbb{R}^n} g(x) \Delta f(x) dx.$$

When the argument is clear, we omit the argument (x) in the formulae for brevity. For example, the last integral above becomes

$$\int f \, \Delta g \, dx = - \int \langle \nabla f, \nabla g \rangle \, dx = \int g \, \Delta f \, dx. \tag{A.1}$$

Derivation of the Fokker-Planck Equation

Consider a stochastic differential equation

$$dX_t = v(X_t) \, dt + \sqrt{2} \, dW_t \tag{A.2}$$

where $v: \mathbb{R}^n \to \mathbb{R}^n$ is a smooth vector field and $(W_t)_{t \geq 0}$ is the Brownian motion on $\mathbb{R}^n$ with $W_0 = 0$.

We will show that if X_t evolves following (A.2), then its probability density function $\rho_t(x)$ evolves following the Fokker-Planck equation:

$$\frac{\partial \rho_t}{\partial t} = -\nabla \cdot (\rho_t v) + \Delta \rho_t. \tag{A.3}$$

We can derive this heuristically as follows; we refer to standard textbooks for rigorous derivation [58].

For any smooth test function $\phi: \mathbb{R}^n \to \mathbb{R}$, let us compute the time derivative of the expectation

$$A(t) = \mathbb{E}_{\rho_t}[\phi] = \mathbb{E}[\phi(X_t)].$$

On the one hand, we can compute this as

$$\dot{A}(t) = \frac{d}{dt} A(t) = \frac{d}{dt} \int_{\mathbb{R}^n} \rho_t(x) \phi(x) \, dx = \int_{\mathbb{R}^n} \frac{\partial \rho_t(x)}{\partial t} \phi(x) \, dx. \tag{A.4}$$

On the other hand, by (A.2), for small $\epsilon > 0$ we have

$$X_{t+\epsilon} = X_t + \int_t^{t+\epsilon} v(X_s)ds + \sqrt{2}(W_{t+\epsilon} - W_t)$$

$$= X_t + \epsilon v(X_t) + \sqrt{2}(W_{t+\epsilon} - W_t) + O(\epsilon^2)$$

$$\overset{d}{=} X_t + \epsilon v(X_t) + \sqrt{2\epsilon}Z + O(\epsilon^2)$$

where $Z \sim \mathcal{N}(0, I)$ is independent of X_t, since $W_{t+\epsilon} - W_t \sim \mathcal{N}(0, \epsilon I)$. Then by Taylor expansion,

$$\phi(X_{t+\epsilon}) \overset{d}{=} \phi\left(X_t + \epsilon v(X_t) + \sqrt{2\epsilon}Z + O(\epsilon^2)\right)$$

$$= \phi(X_t) + \epsilon\langle\nabla\phi(X_t), v(X_t)\rangle + \sqrt{2\epsilon}\langle\nabla\phi(X_t), Z\rangle$$

$$+ \frac{1}{2}2\epsilon\langle Z, \nabla^2\phi(X_t)Z\rangle + O(\epsilon^{\frac{3}{2}}).$$

Now we take expectation on both sides. Since $Z \sim \mathcal{N}(0, I)$ is independent of X_t,

$$A(t + \epsilon) = \mathbb{E}[\phi(X_{t+\epsilon})]$$

$$= \mathbb{E}\left[\phi(X_t) + \epsilon\langle\nabla\phi(X_t), v(X_t)\rangle + \sqrt{2\epsilon}\langle\nabla\phi(X_t), Z\rangle\right.$$

$$\left. + \epsilon\langle Z, \nabla^2\phi(X_t)Z\rangle\right] + O(\epsilon^{\frac{3}{2}})$$

$$= A(t) + \epsilon\left(\mathbb{E}[\langle\nabla\phi(X_t), v(X_t)\rangle] + \mathbb{E}[\Delta\phi(X_t)]\right) + O(\epsilon^{\frac{3}{2}}).$$

Therefore, by integration by parts, this second approach gives

$$\dot{A}(t) = \lim_{\epsilon \to 0} \frac{A(t + \epsilon) - A(t)}{\epsilon}$$

$$= \mathbb{E}[\langle\nabla\phi(X_t), v(X_t)\rangle] + \mathbb{E}[\Delta\phi(X_t)]$$

$$= \int_{\mathbb{R}^n} \langle\nabla\phi(x), \rho_t(x)v(x)\rangle dx + \int_{\mathbb{R}^n} \rho_t(x)\Delta\phi(x)\,dx$$

$$= -\int_{\mathbb{R}^n} \phi(x)\nabla \cdot (\rho_t v)(x)\,dx + \int_{\mathbb{R}^n} \phi(x)\Delta\rho_t(x)\,dx$$

$$= \int_{\mathbb{R}^n} \phi(x)\left(-\nabla \cdot (\rho_t v)(x) + \Delta\rho_t(x)\right)dx. \qquad \text{(A.5)}$$

Comparing (A.4) and (A.5), and since ϕ is arbitrary, we conclude that

$$\frac{\partial \rho_t(x)}{\partial t} = -\nabla \cdot (\rho_t v)(x) + \Delta \rho_t(x)$$

as claimed in (A.3).

When $v = -\nabla f$, the stochastic differential equation (A.2) becomes the Langevin dynamics (7) from Sect. 2.3, and the Fokker-Planck equation (A.3) becomes (8).

In the proof of Lemma 3, we also apply the Fokker-Planck equation (A.3) when $v = -\nabla f(x_0)$ is a constant vector field to derive the evolution equation (30) for one step of ULA.

Remaining Proofs

Proof of Lemma 16

Proof of Lemma 16 Let $g: \mathbb{R}^n \to \mathbb{R}$ be a smooth function, and let $\tilde{g}: \mathbb{R}^n \to \mathbb{R}$ be the function $\tilde{g}(x) = g(T(x))$. Let $X \sim v$, so $T(X) \sim \tilde{v}$. Note that

$$\mathbb{E}_{\tilde{v}}[g^2] = \mathbb{E}_{X \sim v}[g(T(X))^2] = \mathbb{E}_v[\tilde{g}^2],$$

$$\mathbb{E}_{\tilde{v}}[g^2 \log g^2] = \mathbb{E}_{X \sim v}[g(T(X))^2 \log g(T(X))^2] = \mathbb{E}_v[\tilde{g}^2 \log \tilde{g}^2].$$

Furthermore, we have $\nabla \tilde{g}(x) = \nabla T(x) \nabla g(T(x))$. Since T is L-Lipschitz, $\|\nabla T(x)\| \le L$. Then

$$\|\nabla \tilde{g}(x)\| \le \|\nabla T(x)\| \, \|\nabla g(T(x))\| \le L \|\nabla g(T(x))\|.$$

This implies

$$\mathbb{E}_{\tilde{v}}[\|\nabla g\|^2] = \mathbb{E}_{X \sim v}[\|\nabla g(T(X))\|^2] \ge \frac{\mathbb{E}_v[\|\nabla \tilde{g}\|^2]}{L^2}.$$

Therefore,

$$\frac{\mathbb{E}_{\tilde{v}}[\|\nabla g\|^2]}{\mathbb{E}_{\tilde{v}}[g^2 \log g^2] - \mathbb{E}_{\tilde{v}}[g^2] \log \mathbb{E}_{\tilde{v}}[g^2]} \ge \frac{1}{L^2} \frac{\mathbb{E}_v[\|\nabla \tilde{g}\|^2]}{\left(\mathbb{E}_v[\tilde{g}^2 \log \tilde{g}^2] - \mathbb{E}_v[\tilde{g}^2] \log \mathbb{E}_v[\tilde{g}^2]\right)}$$

$$\ge \frac{\alpha}{2L^2}$$

where the last inequality follows from the assumption that v satisfies LSI with constant α. This shows that $\tilde{v}$ satisfies LSI with constant α/L^2, as desired. $\qquad \square$

Proof of Lemma 17

Proof of Lemma 17 We recall the following convolution property of LSI [15]: If $\nu, \tilde{\nu}$ satisfy LSI with constants $\alpha, \tilde{\alpha} > 0$, respectively, then $\nu * \tilde{\nu}$ satisfies LSI with constant $\left(\frac{1}{\alpha} + \frac{1}{\tilde{\alpha}}\right)^{-1}$. Since $\mathcal{N}(0, 2tI)$ satisfies LSI with constant $\frac{1}{2t}$, the claim follows. $\qquad\square$

Proof of Lemma 19

Proof of Lemma 19 Let $g: \mathbb{R}^n \to \mathbb{R}$ be a smooth function, and let $\tilde{g}: \mathbb{R}^n \to \mathbb{R}$ be the function $\tilde{g}(x) = g(T(x))$. Let $X \sim \nu$, so $T(X) \sim \tilde{\nu}$. Note that

$$\mathrm{Var}_{\tilde{\nu}}(g) = \mathrm{Var}_{X \sim \nu}(g(T(X))) = \mathrm{Var}_{\nu}(\tilde{g}).$$

Furthermore, we have $\nabla \tilde{g}(x) = \nabla T(x) \nabla g(T(x))$. Since T is L-Lipschitz, $\|\nabla T(x)\| \le L$. Then

$$\|\nabla \tilde{g}(x)\| \le \|\nabla T(x)\| \|\nabla g(T(x))\| \le L \|\nabla g(T(x))\|.$$

This implies

$$\mathbb{E}_{\tilde{\nu}}[\|\nabla g\|^2] = \mathbb{E}_{X \sim \nu}[\|\nabla g(T(X))\|^2] \ge \frac{\mathbb{E}_{\nu}[\|\nabla \tilde{g}\|^2]}{L^2}.$$

Therefore,

$$\frac{\mathbb{E}_{\tilde{\nu}}[\|\nabla g\|^2]}{\mathrm{Var}_{\tilde{\nu}}(g)} \ge \frac{1}{L^2} \frac{\mathbb{E}_{\nu}[\|\nabla \tilde{g}\|^2]}{\mathrm{Var}_{\nu}(\tilde{g})} \ge \frac{\alpha}{L^2}$$

where the last inequality follows from the assumption that ν satisfies Poincaré inequality with constant α. This shows that $\tilde{\nu}$ satisfies Poincaré inequality with constant α/L^2, as desired. $\qquad\square$

Proof of Lemma 20

Proof of Lemma 20 We recall the following convolution property of Poincaré inequality [23]: If $\nu, \tilde{\nu}$ satisfy Poincaré inequality with constants $\alpha, \tilde{\alpha} > 0$, respectively, then $\nu * \tilde{\nu}$ satisfies Poincaré inequality with constant $\left(\frac{1}{\alpha} + \frac{1}{\tilde{\alpha}}\right)^{-1}$. Since $\mathcal{N}(0, 2tI)$ satisfies Poincaré inequality with constant $\frac{1}{2t}$, the claim follows. $\qquad\square$

Acknowledgments The authors thank Kunal Talwar for explaining the privacy motivation and application of Rényi divergence to data privacy; Yu Cao, Jianfeng Lu, and Yulong Lu for alerting us to their work [14] on Rényi divergence; Xiang Cheng and Peter Bartlett for helpful comments on an earlier version of this paper; and Sinho Chewi for communicating Theorem 8 to us.

References

1. M. Abadi, A. Chu, I. Goodfellow, H.B. McMahan, I. Mironov, K. Talwar, L. Zhang, Deep learning with differential privacy, in *Proceedings of the 2016 ACM SIGSAC Conference on Computer and Communications Security* (ACM, 2016), pp. 308–318
2. D. Applegate, R. Kannan, Sampling and integration of near log-concave functions, in *Proceedings of the Twenty-third Annual ACM Symposium on Theory of Computing*, STOC '91, New York, NY, USA (ACM, 1991), pp. 156–163
3. J.C. Baez, Rényi entropy and free energy. Preprint. arXiv:1102.2098 (2011)
4. S. Bai, T. Lepoint, A. Roux-Langlois, A. Sakzad, D. Stehlé, R. Steinfeld, Improved security proofs in lattice-based cryptography: using the Rényi divergence rather than the statistical distance. J. Cryptol. **31**(2), 610–640 (2018)
5. D. Bakry, M. Émery, Diffusions hypercontractives, in *Séminaire de Probabilités XIX 1983/84* (Springer, 1985), pp. 177–206
6. D. Bakry, F. Barthe, P. Cattiaux, A. Guillin et al., A simple proof of the Poincaré inequality for a large class of probability measures. Electron. Commun. Probab. **13**, 60–66 (2008)
7. K. Balasubramanian, S. Chewi, M.A. Erdogdu, A. Salim, M. Zhang, Towards a theory of non-log-concave sampling: first-order stationarity guarantees for Langevin Monte Carlo, in *Proceedings of the 2022 Conference on Learning Theory*. PMLR (2022)
8. J.B. Bardet, N. Gozlan, F. Malrieu, P.A. Zitt, Functional inequalities for Gaussian convolutions of compactly supported measures: explicit bounds and dimension dependence. Bernoulli **24**(1), 333–353 (2018)
9. E. Bernton, Langevin Monte Carlo and JKO splitting, in *Conference on Learning Theory, COLT 2018, Stockholm, Sweden, 6-9 July 2018* (2018), pp. 1777–1798
10. S.G. Bobkov, F. Götze, Exponential integrability and transportation cost related to logarithmic Sobolev inequalities. J. Funct. Anal. **163**(1), 1–28 (1999)
11. S.G. Bobkov, I. Gentil, M. Ledoux, Hypercontractivity of Hamilton–Jacobi equations. J. Math. Pures Appl. **80**(7), 669–696 (2001)
12. S.G. Bobkov, G.P. Chistyakov, F. Götze, Rényi divergence and the central limit theorem. Ann. Probab. **47**(1), 270–323 (2019)
13. M. Bun, T. Steinke, Concentrated differential privacy: simplifications, extensions, and lower bounds, in *Theory of Cryptography Conference* (Springer, 2016), pp. 635–658
14. Y. Cao, J. Lu, Y. Lu, Exponential decay of Rényi divergence under Fokker–Planck equations. J. Stat. Phys. **176**, 1172–1184 (2019)
15. D. Chafaï, Entropies, convexity, and functional inequalities: on ϕ-entropies and ϕ-Sobolev inequalities. J. Math. Kyoto Univ. **44**(2), 325–363 (2004)
16. D. Chafai, F. Malrieu, On fine properties of mixtures with respect to concentration of measure and Sobolev type inequalities, in *Annales de l'IHP Probabilités et statistiques*, vol. 46 (2010), pp. 72–96
17. Z. Chen, S.S. Vempala, Optimal convergence rate of Hamiltonian Monte Carlo for strongly logconcave distributions. Theory Comput. **18**(9), 1–18 (2022)
18. Y. Chen, S. Chewi, A. Salim, A. Wibisono, Improved analysis for a proximal algorithm for sampling, in *Proceedings of the 2022 Conference on Learning Theory*. PMLR (2022)
19. X. Cheng, P. Bartlett, Convergence of Langevin MCMC in KL-divergence, in F. Janoos, M. Mohri, K. Sridharan, ed. by *Proceedings of Algorithmic Learning Theory*, volume 83 of *Proceedings of Machine Learning Research*. PMLR, 07–09 Apr (2018), pp. 186–211

20. X. Cheng, N.S. Chatterji, Y. Abbasi-Yadkori, P.L. Bartlett, M.I. Jordan, Sharp convergence rates for Langevin dynamics in the nonconvex setting. Preprint. arXiv:1805.01648 (2018)
21. S. Chewi, T. Le Gouic, C. Lu, T. Maunu, P. Rigollet, A. Stromme, Exponential ergodicity of mirror-Langevin diffusions, in *Advances in Neural Information Processing Systems*, vol. 33 (2020), pp. 19573–19585
22. S. Chewi, M.A. Erdogdu, M.B. Li, R. Shen, M. Zhang, Analysis of Langevin Monte Carlo from Poincaré to log-Sobolev, in *Proceedings of the 2022 Conference on Learning Theory*. PMLR (2022)
23. T.A. Courtade, Bounds on the Poincaré constant for convolution measures. Ann. l'Inst. Henri Poincaré Probab. Stat. **56**(1), 566–579 (2020)
24. I. Csiszár, Generalized cutoff rates and Rényi's information measures. IEEE Trans. Inf. Theory **41**(1), 26–34 (1995)
25. A. Dalalyan, Further and stronger analogy between sampling and optimization: Langevin Monte Carlo and gradient descent, in *Proceedings of the 2017 Conference on Learning Theory*, volume 65 of *Proceedings of Machine Learning Research*. PMLR, 07–10 Jul (2017), pp. 678–689
26. A.S. Dalalyan, A. Karagulyan, User-friendly guarantees for the Langevin Monte Carlo with inaccurate gradient, in *Stochastic Processes and their Applications* (2019)
27. X. Dong, The gravity dual of Rényi entropy. Nat. Commun. **7**, 12472 (2016)
28. A. Durmus, E. Moulines, E. Saksman, On the convergence of Hamiltonian Monte Carlo. Preprint. arXiv:1705.00166 (2017)
29. A. Durmus, S. Majewski, B. Miasojedow, Analysis of Langevin Monte Carlo via convex optimization. J. Mach. Learn. Res. **20**(1), 2666–2711 (2019)
30. R. Dwivedi, Y. Chen, M.J. Wainwright, B. Yu, Log-concave sampling: Metropolis-Hastings algorithms are fast!, in *Conference on Learning Theory, COLT 2018, Stockholm, Sweden, 6–9 July* (2018), pp. 793–797
31. C. Dwork, G.N. Rothblum, Concentrated differential privacy. Preprint. arXiv:1603.01887 (2016)
32. A. Eberle, A. Guillin, R. Zimmer, Couplings and quantitative contraction rates for Langevin dynamics. Ann. Probab. **47**(4), 1982–2010 (2019)
33. M.A. Erdogdu, R. Hosseinzadeh, On the convergence of Langevin Monte Carlo: the interplay between tail growth and smoothness, in *Proceedings of Thirty Fourth Conference on Learning Theory*, ed. by M. Belkin, S. Kpotufe, volume 134 of *Proceedings of Machine Learning Research*. PMLR, 15–19 Aug (2021), pp. 1776–1822
34. M.A. Erdogdu, R. Hosseinzadeh, M.S. Zhang, Convergence of Langevin Monte Carlo in chi-squared and Rényi divergence, in *International Conference on Artificial Intelligence and Statistics*. PMLR (2022), pp. 8151–8175
35. A. Ganesh, K. Talwar, Faster differentially private samplers via Rényi divergence analysis of discretized Langevin MCMC, in *Advances in Neural Information Processing Systems*, ed. by H. Larochelle, M. Ranzato, R. Hadsell, M.F. Balcan, H. Lin, vol. 33 (Curran Associates, 2020), pp. 7222–7233
36. A. Garbuno-Inigo, N. Nüsken, S. Reich, Affine invariant interacting Langevin dynamics for Bayesian inference. SIAM J. Appl. Dynam. Syst. **19**(3), 1633–1658 (2020)
37. K. Gatmiry, S.S. Vempala, Convergence of the Riemannian Langevin algorithm. Preprint. arXiv:2204.10818 (2022)
38. J. Gorham, L. Mackey, Measuring sample quality with kernels, in *Proceedings of the 34th International Conference on Machine Learning*, ed. by D. Precup, Y.W. Teh, volume 70 of *Proceedings of Machine Learning Research*, International Convention Centre, Sydney, Australia, 06–11 Aug 2017. PMLR (2017), pp. 1292–1301
39. L. Gross, Logarithmic Sobolev inequalities. Am. J. Math. **97**(4), 1061–1083 (1975)
40. P. Harremoës, Interpretations of Rényi entropies and divergences. Physica A Stat. Mech. Appl. **365**(1), 57–62 (2006)
41. Y. He, A.B. Hamza, H. Krim, A generalized divergence measure for robust image registration. IEEE Trans. Signal Process. **51**(5), 1211–1220 (2003)

42. Y. He, K. Balasubramanian, M.A. Erdogdu, Heavy-tailed sampling via transformed unadjusted Langevin algorithm. Preprint. arXiv:2201.08349 (2022)
43. R. Holley, D. Stroock, Logarithmic Sobolev inequalities and stochastic Ising models. J. Stat. Phys. **46**(5), 1159–1194 (1987)
44. R. Holley, D. Stroock, Simulated annealing via Sobolev inequalities. Commun. Math. Phys. **115**(4), 553–569 (1988)
45. M. Iwamoto, J. Shikata, Information theoretic security for encryption based on conditional Rényi entropies, in *International Conference on Information Theoretic Security* (Springer, 2013), pp. 103–121
46. Q. Jiang, Mirror Langevin Monte Carlo: the case under isoperimetry, in *Advances in Neural Information Processing Systems*, ed. by M. Ranzato, A. Beygelzimer, K. Nguyen, P.S. Liang, J.W. Vaughan, Y. Dauphin, vol. 34 (Curran Associates, 2021)
47. R. Jordan, D. Kinderlehrer, F. Otto, The variational formulation of the Fokker–Planck equation. SIAM J. Math. Anal. **29**(1), 1–17 (1998)
48. R. Kannan, L. Lovász, M. Simonovits, Random walks and an $O^*(n^5)$ volume algorithm for convex bodies. Random Struct. Algorithms **11**, 1–50 (1997)
49. M. Ledoux, Concentration of measure and logarithmic Sobolev inequalities. Sémin. Probab. Strasbourg **33**, 120–216 (1999)
50. Y.T. Lee, S.S. Vempala, Convergence rate of Riemannian Hamiltonian Monte Carlo and faster polytope volume computation, in *Proceedings of the 50th Annual ACM SIGACT Symposium on Theory of Computing* (ACM, 2018), pp. 1115–1121
51. M.B. Li, M.A. Erdogdu, Riemannian Langevin algorithm for solving semidefinite programs. Preprint. arXiv:2010.11176 (2020)
52. Y. Li, R.E. Turner, Rényi divergence variational inference, in *Advances in Neural Information Processing Systems*, vol. 29, ed. by D.D. Lee, M. Sugiyama, U.V. Luxburg, I. Guyon, R. Garnett (Curran Associates, 2016), pp. 1073–1081
53. X. Li, Y. Wu, L. Mackey, M.A. Erdogdu, Stochastic Runge–Kutta accelerates Langevin Monte Carlo and beyond, in *Advances in Neural Information Processing Systems*, vol. 32 (2019)
54. L. Lovász, S. Vempala, Fast algorithms for logconcave functions: Sampling, rounding, integration and optimization, in *FOCS* (2006), pp. 57–68
55. L. Lovász, S.S. Vempala, Hit-and-run from a corner. SIAM J. Comput. **35**(4), 985–1005 (2006)
56. L. Lovász, S. Vempala, The geometry of logconcave functions and sampling algorithms. Random Struct. Algorithms **30**(3), 307–358 (2007)
57. Y.A. Ma, N.S. Chatterji, X. Cheng, N. Flammarion, P.L. Bartlett, M.I. Jordan, Is there an analog of Nesterov acceleration for gradient-based MCMC? Bernoulli **27**(3), 1942–1992 (2021)
58. M.C. Mackey, *Time's Arrow: The Origins of Thermodynamics Behavior* (Springer, 1992)
59. O. Mangoubi, A. Smith, Rapid mixing of Hamiltonian Monte Carlo on strongly log-concave distributions. Preprint. arXiv:1708.07114 (2017)
60. O. Mangoubi, N. Vishnoi, Dimensionally tight bounds for second-order Hamiltonian Monte Carlo, in *Advances in Neural Information Processing Systems*, vol. 31 (Curran Associates, 2018), pp. 6027–6037
61. O. Mangoubi, N.K. Vishnoi, Nonconvex sampling with the Metropolis-adjusted Langevin algorithm, in *Conference on Learning Theory*. PMLR (2019), pp. 2259–2293
62. Y. Mansour, M. Mohri, A. Rostamizadeh, Multiple source adaptation and the Rényi divergence, in *Proceedings of the Twenty-Fifth Conference on Uncertainty in Artificial Intelligence* (AUAI Press, 2009), pp. 367–374
63. G. Menz, A. Schlichting, Poincaré and logarithmic Sobolev inequalities by decomposition of the energy landscape. Ann. Probab. **42**(5), 1809–1884 (2014)
64. I. Mironov, Rényi differential privacy, in *2017 IEEE 30th Computer Security Foundations Symposium (CSF)* (IEEE, 2017), pp. 263–275
65. D. Morales, L. Pardo, I. Vajda, Rényi statistics in directed families of exponential experiments. Stat. J. Theor. Appl. Stat. **34**(2), 151–174 (2000)
66. F. Otto, C. Villani, Generalization of an inequality by Talagrand and links with the logarithmic Sobolev inequality. J. Funct. Anal. **173**(2), 361–400 (2000)

67. M. Raginsky, A. Rakhlin, M. Telgarsky, Non-convex learning via stochastic gradient Langevin dynamics: a nonasymptotic analysis, in *Proceedings of the 2017 Conference on Learning Theory*, ed. by S. Kale, O. Shamir, volume 65 of *Proceedings of Machine Learning Research*, Amsterdam, Netherlands, 07–10 Jul 2017. PMLR (2017), pp. 1674–1703
68. A. Rényi et al., On measures of entropy and information, in *Proceedings of the Fourth Berkeley Symposium on Mathematical Statistics and Probability, Volume 1: Contributions to the Theory of Statistics* (The Regents of the University of California, 1961)
69. O.S. Rothaus, Diffusion on compact Riemannian manifolds and logarithmic Sobolev inequalities. J. Funct. Anal. **42**(1), 102–109 (1981)
70. M. Talagrand, Transportation cost for Gaussian and other product measures. Geom. Funct. Anal. **6**, 587–600 (1996)
71. T. Van Erven, P. Harremos, Rényi divergence and Kullback-Leibler divergence. IEEE Trans. Inf. Theory **60**(7), 3797–3820 (2014)
72. S. Vempala, A. Wibisono, Rapid convergence of the unadjusted Langevin algorithm: isoperimetry suffices, in *Advances in Neural Information Processing Systems*, vol. 32 (Curran Associates, 2019)
73. C. Villani, A short proof of the concavity of entropy power. IEEE Trans. Inf. Theory **46**(4), 1695–1696 (2000)
74. C. Villani, *Topics in Optimal Transportation*. Number 58 in Graduate Studies in Mathematics (American Mathematical Society, 2003)
75. F.Y. Wang, J. Wang, Functional inequalities for convolution of probability measures, in *Annales de l'Institut Henri Poincaré, Probabilités et Statistiques*, vol. 52 (Institut Henri Poincaré, 2016), pp. 898–914
76. X. Wang, Q. Lei, I. Panageas, Fast convergence of Langevin dynamics on manifold: Geodesics meet log-Sobolev, in *Advances in Neural Information Processing Systems*, ed. by H. Larochelle, M. Ranzato, R. Hadsell, M. F. Balcan, H. Lin, vol. 33 (Curran Associates, 2020), pp. 18894–18904
77. A. Wibisono, Sampling as optimization in the space of measures: the Langevin dynamics as a composite optimization problem, in *Conference on Learning Theory, COLT 2018, Stockholm, Sweden, 6–9 July 2018* (2018), pp. 2093–3027
78. A. Wibisono, Proximal Langevin algorithm: rapid convergence under isoperimetry. e-prints arXiv:1911.01469 (2019)

LECTURE NOTES IN MATHEMATICS  Springer

Editors in Chief: J.-M. Morel, B. Teissier;

Editorial Policy

1. Lecture Notes aim to report new developments in all areas of mathematics and their applications – quickly, informally and at a high level. Mathematical texts analysing new developments in modelling and numerical simulation are welcome.

   Manuscripts should be reasonably self-contained and rounded off. Thus they may, and often will, present not only results of the author but also related work by other people. They may be based on specialised lecture courses. Furthermore, the manuscripts should provide sufficient motivation, examples and applications. This clearly distinguishes Lecture Notes from journal articles or technical reports which normally are very concise. Articles intended for a journal but too long to be accepted by most journals, usually do not have this "lecture notes" character. For similar reasons it is unusual for doctoral theses to be accepted for the Lecture Notes series, though habilitation theses may be appropriate.

2. Besides monographs, multi-author manuscripts resulting from SUMMER SCHOOLS or similar INTENSIVE COURSES are welcome, provided their objective was held to present an active mathematical topic to an audience at the beginning or intermediate graduate level (a list of participants should be provided).

   The resulting manuscript should not be just a collection of course notes, but should require advance planning and coordination among the main lecturers. The subject matter should dictate the structure of the book. This structure should be motivated and explained in a scientific introduction, and the notation, references, index and formulation of results should be, if possible, unified by the editors. Each contribution should have an abstract and an introduction referring to the other contributions. In other words, more preparatory work must go into a multi-authored volume than simply assembling a disparate collection of papers, communicated at the event.

3. Manuscripts should be submitted either online at www.editorialmanager.com/lnm to Springer's mathematics editorial in Heidelberg, or electronically to one of the series editors. Authors should be aware that incomplete or insufficiently close-to-final manuscripts almost always result in longer refereeing times and nevertheless unclear referees' recommendations, making further refereeing of a final draft necessary. The strict minimum amount of material that will be considered should include a detailed outline describing the planned contents of each chapter, a bibliography and several sample chapters. Parallel submission of a manuscript to another publisher while under consideration for LNM is not acceptable and can lead to rejection.

4. In general, **monographs** will be sent out to at least 2 external referees for evaluation.

   A final decision to publish can be made only on the basis of the complete manuscript, however a refereeing process leading to a preliminary decision can be based on a pre-final or incomplete manuscript.

   Volume Editors of **multi-author works** are expected to arrange for the refereeing, to the usual scientific standards, of the individual contributions. If the resulting reports can be forwarded to the LNM Editorial Board, this is very helpful. If no reports are forwarded or if other questions remain unclear in respect of homogeneity etc, the series editors may wish to consult external referees for an overall evaluation of the volume.

5. Manuscripts should in general be submitted in English. Final manuscripts should contain at least 100 pages of mathematical text and should always include

    - a table of contents;
    - an informative introduction, with adequate motivation and perhaps some historical remarks: it should be accessible to a reader not intimately familiar with the topic treated;
    - a subject index: as a rule this is genuinely helpful for the reader.
    - For evaluation purposes, manuscripts should be submitted as pdf files.

6. Careful preparation of the manuscripts will help keep production time short besides ensuring satisfactory appearance of the finished book in print and online. After acceptance of the manuscript authors will be asked to prepare the final LaTeX source files (see LaTeX templates online: https://www.springer.com/gb/authors-editors/book-authors-editors/manuscriptpreparation/5636) plus the corresponding pdf- or zipped ps-file. The LaTeX source files are essential for producing the full-text online version of the book, see http://link.springer.com/bookseries/304 for the existing online volumes of LNM). The technical production of a Lecture Notes volume takes approximately 12 weeks. Additional instructions, if necessary, are available on request from lnm@springer.com.

7. Authors receive a total of 30 free copies of their volume and free access to their book on SpringerLink, but no royalties. They are entitled to a discount of 33.3 % on the price of Springer books purchased for their personal use, if ordering directly from Springer.

8. Commitment to publish is made by a *Publishing Agreement*; contributing authors of multiauthor books are requested to sign a *Consent to Publish form*. Springer-Verlag registers the copyright for each volume. Authors are free to reuse material contained in their LNM volumes in later publications: a brief written (or e-mail) request for formal permission is sufficient.

Addresses:
Professor Jean-Michel Morel, CMLA, École Normale Supérieure de Cachan, France
E-mail: moreljeanmichel@gmail.com

Professor Bernard Teissier, Equipe Géométrie et Dynamique,
Institut de Mathématiques de Jussieu – Paris Rive Gauche, Paris, France
E-mail: bernard.teissier@imj-prg.fr

Springer: Ute McCrory, Mathematics, Heidelberg, Germany,
E-mail: lnm@springer.com

Printed in the United States
by Baker & Taylor Publisher Services

Printed in the United States
by Baker & Taylor Publisher Services